Advanced Maintenance Modelling for Asset Management

Adolfo Crespo Márquez
Vicente González-Prida Díaz
Juan Francisco Gómez Fernández
Editors

Advanced Maintenance Modelling for Asset Management

Techniques and Methods for Complex Industrial Systems

 Springer

Editors
Adolfo Crespo Márquez
Department of Industrial Management,
  School of Engineering
University of Seville
Seville
Spain

Juan Francisco Gómez Fernández
Department of Industrial Management,
  School of Engineering
University of Seville
Seville
Spain

Vicente González-Prida Díaz
Department of Industrial Management,
  School of Engineering
University of Seville
Seville
Spain

ISBN 978-3-319-86309-2   ISBN 978-3-319-58045-6   (eBook)
DOI 10.1007/978-3-319-58045-6

Printed on acid-free paper

This Springer imprint is published by Springer Nature
The registered company is Springer International Publishing AG
The registered company address is: Gewerbestrasse 11, 6330 Cham, Switzerland

Foreword

It is humbling to write a foreword on a book edited by colleagues Adolfo Crespo Márquez, Vicente González-Prida Díaz and Juan Francisco Gómez Fernández, who have collated the wealth of experience of the research group SIM (Sistemas Inteligentes de Mantenimiento/Intelligent Maintenance Systems) at the School of Engineering, University of Seville, Spain.

Engineered assets make up our built environment, and include man-made equipment, infrastructure, plant, tools and physical systems that are deployed across all industry sectors of human endeavour. Industrial components, equipment, infrastructure and plant are becoming complex and more sophisticated underpinned by rapid advances in materials technologies, instrumentation and automation. Unprecedented advances in information and communications technologies coupled to globalization and sustainability imperatives are providing new tools and techniques for managing complex industrial systems.

Asset Management involves the strategic tenets of planning, decision-making and control of resources. Maintenance is a critical function for successful implementation of strategy for managing engineered assets. Maintenance is intertwined with operating complex industrial systems inasmuch as it informs the tactical aspects of managing engineered assets.

The emergence of big data analytics, Internet of things, and associated industry platforms and networks significantly accentuates the application of artificial intelligence techniques towards various models of the maintenance function. Such advanced models of the maintenance function are necessary to enhance the tactical and practical implementation of strategies for managing complex industrial systems.

This book extends an earlier publication by Adolfo Crespo Márquez in 2007. It covers recent research work of the SIM research group that focus on the demand for ever-increasing reliability and availability of industrial components, equipment, plants, processes and systems. The numerous industrial case studies discussed in the book constitute an invaluable resource for practitioners, researchers, and academics.

January 2017   Joe Amadi-Echendu DPhil, CEng, PrEng, FISEAM
                                  Professor, Graduate School of Technology Management
                                             University of Pretoria; Editor-in-Chief,
                                   Engineering Asset Management Review Series
                                  Springer; Director, Board of the International Society
                                                of Engineering Asset Management

Preface

Overview

The aim of this book is to continue with the development of a framework for maintenance and assets management that has been promoted by the SIM group over the years (the reader can find the seminal work regarding the referred management framework in Springer: Crespo Márquez 2007). To that end, this manuscript describes new advanced models, methods and techniques, which can be applied at different stages of the originally proposed management process, as well as their practical implementation. During the last 15 years, among other research activities, the SIM group has:

- Published 13 books (4 with Springer-Verlag and 2 with IGI Global, 2 in Chinese with Machinery Industry Press and National Defense Industry Press, 1 in Farsi with the University of Tehran, and 4 in Spanish with AENOR [2], INGEMAN and ETSI Sevilla), and coordinate other 3 books more (two from an international conference and other from the national research network on assets management)
- Authored 75 chapters in scientific books and 104 research articles of which 75 are articles in international journals and 45 are published in the JCR.
- Made more than 80 contributions in congresses of which 65 are international.
- Directed 11 Ph.D. works and over 150 Master Thesis.
- Opened international connections with many universities around the world.
- Evaluated research projects for national and international research agencies (Swedish, Canadian and Italian, among others).
- Developed research lines related to:

  - Asset Management
  - Maintenance Engineering and Management
  - Supply Chain Management and Logistics
  - Simulation and Analysis tools.

In the SIM group we realize that asset management, once considered a tactical area, is now a matter of strategy, given the implications it has for the proper development of the business policy. In addition, the introduction of advanced manufacturing techniques and new production management systems, which lead to increased automation and reduced delivery times, has given great importance to asset management. In manufacturing, production, finance, etc., decisions are increasingly taken based on models or techniques which provide satisfactory, objective decision making, which guarantees improved competitiveness, reducing risk and uncertainty, and that can be justified to management. However, maintenance managers have taken decisions based only on their experience or supported by the advice of system sales staff or consultants. This lack of models and techniques in the area of asset management leads to underperforming maintenance departments characterized by a reactive approach, underutilized maintenance information systems, inaccurately managed costs, no scheduled maintenance hours, feedback on work quality not being provided, etc. Hence, this book looks to promote and address the application of objective and effective decision-making in asset management based on mathematical models and practical techniques that can be easily implemented in organizations.

Summary of Topics and Target Audience

The relevance of maintenance in organizations has increased considerably over the last two decades; this importance is linked to the introduction of a growing number of factors with an influence on the effective and efficient asset management. The existence of increasingly complex equipment and processes, the increase in the number of assets, the speed of technological change, the need to reduce costs in the modern world, together with increases in the level of excellence of commercial goals such as quality and delivery time, and concern for the safety of workers and the environment, make asset management an important source of benefits and competitive advantages for present and future world class enterprises. This book analyses these factors, which are divided into, although not limited to, the following categories:

- Maintenance policy selection.
- After-sales management.
- Knowledge management.
- Critical asset and infrastructure management.
- Asset life cycle management.
- Performance measurement system.
- Sensors and health monitoring systems.
- Reliability centred maintenance.
- Building information modelling.
- Advanced maintenance techniques.
- Set-up processes analysis.

Industrial and manufacturing engineers, managers and plant supervisor, academicians, researchers, advanced-level students (both postgraduate and doctoral), technology developers and managers who take decisions in this field will find in this book a source of ideas, models and techniques which mark out a path for future research in this field and may also serve to encourage original ideas and in many cases practical application in business. This book is aimed at the above-mentioned target audience worldwide and because of the number of chapters it contains and the variety of the subjects analysed, it provides an in-depth look at current global concerns.

Background Material and Origin of Each Chapter

The content of the book is divided into seven parts. Briefly, each part deals with the following matters:

- The first part is an introduction to the topic and to the manuscript.
- The second part presents new possible evolutions in the current assets management framework, according to new standards, techniques and technologies.
- The third part contains advanced tools to improve effectiveness of management, especially under modern dynamic scenario considerations.
- The fourth part includes methods for the improvement of management efficiency, which benefit of a more affordable online information availability regarding assets' conditions.
- The fifth part present innovative techniques to easy management control, providing also a more practical approach to maintenance activities accountability.
- The sixth part compiles new efforts in continuous improvements using artificial intelligence tools mixed with advanced interoperability of the information systems. At the same time explore advance analysis of different operational possibilities to improve assets management.
- Finally, the seventh part is devoted to summarize conclusions and to infer future developments.

Different research results of the SIM group, over the last 5 years, are serving as the main basis and background for the mentioned parts.

Table 1 explicitly mentions the publication linked to each book chapter.

At the same time, each chapter has been developed by a group of authors (some belonging properly to the SIM Research Group, and other assiduous collaborators with the group), whose relevance in the field of asset management has been manifested for years. A brief biographical note of each one of them is shown in the section List of Contributors. Additional information about the contributions in the book can be found in the Intelligent Maintenance Systems Group (SIM) web site (University of Seville) at http://taylor.us.es/sim.

Table 1 Link between chapters and published references

Chapter	Title	Original reference
1	On the Family of Standards UNE-ISO 55000 and How to Effectively Manage Assets	Sola Rosique et al. (2015)
2	A Maintenance Management Framework Based on PAS 55	López-Campos and Crespo Márquez (2011)
3	The Integration of Open Reliability, Maintenance and Condition Monitoring Management Systems	López-Campos et al. (2013)
4	Prognostics and Health Management in Advanced Maintenance Systems	Guillén et al. (2016a)
5	A Framework for Effective Management of CBM Programs	Guillén et al. (2016b)
6	Criticality Analysis for Maintenance Purposes	Crespo Márquez et al. (2015)
7	AHP Method According to a Changing Environment	González-Prida et al. (2014)
8	Reliability Stochastic Modelling for Repairable Physical Assets	Viveros et al. (2016)
9	Economic Impact of a Failure Using Life-Cycle Cost Analysis	Parra et al. (2012)
10	Online Reliability and Risk to Schedule the Preventive Maintenance in Network Utilities	Crespo et al. (2013)
11	Customer-oriented Risk Assessment in Network Utilities	Gómez et al. (2016a)
12	Dynamic Reliability Prediction of Asset Failure Modes	Gómez et al. (2016b)
13	A Quantitative Graphical Analysis to Support Maintenance	Barberá et al. (2012)
14	Case Study of Graphical Analysis for Maintenance Management	Barberá et al. (2013)
15	A Graphical Method to Support Operation Performance Assessment	Viveros et al. (2015)
16	Value-Driven Engineering of e-maintenance Platforms	Macchi et al. (2014)
17	Assistance to Dynamic Maintenance Tasks by Ann-Based Models	Olivencia et al. (2015)
18	Expected Impact Quantification Based on Reliability Assessment	Kristjampoller et al. (2016)
19	Influence of the Input Load on the Reliability of the Grinding Line	Barberá et al. (2014)

The a.m. references are detailed in Chapter "On the Family of Standards UNE-ISO 55000 and How to Effectively Manage Assets"

Conclusions

As introduced at the beginning of this preface, this book looks to promote and address the application of objective and effective decision-making in asset management based on mathematical models and practical techniques that can be easily implemented in organizations. This comprehensive and timely publication aims to be an essential reference source, building on the available literature in the field of asset management while providing for further research breakthroughs in this field. This text provides the necessary resources for managers, technology developers, scientists and engineers to adopt and implement optimum decision-making based on models and techniques that contribute to recognizing risks and uncertainties and, in general terms, to the important role of asset management to increase competitiveness in organizations.

Seville, Spain   Adolfo Crespo Márquez
                                                            Vicente González-Prida Díaz
                                                      Juan Francisco Gómez Fernández

Reference

Crespo Márquez A (2007) The maintenance management framework. Models and methods for complex systems maintenance. Springer Series in Reliability Engineering. ISBN: 978-1-84628-820-3

Acknowledgements

The editors wish to thank specific people and institutions for providing their help during the year 2016 and 2017, making the publication of this book possible.

The 5th of April 2016, Professor Marco Garetti passed away. Marco was Full Professor of Industrial Technology at the Department of Management, Economics and Industrial Engineering of Politecnico di Milano, Italy. As a visionary in the area of asset management and maintenance, he devoted a great deal of effort to the creation of an international community of scholars and researchers in this area. Marco was always a person very close to our research group in Spain and helped very significantly in the international dimension of our work, which can be appreciated in this book. At the time of publishing this book we want to leave in writing our deep recognition of Marco's professional work and effort, and our profound appreciation for his exceptional support and friendship for so many years.

This research work was performed within the context of Sustain Owner ('Sustainable Design and Management of Industrial Assets through Total Value and Cost of Ownership'), a project sponsored by the EU Framework Programme Horizon 2020, MSCA-RISE-2014: Marie Skłodowska-Curie Research and Innovation Staff Exchange (RISE) (grant agreement number 645733—Sustain-Owner—H2020-MSCA-RISE-2014).

At the same time, the funding from the Spanish Ministry of Economy and Competitiveness, as well as from European Regional Development Funds (ERDF), during the time while this book has been written (Research Project DPI2015-70842-R "Development of advanced operation and maintenance processes using Cyber Physical Systems—CPS—within the scope of Industry 4.0") made possible many research works and contributions related to the content of this book.

To all of them, thanks.

Contents

Editors and Contributors

About the Editors

Adolfo Crespo Márquez is currently Full Professor at the School of Engineering of the University of Seville, and Head of the Department of Industrial Management. He holds a Ph.D. with Honours in Industrial Engineering from this same University. His research works have been published in journals such as Reliability Engineering and System Safety, International Journal of Production Research, International Journal of Production Economics, European Journal of Operations Research, Omega, Decision Support Systems and Computers in Industry, among others. Prof. Crespo is the author of eight books, the last five with Springer-Verlag (2007, 2010, 2012, 2014) and Aenor (2016) about maintenance, warranty, supply chain and assets management. Professor Crespo is Fellow of ISEAM (International Society of Engineering Assets Management) and leads the Spanish Research Network on Assets Management and the Spanish Committee for Maintenance Standardization (1995–2003). He also leads the SIM (Sistemas Inteligentes de Mantenimiento) research group related to maintenance and dependability management and has extensively participated in many engineering and consulting projects for different companies, for the Spanish Departments of Defense, Science and Education as well as for the European Commission (IPTS). He is the President of INGEMAN (a National Association for the Development of Maintenance Engineering in Spain) since 2002.

Vicente González-Prida Díaz holds a Ph.D. with Honours in Industrial Engineering by the University of Seville, and Executive MBA (First Class Honours) by the Seville Chamber of Commerce. He also has been awarded with the National Award for Ph.D. Thesis on Dependability by the Spanish Association for Quality; the National Award for Ph.D. Thesis on Maintenance by the Spanish Association for Maintenance; and the Best Nomination from Spain for the Excellence Master Thesis Award bestowed by the EFNSM (European Federation of National Maintenance Societies). Dr. González-Prida has authored a book with Springer Verlag about Warranty and After Sales Assets Management (2014) and many other publications in relevant journals, books and conferences, nationally and internationally. His main interest is related to industrial asset management, specifically the reliability, maintenance and life cycle organization. He currently works as Project Manager in the company General Dynamics-European Land Systems and shares his professional performance with the development of research projects within the SIM (Sistemas Inteligentes de Mantenimiento) research group in the Department of Industrial Organization and Management at the University of Seville and teaching activities in Spain and Latin-America.

Juan Francisco Gómez Fernández is Ph.D. in Industrial Management and Executive MBA. He is currently part of the SIM research group of the University of Seville and a member in knowledge sharing networks about Dependability and Service Quality. He has authored a book with Springer Verlag about Maintenance Management in Network Utilities (2012) and many other publications in relevant journals, books and conferences, nationally and internationally. In relation to the practical application and experience, he has managed network maintenance and deployment departments in various national distribution network companies, both from private and public sector. He has conducted and participated in engineering and consulting projects for different international companies, related to Information and Communications Technologies, Maintenance and Asset Management, Reliability Assessment, and Outsourcing services in Utilities companies. He has combined his professional activity, in telecommunications networks development and maintenance, with academic life as an associate professor (PSI) in Seville University, and has been awarded as Best Master Thesis on Dependability by National and International Associations such as EFNSM (European Federation of National Maintenance Societies) and Spanish Association for Quality.

Contributors

Adolfo Arata Andreani Escuela de Ingeniería Industrial, Pontificia Universidad Católica de Valparaíso, Valparaíso, Chile

Luis Barberá Martínez Department of Industrial Management, School of Engineering, University of Seville, Seville, Spain

Gonzalo Cerruela García Department of Computer and Numerical Analysis, University of Cordoba, Córdoba, Spain

Adolfo Crespo Márquez Department of Industrial Management, School of Engineering, University of Seville, Seville, Spain

Pedro Moreu de León Department of Industrial Management, School of Engineering, University of Seville, Seville, Spain

Jesús Ferrero Bermejo Magtel Systems, Seville, Spain

Luca Fumagalli Department of Management, Economics and Industrial Engineering, Politecnico di Milano, Milan, Italy

Vicente González-Prida Díaz General Dynamics—European Land Systems, Seville, Spain; Department of Industrial Management, School of Engineering, University of Seville, Seville, Spain

Juan Pablo González Department of Industrial Engineering, Universidad Técnica Federico Santa María, Valparaíso, Chile

Juan Francisco Gómez Fernández Department of Industrial Management, School of Engineering, University of Seville, Seville, Spain

Antonio Jesús Guillén López Department of Industrial Management, School of Engineering, University of Seville, Seville, Spain; Intelligent Maintenance System research group (SIM), University of Seville (USE), Seville, Spain

María Holgado Granados Department of Engineering, Institute for Manufacturing, University of Cambridge, Cambridge, UK

Fredy Kristjanpoller Rodríguez Department of Industrial Engineering, Universidad Técnica Federico Santa María, Valparaíso, Chile; Department of Industrial Management, School of Engineering, University of Seville, Seville, Spain

Mónica Alejandra López-Campos Department of Industrial Engineering, Universidad Técnica Federico Santa María, Valparaíso, Chile

Marco Macchi Department of Management, Economics and Industrial Engineering, Politecnico di Milano, Milan, Italy

Rodrigo Mena Department of Industrial Engineering, University Federico Santa María, Valparaíso, Chile

Fernando Agustín Olivencia Polo Magtel Systems, Seville, Spain

Carlos Parra Márquez Department of Industrial Management, School of Engineering, University of Seville, Seville, Spain; IngeCon, Ingenieria de Confiabilidad, Caracas, Venezuela

Antonio Sola Rosique Department Industrial Management, School of Engineering, University of Seville, Seville, Spain; Technical Services, Iberdrola Generación S.A. Madrid, Madrid, Spain; INGEMAN. Association for the Development of Maintenance Engineering, School of Engineering, Seville, Spain

Raúl Stegmaier Department of Industrial Engineering, Universidad Técnica Federico Santa María, Valf Paraíso, Chile

René Tapia Peñaloza RelPro SpA, Valparaíso, Chile

Pablo Viveros Gunckel Department of Industrial Engineering, Universidad Técnica Federico Santa María, Valparaíso, Chile; Department of Industrial Management, School of Engineering, University of Seville, Seville, Spain

On the Family of Standards UNE-ISO 55000 and How to Effectively Manage Assets

Adolfo Crespo Márquez, Antonio Jesús Guillén López, Antonio Sola
Rosique and Carlos Parra Márquez

Abstract The publication of the ISO 55000 family of standards on asset management is surely a very important event for many economic activity sectors, especially for those that are very intensive in capital investments devoted to physical assets. Although the standards set a framework for the requirements to fulfill in order to manage assets properly, companies, and organizations in general, need to know how to reach those requirements. What are the necessary steps to follow and the supporting structure that needs to be built in order to develop a proper, consistent and competitive assets management process and system? This chapter links ISO 55000 requirements to the assets management framework promoted by the authors, and at the same time, links the models presented in the different chapters of the book, with specific elements of the standards.

Keywords Assets management · Management framework · Management supporting structure · Maintenance engineering techniques

A. Crespo Márquez (✉) · C. Parra Márquez
Department of Industrial Management, School of Engineering,
University of Seville, Camino de los Descubrimientos s/n, 41092 Seville, Spain
e-mail: adolfo@us.es

A.J. Guillén López
Intelligent Maintenance System research group (SIM),
University of Seville (USE), Seville, Spain

A. Sola Rosique
INGEMAN. Association for the Development of Maintenance Engineering,
School of Engineering, Camino de los Descubrimientos, Seville 41092, Spain

C. Parra Márquez
IngeCon, Ingenieria de Confiabilidad, Caracas, Venezuela

© Springer International Publishing AG 2018   1
A. Crespo Márquez et al. (eds.), *Advanced Maintenance Modelling
for Asset Management*, DOI 10.1007/978-3-319-58045-6_1

1 Introduction

The ISO 55000 family of standards on asset management has just been published. These are three documents that present the minimum requirements on good practices to establish, implement, maintain and improve the management of any type of asset in organizations. They also offer a strategic approach to incorporate operations and maintenance applications, and thus improve asset availability and utilization. The benefits of asset management in organizations, with a focus on achieving value over the asset life cycle, are solidly proven in many industries and business environments. In addition, it demonstrates organizations' commitment to quality, performance or safety, helping to mitigate the legal, social and environmental risks associated with accidents in industrial facilities.

The recent publication of the family of ISO 55000 Standards on Asset Management (AM) aims to support a management oriented to obtain value of the assets. This is the ISO 55000 Asset Management. General aspects, principles and terminology, which provides a broad view of what AM represents; ISO 55001 Asset management. Management systems. Requirements, which specifies the requirements for establishing an AM system (see Fig. 1); and ISO 55002 Asset Management. Management systems. Guidelines for the implementation of ISO 55001, which provides guidance for the application of that standard. Thus, it presents in a generic way the minimum requirements on good practices to establish, implement, maintain and improve the management of any type of asset, establishing a strategic approach to incorporate operations and maintenance applications to improve the availability and use of assets. These requirements apply to all stakeholders, allowing to measure and show the organization's ability to meet legal, regulatory and contractual requirements, as well as those of the organization.

Standard ISO 55001 does not define "how" to carry out such good practices. And this will depend on the context of the organization itself and the assets to be managed. In addition, it will in the future be a source of development for the different business areas and types of assets in the interpretation and application of the requirements established by the standard. The formal recognition through the standard of what needs to be done (elements and requirements), for the coordination and maintenance of good practices, is the basis for organizing the processes and achieving the goals set.

It helps organizations realize even more the value of their assets, enabling them to demonstrate their ability to control risks, reliability of their plants, loss mitigation and unplanned outages. In short, the purpose of this series of standards is to provide a model for the creation and operation of an Asset Management System (AMS). This system can be integrated (see Fig. 2) with other management systems, such as quality, environment or safety.

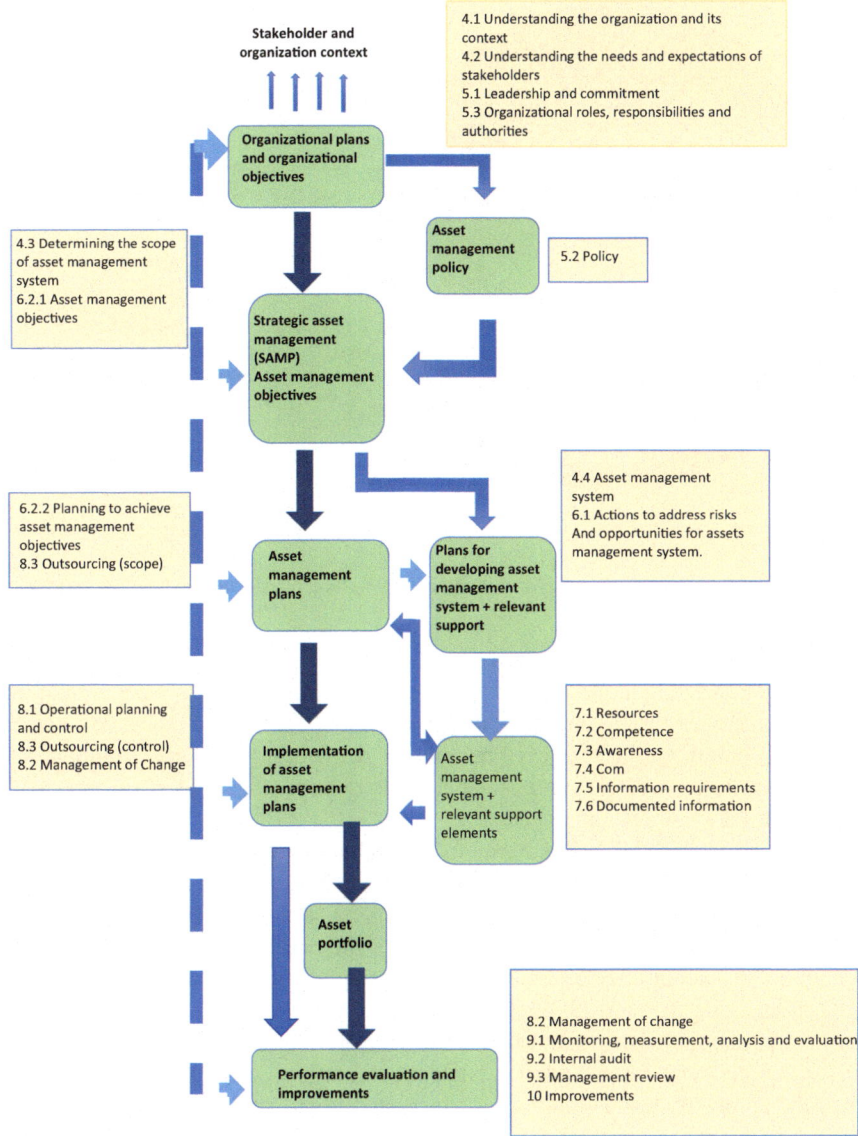

Fig. 1 Relationship between key elements of an asset management system

2   Principles Promoted by the Standards

Among the central themes of the ISO 55000 Standards family are the concepts of value creation and risk-based decision-making, considering four fundamental principles. That is, alignment of the objectives of the company, from the top leaders

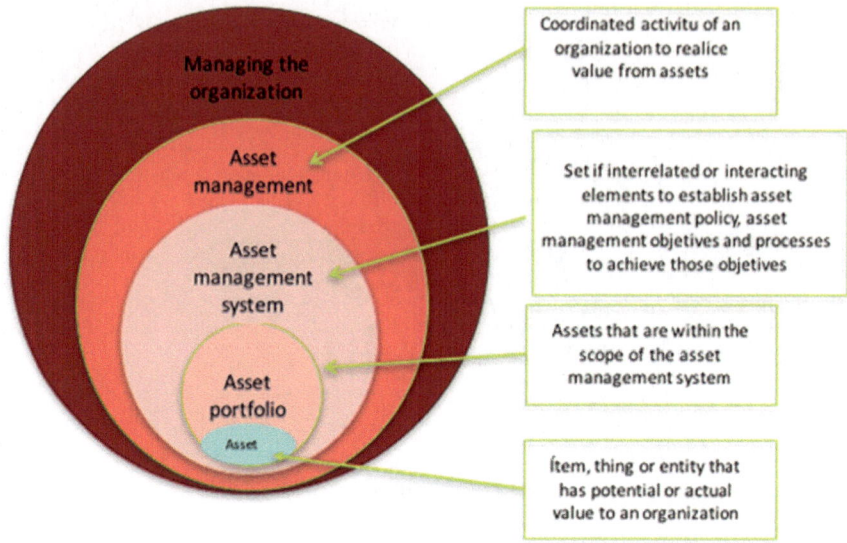

Fig. 2 Relationships between key terms: Integration the assets management with other systems of the organization

of the organization to the technicians responsible for the day-to-day operation of the assets; transparent and consistent decision-making, seeking a balance between potentially conflicting initiatives and limited resources; risk participation in the decision-making process; and balancing long-term asset requirements with short-term business planning cycles.

Emphasis is placed on value creation, but with a focus on the idea and long-term strategy since the duration of the assets can be much greater than the strategic plan of the corporation. Better knowledge of assets helps in making operational decisions and in understanding the performance of the organization in general. It also emphasizes stakeholder engagement as well as alignment with business finance and accounting. The standard ISO 55001: 2014 describes the requirements of the system emphasizing "what needs to be done" and not "how to do it", based on the following elements: organization, context, leadership, planning, resources and support, operation, evaluation, performance, and improvement.

It is important to consider that the application and use of this standard presents certain challenges for its implementation in the different organizations. To a great extent, these challenges are linked to the maturity of their systems and processes of the organization, since a high level of integration, harmonization, and coordination of functions between engineering, operations, maintenance, and the commercial part of the business is required for good management. In this sense, it is necessary to emphasize the "culture" in the use of Management Systems. This will be the main challenge in organizations with no prior experience in these systems. Also, if the organization has competent personnel, since even mature organizations will have to acquire competencies to evaluate the processes and the Asset Management System. In addition, it is necessary to interpret the standard to adapt its generalist approaches to the particularities of the business to which it applies. Finally, leadership commitment is essential, requiring the involvement of top management from the beginning of the implementation process of an Asset Management System.

As a result, the ISO 55000 family of standards offers important opportunities for asset owners to re-examine and refine their management model. It also helps to improve relationships with service providers and customers, governance of the management model, regulatory frameworks, and stakeholder trust. The benefits that the improved of asset management contribute, with an integrated approach to value across the asset life cycle, are solidly proven in many industries and business environments, improving the quality of life through contributing to safety, human health and environmental protection while demonstrating the organization's commitment to quality, performance or safety and helping to mitigate the legal, social, and environmental risks associated with accidents in industrial facilities.

3 Integration of a Maintenance Management Model (MMM) with the Asset Management Standard ISO 55000

Although there are no simple formulas for the implementation of an integral model of management of asset, nor fixed or immutable rules with validity and applicability for all the assets of production, the requirements needed by the proposal of standard ISO 55000 can be covered by the integral maintenance management model (Fig. 3) proposed at the beginning of this report. In the MMM, composed of eight phases, specific actions are described to follow in different steps of the process of management of maintenance that are integrated in a direct form within a process of management of assets [4]. The MMM offers a dynamic, sequential process and in a closed loop that tries to accurately characterize the course of actions to be carried out to ensure the efficiency, effectiveness and continuous improvement of the management of assets from the use and integration of techniques of engineering and maintenance management and reliability.

In particular, in Table 1, a relationship is made between the 8 phases of the model proposed and the general points of the standard ISO 55000, so that the gradual implementation of the generic model progressively covering the requirements of the standard ISO 55000 may be looked at. According to 1, the activities to be developed within the eight stages of the MMM can help organizations, to meet with the 24 requirements demanded by the standard ISO 55000. The following describes in more detail the relationship between the phases of the MMM and the requirements of ISO 55000.

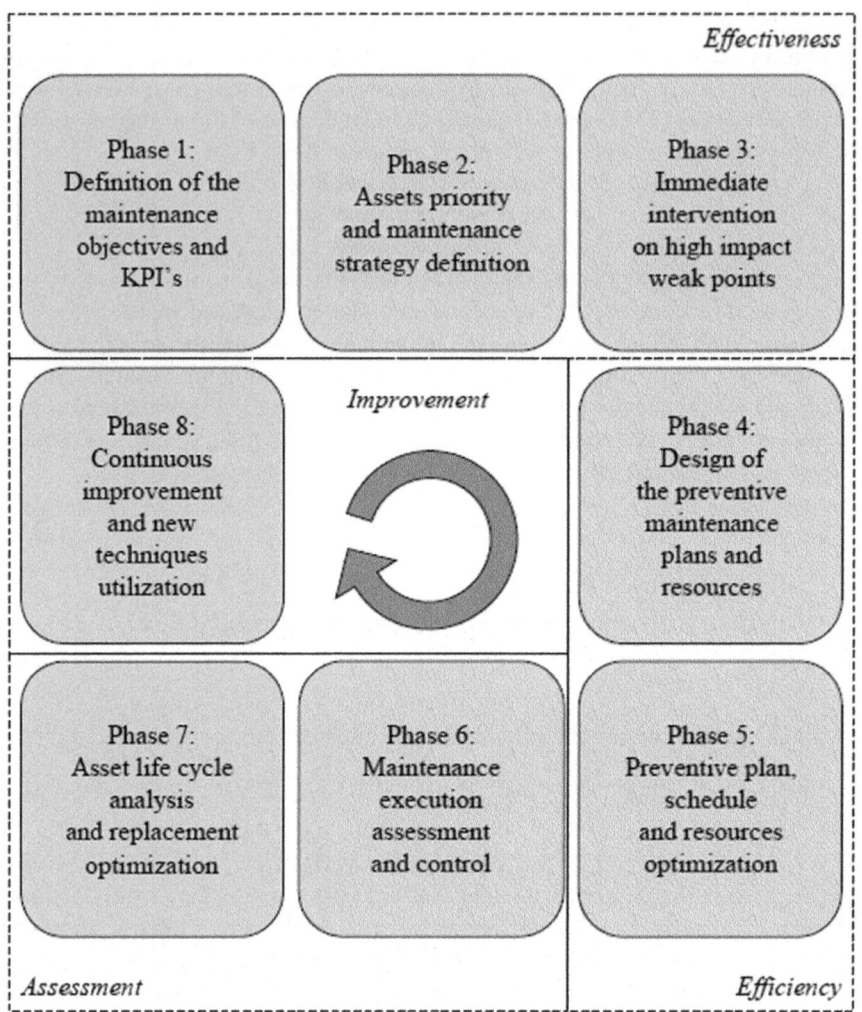

Fig. 3 Model of the process of maintenance management (MMM) integrated into ISO 55000 Crespo Márquez [4]

According to Table 2, out of the 24 requirements defined by the standard ISO 55000, the maintenance management model (MMM) can help us totally or partially meet the demands of the requirements expected by this standard (the proposal of standard PAS 55 represents the most important background of standard ISO 55000).

Table 1 Relationship between the phases of the maintenance management model (MMM) proposed and the requirements of ISO 55000 [18]

ISO 55000 requirements	Integration of the phases of the MMM proposed with standard ISO 55000

4. Context of the organization
    4.1. Understanding the Organization and its context
    4.2. Understanding the needs and expectations of interested parties
    4.3. Determining the extent of the asset management system
    4.4. system of management of assets
5. Leadership
    5.1. Leadership and commitment
    5.2. Policies
    5.3. Roles, organizational responsibilities and authorities
	Phase 1. Proposes the use of the scorecard (balanced Scorecard—BSC), proposed by Kaplan and Norton, model that translates the mission of a business unit into its strategy in a set of objectives and quantifiable measures. By implementing the BSC, organizations get to:
	1. Formulate policies and strategies for the operation and performance of the maintenance of assets throughout their lifecycle
	2. Put into practice the strategies of maintenance and operation, which is translated into objectives at short, medium and long term
	3. Develop the plans of action. These are the means to get to the purposes stipulated in the objectives set out in step (2)
	4. Establish leadership in the different processes to improve in all areas of the Organization
	5. Review and periodically audit the performance of implemented strategies. Monitoring will be made and the casual relations between the measures will be investigated what will be validated at intervals previously established and plans of contingency will be defined
	Additionally in phase 1, the MMMC model proposes that an cohesive organization is designed which supports the process of asset management and is able to implement a holistic process optimization based on the application of techniques of reliability and maintenance, with the assignment of roles, responsibilities, and definition of the leadership of all the activities to be developed during the lifecycle of the asset
	Phase 2. Proposes the use of models of prioritization, which must comply and align with the expectations of stakeholders (interested parties) and at the same time, cover the legal requirements demanded by the environment of the asset

(continued)

Table 1 (continued)

ISO 55000 requirements	Integration of the phases of the MMM proposed with standard ISO 55000
6. Planning	Phase 2. Proposes at the beginning of a process of improvement, the development and the application of basic models of prioritization of assets based on the analysis of the risk factor (example: qualitative and technical matrix of risks AHP: Analytics Hierarchy, Process, etc.)
6.1. Actions to address the risks and the opportunities in the system of management of assets	Phase 3. Proposes the use of the methodology of root cause analysis (RCA) to assess the failures of major impact events, taking as a basis for the definition of solutions, the level of risk caused by failure events to be analyzed
6.2. Objectives for the management of assets and planning to achieve them	Phase 4. Proposes the use of methodology of reliability-centered maintenance (RCM) to optimize maintenance and operation depending on the level of risk plans that generate failures within the context of the operational modes
7 Support	Phase 5. Proposes the use of methods of optimization to be used in the programming and allocation of resources for maintenance and operations. Within the selected methods are the techniques related to processes such as risk analysis: theory of queues, Monte Carlo simulation and probabilistic techniques of point of order from inventory
7.1. Resources	Additionally, at this stage, using continuous improvement methods is proposed in the programming, planning and allocation of resources for maintenance and operations, risk management-based
7.2. Competencies	
7.3. Awareness	Phase 8. Proposes the use of the systems of information support (ERP, EAM, software of reliability, etc.), to manage and disclose all the documentation and information to be generated by the different assets in their processes of operation and maintenance. The information systems for the management of assets are key tools for their ability to support and facilitate their management, thanks to the transmission and processing of information at high speeds and quantities exceeding the organizations' own borders and strengthening the convergence among sectors. The need for a correct implementation of the support for the management of information systems is the basis for the development of programs to improve reliability, maintenance and operations
7.4. Communication	
7.5. Requirements of information	
7.6. Documented information	

(continued)

Table 1 (continued)

ISO 55000 requirements	Integration of the phases of the MMM proposed with standard ISO 55000
8. Operation 8.1. Operational planning and control 8.2. Change management 8.3. Outsourcing 9. Evaluation of performance 9.1. Monitoring, measurement, analysis and evaluation 9.2. Internal audit 9.3. Revision of the management	Phase 1. Proposes the use of the Balanced Scorecard-BSC table to measure and review the indicators of economic performance of the Organization and subsequently, integrate them with the technical indicators of operation and maintenance (technical indicators that are developed in phase 6). Additionally, in this phase 1, the use of audits of control and continuous improvement was proposed among which is found: MES (Maintenance Effectiveness Survey), QMEM (Qualitative Matrix of Excellent in Maintenance), etc. Phases 3 and 4. Propose the application of reliability as the RCA and the RCM methods that allow evaluating modes of failure and determine their causes. These methods help to determine the incidents and non-conformities, allow to evaluate the consequences that the failures can cause on safety, the environment and operations and additionally, these techniques propose procedures that help to define actions of improvement and control: corrective, preventive, of redesign and by condition Phases 5. Proposes the application of methods of optimization of maintenance and reliability engineering, which would help to define the processes of planning, programming, outsourcing and the level of training necessary to improve the management of assets in their lifecycle Phase 6. Offers a comprehensive process of measurement, analysis and evaluation of indicators of performance and improvement (indicators of probabilistic assessment: reliability, maintainability, availability, cost and risk) Phase 8. Proposes to establish a process of continuous improvement which should be able to register and to adjust to the constant changes related to techniques and emerging technologies in areas that are considered of high impact as a result of the studies carried out in the previous 8 phases of the proposed maintenance management model

(continued)

Table 1 (continued)

ISO 55000 requirements	Integration of the phases of the MMM proposed with standard ISO 55000
10 Improvement 10.1. Non-conformity and corrective action 10.2. Preventive action 10.3. Continuous improvement	Phase 2. Proposes at the beginning of a process of improvement, the development and application of basic models of prioritization of assets based on the analysis of the risk factor (example: technical and qualitative risk matrix AHP: Analytics, Hierarchy, Process, etc.) Phase 3. Proposes the use of the methodology of analysis cause root (RCA: Root Cause Analysis) to evaluate them events of failures of greater impact, taking as base for the definition of solutions, the level of risk caused by them events of failures to be analyzed (processes of not conformity and actions corrective) Phase 4. Proposes the use of the reliability-centered (RCM) maintenance methodology, to optimize maintenance and operation depending on the level of risk plans that generate the modes of failures within the operational context (preventive action) Phase 5. Proposes the use of methods of optimization to be used in the programming and allocation of resources for maintenance and operations. Selected methods techniques include related processes such as risk analysis: theory of queues, Monte Carlo simulation and probabilistic techniques of point of order from inventory Phase 6. Proposes a holistic process of probabilistic evaluation of the indicators of: reliability, maintainability, availability, cost and risk Additionally, in this phase a procedure is explained that allows to relate the indicators of reliability and maintainability, with decisions of optimization in the areas of maintenance and operation based on techniques of cost risk benefit analysis (continuous improvement) Phase 7. Proposes a process of cost analysis of life cycle that allows optimizing decision-making associated with the processes of design, selection, development and replacement of assets that make up a production system. The process of life cycle begins with the definition of the different tasks of production for the preliminary design. Then activities are developed such as: plan of production, layout of plant, selection of equipment, definition of processes of manufacturing and other similar activities. Subsequently, prior to the design phase logistics is considered. This phase involves the development of the necessary support for the design and the different stages of production, the possible user support, maintenance plan intended for the use of the asset and the process of divestiture of assets (continuous improvement) Phase 8. Proposes establishing a process of continuous improvement which must be capable of reviewing and evaluating the technical and economic performance of the Organization in a continuous way

Table 2 Link between parts and published references

Part	References
Part I: Introduction	Sola Rosique, Antonio, Crespo Marquez, Adolfo, Guillen Lopez, Antonio Jesus [19]. Bases para la mejora de la gestión de activos en las organizaciones. Industria Química
Part II: A changing asset management framework	López Campos, Mónica Alejandra, Crespo Marquez, Adolfo [13]. Modelling a maintenance management framework based on PAS 55 standard. Quality and Reliability Engineering International
	López-Campos, Mónica Alejandra; Crespo-Marquez, Adolfo; Gómez-Fernández, Juan Francisco [14]. Modelling using UML and BPMN the integration of open reliability, maintenance and condition monitoring management systems: An application in an electric transformer system. Computers in industry
	Guillén A.J., Crespo A., Macchi M., Gómez J [10]. On the role of Prognostics and Health Management in advanced maintenance systems, Production Planning and Control (In printed)
	Antonio J. Guillén, Juan Francisco. Gómez, Adolfo Crespo [11]. Framework for effective management of CBM programs. Computers in Industry
Part III: Pursuing high management effectiveness in a dynamic environment	Adolfo Crespo Márquez, Pedro Moreu de León, Antonio Sola Rosique, Juan F. Gómez Fernández [5]. Criticality Analysis for Maintenance Purposes: A Study for Complex In-service Engineering Assets. Quality and reliability engineering international
	González-Prida, Vicente; Viveros, Pablo; Barberá-Martínez, Luis; Crespo-Marquez, Adolfo [9]. Dynamic Analytic Hierarchy Process: Ahp Method Adapted To A Changing Environment. Journal of Manufacturing Technology Management
	P. Viveros, A. Crespo, R. Tapia, F. Kristjanpoller, V. González-Prida [20]. Reliability Stochastic Modeling for Repairable Physical Assets. Case study applied to the Chilean Mining. DYNA, 91(4). 423–431. doi:http://dx.doi.org/10.6036/7863
	Carlos Parra, Adolfo Crespo, Fredy Kristjanpoller, Pablo Viveros [17]. Reliability stochastic model applied to evaluate the economic impact of the failure in the life cycle cost analysis (LCCA). Case Study for the Rail Freight and Oil Industries. Proc IMechE Part O: Journal of Risk and Reliability. 226(4) 392–405, DOI:10.1177/1748006X12441880

(continued)

Table 2 (continued)

Part	References
Part IV: Advanced methods and techniques to improve management efficiency	Crespo-Marquez, Adolfo; Gómez-Fernández, Juan Francisco; Moreu-De Leon, Pedro; Sola-rosique, Antonio [6]. Modelling on-line reliability and risk to schedule the preventive maintenance of repairable assets in network utilities. IMA journal of management mathematics
	Juan F. Gómez Fernández, Adolfo Crespo Márquez, Mónica A. López-Campos [8]. Customer-oriented risk assessment in network utilities Reliability Engineering and System Safety 147 (2016) 72–83
	J.F. Gómez Fernández, F. Olivencia, J. Ferrero, A. Crespo Márquez, G. Cerruela García [7] Analysis of Dynamic Reliability Surveillance: a Case Study. IMA Journal of management Mathematics
Part V: The need for innovation in assessment and control	Barberá Martinez, Luis; Crespo Marquez, Adolfo; Viveros Gunckel, Pablo; Arata Andreani, Adolfo [2]. The Graphical Analysis for Maintenance Management Method: A Quantitative Graphical Analysis to Support Maintenance Management Decision Making. Quality and Reliability Engineering International
	Barberá-Martinez, Luis; Crespo-Marquez, Adolfo; Viveros, Pablo; Stegmaier, Raúl [3]. Case Study of GAMM (Graphical Analysis For Maintenance Management) in the Mining Industry. Reliability Engineering and System Safety
	Viveros, Pablo, Crespo Marquez, Adolfo, Barberá Martínez, Luis, Gonzalez, Juan Pablo [21]. Graphical Analysis for Operation Management: A Graphical Method to Support Operation Decision Making. Quality and Reliability Engineering International
Part VI: Continuous improvement through emergent process and technologies	Macchi, Marco; Crespo-Marquez, Adolfo; Holgado-Granados, María; Fumagalli, Luca; Barberá-Martinez, Luis [15]. Value-Driven Engineering of E-Maintenance Platforms. Journal Of Manufacturing Technology Management
	Olivencia, Fernando, Ferrero, Jesus, Gómez Fernández, Juan Francisco, Crespo Marquez, Adolfo [16]. Failure mode prediction and Energy forecasting of PV plants to assist dynamic Maintenance tasks by ANN based models. Renewable Energy
	F. Kristjanpoller, A. Crespo, P Viveros, L. Barberá [12]. Expected Impact Quantification based Reliability Assessment Methodology for Chilean Copper Smelting Process—A Case Study. Advances in Mechanical Engineering. Vol 8(10):1–13. DOI 10.1177/1687814016674845
	Barberá Martinez, Luis, Viveros, Pablo, Mena, Rodrigo, Gonzalez-Prida Diaz, Vicente [1]. Influence of the Input Load on the Reliability of the Grinding Line. Dyna

4 Link Between the Book Parts and Published References

The present book connects the phases of the Maintenance Management Model with the ISO 55000 [18], integrating new technologies, methods and concept provided by the so-called context Industry 4.0. There are different research results of the last 5 years serving as the main basis and background for the different parts of this book. Particularly, each part gathers the following papers, previously published.

As commented, the book is a compilation of the most relevant and recent contributions of our research group. Each chapter corresponds to an article published previously.

5 Conclusions

A better knowledge of its assets helps an organization to make operational decisions and, in general terms, will improve organizational performance. In this Chapter we have explained how an MMM can help in the process of establishing an asset management model fulfilling the requirements of the ISO 55000 family of standards. Also, each Chapter of the book has been declared as related to, or linked to, a certain standard requirement. Nevertheless, the reader must be aware that concerning standard ISO 55000, key factors that influence an organization to achieve its goals are the following: (i) Nature and purpose of the organization; (ii) Its operational context; (iii) Its financial restrictions and regulatory requirements; (iv) The needs and expectations of the organization and interested parties (stakeholders).

References

1. Barberá ML, Viveros, P, Mena R, Gonzalez-Prida DV (2014) Influence of the input load on the reliability of the grinding line. Dyna 89(5):560–568
2. Barberá-Martínez L, Marquez AC, Gunckel PV, Andreani AA (2012) The graphical analysis for maintenance management method: a quantitative graphical analysis to support maintenance management decision making. Qual Reliab Eng Int 29(1):77–87
3. Barberá-Martínez L, Crespo-Marquez A, Viveros P, Stegmaier R (2013) Case study of gamm (graphical analysis for maintenance management) in the mining industry. Reliab Eng Syst Saf
4. Crespo Marquez A (2007) The maintenance management framework. Models and methods for complex systems maintenance. Springer, London. ISBN 978-1-84628-820-3
5. Crespo Márquez A, Moreu de León P, Sola Rosique A, Gómez Fernández JF (2015) Criticality analysis for maintenance purposes: a study for complex in-service engineering assets. Qual Reliab Eng Int 32(2):519–533
6. Crespo-Marquez A, Gómez-Fernández JF, Moreu-De Leon P, Sola-rosique A (2013) Modelling on-line reliability and risk to schedule the preventive maintenance of repairable assets in network utilities. IMA J Manage Math 24(4):437–450

7. Gómez Fernández JF, Olivencia F, Ferrero J, Crespo Márquez A, Cerruela García G (2015) Analysis of dynamic reliability surveillance: a case study. IMA J Manage Math
8. Gómez Fernández JF, Crespo Márquez A, López-Campos MA (2016) Customer-oriented risk assessment in network utilities. Reliab Eng Syst Saf 147:72–83
9. González-Prida V, Viveros P, Barberá-Martínez L, Crespo-Marquez A (2014) Dynamic analytic hierarchy process: Ahp method adapted to a changing environment. J Manuf Technol Manage
10. Guillén AJ, Crespo A, Macchi M, Gómez J (2016) On the role of prognostics and health management in advanced maintenance systems. Prod Plan Control (In printed)
11. Guillén AJ, Gómez Juan F, Crespo A (2015) A Framework for effective management of CBM programs. Comput Ind
12. Kristjanpoller F, Crespo A, Viveros P, Barberá L (2016) Expected impact quantification based reliability assessment methodology for chilean copper smelting process—a case study. Adv Mech Eng 8(10):1–13. doi:10.1177/1687814016674845
13. López C, Mónica A, Crespo Marquez A (2011) Modelling a maintenance management framework based on PAS 55 standard. Qual Reliab Eng Int 27(6):805–820
14. López-Campos MA, Crespo-Marquez A, Gómez-Fernández JF (2013) Modelling using UML and BPMN the integration of open reliability, maintenance and condition monitoring management systems: an application in an electric transformer system. Comput Ind 64 (5):524–542
15. Macchi M, Crespo-Marquez A, Holgado-Granados M, Fumagalli L, Barberá-Martínez L (2014) Value-driven engineering of E-Maintenance platforms. J Manuf Technol Manage 25 (4):568–598
16. Olivencia F, Ferrero J, Gómez Fernández, JF, Crespo Marquez A (2015) Failure mode prediction and energy forecasting of PV plants to assist dynamic maintenance tasks by ANN based models. Renew Energy 81:227–238
17. Parra C, Crespo A, Kristjanpoller F, Viveros P (2012) Reliability stochastic model applied to evaluate the economic impact of the failure in the life cycle cost analysis (LCCA). Case Study for the rail freight and oil industries. Proc IMechE Part O J Risk Reliab. 226(4):392–405. doi:10.1177/1748006X12441880
18. Parra C, Crespo A (2015) "Ingeniería de Mantenimiento y Fiabilidad aplicada en la Gestión de Activos. Desarrollo y aplicación práctica de un Modelo de Gestión del Mantenimiento". Segunda Edición. Editado por INGEMAN, Escuela Superior de Ingenieros Industriales de la Universidad de Sevilla, España
19. Sola RA, Crespo Marquez A, Guillen L, Antonio J (2015) Bases para la mejora de la gestión de activos en las organizaciones. Industria Química
20. Viveros P, Crespo A, Tapia R, Kristjanpoller F, González-Prida V (2016) Reliability stochastic modeling for repairable physical assets. Case study applied to the Chilean Mining. DYNA 91(4):423–431. doi:10.6036/7863
21. Viveros, P, Crespo Marquez A, Barberá Martínez L, Gonzalez JP (2015) Graphical analysis for operation management: a graphical method to support operation decision making. Qual Reliab Eng Int

Author Biographies

Adolfo Crespo Márquez is currently Full Professor at the School of Engineering of the University of Seville, and Head of the Department of Industrial Management. He holds a Ph.D. with Honours in Industrial Engineering from this same University. His research works have been published in journals such as Reliability Engineering and System Safety, International Journal of Production Research, International Journal of Production Economics, European Journal of Operations

Research, Omega, Decision Support Systems, and Computers in Industry, among others. Prof. Crespo is the author of eight books, the last five with Springer-Verlag (2007, 2010, 2012, 2014) and Aenor (2016) about maintenance, warranty, supply chain and assets management. Prof. Crespo is Fellow of ISEAM (International Society of Engineering Assets Management) and leads the Spanish Research Network on Assets Management and the Spanish Committee for Maintenance Standardization (1995–2003). He also leads the SIM (Sistemas Inteligentes de Mantenimiento) research group related to maintenance and dependability management and has extensively participated in many engineering and consulting projects for different companies, for the Spanish Departments of Defense, Science and Education as well as for the European Commission (IPTS). He is the President of INGEMAN (a National Association for the Development of Maintenance Engineering in Spain) since 2002.

Antonio Jesús Guillén López is a Ph.D. candidate and contracted researcher in Intelligent Maintenance System research group (SIM) of the University of Seville (USE), focusing his studies in Prognosis Health Management & Condition-Based Maintenance applications and Assets Management; Industrial Engineering and Master in Industrial Organization & Business Management from the University of Seville (USE). From 2003 to 2004, he worked for the Elasticity and Strength of Materials Group (GERM) of the USE in aeronautical materials tests. From 2006 to 2009, he was a member of the Department of Electronic Engineering of USE, and worked in numerous public–private international R&D project, developing new thermo-mechanical design applications for improving the performance and the life cycle of power electronics system. From 2009 to 2010, he was a Project Manager of "Solarkit" project, representing Andalucia's Government and USE in the international competition Solar Decathlon 2010. He was a foundational partner of Win Inertia Tech., technological based company specializing in electronic R&D for Smart Grids and Energy Store fields, where he has carried out different roles, from R&D engineer to General Manager until September 2012. Currently, he is a coordinator of the Spanish National Research Network of Asset Management.

Antonio Sola Rosique is a civil engineer currently acting as Vicepresident of INGEMAN, (a National Association for the Development of Maintenance Engineering in Spain). His professional experience (34 years) is very much related to the field of dependability and maintenance engineering in different types of power generation plants for Iberdrola Generación (nuclear, thermal, hydro and combined cycles gas-steam). Antonio has coordinated various collaborative projects between Iberdrola and the University of Seville, and is active member in the board of European Safety,Reliability and Data Association (ESREDA), Asociación Española de Mantenimiento (AEM), Asociación Española de la Calidad (AEC) and Asociación Española de Normalización (AENOR). At present Antonio is about to finish a Ph.D. within the field of Maintenance and Risk Management in the University of Seville.

Carlos Parra Márquez Rewarded Titles:—Naval Engineering, Polytechnic Institute of the National Armed Forces (IUPFAN), 1986–1991, Caracas, Venezuela.—Master in Maintenance Engineering, Universidad de los Andes, School of Mechanical Engineering, Master Program in Maintenance Engineering, 1994–1996, Merida, Venezuela.—Reliability Specialist Engineering, PDVSA Convention—University Maryland—ASME, 2002–2003, United States.—Specialist in Industrial Organization Engineering, School of Industrial Engineering, University of Seville, 2004–2006, Seville, Spain.—Diploma of Advanced Studies, Industrial Engineering area Organization, PhD in Industrial Engineering, School of Industrial Engineering, University of Seville, 2006–2008, Seville, Spain.—Doctor (Ph.D.) in Industrial Organization Engineering, University of Sevilla, School of Industrial Engineering, Industrial Engineering Department Organization, 2004–2009, Sevilla, Spain. AWARDS/HONORS or Gran Mariscal de Ayacucho

Scholarship to study: Master Maintenance Engineering at the University of the Andes, Merida, Venezuela 1994 or OAS Scholarship ALBAN ESI (Organization of American States—European Community—Association for the Development of Maintenance Engineering—School of Industrial Engineering at the University of Sevilla) to study: Doctorate in Industrial Organization Engineering, University of Seville, Spain, 2004. o Award for best technical work in the 1st. World Congress of Maintenance Engineering, Bahia/Brazil, September 2005. Presentation/Publication: "Optimizing Maintenance Management process in the Venezuelan oil industry from the use of Reliability Engineering Methodologies".

A Maintenance Management Framework Based on PAS 55

Mónica Alejandra López-Campos and Adolfo Crespo Márquez

Abstract This chapter shows the process of modelling a reference maintenance management framework (MMF) that represents the general requirements of the asset management specification PAS 55. The modelled MMF is expressed using the standardized and publicly available Business Process Modelling languages UML 2.1 (Unified Modelling Language) and BPMN 1.0 (Business Process Modelling Notation). The features of these notations allow to easily integrate the modelled processes into the general information system of an organization and to create a flexible structure that can be quickly and even automatically adapted to new necessities. This chapter presents a brief review about the use of UML in maintenance projects, general characteristics of PAS 55, modelling concepts and theirs applications in the project of modelling the MMF. The arguments underlying the methodology and the choice of UML and BPMN are exposed. The general architecture of the suggested MMF is described and modelled through diagrams elucidating the general operation of PAS 55. From this development is appreciated the operation structure of a software tool that can incorporate MIMOSA standards and that can be made suitable to e-maintenance functions, as an alternative from the commercial systems. Finally, some conclusions about the modelled framework are presented.

Keywords Maintenance · Framework · Standard · UML · PAS 55

M.A. López-Campos (✉)
Department of Industrial Engineering, Universidad Técnica
Federico Santa María, Av. España 1680, Valparaíso, Chile
e-mail: monica.lopezc@usm.cl

A. Crespo Márquez
Department of Industrial Management, School of Engineering,
University of Seville, Camino de los Descubrimientos s/n, 41092 Seville, Spain
e-mail: adolfo@us.es

© Springer International Publishing AG 2018
A. Crespo Márquez et al. (eds.), *Advanced Maintenance Modelling for Asset Management*, DOI 10.1007/978-3-319-58045-6_2

1 Introduction

Regarding to asset management, maintenance has been experiencing a slow but constant evolution across years, from the earlier concept of "necessary evil" [1] up to being considered an integral function to the company and a way of competitive advantage [2].

For approximately three decades, companies realized that if they wanted to manage maintenance adequately it would be necessary to include it in the general scheme of the organization and to manage it in interaction with other functions [3].

Implanting a high-quality model to drive maintenance activities, embedded in the general management system of the organization, has become a research topic and a fundamental matter to reach effectiveness and efficiency of maintenance management and to fulfil enterprise objectives [4].

On the other hand, it is known that for a significant number of organizations every activity or important action realized has its reflection on its information system. This means that the enterprise information system is a basic element to consider for the implementation of a maintenance management system. In fact, the most desirable situation is the complete integration of the maintenance management operations into the general information system [5].

To deal with the mentioned integration of maintenance management and enterprise information systems this research proposes the use of the BPM (Business Process Management) methodology, which aim is to improve efficiency through the management of business processes that are modelled, automatized, integrated, controlled and continuously optimized [6]. BPM involves managing change in a complete processes life cycle.

By adopting the BPM methodology it is possible to model a particular maintenance management process and afterwards "connect" this model with a general information system.

In this way, a flexible management process can be created. If it was necessary to modify the management process to adapt its activities to new necessities, it would be quickly and even automatically modified into the enterprise information system [7].

UML and BPMN are the internationally standardized languages used in BPM methodology. A review of the literature of the last 10 years revealed that some maintenance applications expressed using UML already exist, but the majority of those specific applications are designed only for monitoring and/or diagnosis. An integral maintenance management framework (MMF) expressed using an approach to business process modelling (UML and BPMN) is an innovative project. It is also, more innovatory because the approach of the model to the PAS 55 standard.

This chapter presents the process of modelling a MMF, aligned to the asset management specification PAS 55:2008 [8] and expressed using UML and BPMN. Several ICT (information and communication technologies) proposals for the implementation of this project are explained at the end of the paper.

2 The Use of UML and BPMN Languages in Maintenance

The use of UML and BPMN standards in maintenance is relatively new and it is an expanding area. During the past years, the most used modelling standards were IDEF (Integration DEFinition), RAD (Role-Activity Diagrams), EXPRESS-G or STEP (Standard for the Exchange of Product model data).

As a literature review of maintenance developments employing UML and/or BPMN revealed, it is just since 2000 when UML began to be mentioned in maintenance projects [9]. When the noticeable advantages of UML and BPMN as internationally standardized modelling languages got more recognition, the number of maintenance projects using them began to increase. This growth is referred particularly to the use of UML, and specially for developing e-maintenance applications.

This literature review was performed covering the maintenance projects published until September 2010 and it produced the following results:

- The first project found relating UML to a maintenance task appeared in 2000, and it was about a CBM system for e-monitoring applied to an electric system [9].
- The majority of the applications using UML are made in e-monitoring [10–13] and e-diagnosis [14–16].
- The electric and electronic industries, along with the transportation industry, lead the number of maintenance projects modelled using UML.
- Some other applications of UML have to do with several maintenance management areas, as the planning and control of repair operations [17–19], the design of specific information systems [20–22], the generation of optimal maintenance policies and decision-making processes [23], the use of knowledge management in the maintenance function [24, 25], or with the asset management [26].
- Ambitious e-maintenance projects like the fully integrated PROTEUS platform uses UML as well, to model its processes [27].
- The modelling of a MMF based on PAS 55 using UML and BPMN, is a not previously explored assignment in the reviewed literature.

3 General Characteristics of the MMF

In the historical development of maintenance, several authors have proposed what they consider the best practices, steps, sequences of activities or models to manage this function.

Different maintenance management models and frameworks have been developed by researchers as [4, 28–33] among many others, in an effort to create an structure with a set of characteristics that fulfils the maintenance and organizational objectives.

From an analysis made to those MMF proposals [34], some desirable characteristics were identified as necessary for a modern and efficient MMF oriented to
operate with a quality system perspective: input–output processes approach, generation of documents and records, objectives entailment, possibility of incorporation of supporting technologies (CBM for example), orientation to operate
integrated to computer maintenance management systems (CMMS), flexibility to
adopt modern technologies (e-maintenance, expert systems, etc.), management of
material, human and information resources, focus on the constant improvement,
cyclical operation, generation of indicators (economics, efficiency, etc.), orientation
to standards, among others.

Considering that a standard is by itself a norm or model widely recognized by its
excellence, or a compendium of best practices, it is not a surprise that all the
aforementioned factors among others have been identified as existing characteristics
into the PAS 55 standard, basis and model of the MMF developed in this chapter.

PAS 55 is a Publicly Available Specification and it is the only standard available
internationally for asset management. The management of assets deals with the
whole life cycle of the asset, from its design until its final disposal; maintenance
commonly only describes the activities during the operational life of the asset. Then
it is possible to say that PAS 55 is a very complete reference to maintenance
management.

PAS 55 can be applied to any business sector and is independent of asset type,
but it is specially recommended to organizations strongly depending on the performance of the physical assets, as utility networks, power stations, roads, airports,
railways, oil and gas installations, manufacturing and process plants, property and
petrochemical complexes. The first version of PAS 55 was published in 2004 by the
British Standards Institution (BSI). In 2008 a new version of PAS 55 was released,
improving its content.

We can pronounce that the MMF modelled in this paper is a representation of
PAS 55:2008. However, even if the presented MMF is strongly based on
requirements of PAS 55, there are some remarks to make.

Firstly, the MMF does not exactly correspond to PAS 55:2008. Its elements have
been arranged according to the experience of the authors in leading companies that
actually operates with PAS 55 and inspired also, by the operation of the ISO
9001:2008 [35] model. The ISO standard was chosen since its spreading in industry
[36], because it is the international reference for any quality management system,
and hence it can be considered a generic guide for a process operation in which
fulfilment with requirements should be demonstrated, such as the case of the
maintenance function.

Secondly, the MMF suggests original flow processes for performing asset
management (PAS 55 declares *what* have to be done, but not exactly *how* to do it).
Part 2 of PAS 55 contains some recommendations and guidelines for the application
of PAS 55 [37]. These recommendations jointly with the techniques referenced by
other analysed models and previous works published by the authors [38, 39] gave
rise to the internal algorithms and processes of the MMF.

Finally, the most noticeable originality of the MMF is that its structure is formalized in terms of processing models, flow models and data models using BPM techniques, UML and BPMN languages. This brings an important and distinctive feature of the MMF: flexibility to be adapted (for example as a software application) to new requirements.

At this point, it is important to mention that the purpose and requirements of PAS 55 are actually observed in the modelled MMF, regardless of the dissimilar organization of the elements into the proposed model and the use of UML and BPMN diagrams. The proposed MMF is not "an improvement" of PAS 55; it is just a representation of it.

In summary: the chapter proposes a MMF that represents the general requirements of the asset management specification PAS 55, and that it is expressed using the innovative approach to BPM (Business Process Modelling).

The general operation framework of the proposed MMF is presented in Fig. 1.

The model begins and finishes with the requirements and satisfaction of the stakeholders, using the concept proposed in maintenance management by Soderholm et al. [32] that is also in line with the ISO 9004:2000 standard [40]. Furthermore, the proposed model is designed to be efficiently used across the organization levels (reminding Pintelon and Gelders [3] who proposed a model to be executed in three organizational activity levels). This model is composed of four modules or macro-processes, each one containing several processes that are specified in sub-processes and tasks.

The four macro-processes are: System Planning, Resources Management, Implementation & Operation, and Assessment & Continual Improvement.

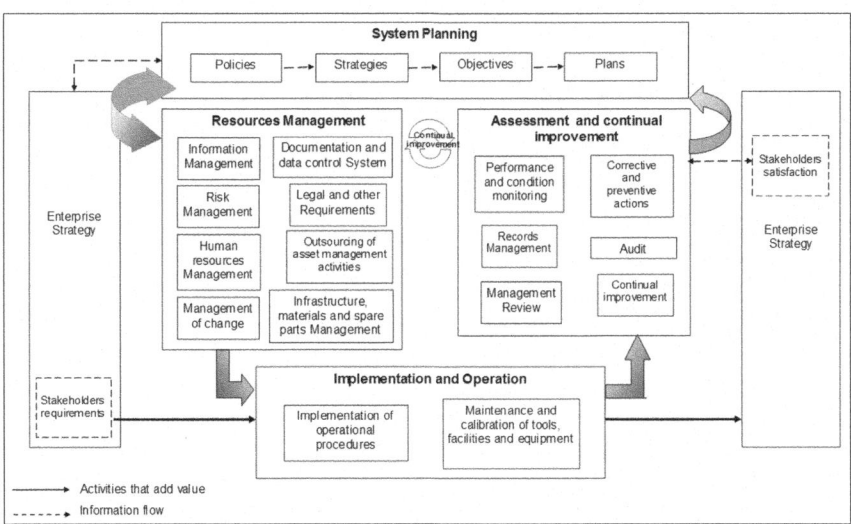

Fig. 1 The proposed maintenance management framework

The System Planning macro-process is constituted by four processes: Policies, Strategies, Objectives and Plans. The Resources Management processes are eight: Information Management, Risk Management, Human Resources Management, Management of Change, Documentation and Data Control System, Legal & Other Requirements, Outsourcing of asset management activities, and Infrastructure, Materials & Spare Parts Management. The Implementation & Operation macro-process is composed by the Implementation of Operational Procedures and by the Maintenance & Calibration of Tools, Facilities and Equipment processes. Finally, the Assessment & Continual Improvement macro-process is constituted by six processes: Performance & Condition Monitoring, Records Management, Management Review, Corrective & Preventive Actions, Audit, and Continual Improvement.

It is noticeable that the System Planning process entails the top Direction of Maintenance. In the presented MMF the medium levels perform the supporting processes (resources management) and control the maintenance execution. The level that executes maintenance also generates data to be used for the continuous improvement of the maintenance function.

The structure of this model enables a link among the maintenance function and the other organizational functions.

In the proposed model, each process (System Planning, Resources Management, Implementation & Operation and Assessment & Continual Improvement) is defined by UML diagrams using the "Eriksson-Penker Business Extensions" and BPMN diagrams that indicate the sequence of activities for the execution of every stage.

4 Business Process Modelling

According to Hammer and Champy [41], a (business process) is "a collection of activities that takes one or more kinds of input and creates an output that is of value to the customer". Davenport [42] defines a (business) process as "a structured, measured set of activities designed to produce a specific output for a particular customer or market. It implies a strong emphasis on how work is done within an organization, in contrast to a product focus's emphasis on what. A process is thus a specific ordering of work activities across time and space, with a beginning and an end, and clearly defined inputs and outputs: a structure for action".

The modelling of business process, understood as the use of methods, techniques and software to design, enact, control and analyse operational processes involving humans, organizations, applications, documents and other sources of information [43], has become an important subject specially since the 1990s, when companies were encouraged to think in "processes" instead of "functions" and "procedures". Process thinking looks horizontally through the company for inducing improvement and measurement [44].

From then on, business process modelling has been used in industry to obtain a global vision of processes by means of support, control and monitoring activities

[45], to facilitate the comprehension of the business key mechanisms, to be a base for the creation of appropriate information systems, to improve the business structure and operations, to show the structure of changes made in business, to identify outsourcing opportunities, to facilitate the alignment of the information and communication technologies with the business needs and strategies [46], and for several other activities such as the automatic processing of documents [47].

The increase in the last years of the quantity of research about business process modelling, and the application of recent technological advances have propitiated the use of business process modelling in other fields such as: planning of managerial resources (ERP), integration of managerial applications (EAI), management of the relations with customers (CRM), management of work flows (WFM) and communication among users to facilitate management requirements [45, 48].

Several benefits deriving from the adoption of business process modelling have been identified in the literature: improvement of the accomplishment speed of business processes, increase of the clients' satisfaction, optimization and elimination of unnecessary tasks, and incorporation of clients and partners in the business processes [49].

Process modelling is object of interest in many different fields, such as the managerial area and software engineering. This is due to the fact that it does not only describe processes, but in addition it represents a preparatory stage for the improvement of business processes, process reengineering, technological transference and processes standardization [50].

Software processes and business processes present certain similarities: they both try to capture the main features of a group of partially ordered activities carried out to achieve a specific goal. However, whereas the aim of a process software is to obtain a software product [51], the aim of a business process is to obtain beneficial results (generally a product or service) for clients or others affected by the process [52, 53].

Actually, the origins of the different business process modelling languages are inspired by software modelling languages. The informatics approach defines modelling as the "designing of software applications before coding" [54]; this focus has contributed to the development of several languages and applications for code generation and processes automation, which have increased in quantity and diversity especially during the last two decades [55].

5 Modelling Language and Software Tool Selection Process

Business processes are modelled using a modelling language, a standard that defines model elements and their meaning, allowing efficient, collaborative business process management across corporate boundaries and disciplines [56].

A large number of business process modelling languages exists and several taxonomies have been proposed [57]. In a general classification, a modelling language can be graphical or textual and in a more detailed taxonomy, Ko et al. [55] classify the BPM languages in relation to the BPM life cycle in: Graphical Standards (BPMN, UML), Execution Standards (BPEL, BPML, WSFL, XLANG), Interchange Standards (XPDL, BPDM) and Diagnosis Standards (BPRI, BPQL).

To select a suitable business modelling language to express the proposed MMF, the present investigation refers to the flowchart proposed by Ko et al. [55]. This flowchart presents a sequential decisional process that leads to define the type of language to be used.

Considering that the objective of this research is to model a new business process (not a web service or an automation application nor a diagnosis), that this model has a private application (for internal BPM, not for collaboration business to business) and that it is desired to work with a graphical representation in order to facilitate the modelling process, the result from Ko's et al. selection procedure indicates that the better choice is a Graphical Standard such as UML, BPMN, Event-driven Process Chains (EPC) [58], Role-Activity Diagrams (RADs) or simple flowcharts.

Among all the mentioned standards, UML 2.1 using the "Eriksson-Penker Business Extensions" and BPMN 1.0 were selected to model the proposed MMF. Both standards are maintained by the OMG (Object Management Group), an "international, open membership, not-for-profit computer industry consortium [...] [that] develops enterprise integration standards for a wide range of technologies, and an even wider range of industries" [54].

A further decision element was the availability to freely access to the OMG website where it is possible to download the latest UML and BPMN specifications, and to consult a variety of resources about those standards.

UML was created in 1997 by Grady Booch, James Rumbaugh and Ivar Jacobson, who developed it from the union of their own methodologies. They proposed UML to consideration of the OMG, being accepted as a standard since the same year it was proposed [59].

For the modelling development of this research, UML will be accompanied by the "Eriksson-Penker Business Extensions" these extensions are a set of specifications about the use of semantics to express the elements of the model in terms of business modelling [6].

UML 2.1 specification is formed by thirteen kinds of diagrams that show a specific static or dynamic aspect of a system.

BPMN first specification was released to the public in May, 2004 with the objective of "provide a notation that is readily understandable by all business users, from the business analysts that create the initial drafts of the processes, to the technical developers responsible for implementing the technology that will perform those processes, and finally, to the business people who will manage and monitor those processes" [60].

BPMN defines a business process diagram (BPD) which is formed by a set of graphical elements to represent activities and their flow [60].

Regarding the software modelling tools there is a large number of applications, some of them non-proprietary and others of proprietary type. The selection of the most appropriate tool depends on the particular modelling requirements and the project scope.

Although a simple graphical tool for diagrams development could be used, a professional software modelling tool including a business process repository offers interesting advantages (storing of elements, simulation, code generation, etc.).

For this research the selected software was Enterprise Architect 7.1; an UML analysis, design, documentation and project management CASE tool, including basic UML models plus testing, metrics, change management, defect tracking and user interface design extensions. This software is developed by Sparx Systems. Enterprise Architect 7.1 was chosen because of its features and its availability to support this research.

6 Business Architecture and Modelling Strategy

The general description of a system that identifies its purpose, vital functions, elements, processes and defines their interaction is called "business architecture" [61]. The OMG [54] provides its definition: "Business architecture is a blueprint of the enterprise that provides a common understanding of the organization and is used to align strategic objectives and tactical demands [...] business architecture defines the structure of the enterprise in terms of its governance structure, business processes, and business information".

The objective of modelling the proposed MMF is to express its business architecture using documents and diagrams known as "artefacts".

The general business architecture can be represented by three principal categories of data: the Business Context (models of the stakeholders relations, mission and vision statements, business goals and physical structure of the "as-is" business), the Business Objects (a domain model of all objects of interest and their respective data), and the Business Workflows (business process diagrams representing the structures and objects defined in the Business Context and in the Business Objects diagrams. These business process diagrams show how objects work together to provide fundamental business activities).

In this paper the Business Context is represented by a Goals Diagram (Fig. 2); the Business Objects by a Model of Classes and Objects (Fig. 9), and the Business Workflows by a set of process diagrams (Figs. 3, 4, 5, 6, 7 and 8).

Generally the business architecture is organized hierarchically so executives can observe how specific processes are aligned to support the organization's strategic aims [62].

The same hierarchical order is used to define the processes and sub-processes for the new MMF proposed in this investigation; the top-down approach will be initially preferred (starting modelling the top value chain process and later modelling the specific processes), not dismissing the possibility of using an inside-out approach in following stages (starting modelling a particular specific process and then extending its influence around the general organization) [63].

There is not a defined way of naming process levels although frequently the smallest process diagram is called an activity (according to UML and BPMN standards).

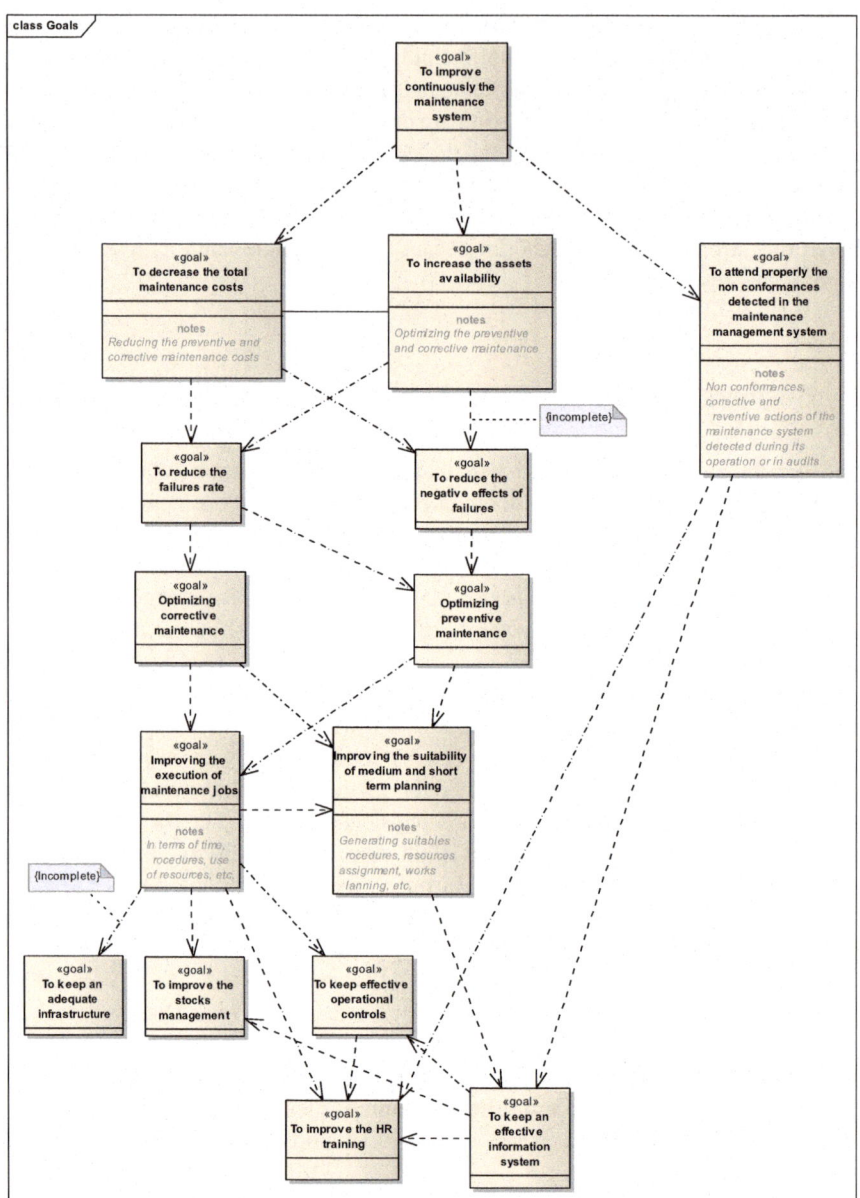

Fig. 2 Goals tree diagram for the proposed MMF

Neither a technical limit exists for a maximum number of processes subdivisions. The most important concept is to keep in mind that processes can be hierarchically arranged [62].

Therefore the proposed MMF has its own nomenclature to refer to its hierarchical levels.

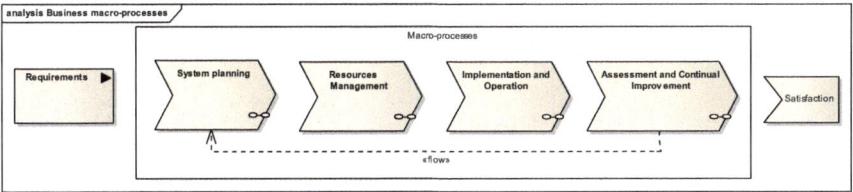

Fig. 3 Macro-processes (top value chain process) of the proposed MMF

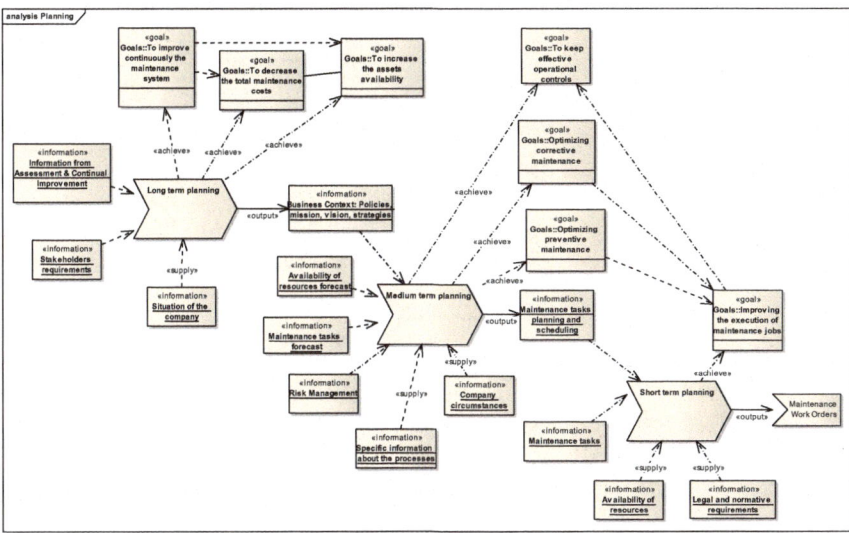

Fig. 4 UML diagram of the system planning macro-process

Once have been defined the operations to be modelled, the boundary of the system and after identifying its mission and vision, is necessary to describe the business strategy to fulfil the goals set. These goals must be achieved through the operation of one or more business processes [64].

In Fig. 2 is represented an example of the goals tree that can be designed for the proposed MMF using a specific type of UML diagram: a class diagram. In this kind of diagrams a goal is described as a class object with the stereotype ≪goal≫. Due in this project a new system is being designed; all the goals presented in the tree diagram are illustrative and of qualitative type.

For this particular MMF the main goal is "to continuously improve the maintenance system". This main goal depends on the fulfilment of other three goals (identified by a dependency line): to decrease the total maintenance cost, to increase the assets availability, and to attend properly the non-conformances detected in the maintenance management system. It is necessary to notice that the first two aims (to decrease the total maintenance cost and to increase the assets availability) are

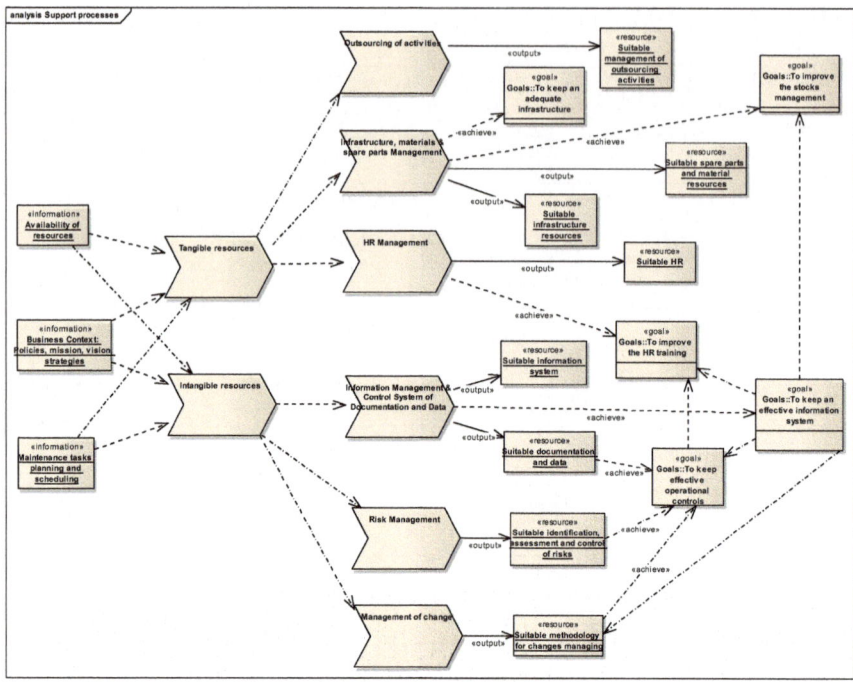

Fig. 5 UML diagram of the resources management macro-process

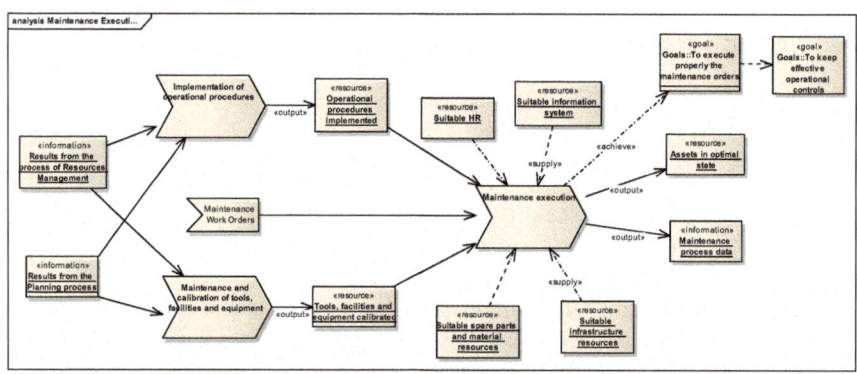

Fig. 6 UML diagram of the implementation and operation macro-process

contradictory goals. This contradictory feature is identified using an association line between the goal objects.

Moreover, the fulfilment of each one of the already mentioned goals depends on another series of hierarchical goals (or sub-goals), which have to be totally or partially achieved. In the diagram, a tag with the legend "incomplete" indicates this condition.

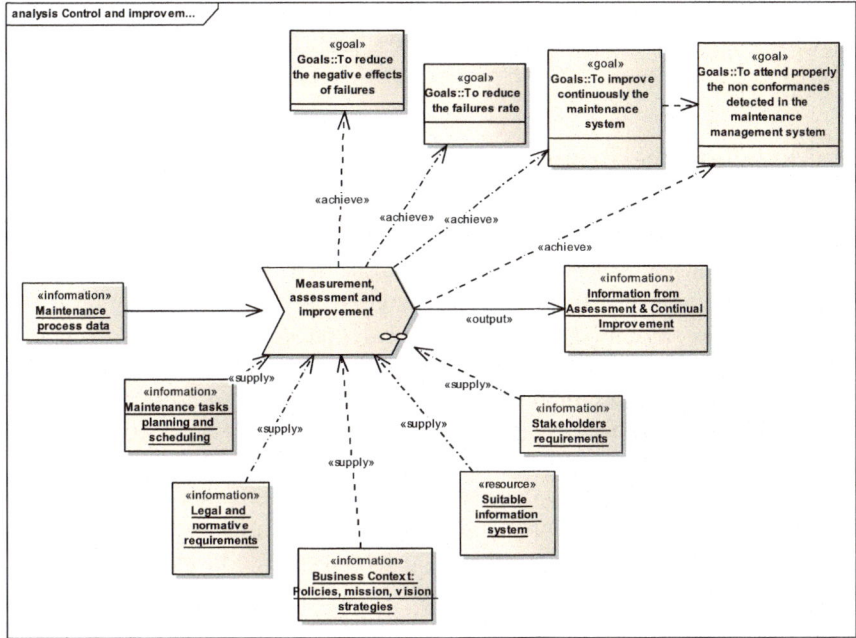

Fig. 7 UML diagram of the macro-process of assessment and continual improvement

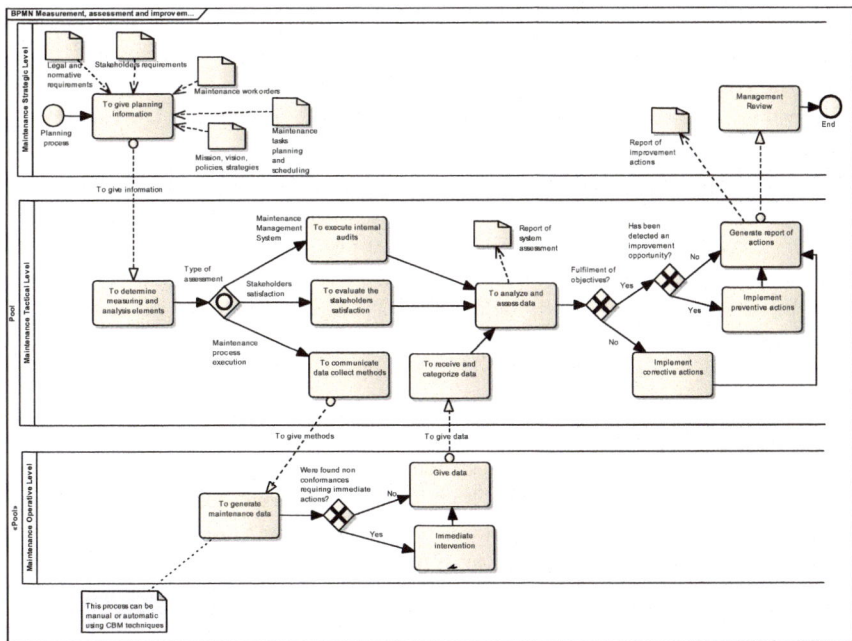

Fig. 8 BPMN diagram of the assessment and continual improvement

Every macro-processes, process and activity described in the model is focused to the satisfaction of the objectives drafted in the goals tree diagram.

7 Modelling the Proposed Maintenance Management System

Following the mentioned top-down approach, the top value chain process of the proposed MMF (or level 0 process) is constituted by the already mentioned four macro-processes: System Planning, Resources Management, Implementation & Operation, and Assessment & Continual Improvement.

Subsequently each macro-process is conformed by processes (level 1 processes) and each process can be subdivided in sub-process (level 2 processes), finally each sub-process can be subdivided in activities (level 3 processes).

In Fig. 3 appears an UML diagram made using the Eriksson-Penker Business Extensions. This diagram represents the top value chain process (or level 0). In a software platform, this diagram can also operates as a main menu to access to the rest of processes.

Every macro-process and process modelled has some invariable related elements: one or several goals associated using a dependence relation with the stereotype «achieve» (these goals are derived from the goals tree); input resources, output resources, both linked using dependence relations, supply resources with a dependence relation and the stereotype «supply» and control resources having the stereotype «control».

The first macro-process to model is System Planning module. Figure 4 shows the level 1 planning diagram, which it was modelled using UML with the Eriksson-Penker Business Extensions. In this diagram it is possible to identify the mentioned elements related to every process (goals, input, output, supply).

Besides, in Fig. 4 it can be observed the three processes composing the total System Planning module: long-term planning, medium-term planning and short-term planning. The information and supplies required are identified in the diagram, as well as the goals to be achieved by each of the three processes.

It is interesting to notice that the output of each process is an input element for the next.

In a general way, this macro-process has defined as start inputs: the maintenance information for improvement (generated by the Assessment & Continual Improvement macro-process), the stakeholder's requirements and information about the situation of the company. Other input elements are going to be needed for the entire planning development, but it is observed that the three mentioned as "start inputs" are the earlier required initiating the process flow.

As a final output of this entire macro-process appears the maintenance work order, which has to be executed by operative personnel. Besides the maintenance work order, there are other essential outputs generated during the planning as: the business context document (policies, mission, vision, agreements, strategies) and the planning and scheduling of maintenance tasks.

The procedures to carry out every planning process belong to level 2 and for this project are named sub-processes. On the whole, a procedure contains a description more detailed of the flow of activities to perform, the required, related and generated documents and the responsibles for the performance.

If it is necessary due the sub-process size or complexity, a level 3 diagram describing activities can be made as well.

Both level 2 diagram and level 3 diagram could be produced using UML (if there is an important quantity of information inside it) or using BPMN (if the procedure is not too long). It is also possible going beyond level 3 if more specific information is required.

The next macro-process to be modelled has to do with Resources Management processes, as Fig. 5 shows using an UML diagram with the Eriksson-Penker Business Extensions.

The Resources Management module classifies the processes into the management of tangible resources, composed by three not sequenced processes (Management of Outsourcing Activities Process; Infrastructure, materials & spare parts Management Process; and Human Resources Management Process) and into the management of intangible resources composed by three not sequenced processes (Information Management & Control System of Documentation and Data Process; Risk Management Process; and Management of Change Process).

These processes are independent in their operation, although they are linked by their goals and by being part of the same general system.

These six supporting processes share the same start input elements as well: information about the availability of resources, about the business context (policies, mission, vision, etc.) and about the planning and scheduling of maintenance tasks.

An appropriate execution of these supporting processes results in having suitable resources to the maintenance development (output elements).

Required procedures for every supporting process can be managed in level 3 diagrams, as previously explained in the System Planning macro-process.

At first glance the macro-process for the maintenance execution, the Implementation & Operation macro-process (Fig. 6) seems to be very simple, since its diagram does not have so many elements as the previous macro-processes. But in fact, this is the core process of the whole system [65].

Beginning from the work order, maintenance tasks are developed according to the particular procedures defined by the organization, using the resources managed in the previous macro-process, and via the outputs supplied by the corresponding level 2 processes: the Implementation of Operational Procedures; and the Maintenance & Calibration of Tools, Facilities and Equipment Process.

From this development, the desirable outputs are: to have and/or to keep the assets in optimal state and, to compile outstanding data about maintenance process.

To have and/or to keep the assets in optimal state is an output that goes directly to satisfy a tangible necessity, generally outside the maintenance function. To compile outstanding data about maintenance process is a required input element in the Assessment & Continual Improvement macro-process.

The particular technical procedures inside the Maintenance execution process, generally involve very specific aspects which can be expressed using UML and BPMN likewise.

These specific technical procedures depend on the kind of organization applying the system and as a core process, its performance is highly supported by the others macro-processes.

The remaining macro-process, Assessment & Continual Improvement is presented in Fig. 7. In its UML diagram it is possible to identify the process start input: data about the maintenance process execution.

Further information is required as well (in diagram expressed as supplies). The desired output of this macro-process is the information for the improvement, which will be used by the following System Planning macro-process.

In this way, the system operates cyclically favouring the continuous improvement approach.

As an example of how the system operation can be detailed in deeper levels, Fig. 8 shows a level 2 BPMN diagram, representing the process inside the Assessment & Continual Improvement module, where the working flows and activities necessary to achieve the corresponding goals are identified.

In this BPMN diagram can also be identified the activities corresponding to the six elements constituting this macro-process (see Fig. 1).

Figure 8 shows the basic elements of a BPMN diagram: flow objects, connection objects, swimlanes and artefacts [66]. Also the six processes inside can be identified as activities.

Inside every activity modelled in the BPMN diagram it is also possible to add more specific procedures, being identified by consecutive levels numeration. Different macro-processes can be conformed by different number of levels, depending on the complexity of the procedures to model.

Besides the diagrams used to symbolize process workflows, there are other kinds of diagrams (called also artefacts) that are useful for having a complete view of the whole system and are indispensables if there is the idea of developing an informatics application.

Those diagrams are categorized by the UML 2.1 standard into structural diagrams (defining the static architecture of a model) and behavioural diagrams (representing the interaction and instantaneous states within a model as it 'executes' over time).

A structural diagram (a class diagram) was used before to symbolize the goals tree (Fig. 2). There are other different kinds of structural diagrams. In order to exemplify, Fig. 9 shows another important structural diagram. This class diagram represents a conceptual model, defining the business concepts about a maintenance management system and how they are related among them.

A conceptual diagram identifies the important concepts related to a specific context and it can be useful to model the business resources, rules and goals [67].

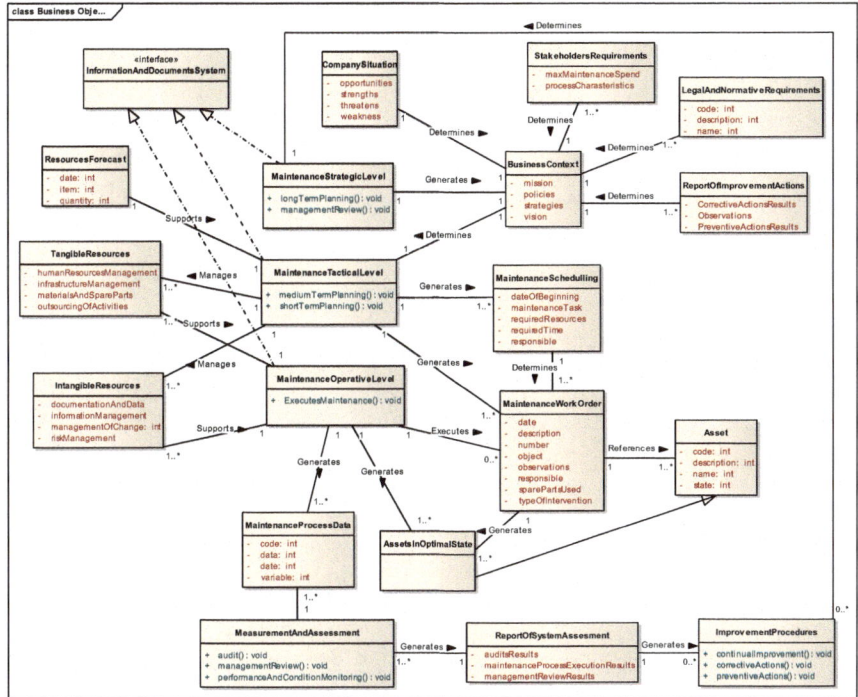

Fig. 9 Conceptual model of classes and objects

Regarding behavioural diagrams, there are also several kinds: use case diagrams, sequence diagrams, state diagrams, etc. Figure 10 shows a state diagram detailing the transitions or changes of state that an object (in this case a maintenance work order) can go through in the system.

State Diagrams show how an object moves from one state to another and the rules that govern that change. State charts typically have a start and an end condition [67].

All diagrams appearing in this chapter were made using Enterprise Architect 7.1. This software was perceived as agile and easy to use, with a variety of online resources.

8 ICT Issues Related to the MMF Implementation

An important distinctive attribute of the proposed MMF is that it has been modelled using the BPM methodology, UML and BPMN languages.

This attribute provides the MMF with integration capacities and flexibility to take advantage of the information and communications technologies (ICT).

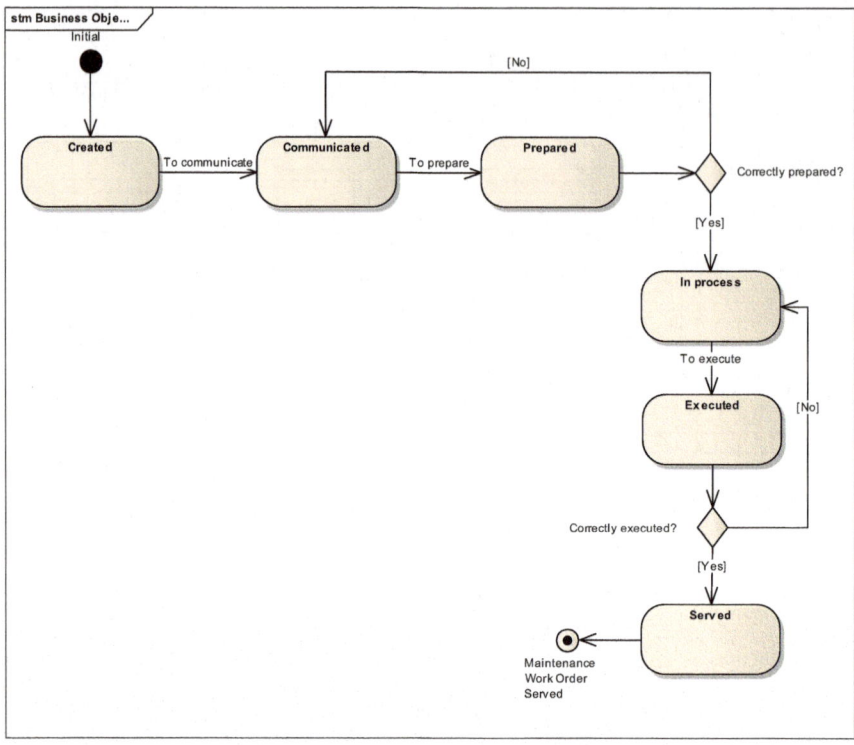

Fig. 10 UML state diagram for a maintenance work order

For instance, from a complete description of the MMF operation algorithms using UML and BPMN diagrams, it is relatively not difficult to generate code for the development of a software application that executes the MMF modules [68, 69].

Some of the ERP systems that currently exist in the market have maintenance management modules, and there are likewise, some software applications specific for asset management [70] (EAM systems); nevertheless the practical experience reveals that there are still several novel functions that can be added in a new system to improve maintenance management. Most of these functions are related to e-maintenance and can be incorporated to the proposed MMF through UML and BPMN models as well, specifically looking for coordination in real time among CMMS, RCM and CBM systems.

Although the e-maintenance term has been used since 2000 as a component of e-manufacturing, at present time there is not yet a standardized definition of e-maintenance given by an official institution [71].

From a pragmatic point of view, we may say that e-maintenance is "the set of maintenance processes that uses the e-technologies to enable proactive decisions in a particular organization" (definition partially derived from Levrat et al. [72]).

Such e-maintenance processes are supported by means of a variety of hardware and software technologies as the wireless and mobile devices, embedded systems, web-based applications, P2P networks, multi-agent applications, specific software architectures, among others.

That variety of technologies implies the existence of multiple communication protocols, data connections, configurations, etc. At this respect, several standards have been developed in order to obtain interconnection and interoperability among the different systems.

The Machinery Information Management Open Systems Alliance or MIMOSA is an important not-for-profit trade association dedicated to developing and encouraging the adoption of open information standards in operations and maintenance (O&M) just to support interoperability [73].

MIMOSA standards are interesting references for the proposed MMF because of two main reasons: the former is that MIMOSA standards are expressed using UML language; the latter is that MIMOSA has developed two types of information-exchanging open standards that are also related to the processes to develop in the proposed MMF: a standard for management applications (OSA-EAI™) and a standard for condition based maintenance (OSA-CBM™). Both standards provide metadata reference libraries and a series of information exchange standards using XML and SQL.

Therefore, it is necessary to consider the adoption of MIMOSA standards to continue modelling deeper levels of the MMF and particularly for the operation of the e-maintenance processes.

Concerning these e-maintenance processes, although e-maintenance can be characterized as a technique, the general idea of this project (based on Iung et al. [74]) is considering e-maintenance as a philosophy supporting the operation of the entire MMF and making possible the information exchange among remote elements. This philosophy allows the decision-making and the fulfilment of the maintenance global objectives depending on collaboration, which implies the use of the ICT.

The majority of the e-maintenance processes to be included in the proposed MMF involve the realization of the classical maintenance management activities but using e-technologies, in a distance environment. The proposed MMF becomes a CMMS system with remote capabilities.

However, the use of e-technologies and large volumes of different data necessarily increases the possibilities to create new emerging e-maintenance processes.

During the development of this MMF, several novel e-maintenance processes have been identified as required, particularly processes related to an integration and exchange of information among CMMS, RCM and CBM systems [75]. This e-maintenance integration is able to optimize the decision-making processes related to the feasibility of the maintenance strategies and programs.

In general terms, this integration works as follows [76]: using the information managed by the CMMS (saved inside each module of the proposed MMF), the RCM methodology is applied to the pre-defined system(s), defining the operational context and the processes involved, doing the FMECA analysis and selecting the appropriate maintenance policies. From the RCM analysis it is possible to detect the necessity of applying CBM in some particular elements in order to generate important economical savings. The real-time CBM signals feed the CMMS and subsequently the RCM, generating an automatic suggestion if a maintenance strategy has to be updated according its behaviour. Then, the RCM is applied again and the improving cycle begins one more time. Moreover, the integration of this information can maximize the effectiveness of the diagnosis: the CBM signals are related to the most critical and frequent failures modes of the RCM analysis, allowing time savings in corrective and preventive actions.

The specific operational characteristics of those e-maintenance processes, the ICT related to them and the additional interoperability standards required to their implementation (i.e. ISO 18435, ISO 62264, OPC standards, etc.) have to be defined according to the special requirements of the specific industrial sector that applies the proposed MMF, and they are material for another paper.

9 Conclusions

In the historical development of maintenance, several models and frameworks looking for the optimal maintenance management structure have been developed [34].

Among all those proposals, PAS 55 standard emerges in 2004, as a complete framework, not only for maintenance but also for management of the entire life cycle of assets.

Besides, PAS 55 involves a set of desirable characteristics and best practices identified as necessaries for the operation of a modern and efficient maintenance management framework (MMF), as the input–output processes approach, the objectives entailment, the orientation to new technologies and the continuous improvement approach.

Then, this chapter shows the process of modelling a MMF that represents the general requirements of PAS 55.

The flow diagrams and processes proposed inside the MMF are a representation of how PAS 55 structure can be implemented in an organization. We have to remember that PAS 55 declares *what* have to be done, but not exactly *how* to do it. For this reason, each company is able to develop its own specific techniques and methodologies to fulfil the PAS 55 requirements.

For the realization of this project, the modelling work involved researching about the basic concepts in the area (business process, modelling, modelling language, business architecture, etc.) to select the most suitable language and software tool to the case.

UML 2.1 and BPMN 1.0 were the modelling languages selected to express the proposed MMF, due to their recognition as international standards, their increasing use in successful maintenance projects (e.g. PROTEUS project [77], among others) and their interesting capabilities.

Later, a modelling methodology was chosen to represent the system architecture, and to develop the structural and behavioural diagrams exposed in this chapter.

Summarizing, the general steps to model the proposed MMF and that have been shown in this chapter are: (i) PAS 55 analysis, (ii) design of the conceptual MMF according to standards, (iii) selection of modelling language and modelling software tool, (iv) definition of the business architecture, (v) modelling of goals tree, (vi) identification of top value chain process, (vii) modelling of involved processes and activities, (viii) tracing several UML and BPMN diagrams to represent specific features of the system, (ix) analysis of the ICT issues related to the implementation of the MMF and conclusions.

The use of processes modelling languages (UML 2.1 and BPMN 1.0) gives to the MMF the interesting possibility of generating code and the subsequent creation of software as an alternative from the ERP, CMMS and EAM commercial systems.

Code generation and the development of a software application involve a hard work detailing the data and operation models for the MMF, a profound working out of algorithms and artefacts describing the use of the tools and techniques required for the operation of the MMF. Also, the identification and interpretation of the interoperability standards required according to the ICT executing of the MMF modules is necessary.

At this respect, the operation of the system through e-maintenance processes [78] is a recommended approach.

Finally, it is important to mention that the activities flow and processes modelled in this paper correspond mainly to the real operation of PAS 55 in a leading Spanish enterprise of the energy sector, and that the project of the e-maintenance integration among CMMS, RCM and CBM for decision-making is actually, being implemented in a transformer and in a water pump [76], equipments of the same energy production and distribution enterprise.

References

1. Sherwin D (2000) A review of overall models for maintenance management. J Qual Maintenance Eng 6(3):138–164
2. Cholasuke C, Bhardwa R, Antony J (2004) The status of maintenance management in UK manufacturing organisations: results from a pilot survey. J Qual Maintenance Eng 10(1):5
3. Pintelon LM, Gelders LF (1992) Maintenance management decision making. Eur J Oper Res 58(3):301–317
4. Prasad Mishra R, Anand D, Kodali R (2006) Development of a framework for world-class maintenance systems. J Adv Manuf Syst 5(2):141–165
5. Vanneste SG, Van Wassenhove LN (1995) An integrated and structured approach to improve maintenance. Eur J Oper Res 82(2):241–257
6. Object Management Group Website. http://www.bpmi.org/. 20 Mar 2009
7. Framiñán J (2008) Introducción a la arquitectura y desarrollo de sistemas de información basados en la web. Secretariado de Publicaciones de la Universidad de Sevilla, Sevilla
8. PAS 55-1:2008 (2008) Asset management. Specification for the optimized management of physical assets. BSI: United Kingdom, 2008

9. Qiu XB, Wimmer W (2000) Applying object orientation and component technology to architecture design of power system monitoring. In: Proceedings of the international conference on power system technology (POWERCON 2000) 2000, vols 1–3:589–594

10. Thurston MG (2001) An open standard for web-based condition-based maintenance systems. In: Proceedings of the IEEE systems readiness technology conference (IEEE AUTOTESTCON 2001): 401–415

11. Huang XQ, Pan HX, Yao ZT, Ma QF, Cai JJ (2005) The study on expert system of state monitoring and fault diagnosis for gearbox. In: Proceedings of the 6th international symposium on test and measurement 2005, vols 1–9:1867–1870

12. Palluat N, Racoceanu D, Zerhouni N (2006) A neuro-fuzzy monitoring system application to flexible production systems. Comput Ind 57(6):528–538

13. Xing W, Jin C, Ruqiang L, Weixiang S, Guicai Z, Fucai L (2006) Modeling a web-based remote monitoring and fault diagnosis system with UML and component technology. J Intell Inf Syst 27(1):5–19

14. Dong X, Liu Y, LoPinto F, Scheibe K, Sheetz S (2002) Information model for power equipment diagnosis and maintenance. In: Proceedings of the IEEE power engineering society winter meeting 2002, vols 1–2, pp 701–706

15. Min-Hsiung H, Rui-Wen H, Fan-Tien C (2004) An e-diagnostics framework with security considerations for semiconductor factories. In: Proceedings of the semiconductor manufacturing technology workshop 2004, pp 37–40

16. Chen B, Gao X, Zhao Z (2006) Research on a remote distributed fault diagnosis system based on UML and CORBA. In: Proceedings of the first international conference on maintenance engineering 2006, pp 363–367

17. Mouritz D (2005) An integrated system for managing ship repair operations. Int J Comput Integr Manuf 18(8):721–733

18. Cerrada M, Cardillo J, Aguilar J, Faneite R (2007) Agents-based design for fault management system in industrial processes. Comput Ind 58(4):313–328

19. Li L, Chen T, Guo B (2010) Simulation modeling for equipment maintenance support system based on stochastic service resource management object. J National Univ Defense Technol 2010

20. Belmokhtar O, Ouabdesselam A, Aoudia M (2004) Conception of an information system for the maintenance management. In: Proceedings of the 5th international conference on quality, reliability and maintenance 2004, pp 141–148

21. Nordstrom L, Cegrell T (2005) Extended UML modeling for risk management of utility information system integration. In: Proceedings of the IEEE power engineering society general meeting 2005, vol 1–3, pp 913–919

22. Keraron Y, Bernard A, Bachimont B (2007) An UML model of the technical information system to enable information handling and recording during the product life cycle. In: Proceedings of the 4th international conference on product lifecycle management 2007, pp 363–372

23. Sadegh P, Concha J, Stricevic S, Thompson A, Kootsookos P (2006) A framework for unified design of fault detection and isolation and optimal maintenance policies. In: Proceedings of the 2006 American control conference, 2006, pp 3749–3756

24. Reiner J, Koch J, Krebs I, Schnabel S, Siech T (2005) Knowledge management issues for maintenance of automated production systems. In: Proceedings of the IFIP international conference on human aspects in production management, 2005, vol 160, pp 229–237

25. Rasovska I, Chebel-Morello B, Zerhouni N (2008) A mix method of knowledge capitalization in maintenance. J Intell Manuf 19(3):347–359

26. Trappey A, Hsiao D, Ma L, Chung YL (2009) Maintenance chain integration using petri-net enabled prometheus MAS modeling methodology. In: Proceedings of the 13th international conference on computer supported cooperative work in design 2009, pp 238–245

27. Bangemann T, Thomesse J, Lepeuple B, Diedrich C (2004) PROTEUS—providing a concept for integrating online data into global maintenance strategies. In: Proceedings of the 2nd IEEE international conference on industrial informatics 2004, pp 120–124

28. Campbell JD, Reyes-Picknell J (1995) Uptime: strategies for excellence in maintenance management. Productivity Press, New York
29. Wireman T (1998) Development performance indicators for managing maintenance. Industrial Press, New York
30. Duffuaa S, Raouf A, Dixon CJ (2000) Planning and control of maintenance systems. Spanish Edition. Limusa: México
31. Waeyenbergh G, Pintelon L (2002) A framework for maintenance concept development. Int J Prod Econ 77(1):299–313
32. Söderholm P, Holmgren M, Klefsjö B (2007) A process view of maintenance and its stakeholders. J Qual Maintenance Eng 13(1):19–32
33. Crespo Márquez A (2007) The maintenance management framework. Models and methods for complex systems maintenance. Springer, United Kingdom
34. López Campos M, Crespo Márquez A (2009) Review, classification and comparative analysis of maintenance management models. J Autom Mobile Robot Intell Syst 3(3):110–115
35. ISO 9001: 2008 (2008) Quality management systems. Requirements. ISO, Geneva
36. Corbett C (2008) Global diffusion of ISO 9000 certification through supply chains. Int Ser Oper Res Manage Sci 119:169–199
37. PAS 55-2:2008 (2008) Asset management. Guidelines for application of PAS 55-1. BSI, United Kingdom
38. Crespo A, Gupta J (2006) Contemporary maintenance management: process, framework and supporting pillars. Omega Int J Manage Sci 34:313–326
39. Crespo Márquez A, Moreu de León P, Gómez Fernández J, Parra Márquez C, López Campos M (2009) The maintenance management framework: a practical view to maintenance management. J Qual Maintenance Eng 2009 15(2):167–178
40. ISO 9004:2000 (2000) Quality management systems—guidelines for performance improvements. ISO, Geneva
41. Hammer M, Champy J (1993) Reenginering the corporation. Harper, New York
42. Davenport T (1993) Process innovation: reengineering work through information technology. Harvard Business School Press, Boston
43. Van der Aalst W (2003) Don't go with the flow: web services composition standards exposed. IEEE Intell Syst 2003 18(1):72–76
44. Rolstadås A (1995) Performance management: a business process benchmarking approach. Kluwer Academic Publishers, England
45. Russel N, VanderAlst W, Hofstede A, Wohed P (2006) On the suitability of UML activity diagrams for business process modelling. In: Proceedings of the third Asia-pacific conference on conceptual modelling (APCCM) 2006, vol 53 of conferences in research and practice information technologies, pp 95–104
46. Beck K, Joseph J, Goldszmidt G (2005) Learn business process modeling basics for the analyst. IBM, 2005 [www-128ibm.com/developersworks/library/wsbpm4analyst]
47. Kalnins A, Vitolins V (2006) Use of UML y model transformations for workflow process definitions. Communications of the Conference Baltic DBIS 2006, Vilnius Technika, pp 3–15
48. Ramzan Ramzan S, Ikram N (2007) Requirement change management process models: an evaluation. In: Proceedings of software engineering conference 2007. Acta Press, Canada
49. Pérez J et al (2007) Model driven engineering Aplicado a business process management. *Informe Técnico UCLM-TSI-002* 2007
50. Succi G, Predonzani P et al (2000) Business process modeling with objects, costs and human resources. Systems modeling for business process improvement. Artech House, London, pp 47–60
51. Acuña S, Ferré X (2001) Software process modelling. In: Proceedings of the 5th World multiconference on systematics, cybernetics and informatics (SCI 2001). Orlando Florida, pp 1–6

52. Sharp A, McDermott P (2000) Workflow modeling: tools for process improvement and application development. Artech House, London
53. Crespo Márquez A (2010) Dynamic modelling for supply chain management. Front-end, back-end and integration issues. Springer, London, p 297
54. OMG Business Architecture Working Group Website (2009). http://bawg.omg.org/. March 2009
55. Ko R, Lee S, Lee E (2009) Business process management (BPM) standards: a survey. Bus Process Manage J 15(5):744–791
56. IDS Scheer Website, http://www.idsscheer.com/en/ARIS/Modeling_Standards/80850.html. March 2009
57. Giaglis G (2001) A taxonomy of business process modelling and information systems techniques. Int J Flex Manuf Syst 13(2):209–228
58. Scheer A (1992) Architecture of integrated information systems: foundations of enterprise modelling. Springer Verlag, New York
59. Schmuller J (2001) Sams teach yourself UML in 24 hours. Macmillan Computer Pub, USA
60. White S (2010) Introduction to BPMN. OMG Website. http://www.bpmn.org/. January 2010
61. Gharajedaghi J (1999) Systems thinking. Managing chaos and complexity: a platform for designing business architecture. Elsevier, USA
62. Harmon P (2007) Business process change: a guide for business managers and BPM and six sigma professionals. Elsevier, Boston
63. Recker J (2006) Process modeling in the 21st century, *BPTrends* 2006 [www.bptrends.com]
64. Vasconcelos A et al (2001) A framework for modeling strategy, business processes and information systems. In: Proceedings of the fifth international enterprise distributed object computing conference (EDOC'2001). IEEE Computer Society, USA
65. Porter ME (1985) Competitive advantage: creating and sustaining superior performance. The Free Press, New York
66. OMG UML Semantics ver. 1.1. ftp://ftp.omg.org/pub/docs/ad/97-08-04.pdf. 20 Mar 2009
67. Consulting M (2001) Modelado de Negocios con UML y BPMN. México, Milestone Consulting Editions
68. Hauser R, Koehler J (2004) Compiling process graphs into executable code. Lect Notes Comput Sci 3286:317–336
69. Ouyang C, Dumas M, Breutel S, Ter Hofstede A (2006) Translating standard process models to BPEL. Lect Notes Comput Sci 4001:417–432
70. Strub J, Jakovljevic P (2010) EAM versus CMMS. *CMMScity*. http://www.cmmscity.com/index.htm. February 2010
71. Muller A, Crespo Márquez A, Iung B (2008) On the concept of e-maintenance: review and current research. Reliab Eng Syst Saf 2008 1165–1187
72. Levrat E, Iung B, Crespo Marquez A (2008) E-maintenance: review and conceptual framework. Prod Plann Control 19(4):408–429
73. MIMOSA (2010) An operations and maintenance information open system alliance. http://www.mimosa.org//. May 2010
74. Iung B, Levrat E, Crespo Márquez A, Erbe H (2009) Conceptual framework for e-Maintenance: illustration by e-Maintenance technologies and platforms. Annu Rev Control 2009 33:220–229
75. Niu G, Yang BS, Pecht M (2010) Development of an optimized condition-based maintenance system by data Fusion and reliability-centered maintenance. Reliab Eng Syst Saf 95(7):786–796
76. López Campos M, Fumagalli L, Gómez Fernández J, Crespo Márquez A, Macchi M (2010) UML model for integration between RCM and CBM in an e-maintenance architecture. In: Proceedings of the 1st IFAC workshop on advanced maintenance engineering services and technology (A-MEST) 2010, pp 133–138

77. Bangemann T, Rebeuf X, Reboul D, Schulze A, Szymanski J, Thomesse J, Thron M, Zerhouni N (2006) PROTEUS-Creating distributed maintenance systems through an integration platform. Comput Ind 57(6):539–551
78. Crespo Márquez A, Iung B (2006) Special issue on e-maintenance. Comput Ind 57(1):473–475

Author Biographies

Mónica Alejandra López-Campos is an Industrial Engineer, and Ph.D. in Industrial Organization from the University of Seville (Spain). Currently she is Professor in the Federico Santa Maria University (Chile), where she teaches the subjects of Maintenance Management, Operations Management and Quality Management. Also, she is member of the National System of Researchers (Mexico). Her research works have been published in journals such as Reliability Engineering and System Safety, International Transactions in Operational Research, Computers in Industry, Quality and Reliability Engineering International, and Journal of Quality in Maintenance Engineering, among others. Her interests in research include maintenance management, modelling of processes, logistics, quality, simulation and educational methodologies for the engineering. She has participated in several research projects in Mexico, Spain and Chile, as well as in international organizations such as the Economic Commission for Latin America and the Caribbean (ECLAC).

Adolfo Crespo Márquez is currently Full Professor at the School of Engineering of the University of Seville, and Head of the Department of Industrial Management. He holds a Ph.D. in Industrial Engineering from this same University. His research works have been published in journals such as the International Journal of Production Research, International Journal of Production Economics, European Journal of Operations Research, Journal of Purchasing and Supply Management, International Journal of Agile Manufacturing, Omega, Journal of Quality in Maintenance Engineering, Decision Support Systems, Computers in Industry, Reliability Engineering and System Safety, and International Journal of Simulation and Process Modeling, among others. Professor Crespo is the author of seven books, the last four with Springer-Verlag in 2007, 2010, 2012, and 2014 about maintenance, warranty, and supply chain management. Professor Crespo leads the Spanish Research Network on Dependability Management and the Spanish Committee for Maintenance Standardization (1995–2003). He also leads a research team related to maintenance and dependability management currently with 5 Ph.D. students and 4 researchers. He has extensively participated in many engineering and consulting projects for different companies, for the Spanish Departments of Defense, Science and Education as well as for the European Commission (IPTS). He is the President of INGEMAN (a National Association for the Development of Maintenance Engineering in Spain) since 2002.

The Integration of Open Reliability, Maintenance, and Condition Monitoring Management Systems

**Mónica Alejandra López-Campos, Adolfo Crespo Márquez
and Juan Francisco Gómez Fernández**

Abstract Maintenance and assets management of an industrial plant has been always a complex activity. Nowadays, Computerized Maintenance Management Systems (CMMS) help to organize information and thus to carry out maintenance activities in a more efficient way. The emergence of new ICT has increased also the use of Condition-Based Maintenance (CBM) systems and the application of Reliability-Centered Maintenance (RCM) analysis. Each system is proved to provide benefits to the maintenance management. However, when all the systems are adopted, the lack of integration among them can prevent the maximum exploitation of their capabilities. This work aims at fulfilling this gap, proposing an e-maintenance integration platform that combines the features of the three main systems. The methodology and the reference open standards used to develop the platform are exposed. UML–BPMN diagrams represent the emerging algorithms of the designed system. The final product, a software demo, is implemented in an electric transformer.

Keywords E-maintenance · Integration · UML · CMMS · CBM · RCM

M.A. López-Campos (✉)
Department of Industrial Engineering, Universidad Técnica
Federico Santa María, Av. España 1680, Valparaíso, Chile
e-mail: monica.lopezc@usm.cl

A. Crespo Márquez · J.F. Gómez Fernández
Department of Industrial Management, School of Engineering,
University of Seville, Camino de los Descubrimientos s/n,
41092 Seville, Spain
e-mail: adolfo@us.es

J.F. Gómez Fernández
e-mail: juan.gomez@iies.es

© Springer International Publishing AG 2018
A. Crespo Márquez et al. (eds.), *Advanced Maintenance Modelling
for Asset Management*, DOI 10.1007/978-3-319-58045-6_3

1 Introduction

Maintenance and assets management of an industrial plant has always been a complex activity that involves handling a large amount of information. Different frameworks, standards, and methodologies have been developed over the years to guide these activities [1].

In the same way, the application of computerized tools and technologies of information and communication (ICT) has always been present in the history of the maintenance function, since the rise of the first personal computers in the 50's decade of the past century [2], until the latest trends involving mobile devices and internet, the so-called e-maintenance processes [3] including the expected future techniques related by Labib [4], as the auto-maintenance systems.

Nowadays, one of the most typical and spread tools used in enterprises around the world is the Computerized Maintenance Management Systems (CMMS) that are basically databases that allow programming and following maintenance activities according to objectives and policies [5]. It is an integrated set of software and data files designed to manage a massive amount of maintenance data in a standardized manner to facilitate the decision-making process [6]. If it is enabled for Web use, it is called e-CMMS [7].

The emergence of new ICT (intelligent sensors, personal digital devices, wireless tools, etc.) has allowed an increase in the efficiency, speed, and proactiveness of maintenance [8]. Prognosis, diagnosis, and monitoring processes are part of the "proactive maintenance" [9]. These three activities involve tracking the item condition. Here is where another maintenance tool stands: the Condition-Based Maintenance (CBM), specifically the CBM developed through remote sensing devices that monitor the condition of assets, where the captured data are transmitted via Web for analysis and decision-making: the denominated e-CBM [7]. As a strategy that offers many benefits, CBM significantly reduces stoppages for failures and if it is well implemented, it is able to reduce by over 20% production losses, to improve quality and reduce spare parts inventories [10]. Campos [11] makes an interesting literature review about e-CBM.

However, in order to obtain all the benefits that e-CBM offers, it is necessary to implement it in an appropriated manner, selecting the most adequate items and the frontiers of the system to be maintained using this policy. Reliability-Centered Maintenance (RCM) can be useful in this sense. The RCM process was developed in the 60's and 70's in order to determine the most appropriate policies to improve the physical assets functioning and to manage the consequences of their failures [12]. The RCM approach contains a variety of methodologies such as FMEA (Failure Mode and Effect Analysis), RBD (Reliability Block Diagram), RP (Reliability Prediction), FTA (Fault Tree Analysis), and ETA (Event Tree Analysis). As pointed out by several authors [13–15], the use of RCM technique is necessary for the proper selection of the CBM processes and technologies. RCM analysis helps in selecting the optimal maintenance policy for every maintainable item.

Each of these three mentioned resources (CMMS, CBM, and RCM) has been shown separately to provide benefits for maintenance and to improve the processes efficiency. However, the lack of integration between them when used simultaneously can avoid the maximum utilization of their advantages. Gabbar et al. [16] mention that despite its benefits, RCM is a process that only consumes time and efforts if it is not automated, due to which it would be using non real-time reliability data to determine the maintenance strategies. This automation relates to a communication CMMS–CBM–RCM platform.

Therefore, this paper aims to fill this gap, proposing an e-maintenance integration platform that combines the features of the three main resources (CMMS–RCM–CBM). Even though CMMS is considered as a tool for managing maintenance information while CBM and RCM are conceived as strategies, putting together the information used and generated by each one of them can create enormous added value and new emerging maintenance processes [8], such as those proposed in this paper.

The purpose of CMMS is to manage, plan, and document the maintenance activities, categorized and related to resources and infrastructure elements. CMMS is the integration between technology and social variables: tasks, resources, personnel and organization, to generate work orders according to costs, risks, equipments, relations, plans, etc. It encompasses the historical knowledge of solutions and learning. Regarding CBM and RCM support systems, they deep into the knowledge based on the physical condition and reliability, respectively.

Hence, we have three sources of maintenance information: (1) the aggregate management, planning, and costs; (2) the real-time status of equipment, using alarms, thresholds, and degradation patterns; and (3) the prediction derived from reliability studies. The benefit is to integrate all three, improving management with real-time information and reliability predictions, controlling and optimizing the processes. Also, as CMMS holds business transactions and technical events information, this historical data can be used as basis for CBM and RCM analysis; therefore, the three resources are clearly defined and integrated.

Until now, there is no evidence in literature of a similar open integration system, referring the data architecture of integration software. Data models and functioning algorithms for this integration are designed using UML and BPMN standards. The final product, a software demo, has been applied to a power transformer owned by a Spanish energy company. Then, the objectives of this paper are as follows: (1) to expose the methodology to develop an e-maintenance CMMS–RCM–CBM integration platform, (2) to provide the UML–BPMN data models and functioning algorithms for the integration platform, and (3) to present the implementation of the platform as a demo software applied to an electric transformer system.

The chapter is structured as follows. Section 2 provides a state-of-art of previous attempts for a CMMS–RCM–CBM integration. Section 3 presents the general architecture and functioning of the proposed integration platform. In Sect. 4, the graphic UML–BPMN models of the platform operation are presented. In Sect. 5 the CBM decisional algorithm is developed. Section 6 describes the implementation of the platform in an electrical transformer system, and finally Sect. 7 concludes the chapter.

2 Previous Integration Attempts

The market has developed and offers powerful software tools that launch the possibility of operating as an integration platform, especially between RCM and CMMS modules. Nevertheless some empirical evidence from industry makes the issue more complicated; for instance, Carnero and Novés [17] report that a high percentage of the CMMS acquired by enterprises is not operational, does not yield the expected benefits, or is eliminated after a short time of use. Even if the CMMS is being used, in many cases its complete functionalities are not exploited by the organization [18], and even less CMMS are integrated with other maintenance tools. In fact, Fumagalli et al. [19] report that just 20% of the enterprises using CMMS use their CMMS to manage CBM activities. These situations may be due to a lack of knowledge [18], or to the fact that the CMMS–RCM–CBM integration is not a common issue, even in the most recognized maintenance applications.

A literature review shows that the CMMS–CBM–RCM integration is a new avenue toward the integration and unification of diagnostic techniques and knowledge about failure modes and use of maintenance data.

Looking in scientific databases for papers related with the integration of RCM–CMMS–CBM, a total of 16 papers have been found. Table 1 summarizes the state-of-art of CBM–RCM–CMMS integration.

From the state-of-art and Table 1, the following can be stated:

As it can be seen, CMMS–RCM–CBM integration is a rarely subject. The idea of using RCM (especially FMECA) as a guide to have a successful implementation of CBM is a common approach already referenced since 1995 [20]. Although most of the related projects establish a clear link between RCM and CBM, the necessity of CMMS–RCM communication has been explicitly recognized only until 2005 [24].

The strength relation between RCM and CBM has created a new term, *CBM+*, presented since the early first years of 2000 [29].

Even if the information coming from CMMS is obviously an input for diagnosis and prognosis processes, the first system that mentions a CMMS–RCM–CBM integration appeared until 2003 [23], a paper available only in Chinese language.

Proposal of Ref. [15], about using MIMOSA standard for the design of the integration system has been followed in this paper, the platform now proposed considers MIMOSA as well.

However, in general, the found contributions deal with very general content; in most cases the articles do not detail the mechanisms of data flow that allows the integration of techniques.

From these observations, the novelty of this paper can be settled as the specification of data models and algorithms for CMMS–RCM–CBM integration, using specific modeling standards that describe the information flow among the three modules.

Moreover, the existence of information exchange among CMMS, RCM, and CBM modules allows the emergence of new e-maintenance processes. These new

Table 1 Previous attempts of integration CBM–RCM–CMMS

Year	Authors	Article title	Main contributions
1995	Tsang [20]	Condition-based maintenance: tools and decision-making	Application of CBM in a RCM context using decision models according to levels of deterioration. It includes a decisional tree as a guide to decide when to apply CBM of another maintenance policy
1998	Deb et al. [21]	Multisignal modeling for diagnosis, FMECA, and reliability	Focused on developing a solution package for Integrated Diagnostics (ID) that addresses Design for Testability (DFT), reliability analysis, failure modes, effects and criticality analysis (FMECA), and test program set (TPS) development and field maintenance
2001	Finley and Schneider [22]	ICAS: the center of diagnostics and prognostics for the US Navy	ICAS (Integrated Condition Assessment System) has the ability to turn the operational data and the information provided by sensor systems into useful information. This information can be used for the diagnosis of failures and the indications of possible future fault conditions
2003	Shum and Gong [23]	Design of an integrated maintenance management system	Overall view of an integrated maintenance management system (IMMS). Its structure consists of three subsystems: the real-time monitoring (RTM), reliability-centered maintenance (RCM), and the computerized maintenance management system (CMMS). These three subsystems work together to form an integrated structure to provide guidance for the maintenance operation. The application in a mold injection machine is described
2005	Penrose [13]	RCM-based motor management	Step-by-step methodology for the integration of RCM and CBM techniques on electric motors Introduction to the selection of electrical motor diagnostic technologies for rotating machinery

(continued)

Table 1 (continued)

Year	Authors	Article title	Main contributions
2005	Huo et al. [24]	CMMS-based reliability-centered maintenance	Analysis of the required characteristics that a CMMS must have to be compatible with RCM methodology. Oriented to a mature power system. This paper proposes that CMMS can be employed to provide original data for the analyses of RCM. The paper also provides guidance for ensuring that the equipment data and history residing in a CMMS are complete and accurate
2006	Lehtonen [25]	On the optimal strategies of condition monitoring and maintenance allocation in distribution systems	This paper first discusses the role of RCM and CBM in distribution systems. Examples are given of typical cases of component condition monitoring and system-level maintenance allocation optimization. In the latter part of the paper, a method is presented for optimal condition monitoring and maintenance strategies for both component and system levels. The method is based on the statistical analysis of component condition data and probabilistic optimization of the overall cost function
2006	Ranganath et al. [26]	System health monitoring and prognostics—a review of current paradigms and practices	Revision of the philosophies and techniques that focus on improving reliability and reducing unscheduled downtime by monitoring and predicting machine health It shows the opportunity to integrate the qualitative information that can be extracted from failure mode and effects analysis (FMEA) or fault tree analysis (FTA) of a process or machine into the quantitative analysis that generates diagnostic recommendations
2006	Han and Yang [27]	Development of an e-maintenance system integrating advanced techniques	Proposal of an e-maintenance system that allows maintenance operations to achieve performance close to zero downtime in a shared platform, through advanced technologies

(continued)

Table 1 (continued)

Year	Authors	Article title	Main contributions
			and integrate distributed resources. This platform has two subsystems: the service center and maintenance sub-local system, both a provider and client relationship
2006	Emmanouilidis et al. [28]	Flexible software for condition monitoring, incorporating novelty detection and diagnostics	Development of software solution for condition monitoring, identification and machinery diagnostics, which can be customized to a range of monitoring scenarios. Its main constituents are a number of independent software modules, such as the fault and symptom tree, the fuzzy classification module, the novelty detection, and the neural network diagnostics subsystems. It is implemented on two different applications, namely machine tool monitoring and gear box monitoring
2008	Jaw and Merrill [29]	CBM+ Research environment—facilitating technology development, experimentation, and maturation	Design and implementation of a CBM+ information system to facilitate CBM and RCM practices in aircraft propulsion systems. CBM+ extends the traditional capabilities of fault detection and isolation with the capabilities of prognostics and logistics. This paper also presents some examples to demonstrate the concept of the CBM+ research environment
2009	Wessels and Sautter [14]	Reliability analysis required to determine CBM condition indicators	A reliability-centered analysis is presented that defines the process to correctly select CBM as the technically and economically feasible maintenance solution. This paper proposes the use of a diagram to decide if CBM is viable or not
2010	Niu et al. [15]	Development of an optimized condition-based maintenance system by data Fusion and reliability-centered maintenance	This paper presents a novel CBM system that uses reliability-centered maintenance mechanism to optimize maintenance cost, and employs

(continued)

Table 1 (continued)

Year	Authors	Article title	Main contributions
			data fusion strategy for improving condition monitoring, health assessment, and prognostics. The proposed CBM+ system is demonstrated by the way of reasoning and case studies. The results show that optimized maintenance performance can be obtained with good generality. The proposed system is based on MIMOSA OSA-CBM
2011	Trappey et al. [30]	Maintenance chain integration using Petri-net enabled multiagent system modeling and implementation approach	This research proposes a collaborative environment integrated by a service center of diagnosis, prognosis, and asset operations. To realize the automation of communication and negotiation, multiagent systems, Petri-net modeling, and UML are applied
2012	Galar et al. [31]	Maintenance decision-making based on different types of data fusion	This paper proposes a combined data mining-based methodology to collect useful data from a CMMS to be used in CBM processes
2012	Chang et al. [32]	Development of an e-operation framework for SoPC-based reconfigurable applications	It presents a conceptual framework for e-operations with the capability of e-maintenance, integrating e-diagnostic, among other processes. UML is used as the tool to accomplish the object-oriented analysis and to design the system

processes can combine reliability data with real-time signals and economical information, generating for example, the emerging process of "economical-maintenance prognosis" afterward exposed in this paper.

The novelty of this paper is that it covers and exposes the systems development since the signal capture, to the formulation of algorithms for advanced analysis of equipments performance. This paper justifies the approach of using multivariable online predictive maintenance strategies with a very high level of accuracy and capability in terms of detectable failure modes. It also allows calculating, monitoring, and predicting the evolution of the failure risks, transferring to a cost estimate. The use of e-maintenance strategies makes possible linking signals of monitored variables to risk levels, relating technical variables (hierarchy) to business values (cost centers), using open information systems for reliability management. In sum, in the analysis for optimizing operational performance suggested in this paper, failure modes are not only the link between technical analysis and functional analysis of systems or equipments, but it is also the nexus for the development of the control systems, with the objective to increase the reliability and efficiency of the network. The proposed system also allows scheduling algorithms and optimization techniques for generating simulation environments that establish the prognosis of future status and evolution of the equipments condition. From them it is possible to quantify and compare magnitudes as relevant as the instantaneous failure probability of equipment and the financial risk if there is a failure. It aims to integrate the use of the most advanced techniques of acquisition and signal processing (network quality, vibration, ultrasound, thermography) with the development of new multivariable algorithms for the prediction of reliability and economic management of the life cycle of the systems.

3 General Integration Architecture

The general architecture of the CMMS–RCM–CBM integration platform proposed is based on the structure and information domain areas defined by the MIMOSA standard and its Open System Architecture for Enterprise Application Integration (OSA-EAI) [33]. MIMOSA standard relates an Open Reliability Management module, an Open Maintenance Management module, and an Open Condition Management module. From this concept we define our integration platform as shown in Fig. 1.

Figure 1 details the three modules composing the integration platform. Within the CMMS module there is a division of four sub-modules: "Systems planning", "Resources Management", "Implementation and Operation", and "Assessment and continual improvement". These sub-modules represent four main activities related with a framework for asset management developed by López-Campos [1] and inspired by the requirements of PAS 55 and ISO 9000 standards [34, 35]. Inside each sub-module there are some items that define more specifically the processes to execute. Inside the "Systems planning" sub-module is the information related to policies, strategies, objectives, and plans of the maintenance function. The "Resources Management" sub-module comprises the management of risks, legal issues, infrastructure, documentation, materials, human resources, etc. The "Implementation and operation" sub-module stores the operational procedures, and data from calibration and preventive maintenance. Finally, the "Assessment and continual improvement" sub-module manages the information related to the system itself: audits, corrective and preventive actions, records, continual improvement, etc. The CMMS module of Fig. 1 has this special structure, due to which it is orientated to the PAS 55 asset management framework, but this can be replaced by any structure of any commercial CMMS. RCM and CBM modules operate as support systems which feed the CMMS, and which in turn are also fed by the CMMS.

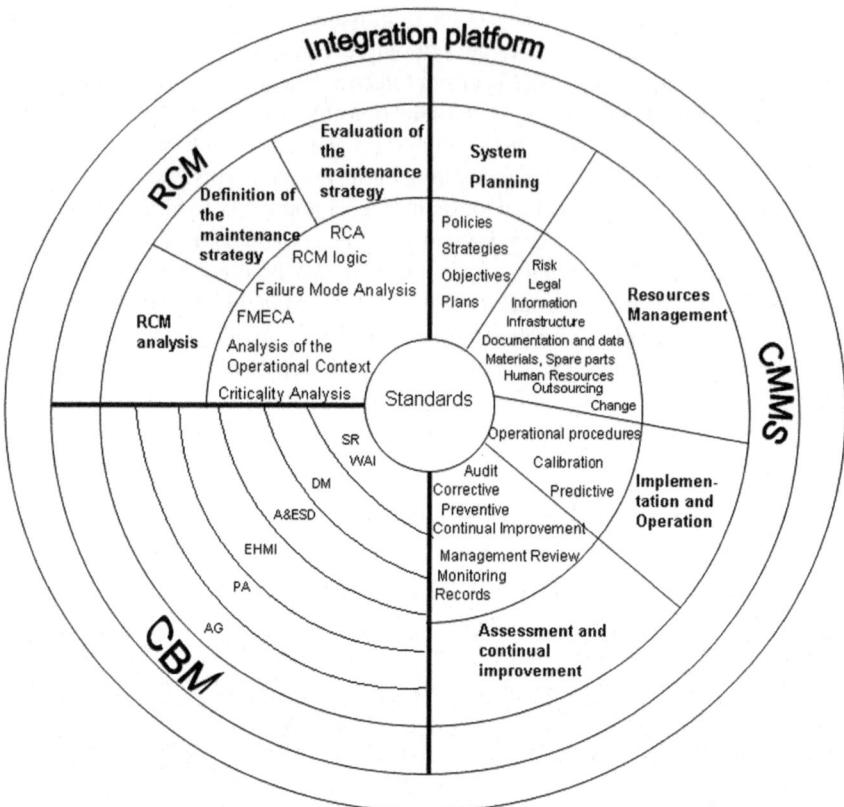

Fig. 1 General architecture of the integration platform CMMS–CBM–RCM

The RCM module comprises three divisions or sub-modules, which correspond to the three main activities to perform the Reliability-Centered Maintenance: "RCM Analysis", "Definition of the maintenance strategy", and "Evaluation of the maintenance strategy". Also, in the part closest to the circle center, there are some of the techniques and tools corresponding to each sub-module, such as Criticality analysis, Analysis of the operational context, Analysis of failure modes, effects and criticality (FMECA), RCM logic, Failure mode analysis, and Root cause analysis (RCA).

The CBM module consists of six layers inspired by the ISO standard 13374: 2003 "Condition monitoring and diagnostics of machines" [36] and the MIMOSA Open Systems Architecture for Condition-Based Maintenance (OSA-CBM) standard [37]. The CBM six levels are represented by its acronyms as follows: SR = "Sensor Registry", WAI = "walk around inspections" (this level is related to the acquisition of data from the controlled variable), DM = "Data manipulation" (it analyzes the signals, computing derived descriptors and sensor readings from the raw measurements), A&ESD = "Alarm and event state detection" (level that facilitates the management of profiles and the detection of abnormalities in new data and alarm generation), EHMI = "Evaluation of the health of the maintainable items" (level that helps to diagnose any faults, and it measures the health of equipments or processes), PA = "Prognostic assessment" (level that analyzes the degradation trend and calculates the convenience of intervention), and AG = "Advisory Generation" (it generates recommendations and alarms).

In the center of the integration platform there is the reference standards used, as standards for ICT interoperability, business process modeling standards, etc.

Figure 2 shows an overview of the hardware architecture of the integration platform, related to the levels indicated by the MIMOSA OSA-CBM standard. It divides a system into seven different layers (partially derived from Ref. [15]). In Fig. 2 it is possible to analyze the information flow to operate the system and the potential access points for the user. In addition, the databases necessary to operate the system are also indicated. The information flow that allows the integration is discussed in detail in the following section.

Figure 3 shows the levels or layers that comprise the ICT platform for supporting the integration of the system [38]. The integration has been done according to ISO 18435 standard [39] and to protocols listed in MIMOSA, generating a comprehensive database as maintenance core that uses the information from the monitoring CBM database, datasheets, and RCM patterns and historic CMMS (as indicated in BPMN diagrams). This has been done to show in parallel the benefits of the integration, adding value without affecting the normal operation of the business.

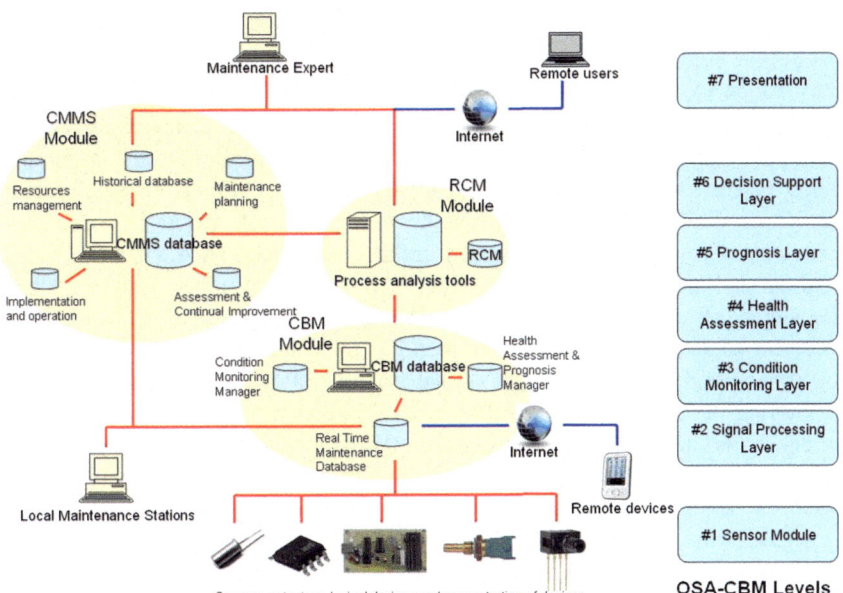

Fig. 2 General view of the integration platform, related with MIMOSA OSA-CBM levels

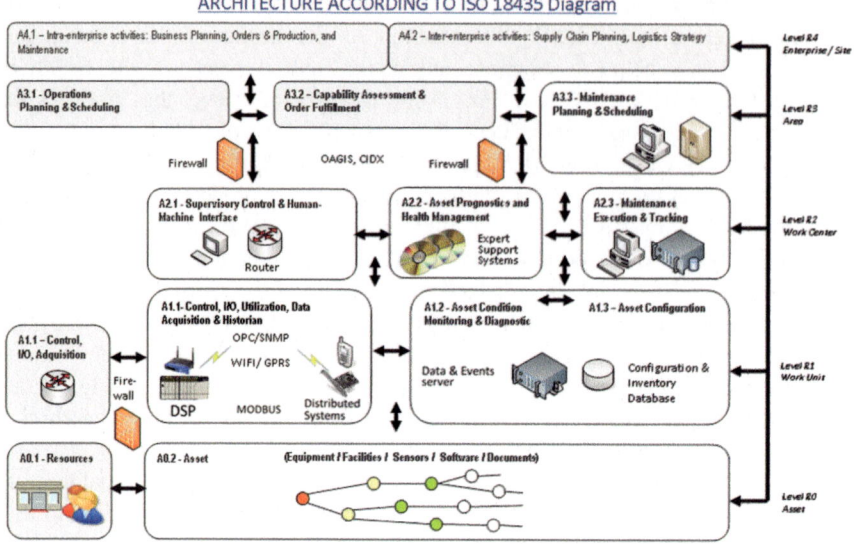

Fig. 3 Levels of the ICT platform for integration

In Fig. 3, white boxes represent the modules developed for the proposed system and gray boxes represent commercial ERP and CMMS modules, which information is exchanged through interoperability. For example, for Level 1 the acquisition and control of assets is made through analog and digital sensors, processed by a system based on DSP and controllers distributed and integrated in the assets. All the processed information is transmitted, concentrated, and transmitted via WI-FI/WIMAX for large bandwidths and through GPRS/UMTS for remote sites and low transmission capacity requirements (this for A1.1). Components A1.2 and A1.3 have been designed integrated into a single software platform, in an application that asks and receives the processed signals of different assets and that configures the warning/alarm thresholds for each asset. Besides, it makes an inventory of the configuration, asset utilization, alarms, and changes control. Furthermore, this part allows working with signals of other systems; signals exchanged using XML formats, as MIMOSA indicates. Level 2 is developed according to the three components of the ISO 18435 standard, a part that allows the supervision and control (A2.1), a second part that allows the implementation and monitoring of maintenance activities, and a third part with specific modules for prognosis and health management for each failure mode. Here is where intelligence is introduced, for the prognosis, for each failure mode and asset. Finally, Level 3 has been developed a planning and scheduling module that can encompass in aggregated way, the management of multiple assets and different time intervals. Besides, considering the adequacy of activities according to other business systems, an interface for online information exchange has been developed, using OAGIS, CIDX format.

Fig. 4 Generic activity
model based on ISO 62264

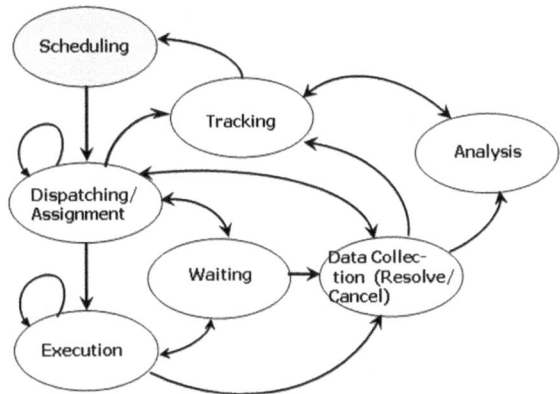

The internal information flows of the proposed system are based on the IEC 62264 standard. Taking as a reference the O&M management domains of the functional enterprise/control model (IEC 62264:2007) [40], a generic activity model has been defined for all the functions for the processes logic implementation (Fig. 4).

The generic activity model of Fig. 4 defines a general request–response cycle of tasks identifying possible data flows. The activity model starts with requests or schedules, generating a detailed schedule whose tasks are dispatched or assigned. The execution management implies that the task could be executed in several parts, considering waiting periods or new assignments. Once the execution ends, the task is resolved collecting data and summarizing reports for the responses. Sometimes the task could be canceled from dispatching, execution or left in a waiting state; in this case the data collection is in charge of documenting the reasons. Inside the tracking state, the maintenance performance and effectiveness information are managed. Finally, the analysis task searches the optimization of resources' operating performance and efficiencies, studying the present–past–future maintenance information to detect problematic areas or areas for improvement.

Standard ISO 14224 [41] is used to organize the elements used by the system, as taxonomy. Figure 5 shows this element organization.

4 Graphic Models of Operation of the Integration Platform

Data model diagrams and operation algorithms for the integration platform developed in this paper are expressed using UML and BPMN standards. UML is a graphical language used for specifying, visualizing, constructing, and documenting systems. UML has proven to be successful in the modeling of many large and complex systems [42], in the modeling of several maintenance applications [1], and it is a basis resource to the further implementation of the designed models as a

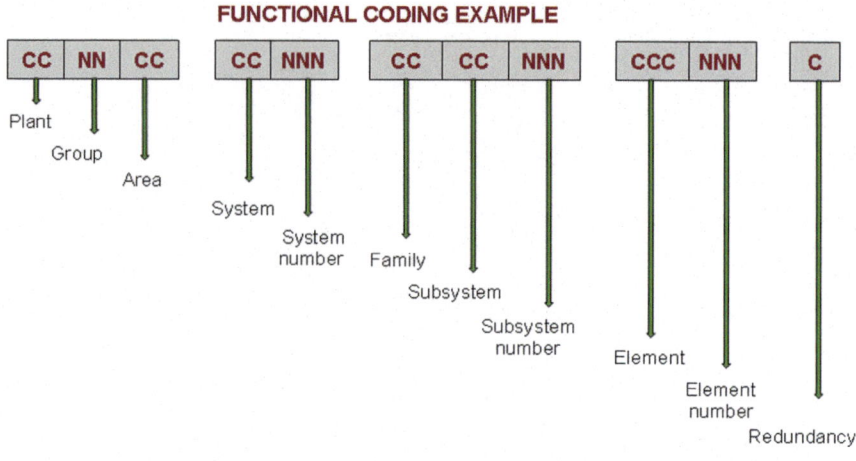

Fig. 5 Taxonomy of elements based on ISO 14224

software program. BPMN is a related graphical standard oriented to business process modeling. Actually BPMN and UML are two of most used and recognized modeling standards around the world in a variety of fields, since informatics science to management [1]. UML 2.1 and BPMN 1.0 are the versions adopted in this paper to model the operation of the designed platform.

The activities flow for the CMMS–RCM–CBM integration platform is presented using BPMN diagrams in Figs. 6, 7, 8, 9, and 10.

A general operation diagram (Fig. 6) and four sub-processes diagrams (Figs. 7, 8, 9, and 10) outline the most significant aspects of the functioning of the integration platform. The input and output documents are indicated by an artefact symbol. As usual in BPMN diagrams, these models exploit the possibility of deepening the definition of processes into sub-processes to be more specific. In the following diagrams, processes including embedded sub-processes are identified by the symbol "∞".

Figure 6 shows the general functioning of the CMMS–RCM–CBM integration, expressed using BPMN standard. Every maintenance tool (CMMS, RCM, and CBM) is represented by a pool division. The tangible output documents are indicated with an artefact figure. The activities flow of Fig. 6 is partially inspired from the functioning of RCMO software developed by Meridium, which integrates RCM with SAP (CMMS), but not yet with a CBM module.

Figure 6 exposes the general functioning of the integration platform as follows: starting from the information of records and CMMS historical data, a RCM analysis is performed to define a maintenance strategy that will be executed later, either through predictive CBM (inspections made by the operator, or sensors real-time monitoring), or proactively, through a planning and subsequent generation of

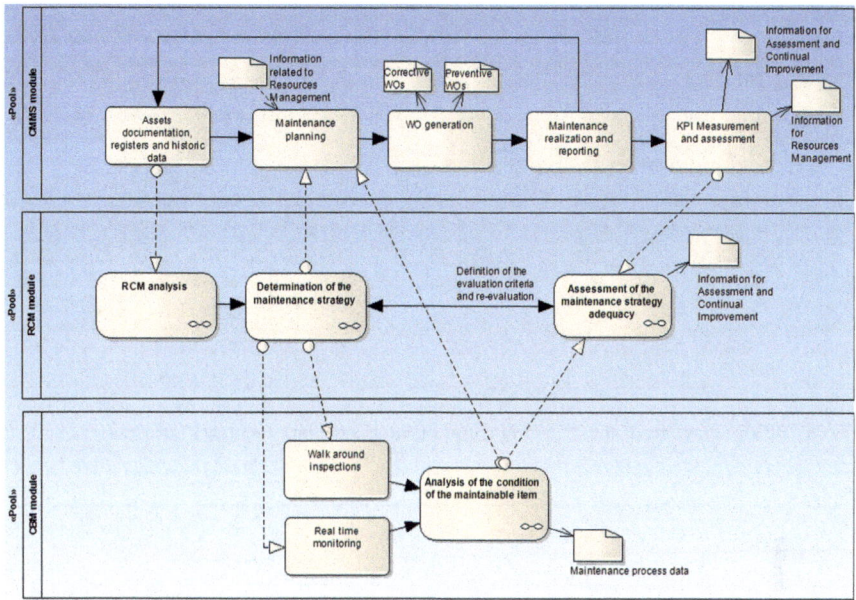

Fig. 6 BPMN diagram of the CMMS–RCM–CBM integration

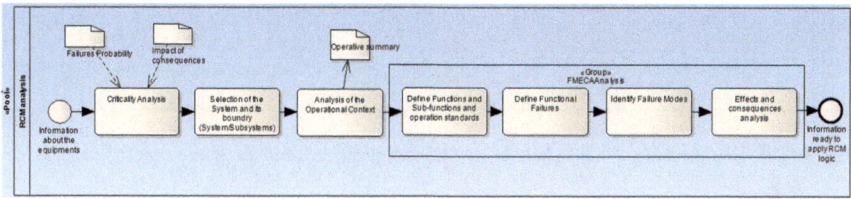

Fig. 7 BMPN diagram of the "RCM Analysis" process

maintenance work orders. It should be mentioned that the maintenance planning comes from the strategy defined by the RCM, from the data yielded by the "Assessment of the maintenance strategy adequacy" process, and from the data about human and material resources available. After the maintenance task mandated by the work order is executed, a report should be generated to be included in the CMMS database. Later this report will serve as an input for the measurement of KPIs and the assessment of the system. An important feature of this system integration is that it allows maintenance strategies to be evaluated, released, and re-evaluated from the results and from the actual state of the maintainable asset. That is, it is intended that the system to be intelligent.

This general functioning diagram (Fig. 6) has four embedded sub-processes: "RCM analysis", "Determination of the maintenance strategy", "Analysis of the condition of the maintainable item", and "Assessment of the maintenance strategy adequacy".

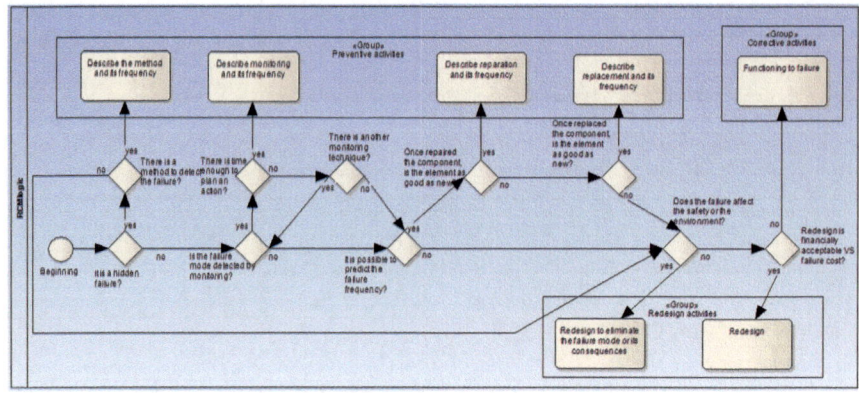

Fig. 8 BPMN diagram of the "Determination of the maintenance strategy" process

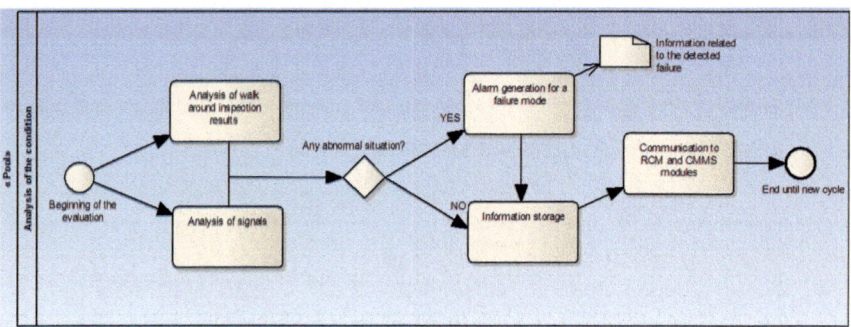

Fig. 9 BPMN diagram of the "Analysis of the condition of the maintainable item" process

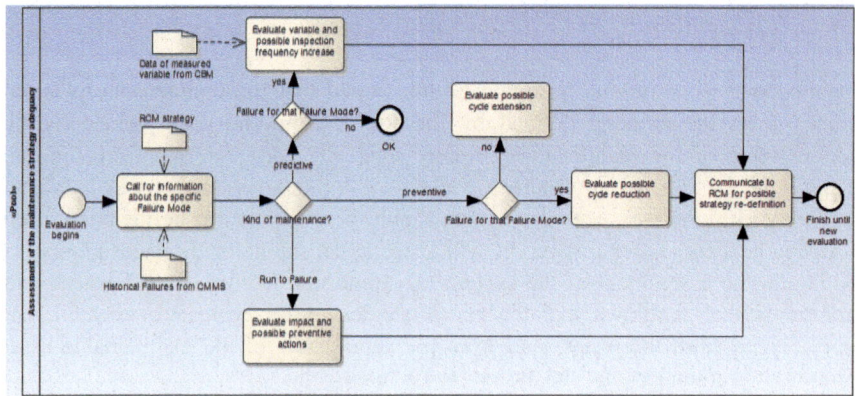

Fig. 10 BPMN diagram of the "Assessment of the maintenance strategy adequacy" process

By clicking on the box corresponding to "RCM analysis" it unfolds its embedded process, which is shown in Fig. 7. Scheme of Fig. 7 describes the RCM activities necessary to identify the effects, consequences, and occurrence probabilities of the main failure modes, in order to assign maintenance strategies that avoid or minimize their occurrence.

The "RCM analysis" sub-process operates as follows: based on the available information of the items involved in the scope of the system, on their failure probabilities and the expected impact of the possible failure consequences, a criticality analysis is generated. This criticality analysis makes possible to select the equipments on which it is necessary having more strict control over maintenance, either because their high failure frequency and/or the major failure consequences.

Once critical equipments are selected, an analysis of their operational context is done and an operative summary is generated. This summary represents the operation guide and the compendium of the main characteristics of the maintainable equipment. Then, a FMECA analysis is made, which defines the main functional failures, failure modes, effects, and consequences. These data are the input information necessary to determine the maintenance policies to assign to each maintainable item through the "RCM logic".

The embedded sub-process that determines the type of maintenance strategy for each failure mode is the "Determination of the maintenance strategy" process that is shown in Fig. 8 under the name of "RCM logic", which is the methodology for making decisions that it is employed. This process determines if for a particular failure mode it should be applied a preventive policy (what kind and how often), an operation until failure policy, or a redesign of the maintainable item.

The "RCM logic" also called "RCM logic tree" runs on a series of questions regarding a specific failure mode. The sequence of questions and their respective decision paths are the following:

Question No. 1: Does the failure mode have hidden or visible consequences? If consequences are hidden but there is a method for detecting the fault, this method has to be described and a frequency has to be assigned (preventive maintenance). If, however, there is no method to detect hidden flaw, question No. 5 has to be answered. If failure consequences are evident, go to question No. 2.

Question No. 2: Does the failure mode is detected by monitoring? If monitoring is feasible and there is enough time to plan an action, the monitoring and its frequency (preventive maintenance) has to be described. If there is not time enough, another monitoring technique has to be selected, and if there is no other technique, go to question No. 3. If it is not possible to detect a failure by monitoring, question No. 3 has to be answered.

Question No. 3: Is it possible to predict the failure frequency? If it is possible make such a prediction, question No. 4 has to be answered. Otherwise, question No. 5 has to be answered.

Question No. 4: Will it be the element "as new" after repairing? If the component can be "as new" the repair method and its frequency plan (preventive maintenance) has to be described. If it is not possible for the component to be "as new" the replacing has to be considered. If replacing, the item will be "as new" the replacement and its frequency (preventive maintenance) has to be described. If replacement is not an option, question No. 5 has to be answered.

Question No. 5: Does the failure affect safety or environment? If the answer is yes, proceed to redesign to eliminate the failure mode or its consequences. If the answer is negative, question No. 6 has to be answered.

Question No. 6: Is redesign economically acceptable versus failure cost? If the answer is yes, then you must redesign. If the answer is no a "run to failure" policy (corrective maintenance) has to be selected.

For the specific case of condition-based maintenance, the sub-process "Analysis of the condition of the maintainable item" (Fig. 9) handles the results of signal detection and inspections made by the operator. It determines the existence of abnormal situations that can trigger an alarm for a specific failure mode.

Detection of abnormal situations from the signals data is not trivial. Behavior of a variable or a set of variables has to be analyzed for a specific given time. From this analysis, the occurrence of abnormal situations has to be inferred, enabling the detection of certain failure modes (diagnosis and prognosis). To accomplish this, typically a methodology of three steps is executed: information acquisition, data processing, and decision-making to recommend the most appropriate maintenance policies [43]. In literature there are a variety of studies that propose various techniques for achieving each of the three stages, depending on the type of technology and the selected solution approach.

There are numerous and varied possible resolution approaches for diagnosis and prognosis. Medjaher et al. [44] suggest a classification: approaches based on mathematical models representing the system behavior including its degradation (for example a set of equations derived from physical laws), approaches based on data obtained from monitoring (data subsequently converted into useful information by means of artificial intelligence techniques and advanced statistical tools, such as Markov chains and Bayesian networks), and approaches based on reliability analysis, using historical data to fit a statistical model (Weibull distributions, exponential, or PHM).

Compared to other approaches, methods based on reliability analysis are the simplest to implement, especially in the case of repairable systems that have a good amount of historical data [45]. Several authors also propose the combined application of two techniques, such as signal analysis and its contrast with historical data [45]. For the proposed integration system, the CBM decisional algorithm is treated in Sect. 5 of this chapter.

Once the defined maintenance policies have been executed, the process that evaluates their appropriateness starts. This process gives intelligence to the system. This is the embedded process "Assessment of the maintenance strategy adequacy", which appears in Fig. 10. This process basically takes the maintenance strategy defined by the RCM analysis and compares it to data obtained from the CMMS for a particular failure mode, and to the condition (in case of predictive maintenance). From this comparison, the system suggests certain tests in order to improve the implemented maintenance policy. For a specific failure mode, the evaluation of the strategy is as follows:

If the maintenance policy for a specific failure mode is "run to failure", it is necessary to evaluate the impact of the failure and the possibility of incorporating preventive actions. Any modification to the strategy, it is necessary to be updated into the RCM module of the system.

If it is a strategy of predictive maintenance (CBM) and so far the failure mode has not occurred in the system, it means the strategy is relevant and it does not need to be changed. If, however, the failure mode has occurred, then it is necessary to evaluate the controlled variable and to consider increasing the frequency of inspection. Any possible change must be reported to the RCM module to redefine the strategy in the system.

If the strategy is preventive maintenance and so far the failure has not existed, a possible extension of the maintenance cycle has to be considered (diminish frequency). If, on the contrary, the failure has already existed the option is evaluating a possible reduction of the maintenance cycle (increase frequency). In any case, any change must be reported to the RCM module. Also in all cases, it is possible to perform the analysis using the tools that appear in the central part corresponding to the RCM module in Fig. 1.

These five main BPMN diagrams (Figs. 6, 7, 8, 9, and 10) show the data flow among the three modules of the integration platform (CMMS–RCM–CBM). Although these diagrams are very illustrative to develop an application software, there are also other types of diagrams that can be helpful to better understand the operation of a system, such as use-case diagrams or object diagrams.

Figure 11 presents another kind of diagram, a UML object diagram that illustrates the relations between the main elements of the integration platform. The interpretation of the diagram is as follows: A system (an electric transformer in this case) is composed by more than one subsystem (magnetic system, dielectric circuit system, protection system, etc.) and each subsystem executes one or more required sub-functions. As part of a preventive analysis (or a FMECA study), each sub-function has one or more identified Functional Failures that may occur. When it happens, the Functional Failure with all its characteristics (description of its related Failure Modes and Failure Effects) is registered as a History Record that later can be used to make health assessments of the system, prognosis analysis, or maintenance activities planning. Obviously, a Work Order can be generated once a Functional Failure is detected. The Work Order describes the observed Failure Mode and matches to a specific Maintainable Element. All this information becomes part of the History Records. This exposed path belongs to the process of detecting a failure through inspection, sudden events, or preventive activities.

The preventive activities are defined by the process of Maintenance Planning that receives information from the History and makes a RCM analysis to program the activities. The designed system has also another path, corresponding to the events detected by CBM processes. In the system, some specific Maintainable Elements have incorporated one or more Sensors. Sensors are continually emitting Signals. A Condition Monitoring Manager receives the Signals and evaluates them. The Condition Monitoring Manager has the specifications in which a Signal is considered normal, out of the boundaries something wrong is occurring.

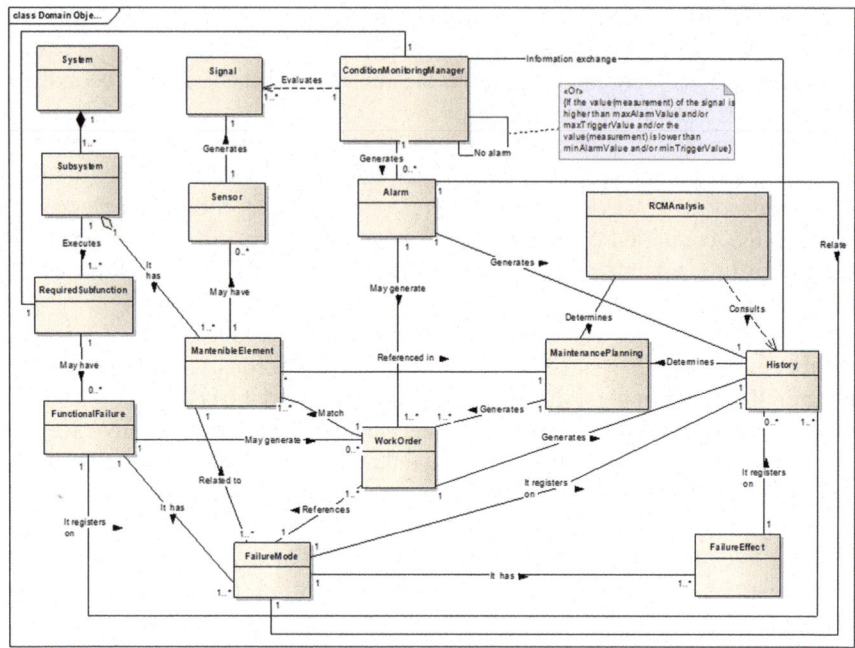

Fig. 11 Object model of the integrated CMMS–RCM–CBM system

The Condition Monitoring Manager can then provide one or more Alarms. The Alarm may generate one or more Work Orders and it also goes to the History Record.

The system tries to relate the generated Alarm to a particular Failure Mode to make easier the corrective actions. Following the described chain of relations, it is possible to affirm that one Maintainable Element can generate multiple Alarms and this seems correct. However, one Alarm has to correspond preferably to just one Failure Mode. This class (Failure Mode) describes the events that are enumerated by a FMECA analysis or a general engineering analysis. One or more Failure Modes are associated with a Required Sub-function and a Maintainable Element can have one or more functions. To be able to check one single Failure Mode by an Alarm, enough information must be collected. It happens, often, that one Alarm is associated with more Failure Modes. We would like to avoid this kind of thing in our system. So, one hypothesis is that there are enough monitored variables (and experienced personnel) to assign one Alarm to each Failure Mode.

The user of the software generated in this project could be called to connect each Failure Mode defined in the system with a History record that represents an event in the real system, related with a failure. The consulted information is based on a Critically Analysis (frequency and consequences) made using the History. When this relation is established, the RCM analysis can be carried out; the frequency of

Table 2 System integration objects related to maintenance tools

CBM	CMMS	RCM
Sensor	System	Required sub-function
Signal	Subsystem	Functional failure
Condition monitoring manager	Mantenible element	Mantenible element
Alarm	Work order	Failure mode
	Maintenance planning	Failure effect
	History	Maintenance planning
		RCM analysis

the Failure Mode is updated in real time by data coming from the records. Also it is possible to define that the software should be able to automatically identify a Signal, through its IP address and associate it to a Maintainable Element through a user interface. The main idea behind this concept is that common sensors are generally related to an IP. This can be the IP of a controller that connect more than one sensors and it is wired with them. There can be also a fieldbus to which all the sensors are connected. In these two cases sub-IP for each sensor is identified. Moreover, due to new development of smart sensors [46] it could be envisioned that each sensor is connected to the overall network through its own IP identifier.

In data model of figure is possible to observe elements from each one of the three tools of the CMMS–RCM–CBM integration. Although an integration platform implies that every object has to do with all the system as a net, there are some objects traditionally used by a defined approach. A classification table is indicated in Table 2. The names of the objects appear as a one word without spaces; this is due to the requirements of the modeling standard and the modeling software.

The objective of diagram in Fig. 11 is to establish the principal elements or objects of the integration platform and their relations to be programmed. This diagram describes the static operation structure of the demo software (different from the BPMN diagrams in Figs. 6, 7, 8, 9, and 10 that show a dynamic information flow of the system). The attributes and operations of classes are omitted due to space restrictions. In general terms, the attributes of each class are defined to allow accurate localization and tracking of each item, for example, for the "system" class it should define its identification code, its operational context, the function required, and the associated subsystems; for the "work order" class it is defined date code, responsible, maintainable element of the intervention, the kind of maintenance, the failure mode, associated subsystem and equipments, etc. The classes that involve a subsystem for decision-making, in addition to attributes, they have defined multiple operations. For example, the "ConditionMonitoringManager" class has operations as signal consulting, alarm generation, estimation of deviation measurement, signal comparing, signal saving, etc., and the "RCMAnalysis" class has operations as the typical executed in a RCM process.

5 CBM Decisional Algorithm

A key module of the integration platform is about CBM decisions: diagnosis and prognosis, understanding diagnosis and the detection, isolation and identification of faults and the prognosis as the "prediction" of the failure before it occurs based on degradation [43]. This means how to interpret the results of CBM, identifying the occurrence of possible failure modes, hazard to equipments and undesirable tendencies.

Tsang [47] states that there are three important decisions to implement CBM: selection of the parameters to be monitored, determination of the inspection frequency, and establishment of warning limits. A prior RCM analysis helps to resolve these issues. At this point, the main failure modes of the transformer equipment are already identified. These failure modes are related to specific signals controlled by probes assembled in various mechanical devices. The CBM system has for each critical failure mode, a range in which the measured variable (temperature, sealing, etc.) operates when the transformer is normal (failure absence). If the installed probes find that the variable is out of the normal range, the system triggers a warning or alarm. This is a diagnosis process related to a specific failure mode. Nevertheless, prognosis processes are more complicated than just the alarm generation.

There are many methodologies for carrying out prognosis. However, methods based on reliability analysis, using historical data to adjust statistical models (Weibull distribution, exponential distribution or proportional hazards model (PHM), among others) stand out for its ease utilization and potential [45]. For these reasons, the statistical procedure PHM (proportional hazards modeling) has been selected to develop the CBM decisional algorithms for the integration system. PHM is used to identify the set of measurements to assess the item condition. Theoretical bases for the application of this procedure are the statistical studies of Cox [48] and the use of the method made by Jardine et al. [45] to estimate the failure risk of equipment when subjected to CBM.

Hence, the objective is to develop a decisional algorithm that relates the most critical failure modes (defined previously through a RCM analysis) with the controlled signals, generating useful information for maintenance decision-making considering reliability, cost, and risk (prognosis).

The methodology for the development of this decisional algorithm consists of the following steps:

Step 1. Determination of the Monitoring Parameters (MP) relevant for maintenance optimization and minimization of risk in the selected equipment.

Step 2. Analysis of the accuracy of the signals and the need for treatment, prior to considering their values in the decision-making process.

Step 3. Identification of the existence of reliable warning and alarm limits for the MPV (Monitoring Parameters Values).

Step 4. Analysis of historical records of failures for each failure mode of the equipment.

Step 5. Analysis of the MPV in the historical failure cases.

Step 6. Analysis of the possible correlation between MPV and failures.

Step 7. Estimation of the instantaneous failure probability and the related reliability function for each failure mode.

Step 8. Calculation of real-time risk for each failure mode.

Step 9. Obtaining a business rule for each failure mode.

Step 10. Implementation of business rules for each failure mode.

Figure 12 presents the BPMN data flow of the algorithm developed, related to each module of the integration system.

The process begins in the CMMS system module, which analyzes the records of each failure mode. A RCM analysis identifies the critical failure modes and monitoring parameters (MP) relevant to diagnosis/ prognosis. The next step is to relate these MP with signals to be generated by sensors, and determining the alarm limits indicating the presence of a failure mode.

The signals generated correspond to $Z(t)$ that is the mathematical representation of the impact in reliability of the $z_n(t)$ variables (Eq. 1):

$$z(t) = \gamma_1 z_1(t) + \gamma_2 z_2(t) + \gamma_3 z_3(t), \tag{1}$$

where

$Zn\ (t)$ real-time measured variables (or function of them) affecting reliability

γ weighting coefficient (importance) of the influence produced by each variable (or function of them).

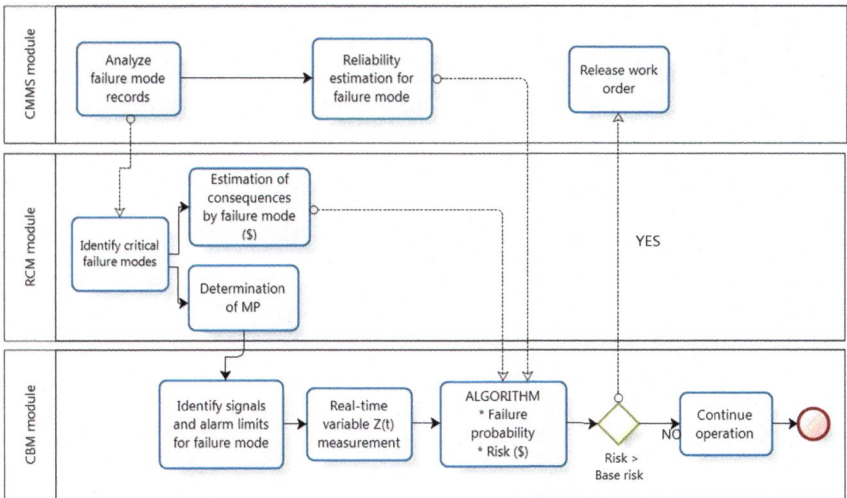

Fig. 12 BPMN process diagram of the CBM decisional algorithm

Subsequently, the system calculates the ideal reliability $Ro(t)$. This reliability only considers the operation time, as indicated in Eq. 2:

$$Ro(t) = \left(e^{-\lambda t}\right), \tag{2}$$

where

λ failure probability, obtained by dividing the number of failures occurrences by the operation time of the equipment (according to information proportioned by the CMMS).

From the value generated by Eq. 1, reliability affected by the real-time signals is calculated using Eq. 3:

$$R(t) = \left(e^{-\lambda t}\right)^{e^{z(t)}}. \tag{3}$$

Then, the algorithm calculates real-time risk in economic terms (Eq. 4). To do this, the algorithm is fed from three inputs: (1) reliability for the specific failure mode (Eq. 2), data obtained from historical records of the CMMS. (2) economical consequences in \$ of the possible failure mode, and (3) real-time measurement of the $Z(t)$ variables generated by the sensors normalized by Eqs. 1 and 3. Then is when the algorithm, based on proportional hazards modeling (PHM), calculates the failure probability as a function of time and risk associated to the failure mode (in \$) for a certain period of time Δt. In this way it is possible to compare in economic terms, the previously calculated risk or base risk (cost of preventive maintenance for the specific failure mode) against the risk identified by the CBM signals (cost calculated by the algorithm expressed by Eq. 4). If the newly calculated risk is greater than the baseline risk, a work order is launched, because if failure appears, the economic impact is greater than the investment required for their regularization. If, however, the newly calculated risk is minor than the baseline cost, the system is allowed to continue operating, since any preventive action represents an extra cost. This designed algorithm produces a new emerging e-maintenance process: "economical-maintenance prognosis" and its innovation rely on the new possibilities offered by the ICT and real-time control applied to maintenance activities.

$$Risk(t, t + \Delta t) = \frac{R(t) - R(t + \Delta t)}{R(t)} C, \tag{4}$$

where

C consequence in economic terms when the failure occurs

6 Implementation of the Integration Platform to an Electrical Transformer System

The CMMS–RCM–CBM integration platform has been developed focused to its initial use in a Spanish power enterprise, specifically over electric transformer equipment. The final objective of this project is the implementation of the integration software, over the most adequate maintainable items. Penrose [13] states as milestones for the implementation of a maintenance software or maintenance informatics application (as the designed in this project) in the following: (1) Identifying the mission of the company; (2) Performing RCM based on analysis of facilities and equipments to determine the systems to be included in the program, and to determine the maintenance practices to be applied; (3) Selection of CBM tools and maintenance practices to meet the analysis; (4) Selection of vendor partners, including discussion of spare parts storage; (5) Selection of personnel to operate the informatics system; (6) Set and communicate goals and metrics of the informatics system; and (7) Periodically review and modify the system as required.

Regarding the implementation process, this paper exclusively focuses on the RCM results, the selected equipment, and on a general view of the created software.

6.1 Description of the Industrial Equipment and RCM Results

Specific equipment has been selected to implement on it the designed integration system. The equipment chosen (according to a previous critically and feasibility analysis) is a three-phase transformer (Fig. 13). The principal function of this equipment is to transform (elevate) the electric energy produced by the generator to 19 kV and transmitting it to the 230 V three-phase external network through the substation. Some features of this transformer are as follows: it has columns in oil coating with two coils; it is installed in the open, operating at 50 Hz frequency, continuous service, with ONAN/ONAF1/ONAF2 cooling system, vacuum transforming ratio $230 \pm 1.5 \pm 3\%/19$ kV and power of 282/376/470 MVA. This transformer system has been built according to the IEC 76, IEC 137, IEC 185, IEC 214, IEC 439, and UNE EN 60551 standards.

A transformer is a device that can have a lifespan of 60 years if having the proper maintenance [49]. Usually transformer failures come from some defined abnormal conditions: (i) general overheating (an abnormal increase in temperature of the oil), (ii) local overheating associated with the main magnetic flux, and (iii) local overheating associated with a flow lost. The main variable that measures the life and condition of a transformer is temperature, especially in relation to the condition of insulation. There is a variety of literature that identifies and analyzes

Fig. 13 Electric transformer, object of the integration software project

the failure modes that can occur during operation of a transformer, and the effectiveness of diagnostic methods (mainly documents from EPRI, CIGRE, and other research institutes).

The RCM analysis defines the operational context of the transformer (a detailed description of the equipment, its processes, performance objectives, operational conditions, etc.), its boundaries, and subsystems (the boundaries of the system and its related subsystems are indicated in Fig. 14, diagram based on the ISO 14224:2004 standard) [41]. Each subsystem has a specific required function, performance standards, and it is made up of several elements. Among these elements, some of them are chosen as maintainable items according to their importance. Then a FMECA analysis determines the possible functional failures and failure modes, establishing the effects and consequences associated to each failure. The final step of RCM logic is to determine the appropriate maintenance activities to prevent the occurrence of the identified failures modes. The relation among each level is expressed in Fig. 15.

A possible maintenance activity defined by RCM analysis is adopting a CBM policy. In this point of the RCM analysis is when the feasibility of CBM is analyzed for every maintainable item. Then, technical work is required, in order to identify the control variables, the inspection methodology, the technology to be utilized, the analysis of data, etc.

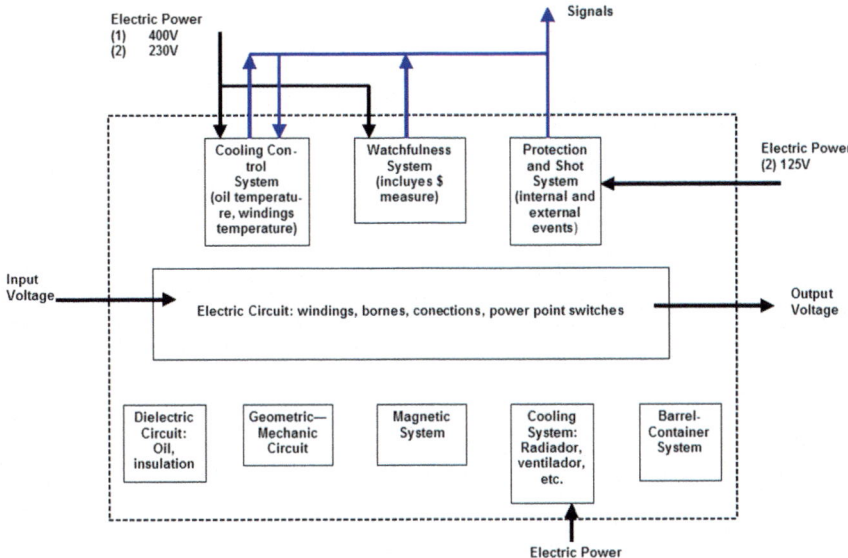

Fig. 14 Transformer's boundary (based on the ISO 14224:2004 recommendations)

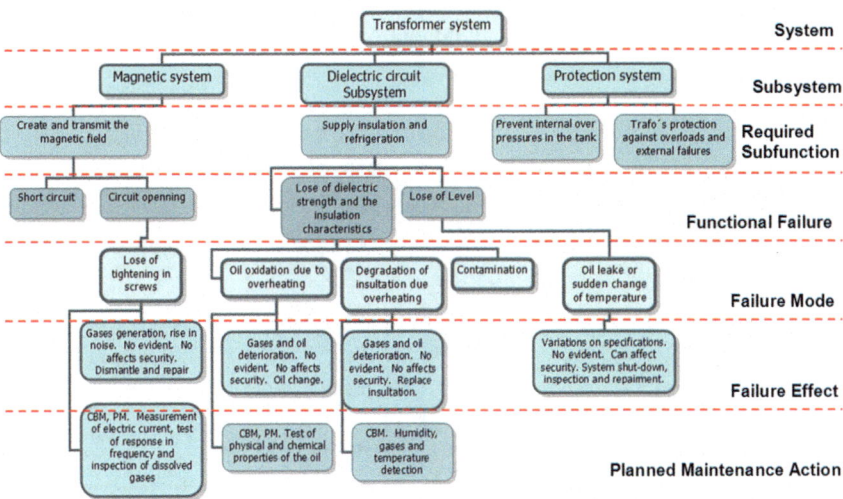

Fig. 15 Part of the structure defined by RCM analysis of the transformer system

6.2 General Description of the Implemented Software System

To validate the designed integration models and from the results of RCM analysis, a software prototype has been built. This software controls the variables related to the main failure modes that occur in the transformer system through probes assembled in various mechanical devices. Those probes are devices that control the processing of analog and digital signals and its conversion to the standard OPC and SNMP for transmission over a TCP/IP. Configuration is done through Web interfaces available. These interfaces include Ethernet, Wi-Fi, and GPRS. The device in its prototype version can handle up to 8 analog/ digital signals and send GPS coordinate information. Then basically, it executes the algorithm exposed in Fig. 12. The rest of informatics processes indicated by Fig. 6 (RCM analysis, Definition of the maintenance strategy, Evaluation of the maintenance strategy, etc.) are not developed as part of the demo software yet.

The created software enables e-maintenance processes by associating failures modes (defined in RCM methodology and supported by CMMS information) with real-time signals obtaining directly from the critical equipment, for a subsequently decision-making process.

The software tool defines three levels for the controlled equipment: physical level, functional level, and monitoring level (see Fig. 16). The lowest element of each level is grouped together as the monitoring elements of the system: maintainable items, parameters, and signals.

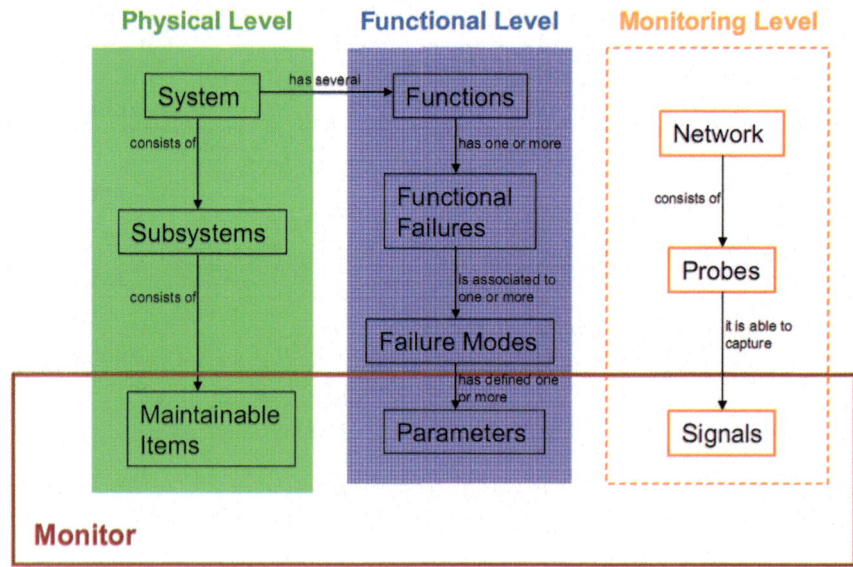

Fig. 16 Levels of the CBM maintainable item

The physical level is defined using ISO 14224:2004 standard ("Petroleum, petrochemical and natural gas industries—Collection and Exchange of reliability and maintenance data for equipment") [41] to break down critical equipment (called system in this context) into subsystems and maintainable items, considering maintainable item as the most basic element on which maintenance actions are executed. It is possible in this level to store for each system, subsystem, and maintainable item, the following information: code, type, subtype, description, manufacturer, manufacture date, installation date, serial number, MTBF, MTTR, availability, etc. This store makes possible to review the most important equipment data in a quick view.

The functional level is defined using the RCM approach. Functions of the equipment must be analyzed and defined. These functions are linked to the functional failures that can occur, and these functional failures are also linked to the failure modes associated. Each failure mode has defined one or more failure parameters that characterize a variable for monitoring, such as temperature, humidity, pressure, vibration, voltage, intensity, and flow. Each failure parameter should have a description and its minimum, medium, and maximum thresholds.

The monitoring level consists of several probes. The probes system is not intrusive as it is based on SNMP, the telecommunication standard. Probes have two parts: one connected to the field sensors and another one connected to the central system by a GPRS communications module. Probes' functions are to get sampling values and in case, sending a warning signal to the central system. It is also possible that probes analyze and make correlations among some signals, sending, if necessary, an alarm that is built by a correlation among them.

Moreover, the developed demo software also includes the following as functionalities:

*Real-time monitorizing panel: as it is shown in Fig. 17, the system can track the variables of interest or monitoring parameters to control their behavior.

Fig. 17 Real-time monitorizing panel

*Panel of MP (monitoring parameters) risk: as shown in Fig. 18, it presents in different colors, the real-time status of the related signals. In case of any probe exceeding the established levels and the percentage of risk allowed, the system will report a warning yellow/red signal.

*Panel of risk comparison (economical-maintenance prognosis): as shown in Fig. 19, the software graphically presents the results of the CBM decisional algorithm exposed in Sect. 5. The left graphic of Fig. 19 shows the increasing trend of the controlled variable (in this case temperature, represented by the yellow line). The blue line (going down with minor slope) represents the reliability decrease only considering the effect of time (Eq. 2); the red line (going down with major slope) represents the reliability decreases considering also the controlled variable (temperature) (Eq. 3). The green lines represent the vertically ascending instantaneous

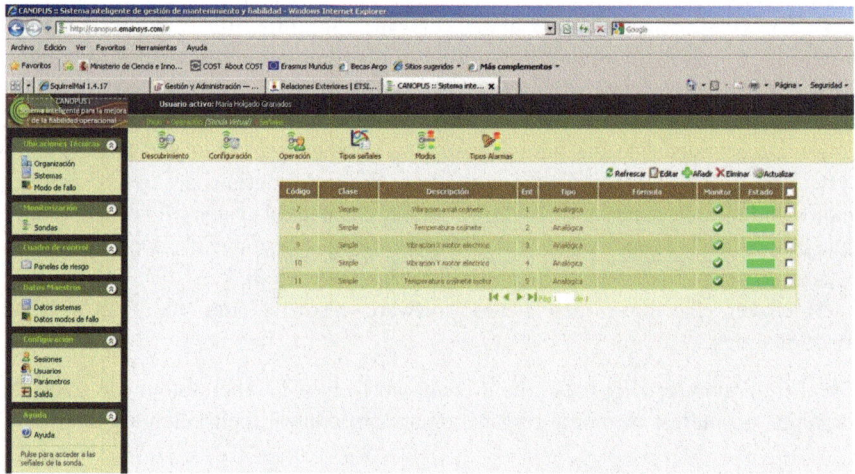

Fig. 18 Panel of MP (monitoring parameters) risk showing some real-time signals status

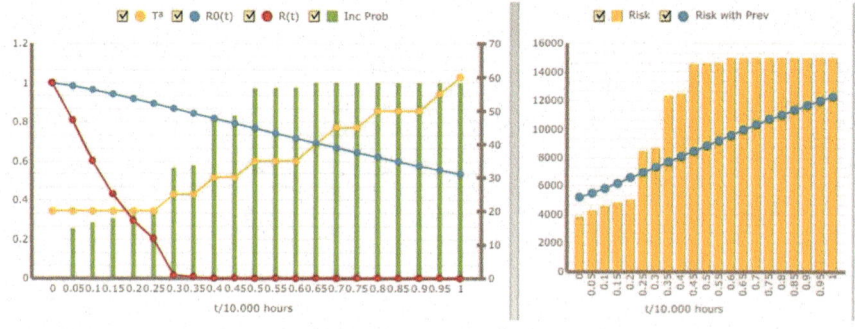

Fig. 19 Risk comparison panel (economical-maintenance prognosis)

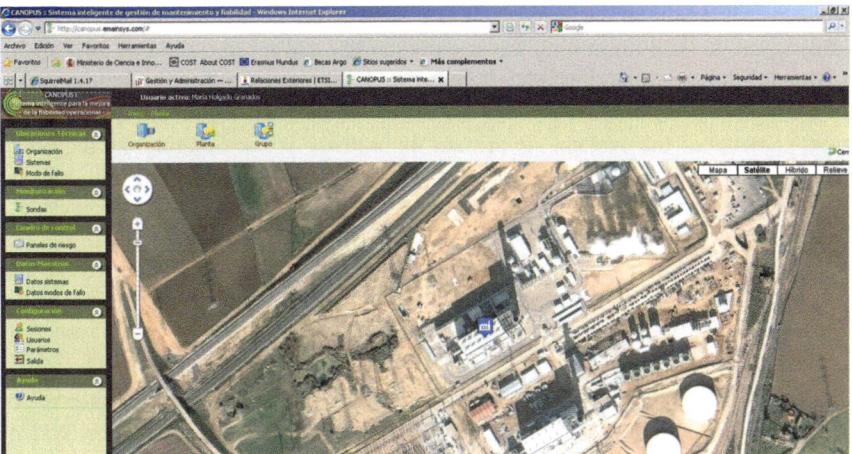

Fig. 20 Screenshot of the GIS software demo

failure probability, which appears in economic terms in the graph on the right (yellow lines) (Eq. 4), where it is compared to the blue line representing the cost of carrying out preventive maintenance. Theoretically at the time that the economic cost corresponding to the current risk (considering controlled signals) outweighs the cost of preventive maintenance, a work order maintenance should be launched.

*Geographic Information System (GIS): as shown in Fig. 20, currently it is possible to georeference each equipment under analysis to study its physical location. GIS would be developed to facilitate, with an intuitive geographical representation, the infrastructure knowledge related to physical location, element interactions, and environment.

*Knowledge Management System: It should be a dynamic and active process. Through the software, knowledge management is facilitated not only storing the equipments information but by implementing a documents repository. This means that it is possible to access the information needed for a maintenance task, any time and place.

At this time, the designed system has been implemented as a demo in the transformer. The software platform is available online at http://canopus.emainsys. com. The distribution has been made to each authorized user providing a username and password. The final software product can be applied to every company that preferably operates in industry sectors with very high investments in productive equipment, relatively high operative and maintenance costs and high failure costs. Some possible related industries are extractive and petroleum industries, energy and water industries, and eventually the general manufacturing sector.

There are ideas for the future of the software system, to expand its scope and possibilities. In this sense a possible future project can be to develop a complete Reliability Analysis module, including complex statistical analysis for improved decision-making in units of time, materials and economical resources. Another point has to do with improving the interoperability and transmission of information between different systems and platforms.

Through the continued use of the tool, the organization will identify the required improvements to the system. At present the outlook is encouraging, as the power company has proposed a new implementation of the integration platform also over other equipments (generators).

7 Conclusions

This paper has exposed a methodology to develop an e-maintenance platform that integrates the database of Computerized Maintenance Management Systems (CMMS), tools for Reliability-Centered Maintenance (RCM), and results from the Condition-Based Maintenance (CBM), an innovating subject according to a literature review. The aim of the integration platform developed is to optimize the synergy of the three maintenance tools (CMMS, RCM, CBM) and to create an intelligent system implemented as a demo software. In this project, the integration approach is preferred over the interoperability because of the software requirements of the enterprise where the system has been applied.

The CMMS–RCM–CBM integration platform created uses MIMOSA standards OSA-EAI, OSA-CBM, and ISO standard 13374:2003 as references to define its general operation architecture. The operation algorithms of the system are modeled using UML and BPMN diagrams.

The paper presents a case study for the implementation of the designed platform in a power transformer. The implementation begins doing a RCM analysis, which determines the operational context of the transformer, the required functions, failure modes and effects, and maintenance policies for each maintainable item. Several technical standards are used as reference at this point (ISO 12224:2004, EPRI, CIGRE).

UML and BPMN models are developed theoretically and then implemented in a prototype hardware + software by integration. Once the prototype has been tested, the models are validated in practice. This application is developed first by integrating databases and by a correlation of information through data warehouse reporting tools, and later through interoperability of proprietary systems of international prestige.

The integration system is implemented as software which controls variables related to the principal failure modes occurring in the transformer system, by means of probes assembled in various mechanical devices. The derived information (signals) is used as an input for decision-making in real time. To achieve this end, a mathematical algorithm has been developed to interpret the measurement results of one or more physical variables, emerging the "economical-maintenance prognosis" process.

The developed demonstration software includes also features as a real-time monitorizing panel, a panel of risk comparison, a geographic information system, and a knowledge management system.

So far, the integration platform has reported favorable results for the power enterprise. Some ideas for the future of the project are oriented to add a complete Reliability Analysis module, and to improve the interoperability and transmission of information between different systems and platforms.

Summarizing, the general steps involved in the process of creating the integration platform are (i) analysis of previous attempts at integration; (ii) analysis of related standards; (iii) design of the general integration structure; (iv) design and modeling of operation algorithms using UML and BPMN diagrams and mathematical equations; (v) equipment critically analysis; (vi) application of RCM methodology to critical equipment, defining the boundaries, functions and maintenance strategies of the system; (vii) analysis of signals in connection to failure modes of critical equipment; (viii) development of the demo software; (ix) customization of the e-maintenance tool to customer needs; (x) implementation of the tool; (xi) validation of the operation and control of the tool with the users; and (xii) audit performance of the tool.

Acknowledgements This research is funded by the Spanish Ministry of Science and Innovation, Project EMAINSYS (DPI2011-22806) "Sistemas Inteligentes de Mantenimiento. Procesos emergentes de E-maintenance para la Sostenibilidad de los Sistemas de Producción", besides FEDER funds. Thanks also to the Mexican Council of Science and Technology (CONACYT) and to the European Project IMAPLA (Integrated Maintenance Planning).

References

1. López-Campos M, Crespo Márquez A (2011) Modelling a maintenance management framework based on PAS 55 standard. Qual Reliab Eng Int 27(6):805–820
2. Kans M (2009) The advancement of maintenance information technology. Literature review. J Qual Maintenance Eng 15(1):5–16
3. Crespo Márquez A, Iung B (2006) Editorial: special issue on e-maintenance. Comput Ind 57:473–475
4. Labib AW (2006) Next generation maintenance systems: towards the design of a self-maintenance machine. In: IEEE international conference on industrial informatics, pp 213–217
5. Mobley K (1999) Creating a diagnostic tool, plant services on the Web, October (1998). In: http://www.plantservices.com [Consulted: June 4 1999]
6. Cato W, Mobley K (2002) Computer-managed maintenance systems: a step-by-step guide to effective management of maintenance, labor and inventory, 2nd edn. Plant Engineering, Butterworth Heinemann
7. Tsang A (2002) Strategic dimensions of maintenance management. J Qual Maintenance Eng 8(1):7–39
8. Muller A, Crespo Márquez A, Iung B (2006) On the concept of e-maintenance: review and current research. Reliab Eng Syst Saf 93:1165–1187

9. Morel G, Suhner M, Iung B, Léger JB (2001) Maintenance holistic framework for optimizing the cost/availability compromise of manufacturing systems. In: Sixth IFAC symposium on cost oriented automation Berlin, Germany, 2001

10. Djurdjanovic D, Lee J, Ni J (2003) Watchdog agent- an infotronics-based prognostics approach for product performance degradation assessment and prediction. Adv Eng Inform 17:109–125

11. Campos J (2009) Development in the application of ICT in condition monitoring and maintenance. Comput Ind 60(1):1–20

12. Moubray J (1997) Reliability-centered maintenance. Butterworth Heinemann, 1997

13. Penrose H (2005) RCM-based motor management. In: Proceedings of the electrical insulation conference and electrical manufacturing expo, 2005

14. Wessels W, Sautter F (2009) Reliability analysis required to determine CBM condition indicators. In: Reliability and maintainability symposium, RAMS Texas, 2009, pp 454–459

15. Niu G, Yang BS, Pecht M (2010) Development of an optimized condition-based maintenance system by data Fusion and reliability-centered maintenance. Reliab Eng Syst Saf 95(7): 786–796

16. Gabbar HA, Yamashita H, Suzuki K, Shimada Y (2003) Computer-aided RCM-based plant maintenance management system. Robot Comput Integr Manufact 19(5):449–458

17. Carnero MC, Novés JL (2006) Selection of computerised maintenance management system by means of multicriteria methods. Prod Plann Control 17(4):335–354

18. Braglia M, Carmignani G, Frosolini M, Grassi A (2006) AHP-based evaluation of CMMS software. J Manuf Technol Manage 17(5):585–602

19. Fumagalli L, Macchi M, Rapaccini M (2009) Computerized maintenance management systems in SMEs: a survey in Italy and some remarks for the implementation of Condition Based Maintenance. In: 13th IFAC Symposium of information control problems in manufacturing, Moscow, 2009

20. Tsang AHC (1995) Condition-based maintenance: tools and decision making. J Qual Maintenance Eng 1(3):1355–2511

21. Deb S, Ghoshal S, Mathur A, Shrestha R, Pattipati KR (1998) Multisignal modeling for diagnosis, FMECA and reliability. In: Proceedings of the IEEE international conference on systems, man and cybernetics 1998, pp 3026–3031

22. Finley B, Schneider E (2001) ICAS: the center of diagnostics and prognostics for the United States navy. In: Proceedings of the SPIE—international society for optics and photonics. 2001, pp 186–193

23. Shum YS, Gong DC (2003) Design of an integrated maintenance management system. J Chin Inst Ind Eng 20(4):337–354

24. Huo Z, Zhang Z, Wang Y, Yan G (2005) CMMS based reliability centered maintenance. In: Proceedings of the IEEE/PES transmission and distribution conference & exhibition: Asia and Pacific, Dalian, China (2005)

25. Lehtonen M (2006) On the optimal strategies of condition monitoring and maintenance allocation in distribution systems. In: International conference on probabilistic methods applied to power systems PMAPS, 2006

26. Kothamasu R, Huang SH, VerDuin WH (2006) System health monitoring and prognostics—a review of current paradigms and practices. Int J Adv Manuf Technol 28:1012–1024

27. Han T, Yang BS (2006) Development of an e-maintenance system integrating advanced techniques. Comput Ind 57:569–580

28. Emmanouilidis C, Jantunen E, MacIntyre J (2006) Flexible software for condition monitoring, incorporating novelty detection and diagnostics. Comput Ind 57:516–527

29. Jaw L, Merrill W (2008) CBM+ research environment—facilitating technology development, experimentation, and maturation. In: Proceedings of the Aerospace Conference 2008, IEEE, 2008

30. Trappey AJC, Hsiao DW, Ma L (2011) Maintenance chain integration using Petri-net enabled multiagent system modeling and implementation approach. IEEE Trans Syst Man Cybernet Part C Appl Rev 41(3):306–315

31. Galar D, Gustafson A, Tormos B, Berges L (2012) Maintenance decision making based on different types of data fusion. Eksploatacja i Niezawodnosc 14(2):135–144

32. Chang Y, Chen Y, Hung M, Chang AY (2012) Development of an e-operation framework for SoPC-based reconfigurable applications. Int J Innovative Comput Inf Control 8 (5 B) (2012) 3639–3660

33. MIMOSA (OSA-EAI™) (2006) Open systems architecture for enterprise application integration, v.3.2.1. 2006

34. PAS 55-1:2008 (2008) Asset management. Specification for the optimized management of physical assets. BSI, UK

35. ISO 9001: 2008 (2008) Quality management systems. Requirements. ISO, Geneva

36. ISO 13374 (2003) Condition monitoring and diagnostics of machines- data processing, communication and presentation, Part 1. General Guidelines, TC 108/SC 5. International Standards for Business, Government and Society. Genove: ISO, 2003

37. MIMOSA (OSA-CBM™) (2006) Open Systems Architecture for Conditioned-Based Maintenance, v.3.2.1. 2006

38. Gómez Fernández J, Crespo Márquez A (2009) Framework for implementation of maintenance management in distribution network service providers. Reliab Eng Syst Saf 94(10):1639–1649

39. ISO 18435-1:2009 (2009) Industrial automation Systems and integration. Diagnostics, capability assessment and maintenance applications integration. Part I: overview and general requirements. ISO, 2009

40. ISO/IEC 62264 (2007) Enterprise-control system integration. International Standards for business, government and society (http://www.iso.org)

41. ISO/DIS 14224:2004 (2004) Petroleum, petrochemical and natural gas industries—collection and exchange of reliability and maintenance data for equipment. ISO, 2004

42. Wu X, Chen J, Li R, Li F (2006) Web-based remote monitoring and fault diagnosis system. Int J Adv Manuf Technol 28:162–175

43. Jardine AK, Lin D, Banjevic D (2006) A review on machinery diagnostics and prognostics implementing condition-based maintenance. Mech Syst Signal Process 20:1483–1510

44. Medjaher K, Tobon-Mejia DA, Zerhouni N, Tripot G Mixure of gaussians hidden markov model based failure prognostic: application on bearings (in press)

45. Jardine AKS, Banjevic D, Wiseman M, Buck S, Joseph T (2001) Optimizing a mine haul truck wheel motors' condition monitoring program. Proportional hazards modelling. J Qual Maintenance Eng 7(4):1355–2511

46. Garetti M, Macchi M, Terzi S, Fumagalli L (2007) Investigating the organizational business models of maintenance when adopting self diagnosing and self healing ICT systems in multi site contexts. In: Proceedings of the IFAC conference on cost effective automation in networked product development and manufacturing IFAC-CEA 07, Monterrey, México, 2007

47. Tsang AHC (1995) Condition-based maintenance: tools and decision making. J Qual Maintenance Eng 1(3):1355–2511

48. Cox DR (1972) Regression models and life tables (with discussion). J Roy Stat Soc Ser B 34:187–220

49. Wang M, Vandermaar AJ, Srivastava KD (2002) Review of condition assessment of power transformers in service. IEEE Electr Insul Mag 18(6):12–25

Author Biographies

Mónica Alejandra López-Campos is an Industrial Engineer, and Ph.D. in Industrial Organization from the University of Seville (Spain). Currently, she is Professor in the Federico Santa Maria University (Chile), where she teaches the subjects of Maintenance Management, Operations Management and Quality Management. Also, she is member of the National System of Researchers (Mexico). Her research works have been published in journals such as Reliability Engineering and System Safety, International Transactions in Operational Research, Computers in Industry, Quality and Reliability Engineering International, and Journal of Quality in Maintenance Engineering, among others. Her interests in research include maintenance management, modeling of processes, logistics, quality, simulation, and educational methodologies for the engineering. She has participated in several research projects in Mexico, Spain, and Chile, as well as in international organizations such as the Economic Commission for Latin America and the Caribbean (ECLAC).

Adolfo Crespo Márquez is currently Full Professor at the School of Engineering of the University of Seville, and Head of the Department of Industrial Management. He holds a Ph.D. in Industrial Engineering from this same University. His research works have been published in journals such as the International Journal of Production Research, International Journal of Production Economics, European Journal of Operations Research, Journal of Purchasing and Supply Management, International Journal of Agile Manufacturing, Omega, Journal of Quality in Maintenance Engineering, Decision Support Systems, Computers in Industry, Reliability Engineering and System Safety, and International Journal of Simulation and Process Modeling, among others. Prof. Crespo is the author of seven books, the last four with Springer-Verlag in 2007, 2010, 2012, and 2014 about maintenance, warranty, and supply chain management. Prof. Crespo leads the Spanish Research Network on Dependability Management and the Spanish Committee for Maintenance Standardization (1995–2003). He also leads a research team related to maintenance and dependability management currently with five Ph.D. students and four researchers. He has extensively participated in many engineering and consulting projects for different companies, for the Spanish Departments of Defense, Science and Education as well as for the European Commission (IPTS). He is the President of INGEMAN (a National Association for the Development of Maintenance Engineering in Spain) since 2002.

Juan Francisco Gómez Fernández is Ph.D. in Industrial Management and Executive MBA. He is currently part of the Spanish Research & Development Group in Industrial Management of the Seville University and a member in knowledge sharing networks about Dependability and Service Quality. He has authored publications and collaborations in journals, books, and conferences, nationally and internationally. In relation to the practical application and experience, he has managed network maintenance and deployment departments in various national distribution network companies, both from private and public sectors. He has conduced and participated in engineering and consulting projects for different international companies, related to Information and Communications Technologies, Maintenance and Asset Management, Reliability Assessment, and Outsourcing services in Utilities companies. He has combined his business activity with academic life as a associate professor (PSI) in Seville University, being awarded as Best Thesis and Master Thesis on Dependability by National and International Associations such as EFNSM (European Federation of National Maintenance Societies) and Spanish Association for Quality.

Prognostics and Health Management in Advanced Maintenance Systems

Antonio Jesús Guillén López, Adolfo Crespo Márquez, Marco Macchi and Juan Francisco Gómez Fernández

Abstract The advanced use of the Information and Communication Technologies is evolving the way that systems are managed and maintained. A great number of techniques and methods have emerged in the light of these advances allowing to have an accurate and knowledge about the systems' condition evolution and remaining useful life. The advances are recognized as outcomes of an innovative discipline, nowadays discussed under the term of Prognostics and Health Management (PHM). In order to analyze how maintenance will change by using PHM, a conceptual model is proposed built upon three views. The model highlights: (i) how PHM may impact the definition of maintenance policies; (ii) how PHM fits within the Condition-Based Maintenance (CBM) and (iii) how PHM can be integrated into Reliability Centered Maintenance (RCM) programs. The conceptual model is the research finding of this review note and helps to discuss the role of PHM in advanced maintenance systems.

Keywords PHM · RCM · Maintenance policies · Proactive maintenance · CBM

A.J. Guillén López (✉) · A. Crespo Márquez · J.F. Gómez Fernández
Department of Industrial Management, School of Engineering,
University of Seville, Camino de los Descubrimientos s/n, Seville, Spain
e-mail: ajguillen@us.es

A. Crespo Márquez
e-mail: adolfo@us.es

J.F. Gómez Fernández
e-mail: juan.gomez@iies.es

M. Macchi
Department of Management, Economics and Industrial Engineering,
Politecnico di Milano, Piazza Leonardo Da Vinci 32, Milan, Italy
e-mail: marco.macchi@polimi.it

© Springer International Publishing AG 2018
A. Crespo Márquez et al. (eds.), *Advanced Maintenance Modelling for Asset Management*, DOI 10.1007/978-3-319-58045-6_4

1 Introduction

Under the term of Prognostics and Health Management (PHM) a body of knowledge is included that nowadays is considered as an engineering discipline [1, 2]. This covers all methods and technologies to assess the reliability of a product in its actual life cycle conditions to determine the advent of failures, and mitigate system risks [3]. Not only prognostics, but also detection and diagnosis of failures are problems addressed by the key skills of this discipline, along with the issues related to the subsequent use of the resulting information (in maintenance, logistics, life cycle control, asset management, etc.). Therefore, the skills are required by the necessities of the entire process, from data collection to interpretation in decision-making [4]. Besides, PHM development and use is linked to Information and Communication Technologies (ICTs) development and its applications inside the maintenance function. It is then interesting to understand the role of PHM in the modern maintenance systems, in particular its contribution to more proactive approaches that allow reaching operational excellence in manufacturing companies.

With the fierce pressure companies are facing, it is nowadays more difficult to compete. Remaining competitiveness, especially in high-tech sectors, requires continuous incorporation of new advances—with higher requirements, among others, of reliability—while optimizing operation and maintenance. Indeed, reliability and maintenance have an increasingly important role in modern engineering systems and manufacturing processes [5], which are becoming increasingly complex and are operating in a highly dynamic environment [6]. Waeyenbergh and Pintelon [7, 8] claim that, in the case of leading-edge systems, characterized by a large number of technical items with great interaction level between them, maintenance is now more important than ever for business goals, not only in terms of cost reduction but regarding decisive contribution to company's performance and efficiency as part of an increasingly integrated business concept. Considering the maintenance department, Macchi and Fumagalli [9] remark the importance of maintenance for the competitiveness of manufacturing companies and, in this regard, assess the maturity of its processes in terms of managerial, organizational, and technological capabilities; especially looking at the technological capability, the maintenance objective is to adopt new technologies and tools in the company's practice to effectively contribute to competitiveness. In short, [10] claim the importance for more efficient maintenance as a key for sustainability and competitiveness of enterprises and production systems, associating the decision-making process with the so called "eco-efficiency" of the production systems: this is an even more comprehensive view of the maintenance role, since eco-efficiency encompasses both the impact on business and on environment. Under this perspective, emphasis on the life cycle of manufacturing assets has caused a redefinition of the role of maintenance as "*a prime method for life cycle management whose objective is to provide society with the required functions while minimizing material and energy consumption*" [10]. On the whole, the changes undergoing for the maintenance function are aligned with the transformation of the current manufacturing models based on the old paradigm of *"unlimited resources and unlimited world's capacity for regeneration"* towards a sustainable manufacturing [11]. Along this vision, manufacturing will be strongly affected by sustainability issues and, what is relevant for the discussion in this paper, *"technology, on which the manufacturing is largely based, is asked to give the tools and options for building new solutions towards a sustainable manufacturing concept"* [11].

PHM will play its role, as it is fundamental in current evolution of maintenance function towards advanced maintenance systems. PHM, along with other trends like E-maintenance—term that serves as conceptual support to general use and applications of ICTs in maintenance [12]—appear as the key factors in achieving higher maintenance efficiency levels and life cycle cost reduction [13, 14]. The current perspective, established within the PHM discipline, can be considered complementary to the vision drawn by E-maintenance, well synthesized by the conceptual framework provided in Levrat et al. [15] with a focus on manufacturing application. This framework facilitates understanding of E-maintenance, by an acknowledgement of its potential through new services, processes, organization, and infrastructure; nonetheless, its contribution is focused on structuring of the technological discipline represented by E-maintenance, without discussing the specific techniques and methods required to fulfill the goals envisioned for the maintenance business functions and processes in future manufacturing. Nowadays, this gap can be completed by the evolution supported thanks to development of PHM discipline.

It is, however, necessary to understand the implications of PHM on maintenance and how to apply the PHM techniques and methods conveniently into a methodological framework for a system maintenance management. Until this moment, the PHM research efforts have been focused on technological issues (models, methods, and algorithms) and their application on very particular system [6], while further discussions with respect to the introduction of PHM into a framework for a system maintenance management are still missing.

In this regard, we believe that there are currently two relevant challenges for the effective design, implementation, and use of PHM solutions in advanced maintenance systems. First of all, the technical profiles, with new skills and capabilities, are far from those that can be found in traditional maintenance engineers or technicians. This is deeply discussed in Bird et al. [4] who introduce how PHM is a multidisciplinary domain that is undergoing rapid evolution especially in the type of demanded skills and capabilities. The industry will then require highly qualified professionals, combining an initial training in PHM techniques and methods with specific expertise in this field; in regard to the work organization, the necessity of simultaneous use of different skills and capabilities—with high-level knowledge and expertise—will also have to be integrated in maintenance work-teams.

The second main challenge can be directly related to the original discussions on PHM, leading to focus the research efforts on models, methods, and algorithms, while missing the management of their potential within a complex engineering system. Hence, it is worth remarking that the maintenance policies design has not been enough studied after considering the potential of PHM solutions.

This challenge is addressed by this paper. Therefore, to effectively introduce PHM in the maintenance function, it is worth considering the existence of different, and well known, frameworks or methodologies that facilitate the design and implementation of maintenance policies.

Some examples of frameworks are RCM II [16], CIBOCOF [17], or MGM-8PH [18]. Although these frameworks consider the Condition Based Maintenance (CBM), none specifically includes the treatment of PHM techniques and methods and the capabilities that it brings to the maintenance function [19]. This is largely owing to the fact that most used frameworks were developed prior to the recent growth of PHM. So, approaches are needed in order to addressing the formal use of PHM within maintenance policies, including as key aspect the integration with these mentioned frameworks which are, in industrial practice, reference tools for designing such policies.

In this sense, the RCM is one of the main references. Its main drawback is that it is a complex and costly implementation methodology, which causes its application limited to equipment or industries with high tech/high risk [7] or, in other words, with high criticality [20]. Nevertheless, even in cases where its full implementation is not advisable, the foundations of its analysis are used as reference for obtaining simpler and more practical maintenance models [7]. Therefore, as a step towards a practical approach to the integration of PHM in maintenance policies, this paper is analyzing how to introduce the PHM within the RCM program, conceived in its simpler and practical versions.

The paper is organized as follows. Section 2 introduces PHM. Section 3 reflects on the link of PHM with maintenance, taking into account its impact on the design of maintenance policies and its relationship with different types of maintenance. Section 4 proposes the integration of PHM within a RCM program. Section 5 presents a case study with the purpose to provide a concrete illustration in real industrial setting of the concepts herein discussed. Eventually, Sect. 6 contains the conclusions.

2 Introduction to PHM

In recent decades traditional maintenance models, that combine run-to-failure (RTF) and time-based preventive maintenance (TBM), are transforming to more proactive types in most industrial sectors, owing to CBM [21]. In this evolution, PHM is considered one of the key factors to achieve system-level efficient maintenance and reduce life cycle costs [14]. Prognosis research field is in fact promising new capabilities to improve the reliability of systems, leveraging both on design and maintenance along the useful lives [22, 23]. Besides, PHM provides capabilities to achieve more proactivity in maintenance: in this regard, it is worth remarking that, as expectation for the future, the equipment data will be transformed by PHM solutions into valuable information to help not only maintenance managers, but also plant managers for optimizing planning, saving cost, and minimizing equipment downtimes [24].

In order to effectively grasp the evolution, it is worth reflecting on the general understanding owing to the terminology currently adopted (Sect. 2.1), before looking at PHM in its process model (Sect. 2.2) and role for competitiveness of enterprises (Sect. 2.3).

2.1 General Understanding of PHM

PHM has been generically understood as the process of determining the current state of a system in view of reliability and prediction of its future state. This is possible by combining the detection and interpretation of different parameters, i.e., environmental, operational, and performance parameters, necessary for assessing the health state of the system and for predicting the remaining useful life (RUL) [25].

The rapid development of PHM techniques and methods, and their applications, is leading to the perception of PHM as engineering discipline [22] based on the use of in situ monitoring and advanced methods for assessing degradation trends of a system and determining the RUL. This further allows to know the state of the system in relation to its life cycle and, thus, to control the risk level with which the system operates.

Haddad et al. [3] present PHM as a discipline that allows different uses: (i) evaluating the reliability of products in each stage of their life cycle; (ii) determining the possible failures occurrence and the levers for risk reduction; (iii) highlighting the RUL estimation to provide accurate lead-time estimation for maintenance implementation, that finally allows greater proactivity within an organization. Lee et al. [6] add that PHM provides information to aid in making scheduled maintenance, or even autonomously triggered maintenance (i.e., self-maintenance); moreover, they mention asset management decisions or actions, even if they limit to advantages mostly related to maintenance management, such as the elimination of unnecessary and costly preventive maintenance, the optimization of maintenance scheduling, and the reduction of lead-time for spare parts and resources.

PHM can be understood also based on the related performance measures. Besides RUL, others terms as ETTF or PD appear in the literature. Estimated Time To Failure (ETTF) is a term equivalent to RUL, that is included and defined by ISO 18331. Another concept that should be considered regarding PHM solutions and their practical use to enable proactivity is the Prognostic Distance (PD). Sandborn and Wilkinson [26] define PD as the time interval of an organization's ability to gather information necessary to predict a failure and forecast a future failure for taking (planning and implementing) appropriate actions. Thus, this term is closely linked with maintenance actions planning and execution, while RUL is only related to failure evolution: if RUL < PD, in a particular moment, the organization would not been able to trigger any maintenance action to prevent the failure. Fritzsche et al. [27] present an interesting case study applied to aircraft logistics maintenance to understand PD concept and its scope.

A more comprehensive view is introduced by Saxena et al. [28, 29], who focus their attention on the performance measurement of prognostics algorithms. This is one of the major problem for PHM practical deployment: it is not enough in developing prognostics methods, it is also essential in evaluating and comparing their performances. Hence, they present a general framework, claiming that they define the most relevant time indexes with the purpose *"to denote important events in a prognostic process"*. This allows not only to develop PHM solutions that calculate the RUL within given probabilistic uncertainty estimates, but also to introduce other relevant concepts for understanding and utilizing the developed solutions. Above all, it is worth remarking the proposal of a terminology to apply metrics for a standardized performance evaluation.

2.2 The Process Model of PHM

PHM solutions implicitly cover a complete process (Fig. 1), from capturing the raw data up to utilizing the information for decision-making (in maintenance, logistics, life cycle control, equipment design, etc.). This process has been originally conceived by ISO 13374 [30, 31], a standard that has been adopted within OSA-CBM [32]: today, ISO 13374/OSA-CBM are jointly being used as main international references for designing particular solutions and general software applications development [33].

Often, the different phases of the process are difficult to interpret and distinguish. Along these phases, the detection, diagnosis, and prognosis are the core issues addressed to develop the PHM solutions. Although the scientific literature treats these three issues separately, they are closely connected. So, for example, in most cases prognosis requires prior detection and diagnosis [28, 29]. Assuming another

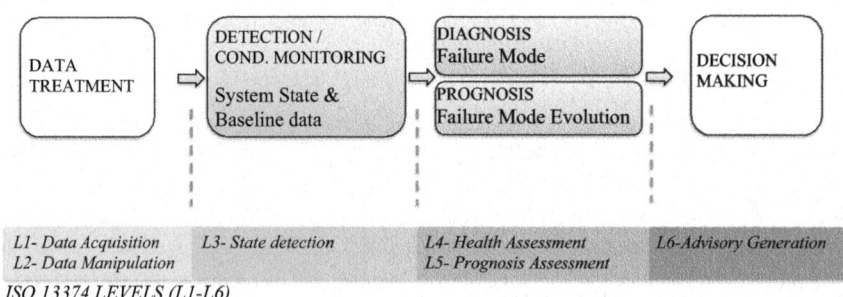

Fig. 1 "From data to business value" process and detection, diagnosis and prognosis problems. Correlation with levels of ISO 13374

perspective, the detection, diagnosis, and prognosis can be considered complementary interpretations of the same gathered data from the system. On the whole, most of techniques and skills that are required when implementing the PHM solutions are similar, or even the same type, so it makes sense to consider the PHM process model as an expression of a unique engineering discipline.

Owing to the fact that an accurate definition of the PHM terms is complicated and, currently, there is no unique standardized and accepted vocabulary within the technical community, it is worth proposing a detailed definition of the core phases of the process, providing an operational perspective.

2.2.1 Detection/Condition Monitoring

It focuses on the state of the machine, equipment or system for which PHM is being developed. It enables to distinguish anomalous behaviors, comparing gathered data against baseline parameters (ISO 13379, [34]), and detecting and reporting abnormal events. In other words, the observed symptoms are related with reference behaviors to determine the condition or state of the system, therefore detecting the abnormality when it happens. This is defined as CM (Condition Monitoring) by ISO 13379 and SD (State Detection) by ISO 13374. Alerts and alarms management is also related with the detection phase. Besides, the operational context determination (current operational state and operational environment) has also to be considered within detection. For what concern performances, one of the detection objectives is to achieve the minimization of false positives and false negatives [23].

2.2.2 Diagnosis

It focuses on the failure modes and its causes. In particular, the ISO 13372 defines diagnosis as the result of diagnostic process, and this as the determination of the "nature" of the failure. This definition can be completed considering the two different stages in the diagnostic process within a complex engineering system

- Isolation: it determines which component, or more accurately, which failure mode is affecting the behavior of the system;
- Identification: it determines or estimates the nature (or causes) and the extent (size and time) of the failures.

Besides, there are two different timings for obtaining the diagnosis

- Before the failure occurs: it corresponds to the diagnosis of incipient failure modes; this enables to prevent the failure and is closely related with prognostic process;
- After the failure occurs: once the failure happens, the failure should be located accurately and as soon as possible, in order to minimize the MTTR (Mean Time to Repair); in this case, the diagnosis can improve the system maintainability

and availability, especially if complex engineering system is considered, with a great level of interaction between system and components.

Some authors consider detection as a part of diagnosis [23], interpreting DII (detection, isolation, and identification) as the three diagnostic phases or diagnosis component. In the interpretation provided by this paper, diagnosis (as Isolation and Identification) is separated from detection owing to two reasons: (i) detection/condition monitoring techniques and methods could be used by both the other process phases, diagnosis and prognosis; (ii) detection/condition monitoring should be implemented before proceeding with any other process phase.

2.2.3 Prognosis

Prognosis is focused on failure mode evolution. The estimation of future behavior of the defined failure mode then allows failure risk assessment and control.

There are different types of outputs from various prognostic algorithms. Some algorithms assess Health Index (HI) or Probability of Failure (PoF) at any given point in time, and others carry out an assessment of the RUL based on a predetermined Failure Threshold (FT) [28, 29]. An interesting aspect to analyze is the different prognostics approaches. Most authors distinguish three basic groups of methods, or approaches, for the prognosis: model-based prognostics, data-driven prognostics, and hybrid prognostics [13, 35].

For what concern performances, the prognosis objectives depend on many factors —time and cost for problem mitigation, system criticality, cost of a failure, and can be related to the minimization of false positives and false negatives [28, 29].

2.3   The Role of PHM for Competitiveness

What is the role of PHM for competitiveness? Why is PHM believed to have high potential for competitive operations? These questions drive the review of this section, extending the discussion of the introduction on maintenance for competitiveness. There are many references that highlight the importance of incorporating new technologies along with diagnosis and prognosis capabilities—in others words, of implementing the CBM supported by PHM solutions, as those levers that can promote a most room for improvement. This review highlights a selected number of cases, ranging from complex products/systems to infrastructures and manufacturing plants; this is then a demonstration of PHM potential in different business context dealing with highly engineered systems.

Vachtsevanos et al. [23] use the term CBM/PHM to represent a new and more powerful CBM (diagnostic and prognostic-enabled CBM), and treats in depth how the area of intelligent maintenance and CBM/PHM of machinery is a vital one for today's complex systems in industry, such as aerospace vehicles, military and merchant ships, automobiles, etc. A major motivation for these developments is the realization that a more powerful CBM is needed to fully reap the benefits of new logistics support concepts, considering the maintenance impact on operations and logistics management.

Jardine et al. [21] focuses on the industry trend to higher cost of preventive maintenance, which has become one of major expenses of many companies. According to Jardine, this is owing to two effects related with rapid technological development: products/systems are more and more complex while the requirements (quality, reliability, etc.) are higher and subjected to a great variability and dynamic changes. More efficient maintenance approaches, such as CBM, are claimed to handle the rapid technological development with cost effectiveness.

In Gómez and Crespo [36], the case of the use of ICTs for asset management on network utilities is analyzed. This may be considered as an example providing a business context currently featuring an high level of integration of ICTs jointly with the inherent problems as physical dispersion of the facilities and availability requirements in terms of customers demand (of a no-interruption service): in this context, a great impact can be expected by using advanced methods based on CBM and E-maintenance.

Focusing more specifically on manufacturing plants, two cases are worth of a citation. Maletič et al. (2014) propose a study that aims to identify the improvement areas of maintenance in relation with their contribution to the profitability of companies. A method of analysis is proposed and a practical case study applied to the textile industry is presented. They conclude that practices related with advanced CBM are introducing the most potential for improvement. Macchi et al. [37] propose a case study in the same industry regarding the investment in E-maintenance solutions developed in the frame of a Supervisory Control and Data Acquisition (SCADA): owing to the study, they underline the link between E-maintenance and CBM as an evidence of the fact that advanced ICT solutions are being adopted in manufacturing to progressively change the maintenance management process, in particular its policies. To this concern, they remark that the automation for controlling the plant's utilities and production lines can become a means to engineer "value" driven solutions based on advanced CBM services available by the plant.

Back to a more generic perspective of the role of PHM for competitiveness, it is worth focusing on "value" as concept for expressing the PHM potential for business. Haddad et al. [3] compare the "value" of different types of maintenance (Fig. 2). They depict the system health as a function of time (top diagram drawn in Fig. 2) inspired in the P-F interval graphics used by classical CBM applications. System health, i.e., the ability of a system to perform its intended functionality, degrades over time and can be managed through maintenance decisions. The goal is to maximize the value of maintenance decisions through time (the bottom diagram of Fig. 2): the most value is provided by CBM/PHM, as advanced evolution of classic CBM, while corrective maintenance (unscheduled maintenance) is the least value option; scheduled preventive maintenance has a low value if the time or usage

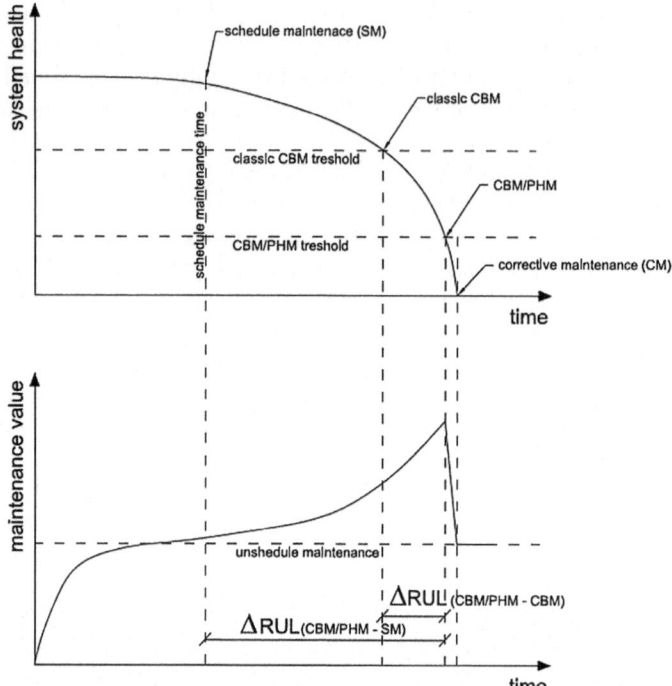

Fig. 2 Value analysis of different maintenance options [3]

to failure is not well characterized, because it wastes substantial part of the actual useful life.

Drawing upon this conceptual perspective of CBM/PHM, it can be concluded that improving maintenance through incorporation of techniques for the study of the condition is a major factor for the competitiveness of companies, being CBM programs, enabled by PHM, often considered to be the paradigm with the highest value, as it minimizes the unused RUL, avoids catastrophic failures, and presents a proper lead-time for logistics management [3]. A good example of this approach of the role of CBM and PHM is the ADS-79D-HDBK standard [33], where CBM is established as priority type of maintenance and provides guides on the use of PHM techniques and methods for the main systems and equipment present in an aircraft.

However, it is necessary to emphasize that, for an optimum design of mainte-nance policies of complex systems, CBM is combined with other maintenance options. In fact, CBM is not always the best type of maintenance, especially from the perspective of cost effectiveness: when failures of machines or components are not critical, we can even allow Run-To-Failure (RTF); when the lives of machines or components can be estimated precisely, scheduled TBM (Time-Based Maintenance) is the most effective means of maintenance [10]; besides, as CBM/PHM opens opportunities to develop the Predictive Maintenance (PdM) policy as CBM evolution [6, 24], a good balance should be found between CBM and PdM.

Now that it has been discussed the role in the "traditional" scope of maintenance management, a further issue is worth of consideration, extending to visions for the future currently drawn by the literature. The specific sub-question driving the search could be the following: is the role of PHM for competitiveness limited only to maintenance function? Based on recent references, it can be asserted that the benefits of prognosis and, by extension, of PHM will go beyond the maintenance function as classically conceived.

PHM application is the foundation for important improvements in all phases of the system lifecycle [22]. In Fig. 3 different life cycle phases, and benefits of every phase, are depicted.

It is evident that the benefits regard a large spectrum of stakeholders. Nonetheless, the major benefits can be achieved based on the role of the maintenance function, even if it is necessary to extend the perspective with respect to the classical one. In fact, in their classification, Sun et al. [22] locate operation and maintenance in different phases of the system life cycle: they are focusing on "where" (what phases) those specific benefits appear and can be counted. All the benefits, especially those listed in the system operation and maintenance phase, are still considered a result of maintenance actions. For example, a PHM solution can reduce the risk of a catastrophic accident during system operation. In this case, the losses avoided, that in this type of event usually have very high impact, can be counted as operating benefit provided by the PHM solution; on the other hand, the cost and effort of solution design and implementation and, finally, of the operational activities to prevent the accident, will be mostly responsibility of maintenance management. This is a clear evidence of how maintenance practices could extend

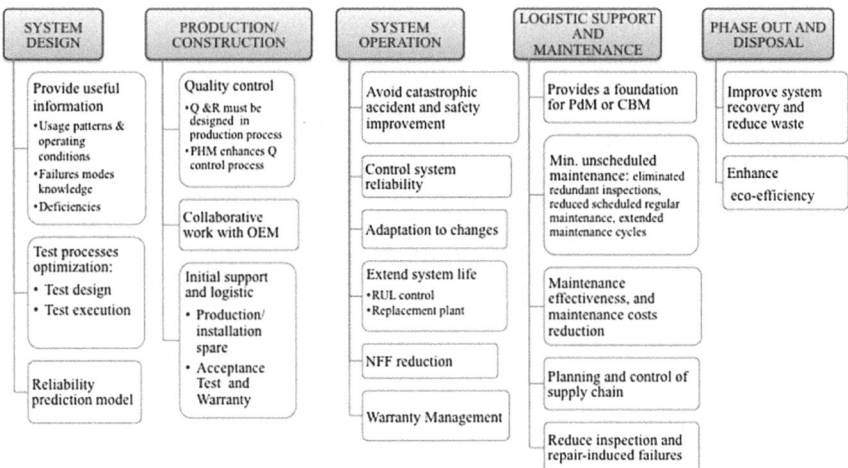

Fig. 3 Benefits of PHM along the life cycle phases

their scope towards system operation phase, bringing a valuable support for safety through PHM potential, which finally confirms the adoption of PHM as a lever for risk reduction.

PHM is also cornerstone in the development of other research lines, like the aforementioned E-maintenance, or the CPS (Cyber Physical System). CPS are physical and engineered systems whose operations are monitored, coordinated, controlled, and integrated by a computing and communication core [38]. Overall, CPS can be considered as adaptive systems with distributed multiple layered feedback loops [39], where the ability to learn, adapt on new situations can be achieved through modeling event-driven techniques. Trends such as CPS, Smart Manufacturing or Self-Maintenance and Engineering Immune System [6, 40, 41, 42], point to a future scenario where the technical content of the equipment, processes, control, and management tools, will reach levels well above current standards. In this development the PHM techniques and methods can have a fundamental role and this will reinforce the importance of maintenance in relation to competitiveness of future models of production systems. The very recent vision presented by Lee et al. [43] proposes a five-level structure as an architecture for developing and deploying a CPS for manufacturing application: within it, PHM is introduced at the second level of the CPS architecture to bring "self-awareness" to machines. Owing to the architecture, the advantage envisioned for PHM is the interconnection between machines through an interface at the cyber level, which would help achieving the benefits of a peer-to-peer monitoring and comparisons in fleet of machines: when a cyber-level infrastructure is made available, the machines could operate in a way "*conceptually similar to social networks*" [43] and PHM will play a relevant role with these networks.

3 The Role of PHM Within Maintenance Types and Policies

The incorporation of PHM, as envisioned in previous Sect. 2, is going to change the way the systems are maintained. To analyze how the PHM capabilities fit within maintenance activities, two complementary perspectives are provided in this section: the relationship of PHM with different types of maintenance is first discussed (Sect. 3.1) before considering the role of PHM within maintenance policies (Sect. 3.2). Concluding remarks are included as a summary (Sect. 3.3).

3.1 PHM and Types of Maintenance

According to Sect. 2, PHM should not be treated as a type of maintenance, as preventive maintenance, condition based maintenance (CBM), or corrective maintenance (see EN 13306). PHM is not a type of maintenance, but a set of tools (techniques and methods) that yield information that can be used as maintenance input. CBM and PdM—as further development of CBM—are consolidated terms in industry: to understand how to introduce PHM in maintenance activities it is necessary to clarify the relation with these terms.

Skills and approaches owing to PHM far outweigh classical techniques that have provided support to CBM. So, in order to mark the difference with the classical concept of CBM, some authors or references have tried to introduce new terms to identify and distinguish these new techniques and methods from traditional approaches of CBM. In this sense, sometimes the term PHM has been used to designate a new maintenance type based on PHM techniques and methods. With the same aim, new terms such as CBM+ [44] or PdM (Predictive Maintenance) [45–47]—identifying predictive maintenance with this concept—have been proposed. Vachtsevanos et al. [23] use the term CBM/PHM to treat this extended CBM approach.

Regarding predictive maintenance, despite both terms—PHM and predictive maintenance—are closely related, it is necessary to understand their differences. PdM is the part of CBM that is focused on the prediction of failure and prevents the failure or degradation, and it has been used profusely by the industry. This makes sometimes PHM would be identified with predictive maintenance. This is not correct, first because the PHM is not a type of maintenance, as it can be deduced from the discussion in previous Sect. 2; second, because PdM can be done without using the PHM techniques and methods, for example it can be actuated based on the experience of the staff.

In conclusion, the type of maintenance that can be performed by using PHM solutions is CBM. In this sense, our vision is that is not necessary to introduce new terms because it can produce confusion or, even worst, the idea that classical techniques are not being useful any more. So we propose to use the expression "CBM enabled by PHM" and the acronym CBM/PHM proposed in Vachtsevanos et al. [23] and adopted also by ADS-79D-HDBK [33].

To give a more comprehensive view and clarify the relationship with different types of maintenance, Fig. 4 locates on the EN 13306 schema of maintenance types other terms usually adopted by the literature: Run-To-Failure (RTF), Breakdown Maintenance (BM), Usage-Based Maintenance (UBM), Time-Based Maintenance (TBM), Failure-Based Maintenance (FBM), Design-Out Maintenance (DOM) and Detection-Based Maintenance (DBM) [48].

The transition from a classical CBM to an extended CBM is part of the evolution of maintenance types based on the progressive introduction of new capabilities serving the maintenance function. This evolution is depicted in Fig. 5: especially, it shows how this CBM evolution is provided by PHM capabilities; likewise, E-maintenance strategies definitely facilitate a degree of proactivity not achieved so far, as they support greater control and capacity to act on the system, including monitoring the efficiency and effectiveness of maintenance plans.

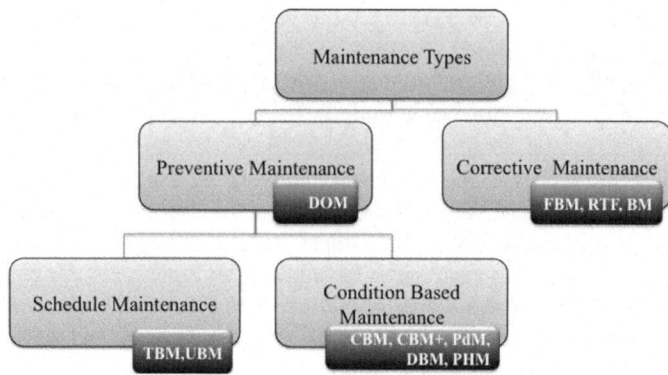

Fig. 4 Maintenance types according to EN 13306 and relation with other terms

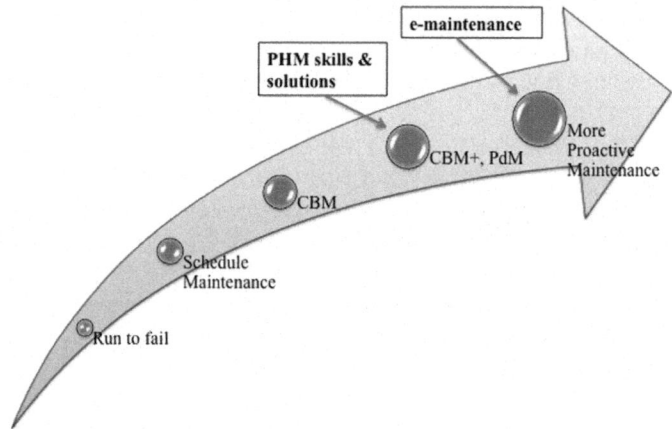

Fig. 5 Maintenance types and incorporation of proactive capabilities to the maintenance function

There are two different views about proactivity. Indeed, exploiting the concept of proactivity, a very important part of the literature adopts an interpretation about maintenance types that is different to the EN 13306 view mentioned above [49]. Moubray [16] considers two fundamental maintenance types: reactive maintenance (after failure occurs) and proactive maintenance (before failure occurs). Nonetheless, the EN 13306 relates the "corrective maintenance" to the "reactive maintenance" defined by Moubray, and the "preventive maintenance" to what Moubray calls "proactive maintenance".

Adopting the EN 13306 interpretation, as in this paper, makes possible preserving the concept of "proactivity", thus giving to this same concept a wider meaning. We in fact prefer that proactivity in maintenance assumes the meaning of "go beyond" and "do more": not only to prevent the failures and its effects in medium or short term, but also to reach an excellent maintenance management in a continuous improvement process. Swanson [49] treats similar ideas when she introduces the term of "aggressive maintenance" as strategy that goes beyond the efforts to avoid equipment failures, seeking to improve the overall equipment operation. Likewise, in our concern, the aim would be to get a proactive maintenance, and the "instruments" to get it are the new capabilities providing a better knowledge about systems states and risks (the PHM solutions are providing such new capabilities) and management tools and strategies based on ICTs (incorporated in E-maintenance). With this consideration, the concept of "proactivity" or "proactive maintenance" is actually driving the evolution of maintenance [50], being PHM and E-maintenance the primary levers of this development [13]. Within the proactive maintenance trend, one of the main aspects to consider is the ability of maintenance adaptation to the dynamic changes of its requirements (considering different issues as technical and operative aspects, owing to the resources limitation, and aligning to the business strategies and external requirements) along the entire life cycle of the system/asset.

The last step of "proactivity" that can be envisioned for the future scenario (in regard to the trends for CPS, Smart Manufacturing, Engineering Immune Systems, etc.) would lead to the integration of PHM and E-maintenance capabilities with operating systems for predictive control strategies. This will enhance the ability to better treat the new system and manufacturing requirements to the maintenance. Indeed, the predictive control (PC) is a research line that is currently creating new opportunities in the field of automation and control: relating it to the maintenance evolution, this would consist of conditioning the basic operation of the system, modifying its control logic according to the observed conditions and estimated RUL. In this way, control decisions impact on efficiency would be accountable, and achievement of excellent performance levels of the system/asset will be possible.

3.2 PHM and Maintenance Policies

The concept of "maintenance policy" according to the IEC 60050-191 will now be used as a framework to facilitate the consideration of PHM within the maintenance function.

The reader may notice that this term, maintenance policy, has had different interpretations or uses in past literature. In some references [17] it is identified with the "type of maintenance" concept defined by EN 13306, i.e., CBM or TBM are sometimes referred as maintenance policies. At the same time, it is also common to find texts where "maintenance policy" is referred as "maintenance strategy", as defined by EN 13306 (i.e., management method used to achieve the objectives of the maintenance function).

IEC 60050-191 definition of maintenance policy is used in this paper. This standard defines the term of maintenance policy as "a description of the interrelationship between the maintenance echelons, the indenture levels, and the levels of maintenance to be applied for the maintenance of an item". This approach is depicted on Fig. 6. So, designing the maintenance policy of a system or asset/installation consists on describing the indenture levels (what elements are subject to the maintenance actions), the level of maintenance for each indenture level (what type of maintenance or maintenance action) and the line of maintenance (who is responsible to execute the maintenance tasks). Figure 6 depicts also the relationship between the result of a RCM program and the components of a maintenance policy. This issue is later treated in Sect. 4. In the next paragraphs of the present section, we review the definition of the different components of a maintenance policy, as given by IEC 60050-191. After clarifying these concepts, according to the standard and offering also examples for their interpretation, PHM will be located in the IEC schema.

3.2.1 Indenture Level

This is the level of decomposition of an item (i.e., system/asset) from the point of view of a maintenance action. Examples of indenture levels could be a subsystem, a circuit board, a component. The indenture level depends on the complexity of the item's construction, the accessibility to sub-items, the skill level of maintenance personnel, the test equipment facilities, safety considerations, etc. For example, in a company without technical capacity for doing corrective maintenance over a circuit board (e.g., repair a certain welding), the indenture level will be established in the next highest level, defining as maintainable item the element that contains the circuit board (a subsystem as a complete electronic module). The failure mechanism

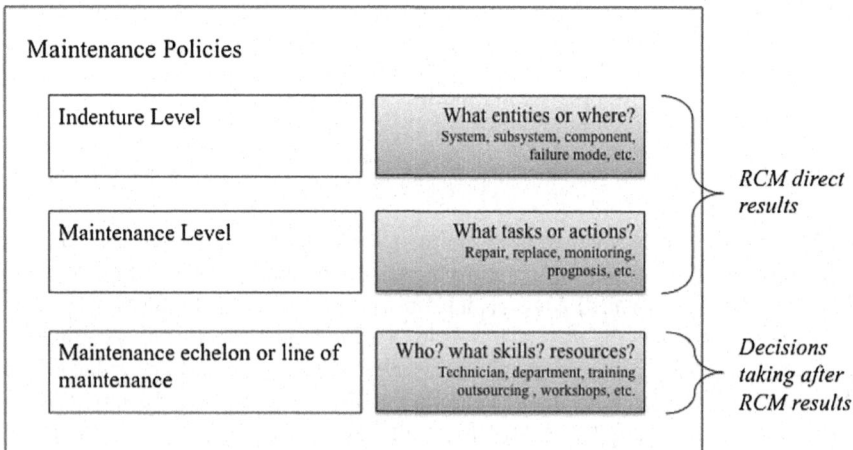

Fig. 6 Components of maintenance policy according to IEC 60050-191 and relation of components with the results of a RCM analysis

should normally be related to the lowest indenture levels (i.e., maintainable item level). In practical terms, the failure mechanism represents a failure mode at maintainable item level. In this regard, the reader is referred to the standard ISO 14224, as a good guideline to establish proper indenture levels in specific equipment, especially within the oil and gas industry, and considering the business technical structure.

3.2.2 Maintenance Level

This is the set of maintenance actions to be carried out at a specified indenture level. Examples of a maintenance action are replacing a component, a printed circuit board, a subsystem, etc. The level of maintenance is closely related to the application of the RCM logic. At that point of the RCM analysis, the maintenance action over each specific failure mode, as depicted by the FMEA/FMECA, is chosen.

3.2.3 Maintenance Echelon or Line of Maintenance

This is the position in an organization where specified levels of maintenance are to be carried out on a maintainable item. This component of the maintenance policy is not explicitly addressed by the RCM analysis, even if it is implicitly present. Maintenance lines description in fact includes the maintenance resources and their availability. Within a general conception, this component includes material, technical, and human resources and required technical knowledge and skills. It is worth noticing how the selection of a particular level of maintenance, for instance a CBM program based on a PHM solution, as better maintenance option over a specific failure mode, implicitly requires analyzing whether the organization can offer a suitable support to carry out this action. In this regard, relevant questions may be, e.g., are existing resources technically capable? Would it be convenient subcontracting?

3.2.4 PHM Contributions and Requirements for Maintenance Policies Design

To understand how PHM can be introduced as fundamental element for advanced maintenance, it is necessary to get a comprehensive view of how maintenance policies—and their components as defined by IEC 60050-191—are affected by the opportunity or the need of using PHM. PHM solutions are in fact complex and, to manage complexity, a simplified interpretation of maintenance policies, and their use within maintenance design methodologies, can help to understand how PHM can affect the definition and execution of maintenance plans.

One of the main problems to address in the maintenance policies design process is to provide a practical interpretation of the different concepts. Of course, it is important to know different concepts meaning. But at the same time it is also needed to know how to interpret them in the most practical way, combining the accurate definitions with some considerations regarding their application on practical cases. In this sense a simplified view of the components of a maintenance policy can be proposed

- Indenture level: failure mode definition for each maintainable item;
- Level of maintenance: maintenance types and tasks/actions selected per failure mode;
- Maintenance echelon or line of maintenance: organizational responsibilities, skills, and maintenance resources required to accomplish the selected maintenance tasks.

Based on this practical interpretation, Table 1 analyses what PHM can provide to each component of the maintenance policy, both in terms of its contributions

Table 1 PHM contributions and requirements for maintenance policies design

Component of the maintenance policy	Interpretation	PHM possible contributions and/or requirements
Indenture level	Failure mode definition	PHM allows more detailed description of the indenture level, owing to its capability to provide accurate information PHM allows reducing NFF (Not Fault Found) and similar events PHM allows reducing hidden failures
Maintenance level	Maintenance types, tasks and/or actions	PHM allows detection, diagnosis and prognosis of a failure before it occurs PHM allows to control the failure risk and the degradation PHM allows diagnosis after a failure occurs PHM allows reducing scheduled maintenance PHM allows effective autonomous maintenance
Maintenance echelon or line of maintenance	Responsibilities, skills and resources	PHM requires new skills and specialized technicians PHM requires new technological resources The shifts of indenture levels definition may introduce changes in the maintenance skills required by PHM as well as in the required logistics support (with changes in the spare parts requirements) PHM allows to better organize responsibilities for an effective autonomous maintenance PHM solutions (design and implementation) can be subcontracted PHM solutions can be a service provided by the OEM (Original Equipment Manufacturer) PHM can be integrated within E-maintenance tools, thus supporting the E-maintenance strategies

(provided capabilities) and its requirements (required resources). It is worth pointing out how the specialization of required resources may lead to opportunities for third parties and the use of high-tech tools.

3.3 PHM Maintenance Types and Policies

Summarizing the role of PHM with respect to the maintenance types and policies, it is worth concluding the following remarks:

- the evolution of maintenance types, based on the progressive incorporation of new capabilities owing to PHM, is leading to more "proactive maintenance";
- maintenance policies will result from properly combining different maintenance levels and lines at different indenture levels; in particular, for a unique system/asset, different maintenance levels can be used for different maintainable items (i.e., maintenance of some items by means of CBM/PHM will be combined with Run-To-Failure in other items).

Thus, using a "proactive maintenance" will designate different matters "to go beyond":

- the maintenance policies are designed and reconfigured, aligning any time with the strategic criteria of the organization and its key strategic factors to achieve competitiveness, profitability, and sustainability of the company (eco-efficiency);
- maintenance policy management promotes progressive growth of E-maintenance tools and CBM/PHM solutions; this introduces the chance of greater value added actions, owing to a better understanding of the system and its internal (between its component items) and external relations (with other systems/assets, the environment, and other maintenance tasks or business areas);
- reconfigurable, adaptive, and evolving maintenance policies means facing better and promptly the uncertain evolution of business requirements, the effect of some disruptive events, or even the changing needs along the asset life cycle.

4 The Role of PHM Within the RCM Framework

PHM techniques and methods have to be conveniently integrated into the maintenance policies [50] to exploit their potential. But, as they are complex, their implementation may generate undesirable costs and side-effects to business. Different frameworks or methodologies facilitating design and implementation of maintenance policies [18] can then help to obtain the expected contributions owing to PHM. Although these frameworks consider CBM, none specifically includes the treatment of PHM techniques and methods and the capabilities that these brings to maintenance [19], largely owing to the fact that the most used frameworks were developed prior to the recent growth of PHM. To address this lack, in this section the RCM methodology is reviewed analyzing how the use of PHM solutions as part of the maintenance function can be facilitated within a RCM program.

RCM, and its evolutions as RCM II, is one of the main maintenance policies design references. Description of RCM is not the scope of this paper. In case of further wishes to deep on this methodology, it is worth consulting Moubray [16] or Parra and Crespo [51] respectively for what concern RCM and RCM II.

Given the definition of maintenance policy in Sect. 3, RCM can be understood as a methodology to define levels of maintenance corresponding to indenture levels of the system/asset to ensure its operational continuity, without occurrence of failure modes in a given operational context. Indeed, the indenture levels are determined when the detailed functional analysis until the possible failure mode of each of the functions is made (FMEA/FMECA); the levels of maintenance are determined by means of the RCM logic. Considering also the organizational aspects, the maintenance lines depend on the maturity of the different actors in each industrial scenario. On the whole, it is possible to link RCM with maintenance policy components as herein illustrated

- Indenture level: it includes the result of RCM operational context identification and FMEA/FMECA;
- Level of maintenance: it includes the result of RCM logic;
- Maintenance Echelon or line of maintenance: it includes the indirect result of RCM logic in regard to the required skills and resources.

If maintenance policies components are the overall results of a RCM analysis, RCM and PHM have a mutually beneficial relationship helpful to lead to such results. In fact, from a system or equipment perspective, PHM without a RCM analysis becomes just a technology insertion without the justified functionality. Conversely, the collection of aggregated health data, without understanding of the failure modes and most effective course-of-action, can lead to wasted effort and unnecessary expenditure of resources [44]. Delving into the areas in which the capabilities of PHM may give support to RCM process steps, the reader is referred to Table 2: the effective achievement of results from a RCM analysis, in regard to the maintenance policies components, can be obtained by benefiting from the support of PHM capabilities.

Table 2 CBM/PHM capabilities and RCM process steps (simplified and adapted from DoD [44])

RCM phases	RCM process steps	CBM based on PHM enabling capabilities
FMEA/FMECA	Functions: the desired capability of the system, how well it is to perform, and under what circumstances	Provides analysis and decision support to determine the maintenance policy to ensure a required system performance; Provides technical data to determine optimal resources to perform maintenance tasks
	Functional failures: the failed state of the system	Provides diagnostic tools to assess degree of system/component degradation; Tracks health and status of installed components
	Failure modes: the specific condition causing a functional failure	Uses sensor and data analysis technology to identify failure physics; Collects, stores and communicates system condition and failure data
	Failure effects: the description of what happens when each failure mode occurs	Uses automated tools and data manipulation software to produce diagnostic information on detected failures; Applies information from Interactive Electronic Technical Manuals to report, troubleshoot, test, and support documentation of failures
	Failure consequences: the description of how the loss of function matters (e.g., safety, environmental, mission, or economics)	Maintains platform hardware and software configuration; Provides data warehouse capability including condition trends, history, and transaction records from business processes; Available to the full range of users
RCM logic	Hidden failures or NFF events	Use of new methods and techniques can make specific types of "hidden failure" no longer being "hidden"
	Maintenance tasks and intervals: the description of the applicable and effective tasks, if any, performed to predict or prevent failures	Incorporates prognostic capabilities to help predict failure causes and timing; Predicts the remaining useful life of equipment/components based on failure predictors derived from composite condition analysis; Includes new task and skills as data gathering task, software actualization, algorithm configuration, performance control
	Default actions: including but not limited to failure finding tasks, run-to-failure, engineering redesigns, and changes/additions to operating procedures or technical manuals	Supports standard graphics and trending displays, user alerts, data mining and analysis, simulation and modeling, enterprise decision-support systems, and advisory generation

5 Case Study

The case study is a further development of a recent publication aimed at presenting the implementation of an E-maintenance tool in an industrial context. The tool extends the functions of a Supervisory Control and Data Acquisition (SCADA) of an Electric Arc Furnace (EAF) for its use in a CBM program [52]. This paper reports the implementation project held in the real industrial setting of an Italian steel-making company. In particular, it provides a thorough explanation of the methodology adopted during the project to implement the E-maintenance tool, with the final purpose to control the degradation of a specific equipment of the EAF, i.e., the burning system. The reader should consult the publication for more details on the methodology and the equipment under control.

It is worth remarking that safety is a relevant objective of EAF's operations. Furthermore, as the furnace runs continuously at high temperatures and in harsh environmental conditions, inspection of many components can occur when the furnace is stopped for a scheduled maintenance; thus, real-time monitoring capabilities are essential to keep under control the health state of the furnace and its components, in this case the burning system. To this end, the E-maintenance tool extends the functionality of an existent SCADA with state detection and diagnosis (observe that these functions are defined in accordance to the ISO 13374, above discussed in Sect. 2.2). State detection compares the gathered data representing the actual functioning of the burning system against baseline parameters (*alias* reference values) of flow rates and pressures built in a statistical model, i.e., a regression model ground on field data available from the SCADA. When a deviation from reference values is detected, abnormal events are reported. Afterwards, diagnosis focuses on the failure modes and its causes: the identification of the causes uses a troubleshooting scheme and, subsequently, an advisory generation task is triggered in order to provide the operator with a check-list of operations to perform to solve the causes. The check-list is a list of counteractions defined as outcome of an HAZOP (Hazard and operability analysis) of the burning system (concerning HAZOP studies, see Lawley [53] for original definition and guide, and Dunjó et al. [54] for a more recent review).

The conceptual model discussed in the present paper is now illustrated in the context of this case study. Table 3 shows the PHM contributions (provided capabilities) and requirements (required resources) for maintenance policies design.

The CBM/PHM capabilities now available are related to the RCM Process Steps where they are bringing contributions (Table 4).

The company is nowadays using the tool as part of the automation running the control of the EAF's operations. This is leading to a greater control and capacity to act on the burning system, with the subsequent benefits for process safety and even for further potentials, in next years, towards process improvement thanks to the use of PHM capabilities within the RCM framework.

Table 3 Case study analysis: PHM contributions and requirements for maintenance policies design

Component of the maintenance policy	PHM possible contributions and/or requirements
Indenture level	PHM allows the capability to provide accurate information for the functioning of the burning system, which is one of the critical equipment for safe and efficient EAF's operations
Maintenance level	PHM allows to extend the functionality of the existent SCADA with the capability of detection and diagnosis of a failure of the burning system before it happens PHM allows to enhance the control of the failure risk and the degradation of the burning system PHM allows shifting maintenance tasks to the CBM type, thus reducing the interventions based on scheduled maintenance of the burning system
Maintenance echelon or line of maintenance	PHM requires to develop some but limited skills for the development, test and setup, and maintenance of the E-maintenance tool PHM requires new technological resources as a software program whose purpose is to extend the functionality of the SCADA (note that no additional hardware was required in the case, as the measures required for state detection and diagnosis were already available for the production process control) PHM is integrated within the E-maintenance tool, thus providing real-time monitoring extended to a maintenance focus

Table 4 CBM/PHM capabilities and RCM process steps in the context of the case study

RCM phases	CBM based on PHM enabling capabilities
FMEA/FMECA	Provides a diagnostic tool to assess degradation of the burning system Tracks, by means of a real-time monitoring, the health state of the burning system
	Uses sensor and data analysis technology to identify failure physics; in this regard, it is worth observing that data analysis—by means of a statistical model (i.e. regression model)—is joined with a model-based approach—thanks to HAZOP application, which finally leads to an hybrid diagnostics as data analysis technology; for what concern sensors, field systems required for operations comprise sensors already installed in the plant for production process control, measuring flow rates, pressures and valves position Collects, stores and communicates system condition data in two ways: a data display conveying, by means of an HMI (Human Machine Interface), dedicated information to the maintenance planner as user; a storage processing, including condition trends and history in a database for future use (i.e. data warehouse capability) Uses automated tools and data manipulation software to produce diagnostic information on detected degradation of failures (it is worth remarking that the implementation is made at different hierarchical levels, as defined by

(continued)

Table 4 (continued)

RCM phases	CBM based on PHM enabling capabilities
	the standard IEC 62264:2003 [55], namely level 1, PLC, and level 2, supervisor computer)
	Applies information built in as a set of HAZOP tables; in particular, the tables form an Interactive Electronic Technical Manual aiding to report different information in relation to the detected deviations (i.e. causes, effects, counteractions, suggestions, for each deviation resulting from the state detection); this is finally an aid for troubleshooting and, more in general, for supporting documentation of failures in relationship to the process of the burning system
RCM logic	Includes new task and skills as data gathering task, software actualization, algorithm configuration, in order to support diagnostics of the burning system
	Supports standard graphics and trending displays, user alerts, and advisory generation; in particular, the advisory generation provides a check-list, as a set of counteractions derived by the HAZOP tables

6 Conclusions

The paper has reflected on the role of PHM in maintenance systems. Thus, PHM correspondence with maintenance policies and types has been analyzed. To better distinguish such concepts, the expression "CBM enabled by PHM" (i.e., CBM/PHM) was considered. This enabled to remark that new technical capabilities have been added to the maintenance function, supported by PHM. It is precisely these new capabilities that open the door to a large room for improvement in terms of competitiveness and profitability: PHM, within E-maintenance strategies, are in fact seen as key points for competitiveness in the future, providing the basis for proactive maintenance management.

PHM impact on maintenance policies can modify levels and lines of maintenance (changing the actions that can be executed on a system and defining who runs the actions). The information actually available by PHM may also modify the assets indenture levels for maintenance definition, and may even render new maintenance actions convenient, contributing to extend the levels of maintenance. Afterwards, it is also important to use the existing methodologies for the design and implementation of maintenance policies: it is indeed essential to consider the capabilities of CBM/PHM in the development of a RCM process, especially in high-tech complex engineering systems, for full exploitation of the PHM potential.

On the whole, the paper provides a handy synthesis of the PHM potential into a framework for a system maintenance management: in fact, we believe that the research finding will contribute fixing the role of PHM in relationship to consolidated matters as the maintenance policies design and RCM framework. For the future work, we consider, as first priority, the need to operationalize the integration of RCM process steps with CBM/PHM capabilities; a second, more advanced issue would be the integration of PHM and E-maintenance capabilities with operating systems for predictive control strategies, which is in line with the expected trends for CPS, Smart Manufacturing, etc.

Acknowledgements This work was supported by the Spanish Ministry of Science and Innovation under Grant DPI2011-22806, besides FEDER funds.

References

1. Cheng S, Azarian M, Pecht M (2010) Sensor systems for prognostics and health management. Sensors 10:5774–5797
2. Pecht M (2008) Prognostics and Health Management of Electronics. Wiley, Hoboken, NJ, p 2008
3. Haddad G, Sandborn P, Pecht M (2012) An options approach for decision support of systems with prognostic capabilities, IEE Trans Reliab 61(4):872–883
4. Bird J, Madge N, Reichard K (2014) Towards a capabilities taxonomy for prognostics and health management. Int J Prognostics Health Manag 2014:2
5. Pinjala S, Pintelon L, Vereecke A (2006) An empirical investigation on the relationship between business and maintenance strategies. Int J Prod Econ 104(1):214–229
6. Lee J, Ghaffari M, Elmeligly S (2011) engineering. Annual Reviews in Control 35:111–122
7. Waeyenbergh G, Pintelon L (2002) A framework for maintenance concept development. Int J Prod Econ 77:299–313
8. Waeyenbergh W, Pintelon L (2004) Maintenance concept development: a case study. Int J Prod Econ 89:395–405
9. Macchi M, Fumagalli L (2013) A maintenance maturity assessment method for the manufacturing industry. J Qual Maintenance Eng 19(3):295–315
10. Takata S, Kirnura F, Van Houten FJAM, Westkamper E, Shpitalni M, Ceglarek D, Lee J (2004) Maintenance: changing role in life cycle management. CIRP Ann Manufact Technol 53(2):643-655
11. Garetti M, Taisch M (2012) Sustainable manufacturing: trends and research challenges. Prod Plan Control 23(2–3):83–104
12. Muller A, Crespo A, Iung B (2008) On the concept of e-maintenance: review and current research. Reliab Eng Syst Saf 93:1165–1187
13. Lee J, Ni J, Djurdjanovic D, Qiu H, Liao H (2006) Intelligent prognostics tools and e-maintenance. Comput Ind 57:476–489
14. Ly C, Tom K, Byington CS, Patrick R, Vatchsevanos GJ (2009) Fault diagnosis an failure prognosis on engineering system: a global perspective. In: 5th annual IEEE conference on automation science and engineering, Bangalore
15. Levrat E, Iung B, Crespo A (2008) E-maintenance: review and conceptual framework. Prod Plan Control 19(4):408–429
16. Moubray J (1997) RCM II: reliability-centred maintenance. Industrial Press Inc, New York
17. Waeyenbergh G, Pintelon L (2009) CIBOCOF: a framework for industrial maintenance concept development. Int J Prod Econ 121:633–6402009
18. Crespo A, Gupta J (2006) Contemporary maintenance management: process, framework and supporting pillars. Omega 34(3):313–326
19. Guillén AJ, Gómez J, Crespo A, Guerrero A (2014) Towards the industrial application of PHM: challenges and methodological approach, PHM Society European conference 2014
20. Crespo A, Moreu P, Sola A, Gómez J (2015) Criticality analysis for maintenance purposes. Qual Reliab Eng Int 32(2):519–533
21. Jardine A, Lin D, Banjevic D (2006) A review on machinery diagnostics and prognostics implementing condition based maintenance. MechSyst Signal Process 20:1483–1510

22. Sun B, Zeng S, Kang R, Pecht M (2012) Benefits and challenges of system prognostics. IEEE Trans Reliab 61(2)
23. Vachtsevanos G, Lewis F, Roemer M, Hess A, Wu B (2006) Intelligent fault diagnosis and prognosis for engineering systems. Wiley, NJ, Hoboken
24. Lee J, Holgado M, Kao H, Macchi M (2014) New thinking paradigm for maintenance innovation design. In: Proceedings of the 19th IFAC world congress, Cape Town, South Africa, IFAC Proceedings Volumes (IFAC-Papers Online), pp 24–29
25. Zio E, Di Maio F (2010) A data-driven fuzzy approach for predicting the remaining useful life in dynamic failure scenarios of a nuclear system. Reliab Eng Syst Safe 95(1):49–57
26. Sandborn PA, Wilkinson C (2007) A maintenance planning and business case development model for the application of prognostics and health management (PHM) to electronic systems. Microlectron Reliab 47(12):1889–1901
27. Fritzsche R, Gupta J, Lasch R (2014) Optimal prognostic distance to minimize total maintenance cost: the case of the airline industry. Int J Prod Econ 151:76–88
28. Saxena A, Roychoudhury I, Celaya JR, Saha S, Saha B, Goebel K (2010) Requirements specifications for prognostics: an overview. AIAA Infotech at Aerospace 2010, art. no. 2010-3398
29. Saxena A, Celaya J, Saha B, Saha S, Goebel K (2010) Metrics for offline evaluation of prognostic performance. Int J Prognostics Health Manag 1
30. International Organization for Standardization (2003) ISO 13374-1:2003—Condition monitoring and diagnostics of machines—data processing, communication and presentation—part 1: general guidelines
31. International Organization for Standardization (2007) ISO 13374-2:2007—Condition monitoring and diagnostics of machines—data processing, communication and presentation—part 2: data processing
32. MIMOSA (Machinery Information Management Open Standards Alliance) (2006) Open systems architecture for condition based maintenance (OSA-CBM) UML specification, v 3.1
33. United States Army (2013) ADS-79D-HDBK—Aeronautical design standard handbook for condition based maintenance systems for US Army Aircraft
34. International Organization for Standardization (2012) ISO 13379-1:2012—condition monitoring and diagnostics of machines—data interpretation and diagnostics techniques—part 1: general guidelines
35. Pecth M, Rubyca J (2010) A prognostics and health management roadmap for information and electronics-rich systems. Microelectron Reliab 50:317–323
36. Gómez J, Crespo A (2012) Maintenance management in network utilities, framework and practical implementation. Springer, London
37. Macchi M, Crespo A, Holgado M, Fumagalli L, Barberá L (2014) Value-driven engineering of E-maintenance platforms. J Manuf Technol Manag 25(4):568–598
38. Rajkumar R, Lee I, Sha L, Stankovic J (2010) Cyber-physical systems: the next computing revolution. Proc Des Autom Conf 2010:731–736
39. Watzoldt S (2012) HPI symposium 2012: reconfiguration in cyber-physical systems. URL: http://www.hpi.unipotsdam.de/fileadmin/hpi/FG_Giese/Slides/waetzoldt_SAP_17_12_2012.pdf
40. Lee J, Lapira E, Bagheri B, Kao H (2013) Recent advances and trends in predictive manufacturing systems in big data environment. Manuf Lett 1(1):38–41
41. Lee J, Ni J (2014) Infotronics-based intelligent maintenance system and its impacts to closed-loop product life cycle systems. Invited keynote paper for IMS'2004—International conference on intelligent maintenance systems, Arles, France
42. Vogl GW, Weiss BA, Donmez MA (2014) Standards for prognostics and health management (PHM) techniques within manufacturing operations. In: Annual conference of the prognostics and health management society 2014
43. Lee J, Bagheri B, Kao H (2015) A cyber-physical systems architecture for industry 4.0-based manufacturing systems. Manuf Lett 3:18–23
44. Department of Defense (USA) (2008) Condition based maintenance plus, DoD guidebook

45. Gupta J, Trinquier C, Lorton A, Feuillard V (2012) Characterization of prognosis methods: an industrial approach. In: European conference of the prognostics and health management society
46. Wang L, Chu J, Wu J (2007) Selection of optimum maintenance strategies based on a fuzzy analytic hierarchy process. Int J Prod Econ 107(1):151–163
47. Tobon-Mejia D, Medjaher K (2010) IEEE prognostics & system health management conference, PHM'2010, Macau (China)
48. Vasseur D, Llory M (1999) International survey on PSA figures of merit. Reliab Eng Syst Safe 66(3):261–274
49. Swanson L (2001) Linking maintenance strategies to performance. Int J Prod Econ 70 (3):237–244
50. López-Campos M, Crespo A, Gómez JF (2013) Modelling using UML and BPMN the integration of open reliability, maintenance and condition monitoring management systems: an application in an electric transformer system. Comput Ind 64:524–542
51. Parra C, Crespo A (2012) Maintenance engineering and reliability for assets management. Ingeman
52. Colace C, Fumagalli L, Pala S, Macchi M, Matarazzo NR, Rondi, M (2015) Implementation of a condition monitoring system on an electric arc furnace through a risk-based methodology. Proc Inst Mech Eng O J Risk Reliab 229(4):327–342
53. Lawley HG (1974) Operability studies and hazard analysis. Chem Eng Prog 70(4):45–56
54. Dunjó J, Fthenakis V, Vílchez JA, Arnaldos J (2010) Hazard and operability (HAZOP) analysis. A literature review. J Hazard Mater 173(1–3):19–32
55. International Electrotechnical Commission (2003) IEC 62264, enterprise-control system integration
56. European Committee for Standadization. EN13306 (2010) Maintenance terminology. Brussels
57. Ferrell BL (1999) JSF prognostics and health management. In: Proceedings of IEEE aerospace conference. 6–13 Mar, Big Sky, MO. doi:10.1109/AERO.1999.793190
58. International Electrotechnical Commission (1990) IEC 60051-901, International electrotechnical vocabulary-part 191: dependability and quality of service
59. International Electrotechnical Commission (2006) IEC 60812—Analysis techniques for system reliability—Procedure for failure mode and effects analysis (FMEA)
60. Sheppard J, Kaufman K, Wilmering T (2008) IEEE standards for prognostics and health management, IEEE AUTOTESTCON 2008

Author Biographies

Antonio Jesús Guillén López is a Ph.D. candidate and contracted researcher in Intelligent Maintenance System research group (SIM) of the University of Seville (USE), focusing his studies in Prognosis Health Management & Condition-Based Maintenance applications and Asset Management. From 2012 collaborates with INGEMAN (a National Association for the Development of Maintenance Engineering in Spain) as consultor for private funding projects for asset management systems and maintenance models designing. Industrial Engineering and Master in Industrial Organization & Business Management from the University of Seville (USE). From 2003 to 2004, he worked for the Elasticity and Strength of Materials Group (GERM) of the USE in aeronautical materials tests. From 2006 to 2009, he was a member of the Department of Electronic Engineering of USE, and worked in numerous public–private international R&D project, developing new thermo-mechanical design applications for improving the performance and the life cycle of power electronics system. From 2009 to 2010, he was a Project Manager of "Solarkit" project, representing Andalucia's Government and USE in the international competition

Solar Decathlon 2010. He was a foundational partner of Win Inertia Tech., technological based company specializing in electronic R&D for Smart Grids and Energy Store fields, where he was General Manager until September 2012. Currently, he is a coordinator of the Spanish National Research Network of Asset Management.

Adolfo Crespo Márquez is currently Full Professor at the School of Engineering of the University of Seville, and Head of the Department of Industrial Management. He holds a Ph.D. in Industrial Engineering from this same University. His research works have been published in journals such as the International Journal of Production Research, International Journal of Production Economics, European Journal of Operations Research, Journal of Purchasing and Supply Management, International Journal of Agile Manufacturing, Omega, Journal of Quality in Maintenance Engineering, Decision Support Systems, Computers in Industry, Reliability Engineering and System Safety, and International Journal of Simulation and Process Modeling, among others. Prof. Crespo is the author of seven books, the last four with Springer-Verlag in 2007, 2010, 2012, and 2014 about maintenance, warranty, and supply chain management. Prof. Crespo leads the Spanish Research Network on Dependability Management and the Spanish Committee for Maintenance Standardization (1995–2003). He also leads a research team related to maintenance and dependability management currently with five Ph.D. students and four researchers. He has extensively participated in many engineering and consulting projects for different companies, for the Spanish Departments of Defense, Science and Education as well as for the European Commission (IPTS). He is the President of INGEMAN (a National Association for the Development of Maintenance Engineering in Spain) since 2002.

Marco Macchi is an Associate Professor at Politecnico di Milano, Department of Management, Economics and Industrial Engineering. He is the chair of the IFAC Working Group on AMEST (Advanced Maintenance Engineering, Services and Technology), the vice-chair of the IFAC Technical Committee 5.1 Manufacturing Plant Control and Book Reviews Editor and Editorial Board Member of the International Journal of Production Planning & Control: The Management of Operations. He is also a member of the IFIP WG 5.7 Advances in Production Management Systems and a Fellow of the International Society of Engineering Asset Management. In his research activity at Politecnico di Milano, he is responsible for several European funded projects in the fields of smart manufacturing, maintenance, and industrial sustainability. Furthermore, he is responsible for the Observatory on Technologies and Services for Maintenance and is Research Co-director of the Observatory on Industria 4.0 of the School of Management at Politecnico di Milano. His research interests are advanced production systems, maintenance management, and asset life cycle management.

Juan Francisco Gómez Fernández is Ph.D. in Industrial Management and Executive MBA. He is currently part of the Spanish Research & Development Group in Industrial Management of the Seville University and a member in knowledge sharing networks about Dependability and Service Quality. He has authored publications and collaborations in journals, books, and conferences, nationally and internationally. In relation to the practical application and experience, he has managed network maintenance and deployment departments in various national distribution network companies, both from private and public sector. He has conduced and participated in engineering and consulting projects for different international companies, related to Information and Communications Technologies, Maintenance and Asset Management, Reliability Assessment, and Outsourcing services in Utilities companies. He has combined his business activity with academic life as an associate professor (PSI) in Seville University, being awarded as Best Thesis and Master Thesis on Dependability by National and International Associations such as EFNSM (European Federation of National Maintenance Societies) and Spanish Association for Quality.

A Framework for Effective Management of CBM Programs

Antonio Jesús Guillén López, Juan Francisco Gómez Fernández and Adolfo Crespo Márquez

Abstract CBM (Condition-Based Maintenance) solutions are increasingly present in industrial systems due to two main circumstances: rapid evolution, without precedents, in the capture and analysis of data and significant cost reduction of supporting technologies. CBM programs in industrial systems can become extremely complex, especially when considering the effective introduction of new capabilities provided by PHM (Prognostics and Health Management) and E-maintenance disciplines. In this scenario, any CBM solution involves the management of numerous technical aspects, that the maintenance manager needs to understand, in order to be implemented properly and effectively, according to the company's strategy. This paper provides a comprehensive representation of the key components of a generic CBM solution, this is presented using a framework or supporting structure for an effective management of the CBM programs. The concept "*symptom of failure*", its corresponding analysis techniques (introduced by ISO 13379-1 and linked with RCM/FMEA analysis), and other international standard for CBM open-software application development (for instance, ISO 13374 and OSA-CBM), are used in the paper for the development of the framework. An original template has been developed, adopting the formal structure of RCM analysis templates, to integrate the information of the PHM techniques used to capture the failure mode behaviour and to manage maintenance. Finally, a case study describes the framework using the referred template.

Keywords CBM management · E-maintenance · Failure mode symptom analysis (FMSA) · Detection · Diagnosis · Prognosis

A.J. Guillén López · J.F. Gómez Fernández (✉) · A. Crespo Márquez
Department of Industrial Management, School of Engineering,
University of Seville, Seville, Spain
e-mail: juan.gomez@iies.es

© Springer International Publishing AG 2018
A. Crespo Márquez et al. (eds.), *Advanced Maintenance Modelling for Asset Management*, DOI 10.1007/978-3-319-58045-6_5

1 Introduction

Condition-Based Maintenance (CBM) is defined by EN 13306:2010 as "Preventive maintenance that includes a combination of condition monitoring and/or inspection and/or testing, analysis and subsequent maintenance actions" [1]. ISO 13372:2012 standard defines CBM as "Maintenance performed as governed by condition monitoring programmes" [2]. CBM monitors the condition of components and systems in order to determine a dynamic preventive schedule [3].

In the literature, it is also possible to find CBM referenced as a system, a program or a solution. The standard ADS-79D-HDBK [4] defines a "CBM system" as that it includes the analytical methods, sensors, data acquisition (DA) hardware, signal processing software and data management standards necessary to support the use of CBM as a maintenance approach to sustain and maintain systems, subsystems and components. A "CBM solution" can be understood as the application of a particular monitoring solution to a specific case (failure mode or element). A "CBM program" comprises the application of the different CBM solutions that have been adopted for a particular system [5], and it involves management and maintenance task planning.

CBM is increasingly becoming common in industrial systems, improving the transition from maintenance approaches that combine run-to-fail and programmed preventive maintenance to more efficient maintenance approaches [6]. In recent decades, the emergence of cheaper and more reliable ICT-Information and Communication Technologies (intelligent sensors, personal digital devices, wireless tools, etc.) has allowed an increase in the efficiency of CBM programs [3]. In automated manufacturing or process plants, CBM is preferred wherever it is technically feasible and financially viable [7].

The classical industrial view of CBM is mainly focused on the use of Condition Monitoring (CM) techniques such as vibration analysis, thermography, acoustic emission, or tribology [8]. The recent development of the PHM discipline (Prognosis and Health Management) is promoting a new CBM, providing powerful capabilities for physical understanding of the useful life of a system through dynamic pattern recognition [9, 10]. These capabilities allow us to treat, efficiently, new maintenance challenges in modern systems and applications [4, 11]. This new CBM, CBM+ [12] or CBM/PHM [10], is the main pillar for the implementation of E-maintenance strategies, where CBM develops its full potential through a more proactive maintenance management.

However, there is still a large gap for effective implementation of these new CBM programs extensively in industry, mainly due to complexity of these solutions and their life cycle. To this end, we propose two practical tools to represent and understand the key points of a CBM solution life cycle

(i) A framework, a basic structure to facilitate the representation of any CBM solution; and

(ii) A template, in table format, that will complement RCM results tables, integrating the information of the CBM solution for a particular existing failure mode).

The paper is organized as follows: Sect. 2 presents and justifies the CBM management approach within the context of E-maintenance strategies, and complexity of its practical implementation. Section 3 develops the proposed framework and its structure, which is depicted with an UML schema. Section 4 introduces the proposed template for CBM solutions compilation in a practical example. Finally, Sect. 5 presents the paper conclusions.

2 CBM Management within an E-maintenance Context

2.1 On the Role of CBM as Pillar of E-maintenance

Information and Communication Technologies (ICTs) are transforming the way systems are maintained; they provide the support to generate more systems behaviour knowledge and to introduce new tools and processes for a more proactive maintenance. This maintenance support has been defined as *E-Maintenance* [13]: "Maintenance support which includes the resources, services and management necessary to enable proactive decision process execution. This support includes e-technologies (i.e. ICT, Web-based, tether-free, wireless, infotronics technologies) but also, e-maintenance activities (operations or processes) such as e-monitoring, e-diagnosis, e-prognosis, etc."

E-maintenance is a broader concept than CBM. Macchi and Garetti [14] claims that E-maintenance provides a new working context extending the service maintenance to a knowledge-driven organization, where the information flows integrating diverse processes (especially those related with monitoring and CBM), knowledge providers (technicians of the service provider, machinery builder/engineers/technicians, and operators on field), and expert/decision support systems (intelligent systems).

Monitoring, diagnosis and prognosis are the basic concepts of CBM [11], three terms appearing in the above definition of E-maintenance. Thus, it is possible to claim that CBM is a basic element of E-Maintenance.

This CBM concept is here understood as an "extended" CBM, where the classical methods of condition monitoring are completed with the new outcomes of an innovative and emerging discipline: PHM. To underline this evolution from the classical CBM view, different terms have been proposed in the literature to name this new concept: CBM+ [15], CBM/PHM (CBM enable by PHM) [10] or the use of the concept PdM (Predictive Maintenance) with this meaning [16].

Sometimes the borders and differences between terms like E-Maintenance, CBM +, PHM and CBM are not clear enough. Simultaneous reference to so many terms can produce great confusion in future practitioners. Figure 1 tries to organize them

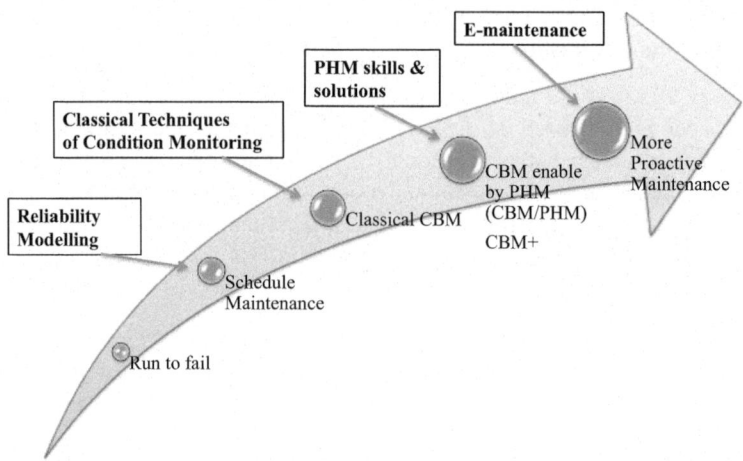

Fig. 1 Positioning CBM with respect to E-maintenance [35]

according to the maintenance types evolution. It shows how CBM evolution is enabled by PHM capabilities; likewise, E-maintenance strategies facilitate a degree of proactivity, supporting greater control and capacity to act on the systems, including efficiency and effectiveness of maintenance plans monitoring [13].

To simplify terminology, in this paper the term CBM should, from now on, be understood as the new global CBM and the E-maintenance strategies its application framework where CBM can provide more value and better results.

Macchi et al. [17] underline this link between E-maintenance and CBM as an evidence of the fact that advanced ICT solutions are being adopted in manufacturing processes in order to progressively change the maintenance management policies. To this concern, automation over available CBM services is crucial to build manufacturing value-driven solutions. This perspective is well synthesized by the conceptual E-maintenance framework provided in Levrat et al. [18], where the role of CBM is acknowledged from a business point of view by its potential to improve services, processes, organization and infrastructure.

Until this moment, the CBM research contributions have been focused on structuring technological issues (models, methods and algorithms) for their application to concrete systems [9], without discussing the specific techniques and methods required to fulfil the envisioned goals in future maintenance processes. It is necessary to understand how to apply CBM techniques and methods conveniently, controlling their implications in a sustainable and efficient way.

The relevance of CBM within E-maintenance justifies the need to analyse the problem of the use, comprehension and applicability of CBM solutions. In fact, there is no CBM/PHM methodological framework covering the management of CBM besides the technical aspects of these solutions. This CBM management approach is the scope of this paper.

The notion "framework", as in standard conceptual computing models Jayaratna [19], used to transmit or address complex issues about some area of knowledge through a generic outline or approach. From a computational context, the reference frameworks serve as templates for the development of specific models and the implementations in a determined scope (ISO15704 2000). In our case, the abstract representation of entities and their relationships will be implemented by UML computer language. Unified Modelling Language (UML) is a well-known graphical language used for specifying, visualizing, constructing and documenting systems. The UML has proven to be successful in the modelling of many large and complex systems [20].

2.2 Complexity Causes and Implementation Challenges of CBM Programs

CBM programs have received several criticisms due to their complexity [21, 22] and to their challenges for practical implementation (see graphic description in Fig. 2):

- Depending on the type of company, the coverage of this program will be more or less complex and so will be the devices required to accomplish this process. It can monitor the entire plant, critical equipment or only their critical functions.

  - The challenge of this cause is to apply the real coverage according to the cost benefit analysis of the program.

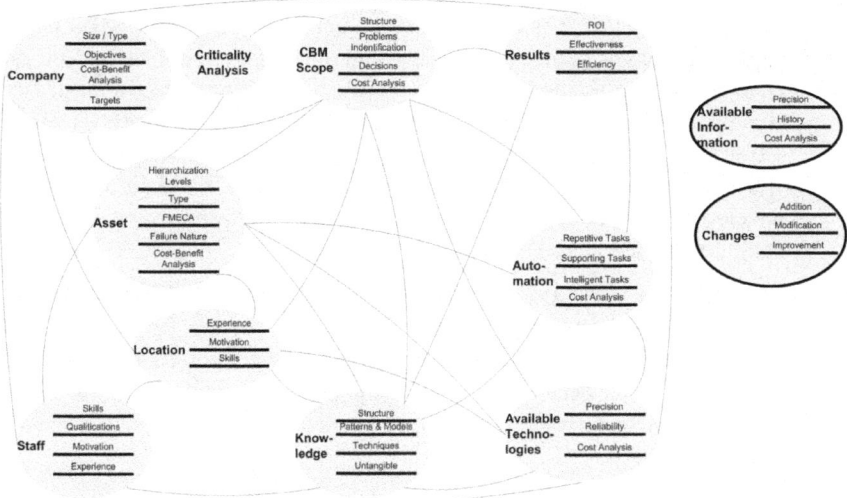

Fig. 2 CBM complexity graph. Entities, attributes and relations

- The system contains a large number of subsystems and components. This case generates a wide variety of maintenance situations that can be handled using different models and methods. Most of the analysis is conducted at a single equipment level, and

  - The challenge is to employ escalating to its surrounding assets of the plant. A generic and scalable prognostic methodology or toolbox does not exist.

- CBM contributes to asset failure reduction thanks to improve the root-cause detection in a short time, providing means for conveying spatial and functional information to operators. The existence of isolated transmission of information and knowledge into islands of specialties or departments normally hampers teamwork and improvements in intergroup activities.

  - The challenge is how to structure the information sustainably and interrelated properly and how to present it in a form which they can assimilate, decreasing mismatches between perceived and real risk, improving rapid decision on critical incidents. That is, defined in a clear and unambiguous way, providing support for decision-making through visual and symbolic representations, simulations and analysis.

- Instrumentation and tools precision, reliability and allocation are crucial frequently unevaluated previously.

  - The challenge is to define these in advance in order to obtain valid information, measurable on homogenous basis, but with the caution of defining for each specific application instead of in a general sense.

- The measurements change according to location, not all measurements are valid in all assets.

  - The related challenge is to define the circumstances in which CBM can be applied, detecting changes that can invalid the measurements. Then, location changes have to be measured continuously. The changeable nature of large technical systems will present constant challenges and many developed CBM programs have been demonstrated in a laboratory environment, but are still without industry validation. All these difficulties highlight the need to develop special computerized systems that can cope with the management of complex engineering systems.

- A CBM program is based on identifying physical changes on equipment conditions, their operation and operation environment. A crucial aspect of this process is to identify equipment patterns triggering warning or alarm messages. The objective is to detect or estimate equipment degradation from normal conditions; consequently to determine the degradation nature and behaviour could be difficult.

  - The challenge is to define the degradation model, whenever possible, in the simplest and most synthetic way easy to use and providing fast feedback.

The models should be according to the problem severity, linking more factors only if the contributions of them are vital.

- Non suitable selection of CBM application according with the problem to solve.

  – The challenge is to have a better problem analysis and identification, providing documented specifications and useful information. These tools may help in reducing the time and resources devoted to decisions to solve repetitive problems instead of singular problems.

- Normally, CBM measurements require specific skills or qualifications, but also the results interpretation. The pattern of degradation depends on the nature of the physical variable and there are diverse international recommendations for each type of variable used: temperatures, pressures, vibrations, amperage, voltage, displacements, humidity, amplitudes, thickness, cracks, the presence of chemicals or particles, etc. There is no expertise regarding modelling and statistical techniques. This expert knowledge can be absent in the company or contracted to an inadequate provider.

  – Then four challenges arise from this, first to verify the technical organizational attitude; second to collect and incorporate expert knowledge (tacit and explicit) to the problem, incorporated after consensus which increases satisfaction and motivation; third to have better analysis consistency with expert knowledge, improving quality and applicability of decisions, in a way that risks decrease, and fourth to train and update the organization in CBM.

- The results should be directly related to the company objectives, they must also adopt financial measures. If this is not adopted at the beginning, managers can lose their faith in the program, and with periodic trends to forsake or reduce the program.

  – The related challenge is to analyse the return-on-investment previously, check the results, visualize the progress and success and share the achievements of the program. It should generate a return on investment that could be between the range from 10:1 to 12:1.

- It is important to determine the items and ultimately the parameters which need to be monitored. This depends on the importance of the item criticality, on the criticality of its failure modes (Failure Mode Effect and Criticality Analysis—FMECA). This information can be obtained by eliciting knowledge from the maintenance staff; of course, this is not a trivial process. The information is often locked away in the heads of domain experts and many times the experts themselves may not be aware of the implicit conceptual models that they use.

  – The challenge is to elicit staff knowledge, drawing out and making explicit all the known knowns, unknown knowns, etc. Another challenge is to develop a more versatile and with a multipurpose staff. Flexible maintenance organizations should be enhanced to facilitate the exchange of knowledge and the teamwork in a confident and motivational environment, avoiding obsolescence and focusing on continuous improvement.

- The patterns have to be reviewed periodically to be properly tuned using the same or different methods and including additional information or knowledge from recent experiences.

  - The related challenge is that CBM programs have to be designed for continuous improvement rather than only monitoring purposes, and also for internal and externally comparison.

- CBM activities execution by hand weighs down the consistency of measurements and analysis.

  - Automation is the challenge. By automation we improve the responsiveness, we reduce complexity, costs and errors in the processes, and also the information is continuously updated, thereby the quality of decision-making increases. Additional challenge is decide the provision of higher levels of intelligence and modelling layers, allowing automatic and fast root cause and weak point analysis.

The complexity of a CBM program is represented in the relationships diagram in Fig. 2. Entities with solid line edges are entities correlated with all the rest of the entities, and they have been drawn separately. This figure shows how, in a CBM program, the concepts have to be considered in a correlational and descriptive way, but not only from a technical point of view, but also from a financial point of view. To minimize the error in the development of a CBM program, to tackle this complexity, the analyst must work at three different levels Russell and Norvig [23]:

  i. In the first level, the objectives and scope of the analysis are defined, delimiting the available technologies and knowledge collection as basis of the next level.
  ii. The second level is related to the expert's knowledge; this will include appropriate criteria for pattern recognition including possible data correlations. The pattern is documented and represented with the intention to make the knowledge explicit and to facilitate simulation and verification. In case of a negative verification, the flow is guided towards an adjustment program where the causes of this negative result are analysed and documented for a first-level refinement. If no pattern is recognized the flow ends.
  iii. The third level chases the generation of new knowledge in the organization based on the verified and explicit pattern, which will be standardized and studied. In the case of potential serious repercussions during the implementation, the flow is redirected towards the first level allowing a modification; otherwise the implementation is carried out searching for the process automation. Automation is developed in proportion to the maximum level of intelligence, as a support system or expert system. Finally, in addition to the

produced prediction and studies, implementation activities could be generated internally or externally as a demanded perfective proposal and depending on the scope of the changes. After the implementation this process has to be reviewed again over time updating information for a sustainable future.

The CBM program requires up-to-date data, information and ultimately knowledge about the assets. The development, management and distribution of assets maintenance knowledge is considered as a foundation for the continuous improvement in CBM program. In consequence, the causes of the complexity of a CBM program can be determined along with the Knowledge Management discipline. Davenport and Prusak [24] define knowledge as "a fluid mix of framed experience, values, contextual information and expert insight that provides a framework for evaluating and incorporating new experiences and information". We have experienced that there is a large amount of dispersed knowledge about assets, which is frequently unprofitable, unknown or inaccessible, and therefore cannot help to process improvement.

2.3 CBM Management Approach

In order to manage the proper adoption of CBM enabling capabilities, each phase of a CBM program life cycle has to be considered: design, implementation and operation; and a continuous review of objectives, on-going and planned activities, and results will introduce changes and new requirement that will modify the CBM plan.

2.3.1 The CBM Management Issue. Principal References

There are references in the literature that somehow try to approximate to the CBM management issue, as ISO 17359:2011 (ISO, 2010), ADS-79D-HDBK [4] or the CBM+ DoD Guide Book [15].

- The ISO 17359:2011 standard in order to establish a condition monitoring program provides a process over the use of the introduced concept of symptom. This standard works with the traditional view of CBM industrial applications, comprising from the equipment analysis to the maintenance action determination point of view.
- ADS-79D-HDBK standard [4] is focused on aircraft CBM applications and provides a very practical approach too, detailing basic concepts and providing practical guides for the application of main measurement techniques for aircraft maintenance. Different processes to manage the implementation of CBM system for new development and legacy system are presented.

- CBM+ DoD Guide Book [15] presents the CBM+ referred above in Sect. 2.1. This is a concept halfway between CBM and E-maintenance. Not only it includes prognosis and diagnosis capabilities and the technological (hardware, software) requirements, but also CBM+ focuses on applying technology that improves maintenance capabilities and business process, complements and enhances reliability analysis efforts, involves the integration of support elements to enable enhanced maintenance-centric logistics system response.

In the case of the first two previous references, the problem of managing the CBM is considered only implicitly, they are focused on the design and/or selection of CBM systems and the choice of CBM versus other types of maintenance. To this aim, both standards introduce the relationship between CBM and RCM (Reliability Centered Maintenance) as a fundamental tool (see Lopez-Campos et al. [25], for an exhaustive revision of the CBM/RCM integration approach), and use the RCM steps (operational context definition, FMEA/FMECA, RCM logic, etc.) as explicit phases in their proposals for CBM design processes.

The CBM+ guide book also treats the RCM as basic tool or reference, but it goes beyond and the CBM management approach is explicitly addressed. Thus, the DoD guide provides a more global vision about the challenges of implementation and operation of CBM systems, showing its real relevance and the need for a specific management of the complete life cycle of CBM solutions.

2.3.2 Linking CBM Management with CBM Data Processing. Understanding the CBM Basic Flow from the Maintenance View

In contrast to the CBM Management approach, technical aspects of CBM have been well studied and characterized in the literature in order to aid the exchange of data in an integrated way, from the devices to the technologies, and together with processes management [26]. Some standards have been developed in order to be used as design guidelines about technical information of CBM solutions and its interoperability from different levels of management (Data processing solutions).

As a result, there are models like OSA-CBM [27] to implement the data processing that allows development complex CBM systems and software [3]. Definitely, these models are leading an important role in the development of CBM, especially as regards the integration between CBM systems with other systems within the organization. The integration and interoperability of systems is a fundamental aspect of the E-Maintenance strategies discussed in the previous point.

Data provides the essential core of CBM, so it is understandable that standards and decisions regarding data and their collection, transmission, storage and processing have dominated until now the requirements for CBM systems development [4]. However, this is not an optimal approach. Sometimes, technical issues hide the most important thing: CBM is a maintenance activity that must be aligned to business maintenance objectives. Figure 3 tries to depict this process introducing three complementary points of views of this same process simultaneously:

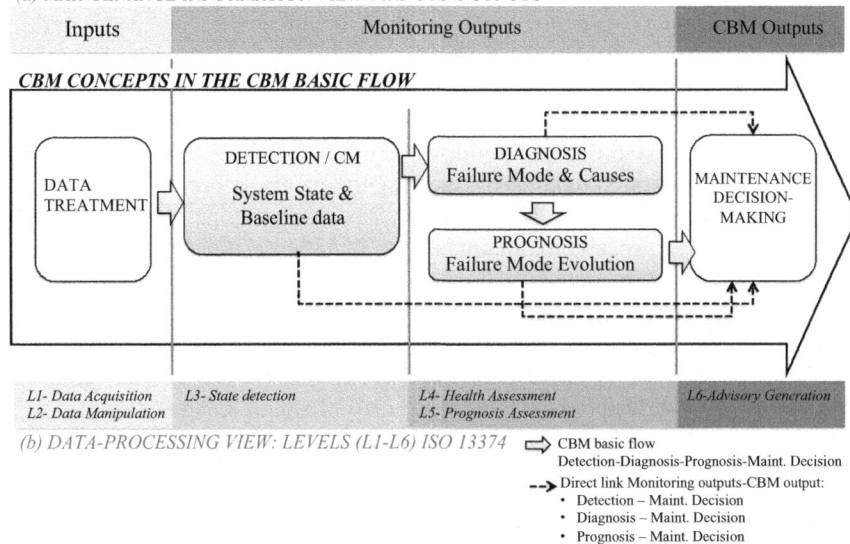

Fig. 3 CBM basic flow and complementary views for its interpretation: **a** maintenance information, **b** data-processing levels (ISO 13374)

- *CBM basic concepts* (*detection, diagnosis, prognosis*) *within the basic CBM flow*. This concept is reinterpreted using two views:
- *The Data-processing view*: CBM flow and concepts reinterpretation within the Data-Processing technical requirements.
- *The Maintenance information view*: maintenance requirements translation.

Managing the potential of CBM implies understanding the concepts detection, diagnosis and prognosis and how to manage them. They are closely connected concepts and sometimes it is difficult distinguish them, due to the lack of a unique and standardized accepted vocabulary within the technical community [10]. Here is presented an interpretation of these concepts from the maintenance function perspective:

- *Detection* is associated with the system states (for example the transition from function state to fault state) and, in general, with normal behaviour-anomalies distinction (in reference to defined baseline data).
- *Diagnosis* is associated with the location of the failure mode and its causes.
- *Prognosis* is associated with the evolution of the failure mode or its future behaviour (risk of failure and remaining useful life in a moment).

They are not independent concepts. In order to understand the connection between them it is also possible to interpret these three terms as sequential phases of a basic flow. For example: prognosis algorithm is triggered by an independent diagnostic algorithm whenever it detects a fault in the system with high certainty probability [28]. This basic flow is the central block in Fig. 3.

The failure mode is the key concept for maintenance management [29]. This definition of diagnosis and prognosis in relation with the failure mode concept makes it clear that the failure mode is the key element of CBM, i.e. the objective of CBM is to control the failure modes.

Regarding the *Data processing view,* the scheme in Fig. 3 is based on the six *layers of functionality* in a condition monitoring system (L1–L6 in Fig. 2) provided by ISO 13374-1:2003 "Condition monitoring and diagnostics of machines—Data processing, communication, and presentation" [30]. This standard (one of the main references) describes the specific requirements for CM&D open-software application, detailing both information model and processing architecture requirements. Its concepts and guidelines are assumed by OSA-CBM, standard that provide processing architecture specification [27] which has been adopted as a reference in multiple application approaches to build CBM systems [4, 10, 15]. OSA-CBM defines an object-oriented data model (defined using Unified Modeling Language, UML) over the provided by ISO 13374-1:2003. There are other proposed models (for example, AI-ESTATE or IEEE 1451.2) but OSA-CBM is one of the most popular.

The *Maintenance Information View* is related to inputs and outputs of the maintenance information system. The inputs are the data (basic data or manipulated data) provide by the different information sources. In Fig. 3, two different *types of outputs* have been distinguished during the process: CBM Outputs and Monitoring Outputs. Monitoring outputs are the basic information to get the maintenance decision-making which is the real CBM output (what, when and how it is necessary to inspect, to repair, to replace, etc.). Thus, the interpretation of Fig. 3 is: in a CBM process, there can be only three possible types of monitoring outputs: Detection, Diagnostic and/or Prognosis; then the interpretation of monitoring outputs drives maintenance decisions (CBM output). Every single-monitoring output can have maintenance interpretations and may support specific maintenance decisions (dashed lines in Fig. 3).

2.3.3 CBM Life Cycle Phases and CBM Management Pillars

Therefore, the needs for CBM programs management actually go beyond simple design guides or application guides, towards a broader approach that includes the two following aspects:

- Integrating both perspectives in the CBM conception: data processing and CBM management,
- Considering CBM life cycle phases: *the design phase and the use phase.*

In order to understand the size of the problem, it is also necessary to consider that the number of data over time, the number of changes of operational modes, decisions taken, re-adaptations of CBM programs, etc., could grow exponentially during the asset life. To show this idea, shaded areas in Fig. 4 represents the magnitude of the problem complexity in the deign phase and the use phase. Therefore, knowledge management becomes another issue to manage. As a result,

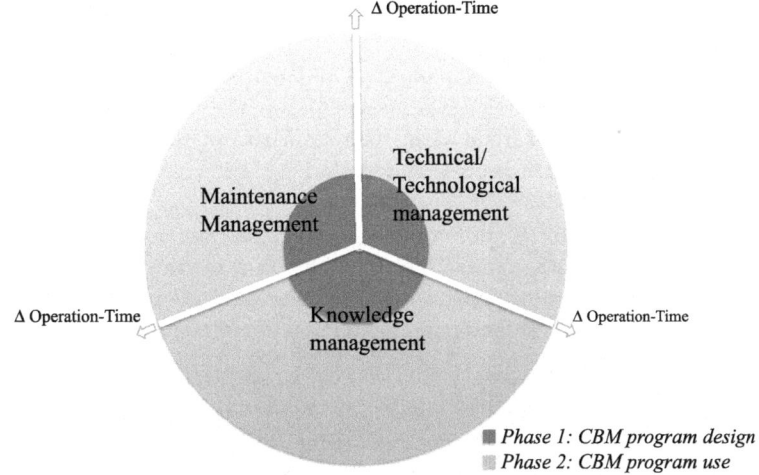

Fig. 4 Phases of CBM management and pillars of CBM management activities

within the life phases of a CBM program, we can find three basic pillars of CBM management activities (also depicted in Fig. 4):

- *Maintenance management:* integration of CBM within the rest of maintenance policies applied over an asset, and according with the available resources and the asset requirements. A reference for this management pillar is the RCM/CBM integration and similar approaches.
- *Technical/Technological management*: supporting of CBM implementation relating this to hardware/software needs. The reference for this pillar is the standard ISO 13374-1:2003 and also OSA-CBM.
- *Knowledge management:* generation and administration of company's knowledge about the asset behaviour and problems (failures, anomalies, etc.), and how they appear and they can be observed. It is materialized with the symptom analysis and description. The reference is the ISO 13379-1:2012. The formal symptom treatment allows standard documentation of the failure modes to be under control using monitoring and the CBM solutions.

3   A Framework to Support CBM Management

3.1   Objectives of the Proposed Framework

In previous sections, we have claimed that the CBM programs can be extremely complex to manage, because they will handle massive information, changing on time, and with complex relationships among them. The maintenance manager needs to cope with all this complexity in an orderly manner previously to implement a CBM solution in a certain software.

Despite the fact that there can be great differences between types of CBM solutions, they can be represented and managed using the same structure. We have depicted this structure in a framework, with main aim of providing an access to CBM knowledge with consistency and uniformity allowing an effective CBM management.

Thus, the central idea of this work is to facilitate the characterization and treatment of all key points of the CBM solutions. In response to the CBM application complexity and its challenges, described above in this paper, the particular objectives of this framework are:

- *The integrated treatment of detection, diagnosis, and prognosis. The maintainer can manage the three types of monitoring outputs simultaneously and in uniform way.* It is needed to understand: (i) the differences between each term, (ii) how they are related between each other and, (iii) what maintenance actions or benefits can be related with each of them.
- *The correct interpretation of the monitoring techniques and their results.* This key element provides enough control over the crucial monitoring and data processing issues. For instance, this allows the performance of the solution to be measured in terms of fulfilling the maintenance goals.
- *The integrated treatment of different possible CBM solutions and different information sources.* Sometimes, different solutions can share the same technical monitoring tools (hardware and software). Other times, more than one technique is needed to observe a unique failure, in order to reach its right interpretation. Technical and human resources must be optimized for this.
- *The integration of CBM results with the rest of maintenance types and strategies (maintenance concept of the company).* This element includes the connection of CBM with the failure mode concept (FMEA process) and the choice of applying CBM instead of any other maintenance possibility (RCM logic). It comprises the approach of CBM-RCM integration [25]. This point also orients the CBM management to the integration within E-maintenance strategies.
- *The definition of a set of groups or blocks of conceptual element, that can then be modelled and easily implemented by software systems.* The design by independent blocks allows decoupling the analysis of different key aspects of a CBM solution, providing a holistic understanding of the problem. Subsequently the framework gives support to the orderly integration of these blocks in an optimized solution and aligned with the maintenance objectives. In order to highlight the orientation to software system implementation, the framework has been depicted using an UML schema.

3.2 Introduction to the Framework

In this work we have identified different blocks for a CBM solution considering that

- Each block introduces a specific perspective or technical area that should be considered for a CBM solution,
- Each block demands specific knowledge and skills and also specific tasks and,
- Each block produces specific results that can be managed and recorded.

The identification of different blocks is fundamental for a formal treatment of CBM solutions, providing a suitable way for interaction between the different disciplines that will have to collaborate in the design and implementation of them.

The proposed structure comprises five blocks (Fig. 5). The five blocks are consistent with the standards that have been included in each block definition (Table 1), and represents different knowledge matters whose information could be correlated in a structure framework. These blocks traditionally are supported by isolated software systems to administrate their information.

We could have chosen other blocks. Using this structure we try to avoid that some important issues will be hidden by others aspects. This happens, for example, if the design of a CBM solution is focused on the monitoring technique acquisition without a previous formal analysis from maintenance management view. In the same way, we introduce the blocks 1 and 2 instead of using a single block in order to give special relevance to the formal definition of the assets. Our experience has shown that this is a lack in many organizations, which can produce important inefficiencies, among other things, it can hinder the information exchanges between different software applications causing reworks and extra costs.

The blocks that we have identified include different abstraction levels. At first glance, it may seem difficult to combine them. But the fact is that all CBM solution involves all these issues and different abstraction levels have to be managed and this is the challenge. Besides we have to think in E-maintenance scenarios, where hundred or thousand CBM solutions will be applied within a complex engineering systems. Thus, formal approaches as it is proposed with this framework are needed.

Fig. 5 Blocks in the proposed framework

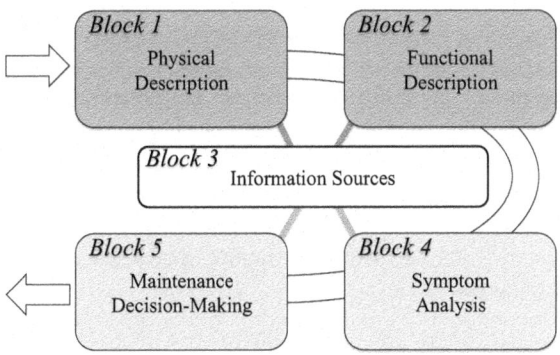

Table 1 Summary of the blocks in the framework for CBM management

Block	Elements	Objective	References: methods & standards
B1	– System – Equipment – Maintainable item	Physical description from system to indenture level	ISO 14224
B2	– Function – Functional failure – Failure mode	Functional description and failure mode definition	RCM/FMEA/FMECA, ISO 17359, IEC 60812
B3	– Sensor – Measurement technique – System variable – Monitoring variable	Information source management and technical resources	ISO 13374, OSA-CBM, ISO 17359
B4	– Symptom – Descriptor – Interpretation rule	Symptom description. Descriptors definition	FMSA, ISO 13379, ADS-79D-HDBK
B5	– Detection – Diagnosis – Prognosis – Maintenance decision	Monitoring outputs and decisions about maintenance (CBM outputs)	ISO 13374, OSA-CBM, ISO 13381, ISO 17359

This approach introduces a formal and standard treatment that allows

- Overall treatment of the solution, assuring that all relevant aspects are considered, avoiding mistakes and low performances.
- Suitable knowledge management regarding CBM applictions.
- *Scalability* and *replicability* of CBM applications.

Four of the blocks are corresponded with maintenance processes (blocks 1, 2, 4 and 5) and one additional block dealing all the information resources (block 3) according to the indicated previously Russell and Norvig [23] ideas.

The sequence of blocks can be interpreted as basic process to analyse any CBM solution. The first two blocks are not only specific to CBM solutions, but they are essentials of any maintenance management application. In the proposed structure, the explicit inclusion of physical and functional elements is mainly due to two reasons: first, it is necessary to connect the CBM with RCM view, as mentioned in previous section; second, the definition of the rest of the elements of the structure depends on the way that failure modes have been defined as a result of these previous blocks.

Table 1 summarizes the blocks elements and the references to deal the concepts of everyone.

From now on, the purposes of all the five blocks are introduced in relation to their functionality inside the framework.

3.3 Blocks and Components of the Framework

From a computational point of view, the elements of the framework and their relationships are represented in a general UML schema (Fig. 6), where for a better understanding of the descriptions, we address the reader to the example in Sect. 4.

3.3.1 Block 1. Physical Description

Physical structure is the most intuitive way to observe the system reality. In the physical structure, the system belongs to a plant, an installation, an industry, etc. and has different subsystems and components that can be physically distinguished. In the structure proposed, this block has been defined hierarchically according with of the ISO 14224:2006 general standard taxonomy [31] by:

- Element 1.1: *System.*
- Element 1.2: *Equipment unit.*
- Element 1.3: *Subunit.*
- Element 1.4: *Maintainable item.*

3.3.2 Block 2. Functional Description

Failure has to be used to indicate functional disorders of elements, of the whole experience surrounding their performance. Information about functional disorders is part of the aim of FMEA/FMECA analysis [5], and has to be obtained by eliciting knowledge from the maintenance staff, but this is not a trivial process. The information is often locked away in the heads of domain experts and many times the experts themselves may not be aware of the implicit conceptual models that they use. Eliciting knowledge consist in drawing out and making explicit all the known knowns, unknown knowns, etc. This block can be summarized in the hierarchical structure through the definition of the following elements:

- Element 2.1: *Function.* The action and activity assigned to, required from or expected from a system [2]. In fact, what the system user wants it to do [29]. The accurate definition of the functions includes the functioning standards determination.

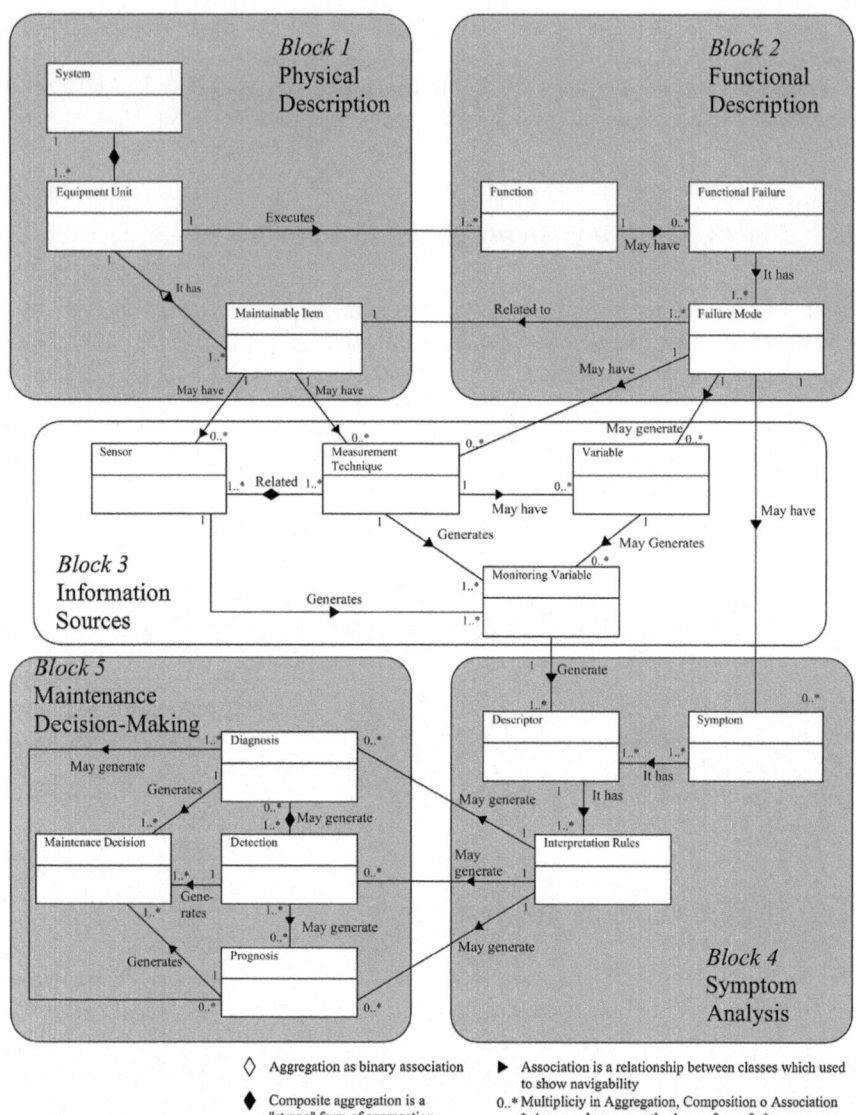

Fig. 6 Basic structure for the CBM solution presented in an UML diagram

- Element 2.2: *Functional failure*: the way in which a system is unable to fulfill a function at the performance standard that is acceptable for the user [29].
- Element 2.3: *Failure mode*: event that is reasonably likely to cause each functional failure. It is also defined as the effect by which a failure is observed [2].

Maintenance management focuses on failure mode in order to preserve system functions [32]. The failure mode is the key concept for maintenance management [29], and its definition is the goal of this block and as central item of CBM process. There can be different ways to define the failure modes: Moubray [29], ISO [31], OREDA [33] or Parra and Crespo [5].

In conclusion, the main output of this block is the failure mode determination. And this step will have great impact on the rest of elements in the following block. Jointly with the symptom (Block 4) is the key component of this analysis structure.

Accordingly, with the ISO14224 standard and RCM methodology, the relation among blocks 1 and 2 are depicted linking the physical structure elements and functional logic elements:

- *Equipment unit* is related to *Function*: *functions* are defined at *Equipment unit* level; an *equipment unit* will have one or more *functions*.
- *Maintainable Item* is related to *Failure Mode*: *failure mode* is defined for a *Maintainable Item*; a *Maintainable Item* will have one or more *failure modes*.

3.3.3 Block 3. Information Sources

One of the main aspects for the maintenance manager control is where the information comes from. Actually, this is the core of the Data-Information process, introduced in previous Sect. 2.3.2. CBM program will have to manage the presence of a great number of different information sources in an industrial system: sensors, software systems (control, operation and maintenance), monitoring devices and techniques, data bases, data warehouses, etc. [26]. The integration of all these information sources is a very complex task, and it is critical for the CBM aims, ensuring the reliability of the information and controlling the performance of the monitoring system [10].

The proposed structure includes a revision and an interpretation of the different terms that can be used to treat the information. This block provides a model to organize and interconnect the different types of information available that will be used in symptoms treatment (Block 5) and considering the separate analysis of three crucial aspects:

- The physical support (sensor determination and location).
- The knowledge for the interpretation of the information (measurement techniques application).

• The relationship with the rest of the organization performance metrics (system variables).

Dealing with ISO 13374 and ISO 17359 jointly with OSA-CBM, the structure defines hierarchically the following concepts in this block:

– Element 3.1: *Sensor*. The "sensor" term is related to the physic measurement process and its communication. A sensor generates a signal and that signal has to be processed and transmitted. The sensors and signals management are related to physical design of the data collection and communication process [10]. In this element it is possible to define the physical characters of the sensor and its location within the system [26]. The location of the sensor can be interpreted as related to a *maintainable item*. The information gathered from a *sensor* can feed one or more *measurement techniques*.
– Element 3.2: *Measurement technique*. It is referred to the technical knowledge and the necessary equipment to observe a particular phenomenon. Techniques as thermography, vibrations or ultrasound analysis have been broadly applied during last decades. Instead of using a simple sensor, these techniques give information that allows to analyse and to interpret the behaviour of an asset. It is possible to refer to them in general with the expression "measurement techniques". There are much more techniques that can be classified as *measurement techniques* [8]. Traditionally, the *measurement techniques* are introduced through periodic inspections, although recently in line with the future factory or the Industry 4.0 models, these measurement techniques are programed in automatic application. The outputs of *measurement techniques* can be included in one or more *system variables*.
– Element 3.3: *System Variable*. This element includes any variable presented in any database related to the system that can model behaviour of good or bad performance of the system. For example, additional information to the obtained variables by measurement techniques: operational variables (from the SCADA), maintenance variables (from the CMMS), economic variables (from ERP), etc.
– Element 3.4: *Monitoring Variable*. The *monitoring variable* list will include the variables that actually are going to be used in the CBM solution. As a result of below elements definition this term includes: (i) variables result of the processing of signals (from *sensors*), (ii) outputs of *measurement techniques* analysis expressed as variables; (ii) the *System Variables* that are used by CBM solutions.

Processed signals from sensors and processed variables from measurement techniques can produce one or more monitoring variables.

3.3.4 Block 4—Symptoms Analysis

The symptoms, how they are managed and interpreted in relation to the failure mode, are the key point of the approach here presented, disaggregating the general concept of symptom into three elements: the symptom, the descriptor and the interpretation rule. Before to describe these elements in details, it is necessary a previous review of the symptom concept.

According to the definition of the ISO 13372 standard, a symptom is the "perception, made by means of human observations and measurements (descriptors), which may indicate the presence of one or more faults with a certain probability" [2]. Thus, a symptom implies that something happen in the system (presence of fault) and that we have the capacity to observe, or measure, some evidence of it (perception). The ISO 13372 definitions use the term "fault". According to the definitions included in Block 2, here the fault concept has been exchanged by the failure mode concept, aggregating author opinions that consider this a more accurate concept. For example, the symptoms can give information of states before the failure (when failure mode is evolving), that allows more detailed interpretations with great relevance in most of CBM applications, while a fault is the state after the failure CEN [1]. On the other hand, the concept of perception is related with terms like sensor, variable, measurement technique, information source, etc.

In addition, this standard introduces the concept of "descriptor", as "feature, data item derived from raw or processed parameters or external observation" [2]. Finally, the measure provided by a descriptor has to be interpreted in order to represent information about the failure mode. So the use of each descriptor implies the definition of *interpretation rules*.

This block is composed by the following elements:

- Element 4.1: *Symptom*. A qualitative description of specific effects or causes that can be measured giving information about the *failure mode*. One *failure mode* can have one or more *symptoms*. On the other hand, a *symptom* can be related to more than one *failure mode*. In the proposed structure if a symptom is related to, for example, two different failure modes, the symptom appears twice in the structure, once with a code that associate the symptom with the first failure mode and other time with a different code related to the second failure mode. It is crucial, that descriptors definition and interpretation rules will be detailed for every single coded symptom-failure mode. A *symptom* will have at least one *descriptor*.
- Element 4.2: *Descriptor*. It is the feature or the specific measurement parameter that actually provides the monitoring of the symptom. A descriptor is related with one coded symptom and one symptom can have one or more descriptors. From the symptom definition, descriptors are measures, so an accurate definition is needed including all the characters of a measure: magnitude, precision, measure frequency, etc. [26]. The difference between descriptor and variable or parameter is that the descriptor is related with a specific failure mode and produce one or more interpretation rules. Descriptor, in this sense, is also referred in the literature as Condition Indicator (CI) [4, 28] or as a feature [10]. Another term sometimes used is the Health Indicators (HI). It is different from CI. HIs are indicators of maintenance action based on the value of one or more

CIs [4]. CI is closer to monitoring while HI to interpretation. Descriptors can be developed recurrently towards sustainable evolution and accuracy in the CBM solutions.

– Element 4.3: *Interpretation rule:* It is the description of how the descriptor values have to be interpreted or treated in order to get the monitoring outputs (detection, diagnosis, prognosis) for a failure mode. A unique descriptor can have more than one interpretation rules, so it has to be considered the necessity of detailing these rules accurately. For example, when the system presents two different possible operational standards, the same values of the same descriptor can be interpreted as normal behaviour on one standard and as failure evidence in the other. For example, the pick power consumed during the engine start is not a failure event, while the same value during regime functioning could be a failure. This can be interpreted as one descriptor with two interpretation rules.

Therefore, in the context of a CBM solution, it has no sense to understand a symptom without, at least, one descriptor and an interpretation rule. They compose a unique entity for the interpretation of the monitoring of a failure mode. However, treating them separately, great advantages can be introduced:

- The use of the *symptom* for introducing maintenance expert knowledge, without giving necessarily details of measures and data process details. Details will be included after with the *descriptor* and *interpretation rule*.
- The possibility to easily improve or adapt the monitoring solutions. Sometimes, improve the solution only require, for example, changes of interpretation rule.
- Introduce new monitoring solutions: by the introduction of new descriptors or new interpretation rules it is possible to obtain new solutions.
- A better understanding of monitoring solutions (detection, diagnosis and/or prognosis), concentring the elements of this block in each of them but relating the knowledge in a sustainable way from detection, to diagnosis and prognosis (see Fig. 3).

The *interpretation rule* element treatment can be extended in a recurrent way, that is, an interpretation rule can be based in one or others interpretation rule. In order to not complicate the paper and to focus it on the element that connects failure mode with information generation, the interpretation rule has not been divided into different elements as in some references. These references include in it other elements in order to concrete the interpretation of the information obtained, such as Health Index, Uncertainty Measurement and evaluation parameter of the performance of the monitoring objectives (detection, diagnosis and prognosis). Interesting references to the use of HI and Performance Metrics can be found by the reader in United States Army [4] and Saxena et al. [28].

Accordingly with the FMSA methodology, ISO 13379 and ADS-79D-HDBK standards, the relation among blocks with this block 4 are depicted next:

- The failure mode element is linked to symptoms as qualitative description of the latter.

- The link to the third block, allows the understanding of the symptom traceability and its descriptors, connected to the maintainable item, or vice versa, the monitoring variables are connected with the *symptom* through the descriptor definition. Then, any variable that is be used by the descriptors have to be distinguished as a monitoring variable.

3.3.5   Block 5—Maintenance Decision-Making

This block supports the two different types of outputs of the CBM process that were defined in Sect. 2 (see Fig. 3): monitoring outputs and CBM outputs.

Considering that it is possible to count and register events of the three types of monitoring outputs respectively (detection, diagnosis and prognosis), it makes sense to treat them as elements. For example, once it is established the *descriptor* and *interpretation rule* to obtain a specific *failure mode detection*, it is possible to register every *detection* event and control the performance of this solution [28] with a performance rate "failure mode event/right detections". Similar consideration can be made with *failure mode diagnosis* and *failure mode prognosis.*

In addition, it is important to understand that detection, diagnosis and prognosis are linked to different maintenance decisions. The maintenance decision depends on the objectives of maintenance. Two illustrative examples can be presented to show the differences between the maintenance decision linked with prognosis or detection:

- In the first example, the maintenance objective is to extend the replacement period of equipment. We will use a prognosis solution since prognosis can provide a measure of the remaining useful life of a critical failure mode of this equipment, thus we would able to programme the replacement of the equipment, from now until the estimated end of useful life.
- In the second example, consider a failure mode where prognosis is not possible (it can happen when the failure mechanism is very fast) and it is critical because it causes the stop of the production of an important part within an industrial plan. In this case the maintenance objective is the maximum reduction of the down time. A detection solution can produce good results within this context, since provides information of the exact moment of the failure, allowing the maintenance resources to be promptly mobilized.

- Element 5.1: *Detection element.* It focuses on the state of the machine or system. It enables to distinguish anomalous behaviours, comparing gathered data against baseline parameters (ISO 13379), detecting and reporting abnormal events. This is defined as CM (Condition Monitoring) by ISO 13379-1:2012 [34] and SD (State Detection) by ISO 13374-1:2003 [30]. Alerts and alarms management also related with the detection issue [8]. Then, as in the example of the previous paragraph, one principal detection objective is to achieve the best possible performance by minimizing false positives and false negatives cases [10].

- Element 5.2: *Diagnosis element.* The ISO 13372 defines diagnosis as the result of diagnosis process and as determination of the nature of failure [2]. In this sense, within a complex system, this definition can be completed considering two different stages in diagnosis process: isolation, determining which component or more accurately which failure mode is affected; and identification, determining or estimating the nature (or causes) and the extent (size and time) of the failures or faults. It is also relevant to note that diagnosis can be done before the failure (the failure mode is evolving but the failure has not happened yet) and after the failure [35]. It is focused on the failure modes and its causes.
- Element 5.3: *Prognosis element.* Prognosis is focused on failure mode evolution. The estimation of future behaviour of the defined failure mode allows failure risk assessment and control. There are different types of outputs from various prognosis algorithms. Some algorithms assess Health Index (HI) or Probability of Failure (PoF) at any given point and others as Remaining Useful Life (RUL) based on a predetermined Failure Threshold (FT) [28].
- Element 5.4: *Maintenance decision.* With this element the CBM outputs are described. The maintenance tasks and general actions that are triggered as consequences of the monitoring outputs can be registered, listed and catalogued to be used within a standard process. They can be connected with the respective monitoring objectives, controlling their implementation balancing cost and performance of them. The knowledge about the decision that a maintainer can take has great value. Actually all the CBM process is founded over this knowledge.

3.4 Basic Structure. UML Diagram

For a more detailed representation of the proposed structure, a being coherent with the CBM-OSA indications [27, 30], an UML diagram is used (Fig. 6), showing a comprehensive view of the structure, elements and blocks, and the different relationships that can be establish among them within CBM solutions [25].

Figure 6 has been developed based on UML Class Diagram Notation. In this figure, classes of entities are represented by rectangular forms where the name of the entity is indicated in the top. A class describes a set of objects that share the same features, constraints and semantics (meaning). The classes can be associated, searching to link certain classes in order to include valuable information related to each other, and it is indicated by a solid line from one class to other (i.e. symptom and descriptor). If there is a solid triangle on the solid line, this indicates navigability in the association instead of bi-directionality, indicating the order to be read the association from the first end to the last end. The association can include two additional information: one expression over the triangle which describes the specific role a class plays in an association and, in both ends the multiplicity notation which indicates the number of instances of a class related to one instance of the other class (i.e. 0..*—zero to any number).

Besides, there are several types of associations among classes, aggregation, composition and reflexive associations. Aggregation is a type of association decorated with an unfilled diamond shape to indicate that one class is a part of another class. Composition is a type of aggregation decorated with a filled black diamond to describe a "strong" dependency of child class with parent class. Reflexive associations indicate that a class instance can also be associated with itself, to another instance of the same class. Example of reflexive associations is descriptor element because it is possible to use a defined descriptor to build another descriptor.

4 The CBM Program Management Template. A Use Case

In this section, the structure of our framework is now translated into a practical business management template, using a table format searching a practical point of view. In order to do so, the typical RCM output table has been adopted as basis, complementing and extending this with the necessary information derived of our framework for a clear description of CBM programs, for management purposes.

In order to illustrate this point, a practical case of the template will be presented, for a CBM program, applied over failure modes of an industrial power transformer. The practical case that has been proposed belongs to a research carried out, in the context of a wider project, about power distribution systems dependability. This study was performed, applying the RCM methodology to a substation of electric power distribution network, and it is partially included in Crespo et al. [36]. In this case, the monitoring objective is the prognosis of a certain failure mode.

Maintenance activities minimization, or their proper execution in time, can increase assets durability and reduce considerably network technical deterioration. With this in mind, we pretend to generate suitable and online estimations of risk of a given failure mode [36]. This is a classical output of prognosis methods [28], where the risk of the failure is the prognosis result and the subsequent decisions is the CBM output, either to do nothing or to release the preventive task.

The table template has been introduced by parts according to the block sequence of the framework.

4.1 Block 1 and Block 2

According to block 1 elements, in the physical description *System* element is the Substation and the Power Transformer is the *Equipment Unit*. The following table summarizes the main results of this functional/physical description of the transformer. In order to simplify the case, due to the wide size of the table, *functional failure* elements have not been included in Table 2.

Table 2 Failure modes of the "power transformer" equipment unit in a substation of power distribution network

Cod	Equipment unit	Cod	Maintainable item	Cod	Failure mode
1	Power transformer	1.1	Core	1.1.1	Short circuit
		1.1	Core	1.1.2	Circuit opening
		1.1	Core	1.1.3	Internal bypass
		1.1	Core	1.1.4	Short circuit between sheets of the core
		1.2	Insulating	1.2.1	Dielectric loss
		1.2	Insulating	1.2.2	Reduction of insulating standard properties
		1.2	Insulating	1.2.3	Insulating level loss
		1.3	Terminal	1.3.1	External defect
		1.3	Terminal	1.3.2	Internal defect
		1.3	Terminal	1.3.3	Bad connection
		1.4	Refrigeration	1.4.1	Lack of outflow
		1.5	Support bench	1.5.1	Support bench collapse
		1.5	Support bench	1.5.2	Transformer movement
		1.6	Chassis tank	1.6.1	Loss of tightness

Referring to Table 2, the failure mode 1.4.1 "Refrigeration-Lack of outflow" has been chosen, to continue the development due to present the higher failure rates in the system and because it produces evaluated consequences as "medium severity".

4.2   Block 3 and 4

Just like in FMEA or RCM analysis, meetings of a group of experts are necessary in order to obtain the knowledge about symptoms of this failure mode. For the symptoms analysis, the expert group has to be compound by specialist of different matters: technologies, operation, corrective/preventive maintenance, and obviously CBM technicians. Any of them can propose system variables which could be on the list of candidates for monitoring variables. About our selected failure mode "Lack of outflow", the team of experts decided that the symptom was the "oil temperature referred to the transformer load". Then, concerning transformer reliability, this failure mode depends on the ratio oil temperature versus load intensity (Table 3).

When this case was elaborated, the only information source was the substation SCADA, while valuable tribology and/or thermography techniques [8] about the failure mode "1.4.1" were not available. So a relevant constraint of the project was

Table 3 Symptom analysis of failure mode 1.4.1 "refrigeration-lack of outflow"

Cod	Failure mode	Cod	Symptom
1.4.1	Lack of outflow	1.4.1.1	Relation of the oil temperature and the current output in the transformer

Table 4 Information sources available: SCADA variables

Cod	System variables
OV1	Upper oil layer temperature (°C)
OV2	Lower oil layer temperature (°C)
OV3	Air temperature (°C)
OV4	Humidity (%)
OV5	Hydrogen (%)
OV6	Load current (intensity)
OV7	Service voltage
OV8	CO (%)
OV9	Oil level
OV10	Lack of fans' feed

Table 5 Monitoring variables obtained from information sources analysis

Cod	Symptom	Information source	Cod	Monitoring variable
1.4.1.1	Relation of the oil temperature and the current output in the transformer	System variables (SCADA)	OV1	Upper oil layer temperature (°C)
			OV2	Lower oil layer temperature (°C)
			OV6	Load current (intensity)
		System variables (CMMS)	MV1	Number of maintenance interventions

to use only the SCADA operational variables for CBM. These operational variables (OV) become *System Variables* within our framework (Table 4).

In Table 4, variables OV1, OV2 and OV6 are the only ones pre-selected in relation to the failure mode under analysis, besides the number of maintenance interventions at a given moment (taken from the CMMS of the company) which could show degradation of the system due to accumulated contribution of bad-executed repairs. The result is the list of monitoring variables included in Table 5.

Due to detection and diagnosis based on measurement techniques were not available, only prognosis analysis were developed as risk estimation for the selected failure mode. Then, the experts search to deduce the transformer reliability at a specified time according to representative stressors for the failure mode. The representative stressors that can reduce transformer reliability are the selected monitoring variable. The value of these variables, besides the number of maintenance interventions at that given moment (taken from the CMMS of the company) could be employed as inputs of the predictive algorithm.

Based on this, only is derived on descriptor: D1, transformer reliability proportional to monitoring variables contribution. For this purpose, a variant of the Cox's proportional hazard model (PHM, published in 1972) is employed. For a detailed description of the used reliability algorithm (see Eqs. 1 and 3), the reader is referred to Crespo et al. [36].

$$R(t,x) = R_o(t)^{e^{-[X(t)]}} = R_o(t)^{e^{\left[-\sum_{j=1}^{k} \gamma_j x_j\right]}}, \tag{1}$$

where

$$R_o(t) = e^{-\lambda t q\left(\frac{-\ln(n+1)}{\ln 2}\right)} \tag{2}$$

Thus,

$$R(t,x) = e^{-\lambda t q\left(\frac{-\ln(n+1)}{\ln 2}\right)^{e^{\left[-\sum_{j=1}^{k} \gamma_j x_j\right]}}} \tag{3}$$

Equations (1 and 3) represent the impact of different monitored parameters on reliability. Here x_i (with $i = 1 \ldots k$) are the representative covariates that contribute to reduce the system reliability, and γ_i are constant coefficients (with $i = 1 \ldots k$) representing each x_i weighted contribution. Equation (2) allows the incorporation of the impact of the number (n) of corrective maintenance activities carried out and their execution quality (q, which is measured as a percentage). For the considered failure mode 1.4.4, experts decide by consensus that $q = 0.9$ over two previous repairs ($n = 2$) according to the historic files presented by the team.

Consequently, D1 is calculated by the reliability function $R(t$, OV1, OV2, OV6, MV1). Although, according to expert group experience, the proportional contribution of the monitoring variables over the reliability is not directly shown, but based on relationships among them. Then, two covariates were considered suitable in order to inform about the presence of the potential failure: x_1 y x_2. Both calculated by the relationship between temperature and current load.

$$x_1(t) = \frac{OV1(t)}{OV6(t)} \text{ and } x_2(t) = \frac{OV2(t)}{OV6(t)} \tag{4}$$

Whenever this ratio increases over time, the failure mode will have a greater risk to show up. The coefficients γ_i are finally two and so the proportional contribution is derived in the function $x(t)$ which has the form of a weighted sum with respective both γ_i equal to 1.

$$x(t) = \gamma_1 x_1(t) + \gamma_2 x_2(t) = 1 \cdot \frac{OV1(t)}{OV6(t)} + 1 \cdot \frac{OV2(t)}{OV6(t)} \qquad (5)$$

As a result, the D1 descriptor is the Eq. (3), producing the value of reliability of the failure mode $R(t)$, proportionally affected by values of x_1 y x_2 raising $R_o(t)$ to $x(t)$, and also reducing $R(t)$ as progressive degradation due to the lack of repair quality $(1 - q)$ by number of maintenance activities carried out $n = $ MV1. As example, Fig. 7 show the evolution of $R(t)$ versus $R_o(t)$ depending on changes in the OV1, OV2, OV6 system variables. $R_o(t)$ is the nominal reliability based only in the number of failures occurrences and the repair quality, following an exponential behaviour. The resulting $R(t)$, including the covariates to $R_o(t)$, shows a scenario where load current intensity would decrease (OV6) while upper and lower layer oil temperature (OV1 and OV2) would increase, which indicates the high reduction in the reliability probability.

Then, the next step is the definition of the interpretation rule, which was agreed by the review team as follows: I1 whenever the condition D1 $\leq$ 0.2 (20%), the risk is very high and a maintenance activity should be detonated in order to avoid or reduce the failure mode consequences with high frequency of occurrence and "medium severity" in each one. Additional descriptor was defined in order to estimate the risk to the next failure mode appearance in time, D2, multiplying the previous D1 by the Mean Time Between this Failures Mode (MTBF). This D2 is an

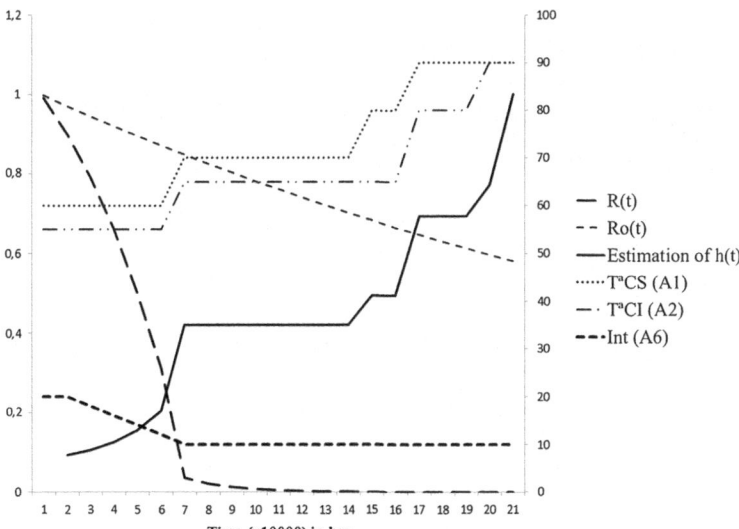

Fig. 7 Evolution of the ideal and actual reliability, failure probability, temperature and load current intensity, for a moment in time (t)

Table 6 Descriptor and interpretation rule of the symptom

Cod	Symptom	Cod	Descriptor	Interpretation rule
1.4.1.1	Relation of the oil temperature and the current output in the transformer	D1	$R(t, OV1, OV2, OV6, MV1)$	If D1 is below 0.2 (20%) it is a warning of a high risk
		D2	$R(t, OV1, OV2, OV6, MV1) \cdot MTBF$	I2 whenever the condition $D2 \leq 0.2 \cdot MTBF$, it is a warning of critical residual life to the next failure

Table 7 Monitoring objective selection

Cod	Failure mode	Cod	Symptom	Detection	Diagnosis	Prognosis
1.4.1	Lack of outflow	1.4.1.1	Relation of the oil temperature and the current output in the transformer			☑

example of the recurrent development of descriptors to obtain an approximation of the residual time to the next failure. The I2 interpretation rule is: I2 whenever the condition $D2 \leq 0.2\,(20\%) \cdot MTBF$, the residual life is critical and it is necessary to react with a maintenance task because the failure occurrence is imminent.

This table template has not considered the two first levels of the block 3 (Information sources), making independent the knowledge management among symptoms and their representative variables of the technology (sensors and measurement techniques or software systems) that makes the variables available.

4.3 Block 5

As we have anticipated, this CBM activity triggers a maintenance task. In this case for both interpretation rules, the expert group decided to dispose a preventive maintenance for this failure mode (see descriptors and interpretation rules in Table 6) in order to avoid the failure mode occurrence. Therefore, maintenance decision-making is automated by a prognosis monitoring output, so the Prognosis cell is selected in Table 7.

Table 8 Example of the proposed template, including the description of the CBM activities, solution adopted, for two failure modes

Cod	Failure mode	Cod	Symptom	Cod	Monitoring variable	Information source	Cod	Descriptor	Detec.	Diag.	Prog.	Interpretation rules
1.4.1	Lack of outflow	1.4.1.1	Relation of the oil temperature and the current output in the power transformer	OV1	Upper oil layer temperature (°C)	SCADA	D1	$R(t,$ OV1, OV2, OV6, MV1)			☑	If $D1 \leq 0.2$ (20%) it is a warning of a high risk
				OV2	Lower oil layer temperature (°C)	SCADA						
				OV6	Load current (A)	SCADA						
				MV1	Number of maintenance interventions	CMMS						
1.1.1	Short circuit	1.1.1.1	Over-current considerably higher than service current, confirmed also by the presence of hydrogen and CO as a result of the arc	OV5	Hydrogen level (%)	SCADA	D5	OV5	☑	☑		When D6 is above 50% standard service values, and $D5 \geq 0$ and $D7 \geq 0$
				OV6	Load current (A)	SCADA	D6	OV6				
				OV8	CO level (%)	SCADA	D7	OV8				

4.4   Final Template for the Use Case

Finally, a template can be arranged to summarize the process followed for the failure mode 1.4.1. "Lack of outflow", and the CBM activity (part of the CBM program) can be properly characterized in a unique table. This type of template is coherent with RCM output tables and completes them with CBM information. As the first columns of the template are coincident with RCM template, Equipment Unit with is code, Maintainable Item and its code, they have been omitted in Table 8. In Table 8 two failure modes are included with the intention to show the versatility of the process and the format adopted by the tables. The failure mode "1.4.1" using only one descriptor D1 and one monitoring output: *prognosis*; and the failure mode 1.1.1 "Core-Short Circuit" that has two monitoring outputs: *detection and diagnosis*. As a result of the expert group work, a very basic algorithm/rule allows to distinguish this failure mode and when it has occurred.

The representation using a single table of all possible CBM activities in the solution facilitates the management of the program by the maintenance staff the decision-making and the sustainable evolution with more available and accuracy measurement techniques, return-on-investment increment on measurement techniques implementation or new available system variables.

5   Conclusions

This paper discusses about the necessity of CBM management approaches in complex context of E-Maintenance strategies. In order to address the CBM management challenge, this paper proposes a framework with a template to clarify the concepts and to structure and to document the knowledge generation for a given condition-based maintenance solution. This framework fulfils, for consistency and robustness purposes, precise standards and well-known methodologies requirements. The CBM framework stresses the importance of concepts (such as fault detection, diagnosis and prognosis), and describes the key building blocks for the characterization of any CBM solution. The five blocks are consistent with the standards that have been included in each block definition, and represent different knowledge matters whose information is correlated in our structure framework. These blocks are traditionally supported by isolated software systems to administrate their information, so the main contribution of our framework is to swing the knowledge using four hinges to join the blocks: failure mode element, symptom element, descriptor element and interpretation rule element. Thanks to structure the relationships among elements, the CBM knowledge can be improved and evolved with the potentiality of capturing new information as monitoring variables inside E-maintenance strategies. Any future change, modification, improvement, management of the solution will be very much facilitated and understood using the provided template. The framework and template have been exemplified for an electrical power transformer CBM solution. An UML model has been developed theoretically and implemented integrating databases in a prototype + software, which has been validated in practice, thanks to proper interoperability of well-known proprietary information systems.

References

1. CEN (2010) Maintenance terminology. European Standard, EN 13306:2010. European Committee for Standardization, Brussels
2. ISO (2012a) ISO 13372:2012—condition monitoring and diagnostics of machines—vocabulary
3. Niu G, Yang BS, Pecht M (2010) Development of an optimized condition-based maintenance system by data fusion and reliability-centered maintenance. Reliab Eng Syst Saf 95(7):786–796
4. United States Army (2013) ADS-79D-HDBK—aeronautical design standard handbook for condition based maintenance systems for United States Army Aircraft
5. Parra C, Crespo A (2012) Maintenance engineering and reliability for assets management. Ingeman
6. Jardine A, Lin D, Banjevic D (2006) A review on machinery diagnosis and prognostics implementing condition based maintenance. MechSyst Signal Process 20:1483–1510
7. Campos J (2009) Development in the application of ICT in condition monitoring and maintenance. Comput Ind 69:1–20
8. ISO (2011) ISO 17359:2011—condition monitoring and diagnosiss of machines—general guidelines
9. Lee J, Ghaffari M, Elmeligy S (2011) Self-maintenance and engineering immune systems: towards smarter machines and manufacturing systems. Ann Rev Control 35:111–122
10. Vachtsevanos G, Lewis F, Roemer M, Hess A, Wu B (2006) Intelligent fault diagnosis and prognosis for engineering systems. John Wiley and Sons, Hoboken, NJ
11. Zio E (2009) Reliability engineering: old problems and new challenges. Reliab Eng Syst Saf 94(2):125–141
12. Jaw L, Merrill W (2008) CBM + research environment—facilitating technology development, experimentation, and maturation. In: Proceedings of the aerospace conference 2008, IEEE
13. Muller A, Crespo A, Iung B (2008) On the concept of e-maintenance: review and current research. Reliab Eng Syst Saf 93:1165–1187
14. Macchi M, Garetti M (2006) Information requirements for e-maintenance strategic planning: a benchmark study in complex production systems. Comput Ind 57(6):581–594
15. DOD (2008) Condition based maintenance Plus, DoD Guidebook. Department of Defense (USA)
16. Gupta J, Trinquier C, Lorton A, Feuillard V (2012) Characterization of prognosis methods: an industrial approach. European conference of the prognostics and health management society
17. Macchi M, Crespo Márquez A, Holgado M, Fumagalli L, Barberá Martínez L (2014) Value-driven engineering of E-maintenance platforms. J Manuf Technol Manage 25(4): 568–598
18. Levrat E, Iung B, Crespo A (2008) E-maintenance: review and conceptual framework. Prod Plann Control 19(4):408–429
19. Jayaratna N (1994) Understanding and evaluating methodologies: NIMSAD, a systematic framework. McGraw-Hill, Inc
20. Wu E, Diao Y, Rizvi S (2006). High-performance complex event processing over streams. In: Proceedings of the ACM SIGMOD international conference on management of data, Chicago, IL, USA, pp 407–420

21. Kobbacy KAH (2008) Artificial intelligence in maintenance in complex system. In: Kobbacy KAH, Murthy DNP (eds) Maintenance handbook. Springer, New York
22. Pintelon L, Parodi-Herz A (2008) Maintenance: an evolutionary perspective in complex system. In: Kobbacy KAH, Murthy DNP (eds) Maintenance handbook. Springer, New York
23. Russell S, Norvig P (2004) Artificial Intelligence: a modern approach
24. Davenport TH, Prusak L (1998) Working knowledge. Harvard Business School Press 2000, Boston
25. López-Campos M, Crespo A, Gómez JF (2013) Modelling using UML and BPMN the integration of open reliability, maintenance and condition monitoring management systems: an application in an electric transformer system. Comput Ind 64:524–542
26. Cheng S, Azarian M, Pecht M (2010) Sensor systems for prognostics and health management. Sensors 10:5774–5797
27. MIMOSA (Machinery Information Management Open Standards Alliance) (2011) Open systems architecture for condition based maintenance (OSA-CBM), v3.2.19
28. Saxena A, Celaya J, Saha B, Saha S, Goebel K (2010) Metrics for offline evaluation of prognostic performance. Int J Prognostics Health Manage 1
29. Moubray J (1991) RCM II: reliability-centred maintenance. Industrial Press Inc, New York
30. ISO (2003). ISO 13374-1:2003—condition monitoring and diagnosiss of machines—data processing, communication and presentation—Part 1: General guidelines
31. ISO (2006). ISO 14224:2006—petroleum, petrochemical and natural gas industries—collection and exchange of reliability and maintenance data for equipment
32. Crespo A (2007) The maintenance management framework. Springer, London Ltd, United Kingdom
33. OREDA (2009) Offshore reliability data handbook, 5th edn. OREDA
34. ISO (2012b) ISO 13379-1:2012—condition monitoring and diagnosiss of machines—data interpretation and diagnosiss techniques—Part 1: general guidelines
35. Guillén AJ, Crespo A, Macchi M, Gómez J (2016) On the role of prognostics and health management in advanced maintenance systems, Production Planning and Control (In printed)
36. Crespo A, Gómez J, Moreu P, Sola A (2012) Modelling on-line reliability and risk to schedule the preventive maintenance of repairable assets in network utilities. IMA J Manage Math 1–14

Author Biographies

Antonio Jesús Guillén López is a Ph.D. candidate and contracted researcher in Intelligent Maintenance System research group (SIM) of the University of Seville (USE), focusing his studies in Prognosis Health Management & Condition-Based Maintenance applications and Assets Management; Industrial Engineering and Master in Industrial Organization & Business Management from the University of Seville (USE). From 2003 to 2004, he was working for the Elasticity and Strength of Materials Group (GERM) of the USE in aeronautical materials tests. From 2006 to 2009, he was a member of the Department of Electronic Engineering of USE, and worked in numerous public–private international R&D project, developing new thermo-mechanical design applications for improving the performance and the life cycle of power electronics system. From 2009 to 2010, he was a Project Manager of "Solarkit" project, representing Andalucia's Government and USE in the international competition Solar Decathlon 2010 and was a foundational partner of Win Inertia Tech., technological based company specializing in electronic R&D for Smart Grids and Energy Store fields, where he has carried out different roles, from R&D engineer to General Manager until September 2012. Currently, he is a coordinator of the Spanish National Research Network of Asset Management.

Juan Francisco Gómez Fernández is Ph.D. in Industrial Management and Executive MBA. He is currently part of the Spanish Research & Development Group in Industrial Management of the Seville University and a member in knowledge sharing networks about Dependability and Service Quality. He has authored publications and collaborations in Journals, books, and conferences, nationally and internationally. In relation to the practical application and experience, he has managed network maintenance and deployment departments in various national distribution network companies, both from private and public sector. He has conduced and participated in engineering and consulting projects for different international companies, related to Information and Communications Technologies, Maintenance and Asset Management, Reliability Assessment, and Outsourcing services in Utilities companies. He has combined his business activity with academic life as a associate professor (PSI) in Seville University, being awarded as Best Thesis and Master Thesis on Dependability by National and International Associations such as EFNSM (European Federation of National Maintenance Societies) and Spanish Association for Quality.

Adolfo Crespo Márquez is currently Full Professor at the School of Engineering of the University of Seville, and Head of the Department of Industrial Management. He holds a Ph.D. in Industrial Engineering from this same University. His research works have been published in journals such as the International Journal of Production Research, International Journal of Production Economics, European Journal of Operations Research, Journal of Purchasing and Supply Management, International Journal of Agile Manufacturing, Omega, Journal of Quality in Maintenance Engineering, Decision Support Systems, Computers in Industry, Reliability Engineering and System Safety, and International Journal of Simulation and Process Modeling, among others. Prof. Crespo is the author of seven books, the last four with Springer-Verlag in 2007, 2010, 2012, and 2014 about maintenance, warranty, and supply chain management. Prof. Crespo leads the Spanish Research Network on Dependability Management and the Spanish Committee for Maintenance Standardization (1995-2003). He also leads a research team related to maintenance and dependability management currently with five Ph.D. students and four researchers. He has extensively participated in many engineering and consulting projects for different companies, for the Spanish Departments of Defense, Science and Education as well as for the European Commission (IPTS). He is the President of INGEMAN (a National Association for the Development of Maintenance Engineering in Spain) since 2002.

Criticality Analysis for Maintenance Purposes

Adolfo Crespo Márquez, Pedro Moreu de León, Antonio Sola Rosique
and Juan Francisco Gómez Fernández

Abstract The purpose of this paper is to establish a basis for a criticality analysis, considered here as a prerequisite, a first required step to review the current maintenance programs, of complex in-service engineering assets. *Review* is understood as a reality check, a testing of whether the current maintenance activities are well aligned to actual business objectives and needs. This paper describes an efficient and rational working process and a model resulting in a hierarchy of assets, based on risk analysis and cost–benefit principles, which will be ranked according to their importance for the business to meet specific goals. Starting from a multi-criteria analysis, the proposed model converts relevant criteria impacting equipment criticality into a single score presenting the criticality level. Although detailed implementation of techniques like root cause failure analysis (RCFA) and reliability centered maintenance (RCM) will be recommended for further optimization of the maintenance activities, the reasons why criticality analysis deserves the attention of the engineers, maintenance and reliability managers are here precisely explained. A case study is presented to help the reader to understand the process and to operationalize the model.

Keywords Criticality · Maintenance management

1 Introduction

In this paper, we deal with the strategic part of the maintenance management definition and process (as in EN 13306:2010 [1]), which is related to the determination of maintenance objectives or priorities and the determination of strategies. A more operational second part of the definition refers to the strategies implementation

A. Crespo Márquez (✉) · P. Moreu de León · A. Sola Rosique · J.F. Gómez Fernández
Department of Industrial Management, School of Engineering,
University of Seville, Seville, Spain
e-mail: adolfo@us.es

© Springer International Publishing AG 2018
A. Crespo Márquez et al. (eds.), *Advanced Maintenance Modelling
for Asset Management*, DOI 10.1007/978-3-319-58045-6_6

(maintenance planning, maintenance control, supervision, and continuous improvement).

Part of this strategy setting process, that we refer to, is devoted to the determination of the maintenance strategies that will be followed for the different types of engineering assets (i.e., specific physical assets such as: production processes, manufacturing facilities, plants, infrastructure, support systems, etc.). In fact, maintenance management can also be considered as "...the management of all assets owned by a company, based on maximizing the return on investment in the asset" [2], also their safety and their respect for the environment. Within this context, criticality analysis is a process providing a systematic basis for deciding what assets should have priority within a maintenance management program [3], and has become a clear business need in order to maximize availability during assets' operational phase.

This type of assets criticality analysis, performed during their operational phase and for maintenance purposes, is therefore different to criticality analysis which is carried out during assets design. At that point, the objective (more linked to asset's reliability assessment) is to identify critical areas so that different design alternatives to achieve a specified availability target can be optimized and compared [4]. Description of specific techniques for criticality analysis during design can be found in MIL-STD 1629A [5], also in [6] and [7]. These techniques use, for instance, the Risk Priority Number (RPN) method, fuzzy logic or approximate reasoning to prioritize failures modes (not assets).

When prioritizing assets for maintenance purposes, and during their operational phase, a large number of quantitative and qualitative techniques can be found in the literature [3]. On some occasions, there is no hard data about historical failure rates, but the maintenance organization may require a certain *gross assessment* of assets priority to be carried out. In these cases, qualitative methods may be used and an initial assets assessment, as a way to start building maintenance operations effectiveness, may be obtained [8].

This paper, however, is about the process to follow on other occasions, lately becoming more frequent, when the maintenance organization has important amounts of data for complex in-service assets for which a certain maintenance strategy has been previously developed and implemented. Therefore, we have evidences of assets behavior for current operational conditions of the asset, and we launch the criticality analysis with the purpose of adjusting assets maintenance strategies to business needs over time.

Most of current quantitative techniques for assets criticality analysis use a weighted scoring method defined as variation of the RPN method used in design [9]. These weighted scoring methods might appear simple, but in order to reach acceptable results, a precise procedure should be considered when determining factors, scores, and combining processes or algorithms [4]. The analysis involves another important issue which is the level of detail required, compromising objective effectiveness (missing focus of subsequent maintenance efforts) and also data collection efforts.

The criticality number to obtain (C), as a measure of risk associated to an asset, is derived by attaching a numerical value to the probability of failure (function loss) of the asset (the higher probability, the higher the value), and attaching another value to the severity of the different categories of asset functional loss consequences (the more serious consequences for each category, the higher the value). The criteria and the relative weighting to assess severity and probability may vary widely for different companies according to their maintenance objectives and KPIs. The two numbers are multiplied to give a third which is the criticality number (C). Of course, assets with the higher (C) will be recognized to be the more critical assets and will deserve special attention from the maintenance (sometimes now called: assets) management organization.

The reader may notice now how the "detectability" factor, used as part of the equation of RPN in design, is not considered now in operations in C. This is because detectability is not an attribute of the assets we are ranking, but of the failure modes for which design alternatives were explored (see [10]).

The inspection and maintenance activities will be prioritized on the basis of quantified risk caused due to failure of the assets [11]. The high-risk assets are inspected and maintained usually with greater frequency and thoroughness and are maintained in a greater manner, to achieve tolerable risk criteria [12].

Although this technique is becoming popular, some authors mention [11] that most of the risk analysis approaches are *deficient in uncertainty and sensitivity analysis*. This may constrain yielding proper results and decisions based on mis-leading results may generate nonessential maintenance efforts spent in less important areas. To avoid this, risk analysis for asset criticality should be evaluated in well-planned manner ensuring that significant sources of risk are reduced or eliminated [13].

In this paper, we propose a criticality analysis taking into account the following process design requirements:

1. The process must be applicable to a large scale of in-service systems within a plant or plants. A reason for this is the fact that PM programs to be evaluated using this criticality analysis are set by plant equipment (placed in a technical location in the plant), and therefore we are forced to deal with this large number of items in the analysis;
2. The scope of the analysis should be, as mentioned in previous point, the same for which the current PM program is developed and implemented;
3. The analysis should support regular changes in the scale adopted for the severity effects of the functional losses of the assets (this is a must to align maintenance strategy in dynamic business environments).
4. The process must allow easy identification of new maintenance needs for assets facing new operating conditions;
5. General guidelines to design possible maintenance strategy to apply to different type of assets according to the results of the analysis (criticality and sources of it) should be provided;
6. Connection with the enterprise asset management system of the company should be possible to automatically reproduce the analysis, with a certain cadence, over time;

7. The process should be tested in industry showing good practical results.

In the sequel the paper is organized as follows: Sect. 2 shows first, briefly, the proposed criticality analysis process description. Then, more precisely and in the different subsections, the notation of the mathematical model supporting the process is introduced, as well as every step of the process to follow, including model equations. Along this second section of the paper, a practical example helps to exemplify the process and model implementation. Section 3 tries to turn, the previous process and model, into a powerful management tool. In order to do so, this section explains how to handle data requirements properly and how to benefit from model outputs and results using suitable graphical representations. Section 4 is devoted to the interpretation of possible results, offering clear guidelines to ensure maximum benefits from the analysis. Conclusions of the paper are finally presented in Sect. 5.

2 Process Description and Rational

In this section, we describe a comprehensive process to be followed by a *criticality review team* (defined as in [14]) in order to generate a consistent criticality analysis based on the use of the PRN method together with multi-criteria techniques to select the weights of factors deriving in the severity of an asset.

The process consists in a series of steps determining the following:

1. Frequency levels and the frequency factors;
2. Criteria and criteria effect levels to assess functional loss severity;
3. Non-admissible functional loss effects;
4. Weights (contribution) of each criteria to the functional loss severity;
5. Severity categories, or levels, per criteria effect;
6. Retrieving data for actual functional loss frequency for an element (r);
7. Retrieving data for maximum possible effects per criteria;
8. Determination of Potential asset criticality at current frequency;
9. Retrieving data for real effects per criteria;
10. Determining observed asset criticality at current frequency;
11. Results and guidelines for maintenance strategy.

The first five steps of the process determine the elements configuring the algorithm in the mathematical model that will be later used to rank assets, once their in-service operational data is retrieved from the enterprise assets management system. Steps 6, 7, and 9 are data gathering related steps. In these sections, the reader will see that assets' data required will be:

- Engineering data concerning maximum possible asset functional loss effects (Step 7); and
- Operational data showing information about current frequency of their functional loss (Step 6) and functional loss effects (Step 9);

Steps (8) and (10) are presenting results for potential and current observed criticality of the assets. Discussion of these two facts will drive to a set of conclusions and action items that will be presented in Step (11).

The process followed to assess the criticality of the different assets considered is supported by a mathematical model, whose notation is now presented:

i: 1 … n criteria to measure severity of an element functional loss,
j: 1 … m levels of possible effects of a functional loss for any criteria,
z: 1 … l levels of functional loss frequency,
e_{ij}:   Effect j of the severity criteria i,
w_i: Weight given to the severity criteria i by experts, with $\sum_{i=1}^{i=n} w_i = 1$,
M_i: Maximum level of admissible effect for criteria i, with $M_i \leq m$, $\forall i$,

MS: Maximum severity value,
v_{ij}:   Fractional value of effect j for the severity criteria i,
S_{ij}:   Severity of the effect j for the severity criteria i,
pe_{rij}:   Potential effect j of criteria i for the functional loss of element r,
f_r: Value for the frequency of the functional loss of element r,
ff_z: Frequency factor for frequency level z,
fe_{rz}:   Boolean variable with value 1 when z is the level of the observed frequency of element r functional loss, 0 otherwise,
af_z: Average frequency of functional loss for frequency level z,
S_r: Severity of the functional loss of element r,
C_r: Criticality of element r,
re_{rij}:   Current probability of the effect j of criteria i for the failure of r,
S'_r: Current observed severity of the functional loss of element r,
C'_r: Current criticality of element r,

Subsequent sections of the paper present precisely the different steps of the process, and introduce the mathematical model supporting them. The model is applied to an example which illustrates a practical industrial scenario.

2.1   Determining Frequency Levels and Frequency Factors

To manage the frequency levels a form of Pareto analysis is used, in which the elements are grouped into z frequency categories according to their estimated functional loss frequency importance. For example, for $z = 4$, the categories could be named: very high, high, medium, and low functional loss frequency. The percentage of elements to fall under each category can be estimated according to business practice and experience for assets of the same sector and operational conditions (for instance, according to existing operating conditions of our assets, in Table 1 the review team has decided to define a category named "Low", including a group of 5 assets having less than 2 failures per year [f/y] and with an average functional loss of 1.2 f/y, easing our corrective maintenance operations, and serving as a reference for the rest of the assets selected categories. Assets with more than 7 f/y are considered to complicate enormously corrective operations, and as soon as severity in consequences of the functional loss increases, current management will consider those assets as very critical). Then, average values for frequencies falling inside each group can be estimated and frequency factors per category calculated (see example in Table 1).

In the model mathematical formulation, if af_z is the average frequency of functional loss for frequency level z, then the frequency factor vector is defined as follows:

$$ff_z = \frac{af_z}{af_1}, \text{for } z = 1 \ldots l \text{ levels of functional loss frequency}$$

2.2 Criteria, and Criteria Effect Levels, to Assess Functional Loss Severity

This part of the analysis should reflect the business drivers recognized by management and shareholders [4]. For the severity classification, this study focuses the attention on both, safety and cost criteria (similarly to [11]). For the safety severity categories, similar hazard severity categories to the ones used in MIL-STD-882C are adopted. This standard proposes four effect categories that can now be reframed as follows:

- catastrophic, could result in multiple fatalities
- critical, resulting in personal injuries or even one single fatality
- marginal, and
- negligible

Table 1 Calculation of frequency factors per selected functional levels

Asset	f/y	Asset	f/y	Category (z)	% (z)	af_z	ff_z
a	1	i	8	Very high	10	8	6.7
b	2	d	7	High	20	6.5	5.4
c	5	h	6				
d	7	c	5	Medium	20	4	3.3
e	3	e	3				
f	1	b	2	Low	50	1.2	1.0
g	1	j	1				
h	6	g	1				
i	8	a	1				
j	1	f	1				

As cost factors, we may selected different criteria for which the functional loss effect can be classified in different levels that can, at the same time, be converted into cost using a certain contract or standard that the company must honor.

In the example for this paper, the following criteria are selected (assuming that we are dealing with the criticality of a collective passenger's transportation fleet):

- Operational reliability: measuring the potential impact of a functional loss to the system where the asset is installed. The effects could be classified in different levels like: No Affection (NA), stopping the system less than x min ($S < x$), stopping the system more than x min ($S > x$) or leaving the system out of order (OO). Each one of these affection levels can be later translated to cost of the functional loss, and the corresponding factors could be obtained.
- Comfort: evaluating whether the functional loss of the element may: have no affection on comfort (NA), affect a passenger (P), a car (C) or the whole train (T). Again, each one of these affection levels can be later translated into cost of the functional loss, and the corresponding factors could be obtained.
- The "corrective maintenance cost" could be selected as another cost-related criteria. Effects could be classified in very high, high, medium, and low corrective maintenance cost, and we could proceed similarly to what has been presented in Table 1, classifying the elements' costs and finding averages costs and the corresponding factors for each effect classification level.

2.3   Determining Non-Admissible Functional Loss Effects

At this point, the process requires the definition of those functional loss effects that would be considered as *"non-admissible"* for each specific criterion. For instance, in Table 2, those categories of effects being considered non-admissible are presented with dark gray inverse video. For this case study, the review team has considered that catastrophic and critical effects on "Safety", besides the "Out of order" condition of the system, are non-admissible effects of a functional loss of an element.

Table 2 Table presenting non-admissible effect categories

Criteria to measure severity			
Safety criteria	Cost related criteria		
	Operational reliability	Comfort	CM cost
Category of effects per criteria			
Cat & Cri (C)	OO	Train (T)	VH (10%)
Marginal (M)	$S > x$	Car (C)	H (20%)
Negligible (N)	$S < x$	Passenger (P)	M (20%)
NA	NA	NA	L (50%)

The model will allocate a maximum value for overall severity (MS) to those assets (elements of the transportation fleet) whose functional loss may produce non-admissible effects for any of the selected criteria. Therefore, those elements will become of maximum severity regardless their functional loss effect on any other criteria under consideration.

In the mathematical model we will use the following notation for this purpose:

M_i: Maximum level of admissible effect for criteria i, with $M_i \leq m$, $\forall i$

MS: Maximum value for overall severity

And in our example,

$$[i] = \text{safety, operational reliability, comfort, CM cost}$$

and $[M_i] = 3, 3, 4, 4$ as maximum levels of admissible effects for each criteria, finally it will be considered MS = 100.

2.4 Criteria Weights in the Functional Loss Severity

To determine these weights various considerations can be taken into account, for instance:

- Criteria correlation to business KPI's.
- Budget allocated to each cost related criteria within the maintenance budget.
- Impact of each criterion on the brand and/or corporate image. For instance, in the previous example the management (or the criticality review team) could consider that "operational reliability" and/or "comfort" criteria could have also impact on the brand image, increasing its weight versus corrective maintenance cost.
- Considerations measuring the importance of the safety factor considering standards, contracts, or market rules.
- Etc.

Regardless all these considerations, assigning criteria weights may contain a certain subjective judgments from the experts involved. In order to make this judgment as much consistent as possible, AHP techniques can be used, and a model presenting the multi-criteria classification problem in a logic decision diagram, can help to solve the multi-criteria decision sub-problem at the highest decision nodes of the diagram (the reader is referred to Bevilacqua et al. [15] for additional information concerning AHP utilization with this purpose). A major advantage of the AHP approach is that both qualitative and quantitative criteria can be included in the classification scheme. In addition, the assignment of weights to the different parameters is considered as a positive characteristic of the method [16]. The reader is referred to [3] (Sect. 9.4.1, steps of the process 6 & 7, pages 121 & 122, concerning the *Quantification of judgments on pair alternative criteria* and the *Determination of the criteria weighting and its consistency*) for a detailed description of the utilization of the AHP in our methodology.

On the other hand, the amount of subjectivity involved in the process of pairwise comparisons is often viewed as the main limitation of this method, another problem arises when the number of alternatives to rank increases forcing to an exponential increase in the number of pairwise comparisons. That is why we just limit the method utilization to the severity criteria level, not to the asset criticality classification level.

In the example of this paper, w_i, weight given to the severity criteria i by experts, resulting from the AHP analysis are assume to be equal to $[w_i] = 10, 30, 20, 40$. This means, for instance, that the review team considers corrective maintenance cost consequences are two times more important than those related to comfort, or that operational reliability consequences are three times more important than admissible safety consequences. Notice that, this is considered, after using AHP, to be a subjective but consistent judgment of the review team.

2.5 Determining Severity Per Criteria Effect

In the mathematical model proposed, an effects severity matrix is defined, for any element included in the analysis (r), as follows:

$$S_{ij} = \begin{cases} MS, & for \quad M_i < j \le m, \quad \forall i \\ w_i v_{ij}, & for \quad 1 \le j \le M_i, \quad \forall i \end{cases} \tag{1}$$

where

$$v_{ij} = \frac{e_{ij}}{e_{ik}}, \text{ with } k = M_i \text{ and } j \le M_i, \text{ and with } v_{ij} = 1 \text{ for } j = M_i \text{ and } \forall i$$

And e_{ij} is the effect j of the severity criteria i, *and* v_{ij} is the fractional value of effect j for the severity criteria i.

In the example, we are following, the effects matrix is included (last 4 rows) in Table 3, where relative values for the different effects for each criteria are presented. Units for these relative values are based on cost (for $i = 2, 3, 4$) or in a dimensionless rule of proportionality of the effect ($i = 1$). The interpretation of Table 3, is as follows: a comfort functional loss, for instance, that in Table 2 is presented as having an effect to the entire train, may have a potential cost of 4500 \$; or, another example, an admissible effect on operational reliability stopping the system more than x minutes ($s > x$), may cause a potential cost of 10,000 \$.

At this point is important to understand that, for a given functional loss, these are maximum possible effects per criteria, but not actual observed effects (later, real observed effects of functional losses, will be considered in the analysis, which are in

Table 3 Effects matrix per functional loss

Criteria to measure severity			
Safety criteria (dmnl) (weight: 10%)	Cost related criteria (e.g., based on penalization cost and CM budget, $)		
	Operational reliability (weight: 30%)	Comfort (weight: 20%)	CM cost (weight: 40%)
Category of effects per criteria and functional loss			
Non-admissible	Non-admissible	4500	300
1.5	10,000	3000	150
1	5000	600	50
0	0	0	10

Table 4 Effects severity matrix per functional loss

Criteria to measure severity (S_{ij})			
Safety criteria (dmnl) (weight: 10%)	Cost related criteria (e.g., based on penalization cost and CM budget, $)		
	Operational reliability (weight: 30%)	Comfort (weight: 20%)	CM cost (weight: 40%)
Category of effects per criteria and functional loss			
100	100	20	40
10	30	13.3	20
6.6	15	4	6.3
0	0	0	1.2

fact conditional probabilities to reach a certain effect once a functional loss takes place). In the example that is presented, the corresponding effects severity matrix (according to Eq. 1, and for MS = 100) is included in last for rows of Table 4.

The interpretation of Table 4, is as follows: a comfort functional loss impacting the train (maximum effect), will count for 20 points of severity (up to 100), while a comfort functional loss impacting only one car will count for 13.3 points of severity (this is calculated proportionally to the functional loss potential cost values). Notice how a non-admissible effect of a functional loss will count for a 100 (maximum value) regardless the effect in any other criteria.

2.6 Retrieving Data for Actual Functional Loss Frequency

Actual data for frequency of functional losses can be retrieved and captured in the variables fe_{rz}, these variables conform, for each asset r, a vector of l elements, once there are $z = 1 \ldots l$ levels of functional loss frequency. Thus, fe_{rz} are Boolean variables with values:

$$fe_{rz} = \begin{cases} 1, & \begin{array}{l} \text{When } z \text{ is the observed frequency} \\ \text{category of element } r \text{ functional loss} \end{array} \\ 0, & \text{Otherwise} \end{cases}$$

Example: For functional loss frequencies expressed in Table 1, the criticality analysis review team could retrieve the asset b functional loss frequency and this would expressed as:

$$[fe_{bz}] = 0, 0, 1, 0$$

The frequency factor to apply to this element would be the result of the following scalar product:

$$f_r = \sum_{z=1}^{z=l} ff_z fe_{rz} \tag{2}$$

In our example:

$$f_b = 1 \times 0 + 3.3 \times 0 + 5.4 \times 1 + 6.7 \times 0 = 5.4$$

And therefore 5.4 would be the frequency to consider for the element when finally calculating its criticality.

2.7 Retrieving Data for Maximum Possible Effects Per Criteria

Data concerning maximum potential effects, when a functional loss of an element happens, can be retrieved and captured in the variables pe_{rij}, these variables conform, for each asset r, a matrix of $n \times m$ elements, once there are i: $1 \ldots n$ criteria to measure severity of an element functional loss, and j: $1 \ldots m$ levels of possible effects of a functional loss for any criteria.. Thus, pe_{rij} are Boolean variables with values:

$$pe_{rij} = \begin{cases} 1, & \begin{array}{l} \text{When } j \text{ is the level of maximum potential effect} \\ \text{of the functional loss of an element } r \text{ and for the} \\ \hspace{4cm} \text{severity criteria } i \end{array} \\ 0, & \text{Otherwise} \end{cases}$$

Assume, as an example, that for the effects severity matrix expressed in Table 4, the criticality analysis review team retrieves potential effects of a functional loss of an element r, this could be represented with the following *potential effects matrix*:

$$[pe_{bij}] = \begin{bmatrix} 1 & 0 & 0 & 0 \\ 0 & 1 & 1 & 0 \\ 0 & 0 & 0 & 1 \\ 0 & 0 & 0 & 0 \end{bmatrix}$$

Then, we can model the severity of the functional loss of element r as follows:

$$S_r = Min(MS, \sum_{i=1}^{i=n} \sum_{j=1}^{j=m} pe_{rij} S_{ij}) \tag{3}$$

In the previous example, the severity of the asset b would result in:

$$S_b = Min(100, 100 + 30 + 13.3 + 6.3) = 100$$

This, in fact, represents a weighted average type of algorithm, where the weights are introduced through the value of the different criteria effects, as calculated in Eq. 1. In this way, consistency in the severity calculation of one element with respect to another is ensured. It has been experienced that how by giving maximum severity to inadmissible effects, like for instance in our previous example, the different roles of actors represented in the review team are safeguarded (for instance, *safety department* people in the review team of our example), discussions in the meetings are reduced and consensus is more easily reached.

Notice how, in case of good data integrity for frequency and functional loss effects of the elements under analysis, the review team can and must concentrate its efforts in the selection of the severity criteria and in establishing proper weights according to business needs.

2.8 Determining Potential Criticality at Current Frequency

The criticality of the element is finally calculated as

$$C_r = f_r \times S_r \tag{4}$$

Thus, for asset b of the example previously introduced:

$$C_b = 1 \times 100 = 100$$

2.9   Retrieving Data for Real Effects Per Criteria

Actual data for real element functional loss effects can be retrieved and captured in the variables re_{rij}, these variables conform, for each asset r, a matrix of $n \times m$ elements, once there are i: $1\ldots n$ criteria to measure severity of an element functional loss, and j: $1\ldots m$ levels of possible effects of a functional loss for any criteria.

re_{rij} = current probability of the effect j of criteria i for the functional loss of element r, with $\sum_{j=1}^{j=m} re_{rij} = 1$.

Assume, as an example, that for the effects severity matrix expressed in Table 4, the criticality analysis review team could retrieve data concerning real element functional loss effects for asset r, this could be represented with the following *real effects matrix*:

$$[re_{rij}] = \begin{bmatrix} 0 & 0 & 0 & 0 \\ 0 & 0.5 & 0.1 & 0 \\ 0.2 & 0.3 & 0.8 & 0.9 \\ 0.8 & 0.2 & 0.1 & 0.1 \end{bmatrix}$$

Then, we can model the severity of the functional loss of element r as follows:

$$S'_r = \text{Min}(MS, \sum_{i=1}^{i=n} \sum_{j=1}^{j=m} re_{rij} S_{ij}) \tag{5}$$

In the previous example, the severity of asset r would result in:

$$S'_b = \text{Min}[100, 6.6 \times 0.2 + (30 \times 0.5 + 15 \times 0.3) + (13.3 \times 0.1 + 4 \times 0.8) + 6.3 \times 0.9]$$
$$= 31.1$$

2.10   Determining Observed Criticality at Current Frequency

The criticality of the element is finally calculated as

$$C'_r = f_r \times S'_r \tag{6}$$

In the previous example presented

$$C_b' = 1 \times 31.1 = 31.1$$

So real criticality is much lower than potential (100).

3 Criticality Analysis as a Practical Management Tool

For complex in-service engineering assets, the implementation of the model must be fast and automatic over time. Also, the results of the analysis, and corresponding maintenance strategic actions carried out as a consequence of it, must be accountable in the future.

In order to easy the data entry process (considering now the need to rank an important amount of elements), and to make the interpretation of the analysis results user friendly, the following process, as a result of previous implementation in complex and large engineering assets, is recommended:

1. Retrieve asset's data for frequency and potential severity effects (Table 5). Data must show assigned frequency and severity criteria categories, per asset. It is frequently convenient, to save time and easy analysis replications, that the list of assets can be directly retrieved (to the scope of the analysis) from the enterprise assets management system database.
2. Convert this qualitative data into quantitative data (Table 6) considering the weights of each criteria and the model explained in previous paragraph. It is very important to separate qualitative and quantitative datasets. A reason for this is related to the possibility to test sensitivity of changes in the criteria weights

Table 5 Frequency and potential effects qualitative matrix

Assets	Frequency	Severity criteria			
	FE	PE_1 (safety)	PE_2 (Op. Reliability)	PE_3 (Comfort)	PE_4 (CM cost)
a	L	L	M	VH	H
b	L	VH	H	H	M
c	M	L	M	L	H
d	H	H	H	H	M
e	M	L	M	M	L
f	L	L	L	H	M
g	L	M	VH	M	L
h	H	H	L	L	L
i	VH	L	VH	M	L
j	L	L	L	L	L

Table 6 Current frequency and potential effects quantitative matrix

Assets	Frequency	Severity criteria				Severity	Criticality
	FE	PE$_1$	PE$_2$	PE$_3$	PE$_4$		
a	1	0	6.6	20	20	46.6	46.6
b	1	100	10	13.3	6.3	100.0	100.0
c	3.3	0	6.6	0	20	26.6	87.8
d	5.4	30	10	13.3	6.3	59.6	321.8
e	3.3	0	6.6	4	1.2	11.8	38.9
f	1	0	0	13.3	6.3	19.6	19.6
g	3.3	15	100	4	1.2	100.0	330.0
h	5.4	30	0	0	1.2	31.2	168.5
i	6.7	0	100	4	1.2	100.0	670.0
j	1	0	0	0	1.2	1.2	1.2

Table 7 Criticality criteria assignment

Criticality level	% of assets	Criticality value interval	Assets	Area color in matrix (Figs. 1 and 2)
Critical	20	326–670	g, i	Dark gray
Semi-critical	30	90–325	b, d, h	Gray
Not critical	50	0–89	a, c, g, e, f, j	White

for criticality assessment, modifying the final ranking of the assets, but not changing data retrieved in Step 1.

3. See the quantitative criteria for the assignment of the category low, mid or high criticality to an asset, like, for instance, in Table 7.

This decision may condition organizational efforts to be dedicated later to the management of the different category of assets. This is a business issue and consensus should be reached within the review team and the management team before any further process development.

4. Populate, with assets, the *potential criticality matrix* representation (in Fig. 1). Notice that this matrix considers current (observed) frequencies and potential severities, for each equipment functional loss.

Notice that shaded areas in the matrices in Figs. 1 and 2, mean minimum criticality level of all elements in that area.

5. Retrieve quantitative data, using a unique matrix, and now for current observed frequency and severity effects per element (see Table 8).
6. Populate, with assets, the *current criticality matrix* representation (Fig. 2). Notice that this matrix considers current (observed) frequencies and severities, for each equipment functional loss.

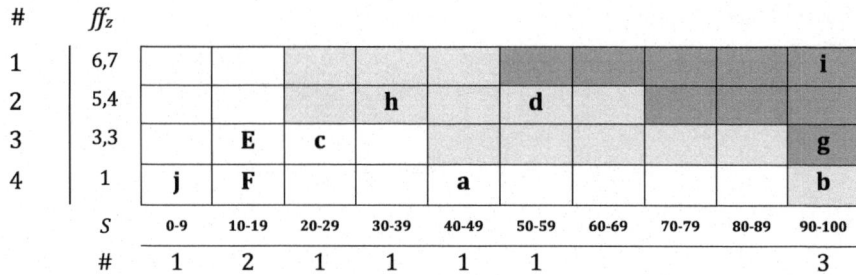

#	ff_z	0-9	10-19	20-29	30-39	40-49	50-59	60-69	70-79	80-89	90-100
1	6,7										**i**
2	5,4				**h**		**d**				
3	3,3		**E**	**c**							**g**
4	1	**j**	**F**			**a**					**b**
S		0-9	10-19	20-29	30-39	40-49	50-59	60-69	70-79	80-89	90-100
#		1	2	1	1	1	1				3

Fig. 1 Potential criticality matrix representation

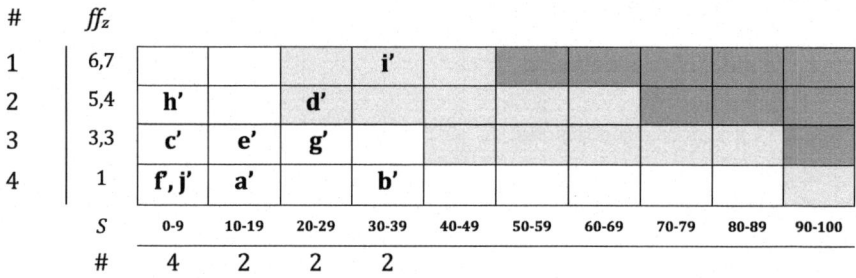

#	ff_z	0-9	10-19	20-29	30-39	40-49	50-59	60-69	70-79	80-89	90-100
1	6,7				**i'**						
2	5,4	**h'**		**d'**							
3	3,3	**c'**	**e'**	**g'**							
4	1	**f', j'**	**a'**		**b'**						
S		0-9	10-19	20-29	30-39	40-49	50-59	60-69	70-79	80-89	90-100
#		4	2	2	2						

Fig. 2 Current criticality matrix representation

7. To easy further analysis, populate with both assets with potential and observed functional loss severity, a criticality matrix representation (Fig. 3). In the matrix in Fig. 3, we can compare results obtained in two previous criticality matrices: potential and current.
8. Automatic generation of the criticality report, listing the assets ranking per criticality levels, and within each level, classify the assets by common frequencies and severities. Rational for this is related to the type of strategy that will be required to manage them. Realize that, for managers, the most important outputs from this analysis are these lists of assets falling under different criticality categories and subcategories.

4 Interpretation of Current and Future Results

Once all these data is available to the analyst, some guidelines for the interpretation of results are the following:

- Assets whose representation in matrix of Fig. 3 are similar to the one in Fig. 4. For all these elements observed criticality (S'_x, ff'_x) is lower than potential

Table 8 Frequency and current observed effects quantitative matrix

r	fe	RE1				RE2				RE3				RE4				Severity				S'r	Criticality
		L	M	H	VH	L	M	H	VH	L	M	H	VH	L	M	H	VH	re1	re2	re3	re4		
a	1	1.0				0.1	0.9			0.7	0.2	0.1		0.5	0.5			0	13.5	2.13	3.75	19.4	19.4
b	1	0.8	0.2			0.2	0.3	0.5		0.1	0.8	0.1		0.1	0.9			1.32	19.5	4.53	5.79	31.1	31.1
c	3.3	1.0				0.5	0.5			1.0				0.8	0.2			0	7.5	0	2.22	9.7	32.1
d	5.4	0.8	0.1	0.1		0.3	0.2	0.5		0.1	0.9			0.0	1.0			1.66	18	3.6	6.3	29.6	159.6
e	3.3	1.0				1.0				0.1	0.2	0.7		1.0				0	0	10.1	1.2	11.3	37.3
f	1	1.0				1.0				0.7	0.2	0.1		1.0				0	0	2.13	1.2	3.3	3.3
g	3.3	0.8	0.2			0.2	0.3	0.5		0.5	0.5			1.0				1.32	19.5	2	1.2	24.0	79.3
h	5.4	0.8	0.1	0.1		1.0				1.0				1.0				1.66	0	0	1.2	2.9	15.4
i	6.7	1.0				0.0	0.0	1.0		0.2	0.8			1.0				0	30	3.2	1.2	34.4	230.5
j	1	1.0				1.0				1.0				1.0				0	0	0	1.2	1.2	1.2

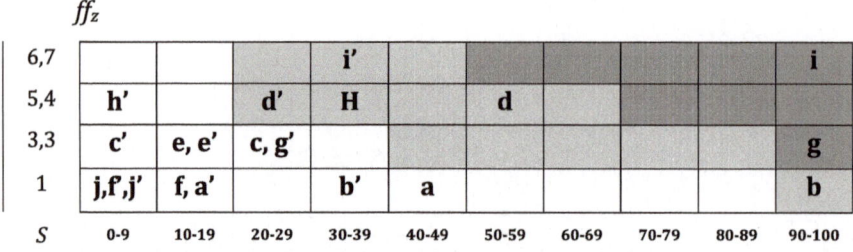

ff_z										
6,7				i'						i
5,4	h'		d'	H		d				
3,3	c'	e, e'	c, g'							g
1	j,f,j'	f, a'		b'	a					b
S	0-9	10-19	20-29	30-39	40-49	50-59	60-69	70-79	80-89	90-100

Fig. 3 Potential and current criticality matrices representation

Fig. 4 Potential and current criticality representation (within the matrix)

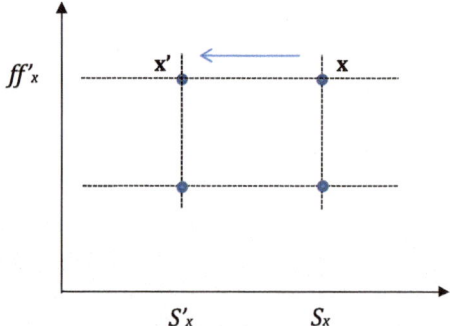

(S_x, ff'_x). Regarding these elements, some of the following statements could be applicable:

(i) Dynamic capabilities to avoid serious failures consequences or fault propagation are somehow in place, and although some of the assets may still have a high frequency of failures (see for instance asset *i* in Fig. 3), consequences are low due to this fact.

(ii) Passive mitigation mechanism to avoid consequences of functional losses has been successfully introduced (redundancy, passive safety and environmental protections, etc.).

(iii) Predictive maintenance programs are mature, and levels of potential and functional failures are properly selected once the consistency of the failure mode PF interval is well studied and known (See Moubray's Maxim 2 formulated in [17]).

- Assets whose representation in current criticality matrix (Fig. 2) over time $(x' \rightarrow x''$ for $t'' = t' + \Delta t)$ are improving in criticality due to a reduction in functional loss frequency over time $(ff_x \rightarrow ff''_x)$, as in Fig. 5. For all these elements observed, criticality is becoming lower and some of the following statements could be applicable:

Fig. 5 Representation of criticality changes, in a matrix, over time

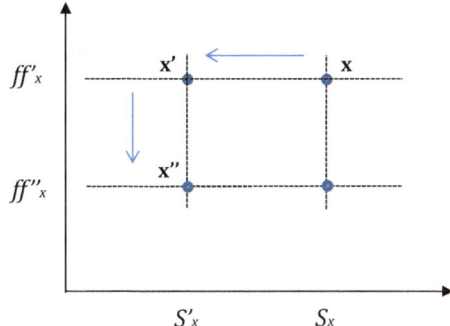

Fig. 6 Risk–cost optimization programs for noncritical (y) assets

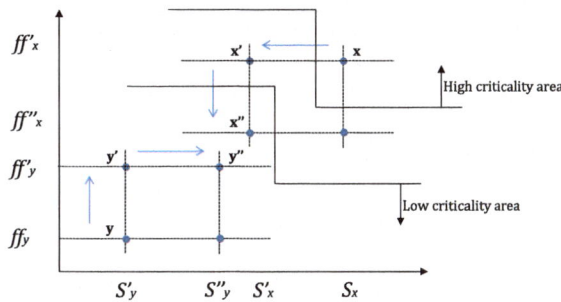

(iv) At time '*t*', the opportunity for an operational reliability improvement was detected. For instance, through a benchmarking of industry standards for current assets operational conditions existing PM programs were optimized; or for example, operational reliability enhancements were discussed with equipment vendors offering their experience and best estimates to consider environmental factors and current operating conditions inside existing PM programs.

(v) An elimination of failure latent root causes in assets was accomplished. Some programmatic and organizational measures were introduced to discard these latent failure causes (Root Cause Failure Analysis Effect).

(vi) There have been proper adjustments of the PM program to existing asset operational conditions by accomplishing a RCM program (reliability centered maintenance effect). RCM programs reached lower failure rates in critical assets, but also possibly lower failure consequences as explained in (ii). RCM programs should be carried out once no abnormal functional loss frequency, in high severity equipment, is found (i.e., it is convenient once point v is accomplished).

- In many occasions, cost–risk–benefit analysis may lead to the discard of PM activities for noncritical assets (for instance, for those in the white area in Fig. 1). Assets with criticality (S'_y, ff_y), as shown in Fig. 6, could become the target of the analysis if they deserve important preventive attention at the moment of the analysis (favorite candidate assets would be: j, f & e in Fig. 1, notice how asset c is in the white area, but could potentially become semi-critical in case of increasing its functional loss frequency, therefore it would not be a favorite candidate asset for this analysis).

The PM tasks to discard should not be those ones avoiding early equipment deterioration; otherwise a significant increase in the LCC of the asset could happen. Nevertheless, even taking into account this consideration, we may expect an increase in functional loss frequency (see Fig. 6, $ff_y \rightarrow ff'_y$) and criticality (S'_y, ff'_y) as a result of less monitoring, inspection, or calibration activities on the asset (activities typically discarded or reduced after the analysis). Severity of the assets failures should also be under control before any discard of PM activities. Functional loss severity could make the asset to exceed the low criticality area, white area in the matrix, increasing business risk. Resulting criticality after discarding PM activities (S'', ff'_y) should remain within the white area in the matrix (see Fig. 6).

It is important to notice that failures with minor consequences tend to be allowed to occur, precisely because they may not matter very much. As a result, large quantities of historical data are normally available concerning these failures, which mean that there will be sufficient material for accurate actuarial analyses if required.

- In other occasions, assets may remain with a high severity and very low functional loss frequency (see asset b in Fig. 1). To these elements, the Resnikoff Conundrum is applicable (this conundrum states that in order to collect failure data, there must be equipment failures, but failures of critical items are considered unacceptable, causing damage, injury and death. This means that the maintenance program for a critical item must be designed without the benefit of failure data which the program is meant to avoid). For a failure with serious consequences, the body of data will never exist, since preventive measures must of necessity be taken after the first failure or even before it. Thus actuarial analysis cannot be used to protect operating safety. This contradiction applies in reverse at the other end of the scale of consequences, as we have discussed in the previous bullet. Therefore, for these elements, maintenance professionals should turn their attention away from counting failures (in the hope that an elegantly constructed scorecard will tell us how to play the game in the future), toward anticipating or preventing failures which matter [14].

5   Validation of the Model and the Model Strengths

Since 2007, authors have implemented this model in a series of critical infrastructures in Spain and South America, through different collaboration and R&D projects with a clear purpose of infrastructure maintenance reengineering and alignment to business needs. The type of infrastructure analyzed were: Electrical power generation plants (all types, including renewable energy plants), network utilities (electricity, gas & water), transportation systems (like the one we use as an example in this paper) and networks, army warships, etc. Over 250,000 assets have been ranked in different type of plants or infrastructure in general. Over these years, the methodology depicted in this paper has been upgraded and different utility models (including software tools) have been developed.

As an example, average number of assets ranked per power plant was over 9900, while for regasification plants over 14,000 assets, or 700 (non-repetitive assets) for a train model. There is no reference in literature to a methodology dealing with such a massive asset criticality assessment.

Again, as an example trying to validate the methodology, Fig. 7 presents a case study where 9921 systems are analyzed in a four years old power plant. Purpose of the study was to audit current maintenance management. For this particular case, after a first round of analysis (prior to the method application) the review team decided to rank only 4816 of those assets, considered now as technical locations (TLs) in the ERP of the company. The only reason for this was the consideration of the rest of the assets (5015, up to the total amount 9921) as auxiliary equipment of the plant that were not relevant for the suggested maintenance study.

After 2 months, the criticality analysis process was finished, the team realized that 2123 technical locations of the plant (Fig. 7) had assets with preventive maintenance plans (MPs) assigned having null or very low severity in consequences and deserving probably less maintenance efforts. It was curios to appreciate (see Fig. 8) how preventive maintenance efforts in low criticality assets, after 4 years of operation of the plant and considering a planning horizon of 5 more years, would be almost 40% higher than in medium and high criticality items. For this case study around 70% of the 41,387.7 h were discarded, directly changing hours assigned to them in the MPs introduced in ERP, and 10% of those hours were dedicated to high criticality equipment preventive MPs. After 5 years of operation, in 2013, same performance of the plant was reached with a 30% savings in overall maintenance cost (direct & indirect cost) only by redirecting and aligning maintenance efforts to

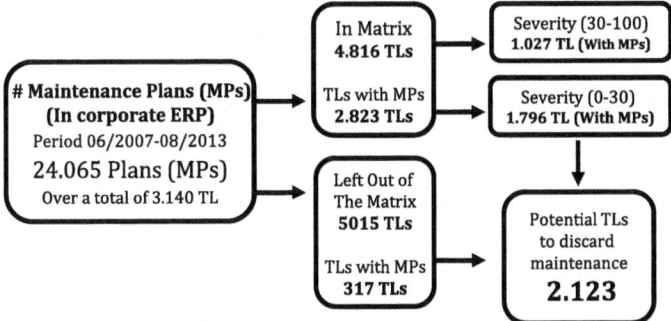

Fig. 7 Case Study: Priority for 9921 systems—power plant—Results 1

Fig. 8 Case Study: Priority for 9921 systems—power plant—Results 2

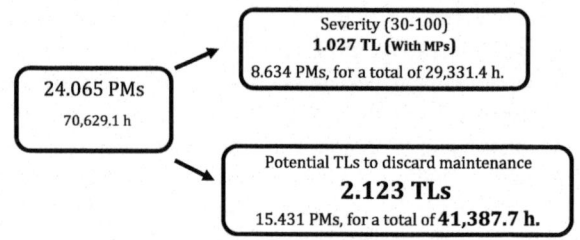

business needs using this criticality analysis. Similar results have been obtained when dealing with this issue in other scenarios, even for companies showing high maturity levels in many in maintenance related topics.

6 Conclusions

This paper contains the design of a process and model for criticality analysis with maintenance purposes and specific design constraints. The methodology ensures analysis consistency to business needs and for existing data of in-service complex engineering assets. At the same time, there is an effort to describe how to turn this process into a practical management tool. Issues arising related to extensive data handling and easy results representation are addressed. Finally, guidelines for results interpretation are offered. The authors believe that this type of analysis will become a must for complex in-service assets maintenance strategy review and redesign. Further research can use this methodology for the improvement of specific operational scenarios, or to refine the different steps of the process presented in this work.

Acknowledgements This research is funded by the Spanish Ministry of Science and Innovation, Project EMAINSYS (DPI2011-22806) "Sistemas Inteligentes de Mantenimiento. Procesos Emergentes de E-maintenance para la Sostenibilidad de los Sistemas de Produccio´n", besides FEDER funds.

References

1. EN 13306:2010 (2010) Maintenance terminology. European Standard. CEN (European Committee for Standardization), Brussels
2. Wireman T (1998) Developing performance indicators for managing maintenance. Industrial Press, New York
3. Crespo Márquez A (2007) The maintenance management framework. Models and methods for complex systems maintenance. Springer Verlag, London
4. Moss TR, Woodhouse J (1999) Criticality analysis revisited. Qual Reliab Eng Int 15:117–121
5. US Department of Defense (1977) Procedures for performing a failure mode and effects analysis, MILSTD 1629A, 1977

6. Palaez CE, Bowles JB (1994) Using fuzzy logic for system criticality analysis. Proc Reliab Maintainability Symp, Anaheim, CA, January, pp 449–455
7. Pillay A, Wang J (2003) Modified failure mode and effects analysis using approximate reasoning. Reliab Eng Syst Saf 79:69–85
8. NORSOK Standards (2001) Criticality analysis for maintenance purposes. NORSOK standard Z-008 Rev. 2, Nov
9. Duffuaa SO, Raouf A, Campbell JD (2000) Planning and control of maintenance systems. Wiley, Indianapolis
10. Taghipour S, Banjevic D, Jardine AKS (2010) Prioritization of medical equipment for maintenance decisions. J Oper Res Soc 62(9):1666–1687
11. Arunraj NS, Maiti J (2007) Risk-based maintenance -techniques and applications. J Hazard Mater 142:653–661
12. Brown SJ, May IL (2000) Risk-based hazardous protection and prevention by inspection and maintenance. Trans ASME J Press Ves Technol 122:362–367
13. Khan FI, Haddara M (2003) Risk-based maintenance (RBM): a new approach for process plant inspecton and maintenance. Process Saf Prog 23(4):252–265
14. Moubray J (1999) Reliability-centered maintenance, 2nd edn. Industrial Press Inc, New York
15. Bevilacqua M, Braglia M (2000) The analytic hierarchy process applied to maintenance strategy selection. Reliab Eng Syst Saf 70:71–83
16. Molenaers A, Baets H, Pintelon L, Waeyenbergh G (2012) Criticality classification of spare parts: a case study. Int J Prod Econ 140:570–578
17. Moubray J (2014) Maintenance management. A new paradigm. Reliability Web.com. http://reliabilityweb.com/index.php/articles/maintenance_management_a_new_paradigm/

Author Biographies

Adolfo Crespo Márquez is currently Full Professor at the School of Engineering of the University of Seville, and Head of the Department of Industrial Management. He holds a Ph.D. in Industrial Engineering from this same University. His research works have been published in journals such as the International Journal of Production Research, International Journal of Production Economics, European Journal of Operations Research, Journal of Purchasing and Supply Management, International Journal of Agile Manufacturing, Omega, Journal of Quality in Maintenance Engineering, Decision Support Systems, Computers in Industry, Reliability Engineering and System Safety, and International Journal of Simulation and Process Modeling, among others. Prof. Crespo is the author of seven books, the last four with Springer-Verlag in 2007, 2010, 2012, and 2014 about maintenance, warranty, and supply chain management. Prof. Crespo leads the Spanish Research Network on Dependability Management and the Spanish Committee for Maintenance Standardization (1995–2003). He also leads a research team related to maintenance and dependability management currently with 5 Ph.D. students and 4 researchers. He has extensively participated in many engineering and consulting projects for different companies, for the Spanish Departments of Defense, Science and Education as well as for the European Commission (IPTS). He is the President of INGEMAN (a National Association for the Development of Maintenance Engineering in Spain) since 2002.

Pedro Moreu de León is an expert in maintenance engineering and maintenance management. He holds a Ph.D. in Industrial Management from the University of Seville. He is author of several books and publications on these topics, is Chairman of the Technical Committee for Standardization of Maintenance CTN AEN-151 "Maintenance" of AENOR (Spanish Association for Standardization) (2002–2014) and Expert Spanish Officer (AENOR) and Secretary Delegation European Committee for Standardization CEN 319 "Maintenance" and

Convenor of the Working Group for Maintenance Management standard (WG7, "Maintenance Management") of the Committee. Dr Moreu is an expert in maintenance of different process plants. He has participated in many engineering and consulting projects for different companies. He is currently member of the board of Directors of INGEMAN (a National Association for the Development of Maintenance Engineering in Spain) since 2002.

Antonio Sola Rosique is a civil engineer currently acting as Vicepresident of INGEMAN, (a National Association for the Development of Maintenance Engineering in Spain). His professional experience (34 years) is very much related to the field of dependability and maintenance engineering in different types of power generation plants for Iberdrola Generación (nuclear, thermal, hydro and combined cycles gas-steam). Antonio has coordinated various collaborative projects between Iberdrola and the University of Seville, and is active member in the board of European Safety, Reliability and Data Association (ESREDA), Asociación Española de Mantenimiento (AEM), Asociación Española de la Calidad (AEC) and Asociación Española de Normalización (AENOR). At present Antonio is about to finish a Ph.D. within the field of Maintenance and Risk Management in the University of Seville.

Juan Francisco Gómez Fernández is Ph.D. in Industrial Management and Executive MBA. He is currently part of the SIM research group of the University of Seville and a member in knowledge sharing networks about Dependability and Service Quality. He has authored a book with Springer Verlag about Maintenance Management in Network Utilities (2012) and many other publications in relevant journals, books and conferences, nationally and internationally. In relation to the practical application and experience, he has managed network maintenance and deployment departments in various national distribution network companies, both from private and public sector. He has conduced and participated in engineering and consulting projects for different international companies, related to Information and Communications Technologies, Maintenance and Asset Management, Reliability Assessment, and Outsourcing services in Utilities companies. He has combined his professional activity, in telecommunications networks development and maintenance, with academic life as an associate professor (PSI) in Seville University, and has been awarded as Best Master Thesis on Dependability by National and International Associations such as EFNSM (European Federation of National Maintenance Societies) and Spanish Association for Quality.

AHP Method According to a Changing Environment

Vicente González-Prida Díaz, Pablo Viveros Gunckel, Luis Barberá Martínez and Adolfo Crespo Márquez

Abstract *Purpose* Actual situations evidence how adopted decisions can change the decision constraints of the system where the AHP is being applied. Therefore, this research is intended to provide a dynamic view of the AHP method, considering the criteria and alternatives as temporary variables and finally obtaining not only one good choice for a specific moment but a more comprehensive picture of those alternatives resulting more important for the business, according to strategy and over time. *Design/methodology/approach* With this purpose this paper starts with a short literature review and the general characteristics of the AHP method. Afterwards, this paper presents the problem that appears frequently in actual situations which justify the development of this research. Once described, the uncertainty appeared after the AHP implementation, the proposed methodology called dynamic analytic hierarchy process (DAHP) is presented. *Findings* Finally, this paper shows a case study and concludes with the main points of the research suggesting applications and further extensions. *Originality/value* The value of this paper is the description of a DAHP as a tool that can facilitate decision-making related to some of the critical aspects in maintenance or post-sales area, permitting the alignment of actions with the business' objectives.

Keywords Post-sales service · Analytic hierarchy process · Decision-making · Dynamic environment · Maintenance · Warranty

V. González-Prida Díaz
General Dynamics—European Land Systems, Seville, Spain

P. Viveros Gunckel
Department of Industrial Management, Universidad Técnica Federico Santa María, Valparaíso, Chile

L. Barberá Martínez · A. Crespo Márquez (✉)
Department of Industrial Management, School of Engineering,
University of Seville, Seville, Spain
e-mail: adolfo@us.es

© Springer International Publishing AG 2018
A. Crespo Márquez et al. (eds.), *Advanced Maintenance Modelling for Asset Management*, DOI 10.1007/978-3-319-58045-6_7

1   Introduction

The general aim of the analytic hierarchy process (AHP) is to show an assessment of the best decision, which includes certain judgment incoherencies when subjective opinions are adopted, due to the fact that human judgment is not always consistent [1, 2]. AHP method establishes a series of scales of comparisons, where inputs can be measured as the price, weight, time, provisioning, etc., or even the subjective opinion on how the satisfaction and preference sentiments can be. The AHP has been applied to multiple business situations, as well as political or personal issues, especially when it is necessary to synthesize the knowledge of different specialists to support decisions:

1. Personal situations, where everything is looking on organizing and reflecting internal preferences, for example acquisition of resources, identification of tracks, etc. [3].
2. Political situations in public administrations. Generally it is orientated to consensus achievement or forecasting the future, for example identification of public transport routes, public services allocation, etc. [4].
3. Business situations in private companies. Orientated mainly by competitiveness and improvement, it is used in all situations to achieve objectives: organization, structuring projects, resources allocation, prediction, etc. [5].

Therefore, AHP provides a picture of a specific situation where decisions are made following alternatives and criteria which are in accordance to the decision constraints of this specific moment. However, the adopted decisions may change the decision maker's point of view and the effects of such implementations can change the decision constraints and, consequently, the evaluation of alternatives and criteria. This paper intends to synthesize in a simple way how the analytic hierarchical process at production and sale stage can help to make more appropriate decisions on strategic actions, for example, with regards to spare parts in order to properly assist customer claims or maintenance activities [6–9], including a dynamic view of the scenario where the weights of criteria and alternatives may change with each adopted decision. Therefore, it is important to highlight that the current development of AHP is clearly static, i.e., it does not generate any feedback nor consider the type of industry. Therefore, one main goal of this document is to refer specifically to those important differences between static traditional decision-making procedures, with highly dynamic contexts where a decision in time t, influences the same decision in time $t + 1$.

The tool to apply should be simple and deal with the problem of a multiple criteria decision-making (MCDM). Among the different tools, AHP has been chosen due to its simplicity, the possibility to consider subjective opinions, and its flexibility to provide feedback from the method with the ranking result obtained from a previous time period. A case study on this field is included at the end of this paper.

2 Problem Statement

The AHP provides a logical framework to determine the benefits of each alternative. Using the conventional AHP method, maintenance or post-sales manager can check the influence, for instance, of:

- Purchasing additional spare parts, which would change the weight of the stock level criterion but also the supply term and the supply cost.
- Repairing a failed piece although it would have a longer repair term, instead of purchasing a new one.
- Performing an opportunistic maintenance during any corrective repair, which increases the reliability criterion.
- Etc.

In order to give a better understanding to the reader, these issues can be found listed and exemplified in more details in references like González-Prida and Crespo [10].

This tool allows us to use it as a sensitivity analysis, showing how projected choices can change with qualitative or quantitative variations in the input of key assumptions on which the decision-making is based, and showing in some way a criticality analysis of the issue.

The problem appears when the AHP is implemented and the decision adopted from it can change the initial conditions under which the method was applied. Consequently, once the decision is adopted, the decision constraints of the system under study can change, and the original hypotheses considered at the beginning are no longer valid. In addition to this, sometimes the adopted decision requires some time to be properly implemented (for example, in terms of provisioning times). Therefore, any possible delay adopting the decision can cause that, once the action is implemented, the proper decision could be another completely different one. These situations appear in actual cases so frequently that they require a specific adaptation of the AHP to this changing environment. This adaptation is what is here called Dynamic Analytic Hierarchy Process (DAHP). In short, the DAHP applies the same AHP methodology but considers the influence of the decisions in the decision constraints. Figure 1 is an attempt to illustrate the complexity of the problem in a simple way, just showing the influence of decisions on alternatives.

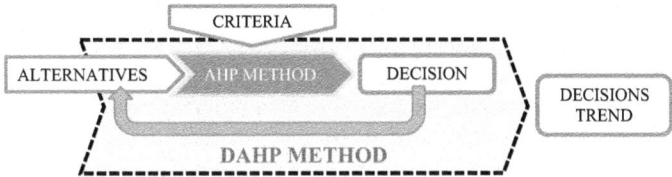

Fig. 1 Relationship between AHP and DAHP

Let's say that, while the AHP provides a fixed picture of a system at a specific moment with its best local decision, the DAHP provides a moving picture of the system where the best decision can be different to the ones calculated at determined moments. It is important here to add that, although costs have been included in AHP and DAHP, in many complex decisions, costs should be set aside until the benefits of the alternatives are evaluated. Otherwise it could happen that the general costs of the maintenance or warranty assistance were too high, not taking care of its benefits. In other words, discussing costs together with benefits can sometimes bring forth many political and emotional results.

3   Short Literature Review and General Characteristics

The AHP is a methodology developed by Thomas Saaty in 1970, based on facilitating the understanding of a complex problem through a breakdown of parts hierarchically ranked (approaches and alternatives), quantifying and comparing alternatives through addition of views with geometrical average to synthesize a solution [11–13]. The process has been used to assist numerous corporate and government decision makers. Some examples of decision problems are: choosing a telecommunication system, choosing a product marketing strategy, etc. In short, problems are decomposed into a hierarchy of criteria and alternatives. AHP is one of the most commonly used MCDM methods. A quick *sciencedirect* search of the term AHP in title, from 2003 to 2013, provides around 300 results. Some recent relevant papers on last developments of AHP are, for example, developed by Ishizaka [14], Ishizaka et al. [15], Ishizaka and Labib [16] and Saaty [17].

A presentation of AHP is not the aim of this paper. For a proper understanding of the whole AHP methodology, authors highly suggest the reader to consult specific literature, such as Saaty [11, 18, 19], González-Prida et al. [20], Ishizaka et al. [21] and Ishizaka and Labib [22]. These references not only make easier the comprehension of the method presented in this paper, but also the following properties of the decision matrix, the axioms considered in AHP, as well as the relationship between criteria and alternatives.

As commented, AHP has been applied to multiple situations. In our case with a dynamic environment, AHP method is orientated to a continuous improvement, also taking into account the following points [18, 19, 23]:

- If the number of alternatives grows, comparisons grow exponentially and the method use can be made cumbersome [24].
- It does not consider criteria variation ranges in a specific t.
- It is more a comparison tool for management than a statistical method [25].
- Valuations of comparisons can be interpreted differently by different subjects.
- Individual comparison may lead to conflicts, because if A > B and B > C, it may not occur A < C.

- Inclusion of a new irrelevant approach may affect management of two relevant criteria [26]. This would contradict axiom of MAVT (Multi Attribute Value Theory) on irrelevant alternatives [27].
- Asymmetrical inconsistency in eigenvectors by Saaty scales [28, 29].

To correct this lack of consistency the Modified AHP Method was developed [28, 29], but also it has its criticisms [30], indicating that improvements in real cases are not as crucial as the original Saaty AHP method. Other published articles about AHP method on decision-making are authored by Belton [31], Bevilacqua and Braglia [32] or González-Prida et al. [20].

4 Proposed Methodology

The methodology is described as follows which could be adapted to each scenario under study. The reader will notice that many parts of this section refer to the classical AHP. Nevertheless, we consider it more useful to present it together with the new version in order to provide a more solid view of the proposed methodology. The DAHP application will be performed following these stages:

(a) State the objective (for instance, it can be to obtain spare parts ranking for maintenance or warranty assistance).
(b) Define the criteria (in the a.m. example this could be the stock level, supply term, repair term, reliability, costs, etc.).
(c) Selection of alternatives or variables in function of time ($f(t)$).
(d) Conventional AHP solving for each value of t.
(e) Analysis of the alternatives trends.

Stages (a) to (c) do not differ to a standard AHP. The novelty here appears in stage (d) when the AHP has to be applied for each value of $t + 1$, taking into consideration the ranking result from t. When defining the criteria, both qualitative and quantitative criteria can be compared using informed judgments to derive weights and priorities. Using pairwise comparisons, the relative importance of one criterion over another can be expressed in c_{ij} according to Table 1. This table will be obtained every time the AHP techniques are applied.

In order to turn this matrix ($n \times n$) into a ranking of weights, the eigenvector solution is the best approach [11]. In order to review its consistency, it is necessary to apply the already stated formula [1] on CI and CR. It is obtained from the

Table 1 Matrix of relative importance between criteria

	C_1	C_2	C_j	C_n
C_1	–	c_{12}	c_{1j}	c_{1n}
C_2	c_{21}	–	c_{2j}	c_{2n}
C_i	c_{i1}	c_{i1}	–	C_{in}
C_n	c_{n1}	c_{n2}	c_{nj}	–

indicated eigenvalues and the matrix of pairwise comparisons, once it is normalized [29]. In terms of the alternatives, a pairwise comparison determines the preference of each alternative over another, according to each criterion. In fact, the aim of the AHP is to rank alternatives, while the aim of criteria pairwise comparison is to derive weights for criteria. It is possible to constitute a 3-D matrix ($m \times m \times n$), where each slide (Fig. 2) refers to a pairwise comparison according to a specific criteria. The novelty here is that values for each matrix cell are not fixed. On the contrary, they depend on those decisions made previously. This dependency on historical data yields finally in a dependency on time. With this point of view, a conventional AHP can be applied for each t, but getting together all these pictures for each t, we finally obtain a dynamical representation of the system behaviour where the system feeds back on itself like in an iterative process.

With this proposal, the new methodology not only combines both qualitative and quantitative information, it also introduces a dynamic view of the solving process due to the fact that the adopted decisions themselves influence in the later system behaviour. The eigenvector of criteria does not have to be computed each time if the criteria pairwise comparison matrix does not change over time. With the mentioned context, the method determines the relative ranking of alternatives under each criterion and time (Table 2).

With the above chart related to the ranking position of each alternative in comparison to the rest of them, it is possible to illustrate graphically these evolutions in order to obtain trends (Fig. 3).

Therefore, the best decision under a dynamic context is not always what the AHP offers at a first glance, but the alternative which remains in prior positions in the ranking of alternatives with the development of time. The ranking obtained with the conventional AHP will refer to good decisions from a local point of view, but not necessarily under a temporary context. Therefore, de DAHP considers the time factor as a crucial aspect for making long-term decisions properly [17]. At each specific stage, the best alternative is chosen as the decision from a local point of view. For instance, in Fig. 3, V_m is chosen from time t_1 to t_4, then V_2 is chosen until t_9. Details will be clarified in the case study.

Fig. 2 Alternative matrices function of time

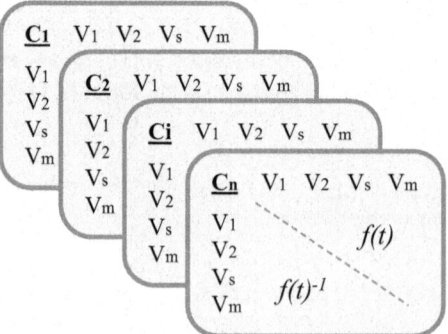

Table 2 Ranking values for each alternative and time

	t_0	t_1	...	t_f
V_1	v_{10}	v_{11}	...	v_{1f}
V_2	v_{20}	v_{21}	...	v_{2f}
V_s	v_{s0}	v_{s1}	...	v_{sf}
V_m	v_{m0}	v_{m1}	...	v_{mf}

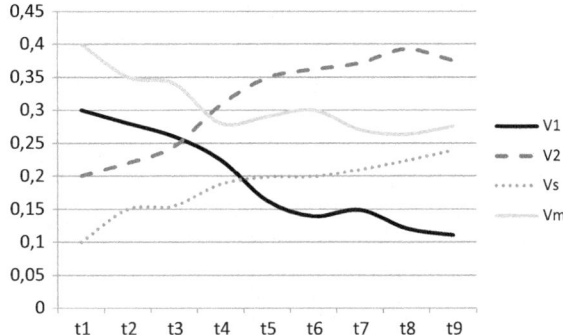

Fig. 3 Example of alternatives scores with time

Next, the methodology presented in this section is going to be detailed in steps. In order to simplify the notation, four generic criteria (A, B, C and D), and three possible alternatives (X, Y and Z) will be considered for a specific objective as illustrated in Fig. 1. Nevertheless, the extrapolation to n criteria and m alternatives would be easy and direct. The dynamic process has been divided into 6 steps (these steps are mainly included in stages (c) and (d) commented in the previous section). The following section (Sect. 5) will present a full case study with data extracted from the mining industry. In order to avoid a lengthy reading of the paper, the amount of equations have been reduced and just the main formulas have been considered, which are included and particularized in matrices for a better understanding of the reader.

4.1 Step 1: The Data Source

In the conventional AHP, specific data are considered for each criterion and alternative. If we deal with four criteria and three alternatives, that means the handling of twelve values. However, with a dynamic context, these 12 values will be time-dependent and can follow different behaviours [17]. For example, values for a particular criterion for an alternative can be pulses [0, 1] or can follow a sinusoidal function between [−1, 1], etc. AHP is a semiquantitative technique, forcing us to consider some subjective important features of the problem as well as the evaluation process and results analysis. Pairwise can take values in different ranges, so the evaluation results are normalized to a range between [0, 1], in order to make the ranking among the different choices more intuitive.

4.2 Step 2: The Evaluation of Each Alternative

Based on the scores from the data sources, it is possible to gather these values in a matrix (Table 3). After the sum of values in columns, the normalized score can be obtained with a first approach of the priority ranking, that is,

$$\text{Alternative X}: \quad \sum V_{ix}(t) / \sum V_{ij}(t) \tag{1}$$

$$\text{Alternative Y}: \quad \sum V_{iy}(t) / \sum V_{ij}(t) \tag{2}$$

$$\text{Alternative Z}: \quad \sum V_{iz}(t) / \sum V_{ij}(t) \tag{3}$$

Alternatives are evaluated based on ranks according to the value of $V_{ij}(t)$ in comparison to the values of the rest of the row, approaching to a priority ranking. As in the classical AHP, the global score of an alternative depends both on its evaluation over the criterion and of the weight of criterion (see next steps). The normalized score for alternative X (for instance) would be: $\sum V_{ix}(t)/\sum V_{ij}(t)$, where $j = X, \ldots, Z$. That means the sum of values of each criterion (for alternative X), divided by the total sum of values for all the alternatives.

4.3 Step 3: Scores Conversion Based on a Normalized Range

Considering that each criterion has values according to a particular range of numbers $[R_{i\_min}, R_{i\_max}]$, it is necessary here to homogenize the different ranges. For that purpose, we consider N_{lower} as the normalized lower bound and N_{upper} as

Table 3 Evaluation based on scores of each alternative

	Alternative X	Alternative Y	Alternative Z
Criterion A	$V_{ax}(t)$	$V_{ay}(t)$	$V_{az}(t)$
Criterion B	$V_{bx}(t)$	$V_{by}(t)$	$V_{bz}(t)$
Criterion C	$V_{cx}(t)$	$V_{cy}(t)$	$V_{cz}(t)$
Criterion D	$V_{dx}(t)$	$V_{dy}(t)$	$V_{dz}(t)$
Sum (i = A, ..., D)	$\sum V_{ix}(t)$	$\sum V_{iy}(t)$	$\sum V_{iz}(t)$

the normalized upper bound. The following formula shows how the different gradient for criterion A is calculated (similar formulation for the rest of criteria):

$$\text{Grad}_A = \left(N_{\text{lower}} - N_{\text{upper}}\right)/\left(R_{A\_\text{min}} - R_{A\_\text{max}}\right), \tag{4}$$

where R_{A_min} and R_{A_max} refers to the minimal and maximal values of the range. Usually, the values for N_{lower} and N_{upper} adopted by experts are:

$$N_{\text{lower}} = 0$$

$$N_{\text{upper}} = 1$$

Once the gradient criterion regarding the new lower and upper bounds is calculated, taking into account the scores matrix from Table 3 it is possible to converse these scores according to a normalized range. Next is shown a formula for Criterion A, Alternative X as an example:

$$\text{Grad}_A x \left[V_{ax}(t) - R_{A\_\text{min}}\right] + N_{\text{lower}} \tag{5}$$

4.4 Step 4: Weighted Scores

Linked to the decision matrix W, we continue here with the concept about weight scores (w_i, where $i = A, ..., D$). The result of the conventional AHP will be expressed as a ranking of alternatives, $\text{Alt}_j(t)$, which is obviously time-dependent. This time dependency is the central issue of the paper. Therefore, independently of the different notation used throughout the diverse literature on AHP, the key aspect is the mentioned time dependency in $\text{Alt}_j(t)$ for the step $t + 1$. In other words, the result in t influences the evaluation in $t + 1$. The manager who applies this method shall decide not only the values for the first iteration, but also the dependency between the scores weight with the alternatives weight resulted as AHP solution (Alt_j, where $j = X, Y, Z$). Consequently,

$$w_i(t+1) = f(\text{Alt}_j(t)) \tag{6}$$

The normalized importance weight for criterion A (and similarly for the rest of criteria) is:

$$\left(w_A^N\right) = w_A\left(\text{Alt}_j\right)\Big/\sum w_i\left(\text{Alt}_j\right) \tag{7}$$

Applying these normalized importance weights, $w_i^N(t)$, to the scores based on a normalized range, we obtain a matrix of normalized and weighted scores which will be used in step 6. For criterion A and Alternative X, the formula is as follows (and similar for the rest of the criteria and alternatives):

$$w_A^N \times [\text{Grad}_A \times [V_{ax}(t) - R_{A\_min}] + N_{lower}]$$ (8)

4.5 Step 5: Priority Vector of Criteria

In the same way as the scores weight, the scale to compare criteria in $t + 1$ will be function of the results in t. These judgements are usually taken intuitively by the manager. Therefore, the manager will consider those rules of dependency as well as the initial conditions (Table 4). Consequently and applying the same notation as in Table 2:

$$c_{ik}\left(\text{Alt}_j\right) = c_{ik}\left(\text{Alt}_j(t)\right) = c_{ik}(t)$$ (9)

where $i, k = A, \dots, D$ and $j = X, Y, Z$.

In order to simplify the notation, the content of the reciprocal matrix (each pairwise judgment) will be expressed as $[p_{ik}]$, where $i, k = A, \dots D$, and the sum of each column and row will be expressed as:

• Columns : $\sum p_{iA}, \sum p_{iB}, \sum p_{iC}, \sum p_{iD}$ $(i = A, \dots, D)$

• Rows : $\sum p_{Ak}, \sum p_{Bk}, \sum p_{Ck}, \sum p_{Dk}$ $(k = A, \dots, D)$

Considering this notation, and normalizing the reciprocal matrix (Table 4), it is possible to obtain the priority vector of criteria. Particularly, for criterion A (and similarly for the rest of criteria) we obtain a scalar applying the following expressions:

Table 4 Reciprocal matrix of pairwise judgments

Criteria	A	B	C	D
A	1.00	c_{AB} (Alt$_j$)	c_{AC} (Alt$_j$)	c_{AD} (Alt$_j$)
B	$1/c_{AB}$(Alt$_j$)	1.00	c_{BC} (Alt$_j$)	c_{BD} (Alt$_j$)
C	$1/c_{AC}$(Alt$_j$)	$1/c_{BC}$ (Alt$_j$)	1.00	c_{CD} (Alt$_j$)
D	$1/c_{AD}$(Alt$_j$)	$1/c_{BD}$ (Alt$_j$)	$1/c_{CD}$ (Alt$_j$)	1.00
Sum	$\sum p_{iA}$	$\sum p_{iB}$	$\sum p_{iC}$	$\sum p_{iD}$

- Sum by row:

$$p_{AA}\Big/\sum p_{iA}+p_{AB}\Big/\sum p_{iB}+p_{AC}\Big/\sum p_{iC}+p_{AD}\Big/\sum p_{iD}$$

- Priority vector:

$$\left[p_{AA}\Big/\sum p_{iA}+p_{AB}\Big/\sum p_{iB}+p_{AC}\Big/\sum p_{iC}+p_{AD}\Big/\sum p_{iD}\right]\Big/\text{Sum}$$

where

$$\text{Sum} = \sum p_{Ak}\Big/\sum p_{iA}+\sum p_{Bk}\Big/\sum p_{iB} \\ + \sum p_{Ck}\Big/\sum p_{iC}+\sum p_{Dk}\Big/\sum p_{iD} \tag{10}$$

In order to analyze the consistency of the pairwise judgments, we calculate now the Lambda max (λ_{max}) as the maximum eigenvalue of the matrix, the consistency index (CI) and the Consistency Ratio (CR):

$$\lambda_{max} = \Big[\sum p_{iA}+\sum p_{iB}\times\Big(p_{BA}\Big/\sum p_{iA}+p_{BB}\Big/\sum p_{iB}+p_{BC}\Big/\sum p_{iC}+p_{BD}\Big/\sum p_{iD}\Big) \\ + \sum p_{iC}\times\Big(p_{CA}\Big/\sum p_{iA}+p_{CB}\Big/\sum p_{iB}+p_{CC}\Big/\sum p_{iC}+p_{CD}\Big/\sum p_{iD}\Big) \\ + \sum p_{iD}\times\Big[\Big(p_{DA}\Big/\sum p_{iA}+p_{DB}\Big/\sum p_{iB}+p_{DC}\Big/\sum p_{iC}+p_{DD}\Big/\sum p_{iD}\Big)\Big] \\ \Big/\Big[\sum p_{Ak}\Big/\sum p_{iA}+\sum p_{Bk}\Big/\sum p_{iB}+\sum p_{Ck}\Big/\sum p_{iC}+\sum p_{Dk}\Big/\sum p_{iD}\Big] \tag{11}$$

$$\text{CI} = (\lambda_{max}-n)/(n-1) \tag{12}$$

$$\text{CR} = \text{CI}/\text{CI}_{random} \le 0,1 \tag{13}$$

where $n = 4$ (in this case, we consider four criteria), and CI_{random} proceeds from the chart titled random consistency index (Table 5) described in the classical AHP [11].

With all those time-dependent functions, it can occur that CR > 0, 1. In that situation, the manager in charge of the decision-making has to manually readapt the pairwise judgments in order to give consistency to the whole process.

Table 5 Random consistency index

n	1	2	3	4	5	...	9	10
RI	0.00	0.00	0.58	0.90	1.12	...	1.45	1.49

4.6   Step 6: Overall Composite Weight of the Alternatives

Multiplying:

$$
\begin{aligned}
&[\text{Matrix of normalized weighted scores}]_{3\times4}\times[\text{Priority vector of criteria}]_{4\times1} \\
&= [\text{Ranking of alternatives}]_{3\times1}
\end{aligned} \quad (14)
$$

This equation takes into account, on one hand, the alternative evaluation (using w_{ij}) and, on the other hand, the priority vector of criteria. The obtained ranking of alternatives is also a priority vector that can be normalized and expressed in percentages, where the highest ranked alternative can be taken. This priority vector of alternatives is a local AHP solution for time t, which feeds back the whole process in $t+1$. Finally, with a specific amount of local solutions, the manager can observe trends which can help make decisions from the general behaviour of the whole system. Therefore, it is important to note that the obtained result is a ranking of alternatives in a specific period, but it is not the global score, which requires the analysis of further periods in order to observe the trends in the whole system.

5   Case Study from the Mining Industry

This section presents a case study that implements the proposed method DAHP. The case covers the main concepts of critical equipment by prioritizing the traditional AHP model, and integrating dynamic phenomena that modify the absolute weighting with the criticality evaluation criteria, which are obviously special and unique to the operational scenario where the methodology is implemented. This case study refers to the criticality evaluation of electric engine components, which drives the main equipment (crushers), applied during the secondary crushing process (Comminution Process) in a copper mine in northern Chile.

5.1   Case Study Scenario

Comminution is a process that is used to reduce the size of an extracted material; it is performed to separate two minerals from each other and to achieve the optimal size for manipulation of a material within an industrial process. The comminution process is used extensively in extractive metallurgy [33]. A mineral extracted from a mine inevitably has a wide particle size distribution, ranging from particles of less than 1 mm to fragments larger than 1 m of diameter: in such a situation, the objective of comminution is to reduce the size of the larger fragments down to a uniform small size. Crushing and grinding are the two primary comminution processes. Crushing is normally carried out on 'run-off-line', whereas grinding (normally carried out after crushing) may be conducted on dry or slurred material. Generally, the extracted mineral is crushed in three stages to a size suitable for the later stages of the production process. The primary crusher reduces the size of the material from the mine, which is then transported by a conveyor belt to a stockpile. After that, some feeders located at the base of the large stockpile load the material onto a conveyor belt that feeds the secondary and tertiary crushers. The equipment selected (Electric Engine) belongs specifically to a mining process denominated: secondary crushing.

- Medium Voltage Electric Engine

The equipment analyzed is a medium voltage engine that performs the operation through pulleys to the secondary crushers crushing plant. The engines used have special treatment due to the 4.6 kV voltage level and its criticality in the crushing process, so any failure generates an operating loss of the main equipment.

From the maintenance point of view, the electric engine can be broken down into the following main items:

- Bearing load side and fan side
- Grease bearings
- Fan cooling system
- Medium voltage junction box
- RTD connections box engine
- Stator
- Rotor.

5.2 DAHP Implementation

The first step of AHP is to define the objective according to the alternatives to be evaluated. In this particular case, the objective for using DAHP is to decide what kind of failure mode has the worst effect on the system, according to some specific maintenance criteria. The result may influence the next period of time by deciding (for instance) a better maintenance policy for the system under study. Based on the main engine components, we proceed to implement the proposed methodology.

- Stage 1: Problem description and decision constraints.

The medium voltage electric engine is critical to the operation of the secondary crushing process, having a functional dependency and logic condition equivalent to a serial system, i.e. any engine failure automatically generates an immediate loss in the production process, which is equivalent to the capacity of each crushing unit. Since 2006, criticality matrices have been developed related to the main subsystems, using as a basis the AHP model, which has been developed by different multidisciplinary teams actively involved in analyses and evaluations. As parallel phenomena to the technical characteristics of the equipment, it is important to show the following data:

1. The organization has a high staff turnover, both at headquarters and at the operational level.
2. Multidisciplinary teams have constantly changed in the organization.
3. Considerable improvements have been developed in the process, such as: increase main component redundancy, acquisition of a replacement engine for the secondary crusher system, change of suppliers with more reliable components, development of digital procedures to improve maintenance performance and equipment overhaul, among others.
4. The Maintenance Department has had three managers in the last four years, which means a continuous variation of maintenance policies and strategies.
5. Policies regarding health, safety and environment are stricter each year. Even since 2010, safety is equivalent to production in terms of weight and importance.

- Stages 2–4: Criteria and alternatives assessed.

  Criteria are shown in the next chart (Table 6):

  Failure Frequency (FF) criterion values the frequency of the phenomenon of functionality failure/loss of the component or failure mode assessed.

  Repair Time (TR) criterion assesses how quickly the component or failure mode that occurs in the system is returned to the operating condition.

  Repair Cost (RC) criterion rates the cost associated with repairing the failure that occurs. It is assumed that 100% of the needed cost to return to the operating condition is a required standard.

  Safety (SE) criterion rates the risks, in terms of physical integrity of people that occurs when a failure takes place [34].

  Next, the alternatives to be evaluated according to the criteria described above and for the main failure mode related to each component, denominator in brackets, are:

- Bearing load side and fan side: Break of the load bearing side and the bearing end shield (ALT1 and ALT2).
- Greases bearings: bearings lubricant loss (ALT3).
- Fan cooling system: engine overheating (ALT4).
- Medium voltage junction box: Loss of engine power (ALT5).
- RTD connections box engine: Loss of engine protection system (ALT6).

Table 6 Weighing criteria for each system component

Analysis criteria		Range	Scale
FF	Failure frequency	1–9	1: Low
TR	Repair time		5: Medium
RC	Repair cost		
SE	Safety		9: High

- Stator: Loss of stator insulation (ALT7).
- Rotor: Magnetization loss (ALT8).

As an example, let us consider that result obtained with this methodology using the presented alternatives may influence the maintenance policies to be applied on the system (as already commented above).

- Stages 5 and 6: Defining Pair Comparison Matrix and Evaluation Matrix.

The evaluation matrices for the three periods (2008, 2010 and 2012) are summarized in the following chart (Table 7).

In this case, each criterion has the same evaluation scale, which considerably simplifies the data processing and later analysis. Table 8 presents the paired comparison matrices for each period, with the corresponding calculation of geometric means and the priority vector, which should be further validated through consistency analysis, identifying the indicators: Lambda max (λ_{max}), the Consistency Index (CI) and the Consistency Ratio (CR). These charts are the key point of the proposed method, since they represent the time dependency. They are obtained applying the conventional AHP (in this particular case) three times. The

Table 7 Evaluation based on scores for each criterion for each period

	Range			Year 2008	Year 2010	Year 2012
	min	max		1	2	3
Criterion FF	1	9	ALT 1	3	3	3
	1	9	ALT 2	3	3	3
	1	9	ALT 3	1	1	1
	1	9	ALT 4	1	1	1
	1	9	ALT 5	2	2	2
	1	9	ALT 6	2	2	2
	1	9	ALT 7	1	1	1
	1	9	ALT 8	1	1	1
Criterion TR	1	9	ALT 1	3	3	3
	1	9	ALT 2	2	2	2
	1	9	ALT 3	1	1	1
	1	9	ALT 4	5	5	5
	1	9	ALT 5	1	1	1
	1	9	ALT 6	5	5	5
	1	9	ALT 7	6	6	6
	1	9	ALT 8	9	6	6
Criterion RC	1	9	ALT 1	9	9	9
	1	9	ALT 2	6	6	6
	1	9	ALT 3	3	3	3
	1	9	ALT 4	6	6	6
	1	9	ALT 5	9	9	9
	1	9	ALT 6	3	3	3
	1	9	ALT 7	4	4	4
	1	9	ALT 8	9	9	9
Criterion SE	1	9	ALT 1	3	3	3
	1	9	ALT 2	3	3	3
	1	9	ALT 3	1	1	1
	1	9	ALT 4	5	3	2
	1	9	ALT 5	2	2	2
	1	9	ALT 6	5	3	2
	1	9	ALT 7	3	3	3
	1	9	ALT 8	3	3	2

Table 8 Reciprocal matrixes of pairwise judgments

2008	FF	TR	RC	SE	G. Mean	P. Vector		
FF	1	2	4	2	2,00	0,44	λ	4,065
TR	0,5	1	2	2	1,19	0,26	CI	0,022
RC	0,25	0,5	1	0,5	0,50	0,11	Ri	0,900
SE	0,5	0,5	2	1	0,84	0,19	CR	2,41%

2010	FF	TR	RC	SE	G. Mean	P. Vector		
FF	1	4	4	1	2,00	0,43	λ	4,121
TR	0,25	1	2	0,5	0,71	0,15	CI	0,040
RC	0,25	0,5	1	0,5	0,50	0,11	Ri	0,900
SE	1	2	2	1	1,41	0,31	CR	4,49%

2012	FF	TR	RC	SE	G. Mean	P. Vector		
FF	1	4	4	0,5	1,68	0,31	λ	4,061
TR	0,25	1	1	0,125	0,42	0,08	CI	0,020
RC	0,25	1	1	0,25	0,50	0,09	Ri	0,900
SE	2	8	4	1	2,83	0,52	CR	2,25%

results of t (for example 2008) influence the following year before the conventional AHP is applied to $t + 1$ (for example 2009), and so on. Consistency results for each pairwise comparison matrix are accepted because they comply with the rule CR $\leq$ 10% (which is of course a questionable threshold as far as it can be 5% or 20%). This condition allows continuing with the result analysis and the ranking, criticality and dynamism search in the last three period's assessments [35].

According to the scale defined at the beginning of the case study, critical equipment for each period are those with the highest percentage in their results.

As noted, these matrices are obtained applying the conventional AHP in each t. Therefore, there should be as many reciprocal matrices as there are periods under study, hence only one matrix is obtained for each t.

5.3 Results and Consequences After Implementing DAHP

The ranking of alternatives during the three analyzed periods are shown in Fig. 4 and Table 9. This table details the results in charts with the hierarchy percentage (from most critical to least critical), and the cumulative for the development of Pareto analysis, where the dynamic behaviour in criticalities is easily identifiable, both in the continuous variations of criticality and in the evaluation of each alternative year after year, see in Figs. 4 and 5. Pareto analysis is only a complementary graphical representation to display hierarchies, allowing readers to compare the variation in years quickly and easily.

Fig. 4 Evolution of
alternative rankings

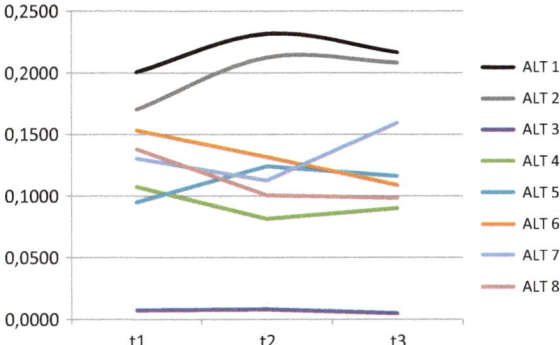

Table 9 Results and ranking of alternatives

2008

Ranking	%	sum %
ALT1	16,18%	16,18%
ALT6	14,80%	30,97%
ALT8	14,79%	45,76%
ALT2	14,68%	60,44%
ALT4	12,32%	72,76%
ALT7	11,20%	83,97%
ALT5	10,64%	94,61%
ALT3	5,39%	100,00%

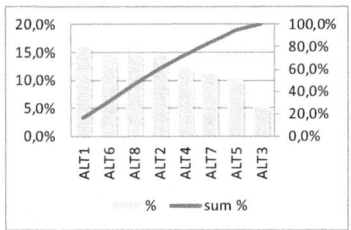

2010

Ranking	%	sum %
ALT1	17,22%	17,22%
ALT2	16,03%	33,24%
ALT6	13,85%	47,10%
ALT8	12,62%	59,71%
ALT5	11,61%	71,32%
ALT7	11,51%	82,84%
ALT4	11,43%	94,26%
ALT3	5,74%	100,00%

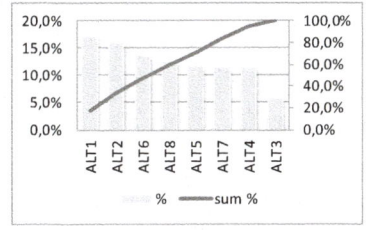

2012

Ranking	%	sum %
ALT1	17,81%	17,81%
ALT2	16,98%	34,79%
ALT7	13,25%	48,03%
ALT5	12,17%	60,20%
ALT6	12,11%	72,31%
ALT8	11,29%	83,60%
ALT4	10,46%	94,06%
ALT3	5,94%	100,00%

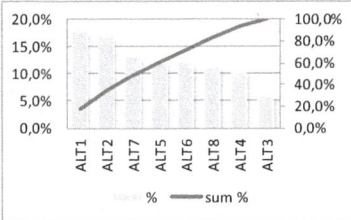

Another important analysis is the direct pattern search, such as the tendency of
each alternative in time, which enables understanding, and design in some cases, of
the evolution of criticality of the components. Figure 5 presents exponential

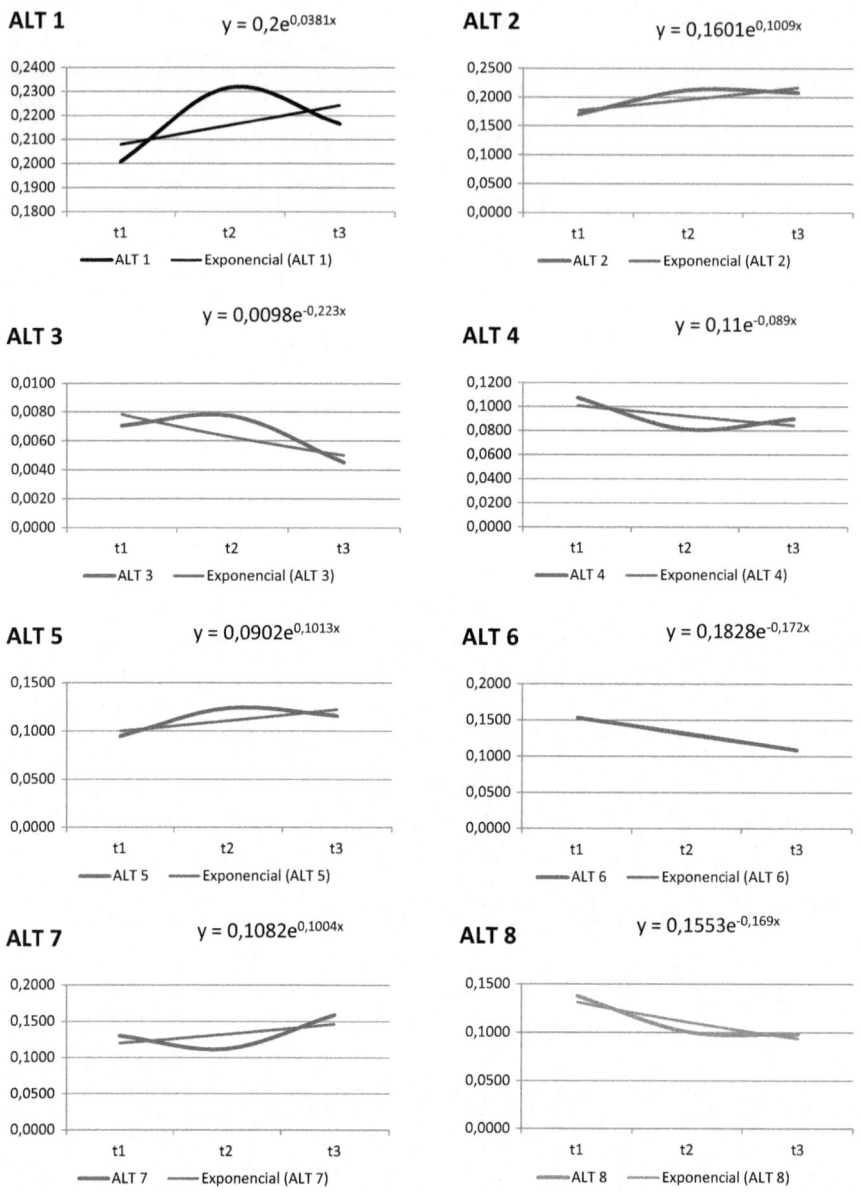

Fig. 5 Exponential approach for alternatives trend

approaches for each alternative, and clearly shows the behaviour that follows each alternative. Future research can examine and assess the parameters of the exponential law, bringing further information for the present analysis.

The mathematical expressions for these approaches are as follows:

$$\text{Alternative 1:} \quad y = 0.1549e^{0.0481\times} \tag{15}$$

$$\text{Alternative 2:} \quad y = 0.1372e^{0.0727\times} \tag{16}$$

$$\text{Alternative 3:} \quad y = 0.0516e^{0.0481\times} \tag{17}$$

$$\text{Alternative 4:} \quad y = 0.134e^{-0.082\times} \tag{18}$$

$$\text{Alternative 5:} \quad y = 0.1002e^{0.0672\times} \tag{19}$$

$$\text{Alternative 6:} \quad y = 0.1655e^{-0.1\times} \tag{20}$$

$$\text{Alternative 7:} \quad y = 0 = .1011e^{0.0837\times} \tag{21}$$

$$\text{Alternative 8:} \quad y = 0.1679e^{-0.135\times} \tag{22}$$

During the evaluated periods, the ranking of components considerably changed. This is a fact that must be understood as part of production systems dynamics, which may depend for instance on: the change of a provider associated to one or more components; the improvement on systems design and/or technology; the proper management of critical spare parts; the implementation of technical training programmws for workers or higher turnover; or the development of criticality analysis by a multidisciplinary team that changes or rotates in time.

Future applications can provide a positive externality of this methodology as the ability to audit existing continuous improvement in organizations, whether with a problem of components, main equipment or failure modes hierarchy [36]. That means, companies must develop sustainable continuous improvement processes in time [6, 7], which are specific and aimed effectively and efficiently to the critical points of the process. That is why the dynamic behaviour in production systems must be understood as an opportunity that highlights the correct or incorrect decisions being made.

6 Conclusions

The traditional AHP method as applied in the area of maintenance tries to answer questions such as: What information should be relevant to make decisions regarding the maintenance or the post-sales service? How to select a policy for maintenance or warranty assistances in order to improve the profit and the image of the company? This paper adds the possibility to track, over time and on a common scale, those variables (alternatives) becoming more relevant to the business when strategy is changing dynamically according to emerging market needs. This can add value to managers by identifying trends of issues becoming critical to the business, and can foster more proactive business behaviour.

A new modified version of the AHP methodology, called DAHP, has been developed and presented in this paper. Using DAHP, the case study is linked to real decisions as far as the obtained tendency of each failure mode in time, enables the manager to choose different maintenance policies over time. The case study shows that the fact of decision constraints that change gradually with each adopted decision is a phenomenon that occurs in highly dynamic systems, so the criticalities resulted affected by changes in the production plan. In other words, the feedback of decision taken at time t, influences the AHP parameters at time $t + 1$.

The proposed DAHP significantly strengthens the vision of how easily it can be to analyse a specific problem, since the dynamism of important qualitative and quantitative characteristics of the organizations can be conceptually and mathematically integrated factors that must be present in making decisions for the present and future. This is particularly true today as organizations must react to the dynamic environment in which they are embedded, either by intrinsic characteristics of the business (e.g. price), as well as by the organizational culture that evolves over time. In the particular case of the current criticality assessment methods, a fundamental step to identify opportunities of improvement and define strategies.

References

1. Crespo A, Parra C, Gómez JF, López M, and González-Prida V (2012) In: Van der Lei T, Herder P, Wijnia Y (eds) Life cycle cost analysis. Asset management. The state of the art in Europe from a life cycle perspective. Springer, Berlin, pp 81–99. ISBN: 978-94-007-2723-6
2. Moreu P, González-Prida V, Barberá L, Crespo A (2012) A practical method for the maintainability assessment using maintenance indicators and specific attributes. Reliability Eng Syst Saf 100:84–92. ISSN: 0951-8320
3. Rezaei J, Ortt R (2013) Multi-criteria supplier segmentation using a fuzzy preference relations based AHP. Eur J Oper Res 225(1):75–84
4. Serrano-Cinca C, Gutiérrez-Nieto B (2013) A decision support system for financial and social investment. Appl Econ 45(28):4060–4070
5. De Bin L, Yong Y, Meng H (2012) The analysis and application of AHP in a construction project evaluation. Adv Mater Res 446:3740–3744
6. Barberá L, Crespo A, Viveros P, Stegmaier R (2012) Advanced model for maintenance management in a continuous improvement cycle: integration into the business strategy. Int J Syst Assur Eng Manag 3(1):47–63. doi:10.1007/s13198-012-0092-y
7. Barberá L, González-Prida V, Parra C, Crespo A (2012) Framework to assess RCM software tools. In: ESREL conference 2012, Helsinki (Finland)
8. Stegmaier R, Viveros P, Nikulín C, Gonzalez-Prida V, Crespo A, Parra C (2011) Generación de solución iinovadora y sustentable: uso de la metodología RCA y la teoría inventiva TRIZ. In: Babarovich V, Endo A, Pascual R, Stegmaier R (eds) MAPLA conference 2011, 8º Encuentro Internacional de Mantenedores de Plantas Mineras. Antofagasta, Chile
9. Viveros P, Zio E, Nikulín C, Stegmaier R, Bravo G (2013) Resolving equipment failure causes by root cause analysis and theory of inventive problem solving. Proc Inst Mech Eng Part O J Risk Reliability (in press)

10. González-Prida V, Crespo A (2012) A framework for warranty management in industrial assets. Comput Ind 63(9):960–971
11. Saaty TL (1990) How to make a decision: the analytic hierarchy process. Eur J Oper Res 48:9–26
12. Saaty TL (1995) Decision making for leaders. RWS Publications, New York
13. Saaty TL, Vargas LG (1982) Logic of priorities. Kluwer-Nijhoff Publishing, Boston
14. Alessio Ishizaka (2012) A multicriteria approach with AHP and clusters for the selection among a large number of suppliers. Pesquisa Operacional 32(1):1–15
15. Ishizaka A, Balkenborg D, Kaplan T (2011) Does AHP help us make a choice? An experimental evaluation. J Oper Res Soc 62(10):1801–1812
16. Alessio Ishizaka, Ashraf Labib (2011) Review of the main developments in the analytic hierarchy process. Expert Syst Appl 38(11):14336–14345
17. Saaty TL (2007) Time dependent decision-making; dynamic priorities in the AHP/ANP: generalizing from points to functions and from real to complex variables. Math Comput Model 46:860–891
18. Saaty TL (1994) Fundamentals of decision making. RWS Publications, Pittsburgh
19. Saaty TL (1994) Highlights and critical points in the theory and application of the analytic hierarchy process. Eur J Oper Res 74:426–447
20. González-Prida V, Gómez J, Crespo A (2011) Practical applications of AHP for the improvement of waranty management. J Qual Maint Eng 17(2):163–182 (Emerald Grsoup Publishing Limited, pp 1355–2511)
21. Ishizaka A, Balkenborg D, Kaplan T (2011) Influence of aggregation and measurement scale on ranking a compromise alternative in AHP. J Oper Res Soc 62(4):700–710
22. Ishizaka A, Labib A (2011) Selection of new production facilities with the group analytic hierarchy process ordering method. Expert Syst Appl 38(6):7317–7325
23. Harker PT, Vargas LG (1990) Reply to remarks on the analytic hierarchy process. J.S. Dyer. Manage Sci 36(3):269–273
24. Moffett A, Garson J, Sarkar S (2005) A software package for incorporating multiple criteria in conservation planning. Environ Model Softw 20:1315–1322
25. Dyer J (1990) Remarks on the analytic hierarchy process. Manage Sci 36:249–258
26. Dyer J (2005) MAUT-multiattribute utility theory. In: Figueira J, Greco S, Ehrgott M (eds) Multiple criteria decision analysis: state of the art surveys. Kluwer, Dordrecht, pp 265–294
27. Arrow K, Raynaud H (1986) Social choice and multicriterion decision-making. The MIT Press, Cambridge
28. Donegan HA, Dodd FJ, McMaster TBM (1992) A new approach to AHP decision-making. Statistician 41:295–302
29. Donegan HA, Dodd FJ, McMaster TBM (1995) Theory and methodology inverse inconsistency in analytic hierarchies. Eur J Oper Res 80:86–93
30. Zanakis SH, Solomon A, Wishart N, Dublish S (1998) Multi-attribute decision making: a simulation comparison of select methods. Eur J Oper Res 107(3):507–529
31. Belton V (1986) A comparison of the analytic hierarchy process and a simple multi-attribute value function. Eur J Oper Res 26(1):7–21
32. Bevilacqua M, Braglia M (2000) The analytical hierarchy process applied to maintenance strategy selection. Reliability Eng Syst Saf 70(1):71–83
33. Viveros P, Zio E, Kristjanpoller F, Arata A (2011) Integrated system reliability and productive capacity analysis of a production line. A case study for a Chilean mining process. Proc Inst Mech Eng Part O J Risk Reliability
34. Costantino F, De Minicis M, González-Prida V, Crespo A (2012) On the use of quality function deployment (QFD) for the identification of risks associated to warranty programs. In: ESREL conference 2012, Helsinki, Finland
35. González-Prida V, Barberá L, Gómez JF, Crespo A (2012) Contractual and quality aspects on warranty: best practices for the warranty management and its maturity assessment. Int J Qual Reliability Manag 29(3). ISSN: 0265-671X, Emerald

36. González-Prida V, Parra C, Gómez JF, Crespo A (2012) Audit to a specific study scenario according to a reference framework for the improvement of the warranty management. In: Bérenguer G, Guedes S (eds) Advances in safety, reliability and risk management. Taylor & Francis Group, London, pp 2757–2767. ISBN 978-0-415-68379-1

Author Biographies

Vicente González-Prida Díaz has PhD (Summa Cum Laude) in Industrial Engineering from the University of Seville and Executive MBA (First Class Honors) by the Chamber of Commerce. He has been honoured with the following awards and recognitions: Extraordinary Prize of Doctorate by the University of Seville; National Award for PhD Thesis on Dependability by the Spanish Association for Quality; National Award for PhD Thesis on Maintenance by the Spanish Association for Maintenance; and Best Nomination from Spain for the Excellence Master Thesis Award bestowed by the EFNSM (European Federation of National Maintenance Societies). Dr. Gonzalez-Prida is a member of the Club of Rome (Spanish Chapter) and has written multitude of articles for international conferences and publications. His main interest is related to industrial asset management, specifically the reliability, maintenance and after sales organization. He currently works as Program Manager in the company General Dynamics—European Land Systems and shares his professional performance with the development of research projects in the Department of Industrial Organization and Management at the University of Seville.

Pablo Viveros Gunckel Researcher and Academic at Technical University Federico Santa María, Chile, has been active in national and international research, both for important journals and conferences. He has also developed projects in the Chilean industry and consultant specialist in the area of reliability, asset management, system modelling and evaluation of engineering projects. He is an Industrial Engineer and has Master in Asset Management and Maintenance.

Luis Barberá Martínez has PhD (Summa Cum Laude) in industrial Engineering from the University of Seville, and Mining Engineer by UPC (Spain). He is a researcher at the School of Engineering of the University of Seville and author of more than 50 articles, national and main internationals. He has worked at different international Universities: Politecnico di Milano (Italy), CRAN (France), UTFSM (Chile), C-MORE (Canada), EPFL (Switzerland), University of Salford (UK), or FIR (Germany), among others. His line of research is industrial asset management, maintenance optimization and risk management. Currently, he is Spain Operations Manager at MAXAM Europe. He has three Master's degrees: Master of Industrial Organization and Management (2008/2009) (School of Engineering, University of Seville), Master of Economy (2008/2009) (University of Seville) and Master of Prevention Risk Work (2005/2006). He has been honoured with the following awards and recognitions: Extraordinary Prize of Doctorate by the University of Seville; three consecutive awards for Academic Engineering Performance in Spain by AMIC. (2004/2005), (2005/2006) and (2006/2007); honour diploma as a Ten Competences Student by the University of Huelva and CEPSA Company (2007); graduated as number one of his class.

Adolfo Crespo Márquez is currently Full Professor at the School of Engineering of the University of Seville, and Head of the Department of Industrial Management. He holds a PhD in Industrial Engineering from the same university. His research works have been published in journals such as the International Journal of Production Research, International Journal of Production Economics, European Journal of Operations Research, Journal of Purchasing and Supply Management, International Journal of Agile Manufacturing, Omega, Journal of Quality in Maintenance Engineering, Decision Support Systems, Computers in Industry, Reliability Engineering and

System Safety and International Journal of Simulation and Process Modeling, among others. Professor Crespo is the author of seven books, the last four with Springer-Verlag in 2007, 2010, 2012 and 2014 about maintenance, warranty and supply chain management. Professor Crespo leads the Spanish Research Network on Dependability Management and the Spanish Committee for Maintenance Standardization (1995–2003). He also leads a research team related to maintenance and dependability management currently with five PhD students and four researchers. He has extensively participated in many engineering and consulting projects for different companies, for the Spanish Departments of Defense, Science and Education as well as for the European Commission (IPTS). He is the President of INGEMAN (a National Association for the Development of Maintenance Engineering in Spain) since 2002.

Reliability Stochastic Modeling for Repairable Physical Assets

Pablo Viveros Gunckel, Adolfo Crespo Márquez,
René Tapia Peñaloza, Fredy Kristjanpoller Rodríguez
and Vicente González-Prida Díaz

Abstract The reliability modeling, calculating, and projecting for industrial equipment and systems are today a basic and fundamental task for reliability and maintenance engineers, regardless of the nature or genetics of those industrial assets. In this paper, the stochastic models Perfect Renewal Process (PRP), Nonhomogeneous Processes of Poison (NHPP), and GRP are explained in detail with the corresponding conceptual, mathematical, and stochastic development. For each model, the respective conceptualization and parameterization is analyzed in detail. The practical application is developed for a real case in the mining industry, which shows step by step the appropriate stochastic and mathematical development. Finally, this research becomes an analytical and explanatory procedure on the definition, calculation, methodology, and criteria to be considered for industrial assets parameterization with partial or null post maintenance degradation.

Keywords Reliability · Degradation · Repairable assets · Simulation

P. Viveros Gunckel · A. Crespo Márquez (✉) · F. Kristjanpoller Rodríguez ·
V. González-Prida Díaz
Department of Industrial Management, School of Engineering,
University of Seville, Avda, Camino de los Descubrimientos,
s/n. Isla de la Cartuja, 41092 Seville, Spain
e-mail: adolfo@us.es

V. González-Prida Díaz
e-mail: vicente.gonzalezprida@gdels.com

P. Viveros Gunckel · F. Kristjanpoller Rodríguez
Department of Industrial Engineering, Universidad Técnica Federico Santa María,
Valparaíso, Chile

R. Tapia Peñaloza
RelPro SpA, Valparaíso, Chile

© Springer International Publishing AG 2018
A. Crespo Márquez et al. (eds.), *Advanced Maintenance Modelling
for Asset Management*, DOI 10.1007/978-3-319-58045-6_8

1   Introduction

The model and analysis of repairable equipment are of great importance, mainly in order to increase the performance oriented to reliability and maintenance as part of the cost reduction in this last item. A reparable system is defined as follows:

> A system that, after failing to perform one or more of its functions satisfactorily, can be restored to fully satisfactory performance by any method other than replacement of the entire system [1].

Depending on the type of maintenance given to equipment, it is possible to find five cases [2]:

(a) Perfect maintenance or reparation: Maintenance operation that restores the equipment to the condition "as good as new".
(b) Minimum maintenance or reparation: Maintenance operation that restores the equipment to the condition "as bad as old".
(c) Imperfect maintenance or reparation: Maintenance operation that restores the equipment to the condition "worse than new but better than old".
(d) Over-perfect maintenance or reparation: Maintenance operation that restores the equipment to the condition "better than new".
(e) Destructive Maintenance or reparation: Maintenance operation that restores the equipment to the condition "worse that old".

For a perfect maintenance, the most common developed model corresponds to the Perfect Renewal Process (PRP). In it, we assume that repairing action restores the equipment to a condition as good as new and assumes that times between failures in the equipment are distributed by an identical and independent way. The most used and common model PRP is the Homogeneous Processes of Poison (HPP), which considers that the system not ages neither spoils, independently of the previous pattern of failures. That is to say, it is a process without memory. Regarding case (b), "as bad as old" is the opposite case to what happens in case (a) "as good as new", since it is assumed that the equipment will stay after the maintenance intervention in the same state than before each failure. This consideration is based that the equipment is complex, composed by hundreds of components, with many failure modes and the fact that replacing or repairing a determined component will not affect significantly the global state and age of the equipment. In other words, the system is subject to minimum repairs, which does not cause any change or considerable improvement. The most common model to represent this case is through Nonhomogeneous Processes of Poison (NHPP); in this case the most used model to represent NHPP is called "Power Law". In this model, it is assumed a Weibull distribution for the first failure, and that later it is modified over time.

Although the models HPP and NHPP are the most used, they have a practical restriction regarding its application, since a more realistic condition after a repairing action is what we find between both: "worse than new but better than old". In order to find a generalization to this situation and not distinguish between HPP and NHPP, it was necessary to create the Generalized Renewal Process (GRP) [3], which establishes an improvement ratio. Unfortunately, the incorporation of this variable can complicate the analytic calculation of parameters and adjustments of probability. Therefore, its applicability in mathematic terms is complex. For this reason, it has been considered solutions through the Monte Carlo simulation (MC) being one of the most validated methods according the proposal developed by author Krivstov [4] where time series of good functioning is generated through the use of the inverse function of the probability distribution (pdf) that has as a base a random variable.

Understanding the importance and applicability of methods PRP, NHPP, and GRP, this paper introduces the conceptual, mathematical, and stochastic development for each one, as explained and presented briefly in the previous paragraphs. Each model is explained and developed in the following way: conceptualization and parameterizing. In addition, each model will be complemented with a numerical application, specifically it corresponds to two pulp pumps (water, copper concentrate and inert material) used in the mining industry of Chile, which suffer different levels of erosion due to use intensity, geographic height, and of course according to the maintenance type developed in its life, planned or not planned (preventive or corrective maintenance).

With the above, this article begins by introducing stochastic models (PRP, NHPP, and GRP), and then is presented a brief of parameterization processes and a numerical analysis application. Finally, the paper concludes by summarizing the main lines provided by the article and its application to the industrial sector. Note the innovative contribution of this article involving the applicability of IT tools in resolving existing models. Such applicability is presented here as a sample, and with a specific real case of Chilean mining.

With the intention of highlighting the scientific and technical contribution of this article it is necessary to emphasize in this introduction by the following considerations: The wide range and variability of their behavior requires the application of techniques of varying complexity and depth, which can adapt to the best way to each of the realities. The variable that defines and conditions the use of techniques is the state assets remaining after repair.

In this regard, there are five classifications of repair: Perfect, minimal, imperfect, over-perfect, and destructive. For perfect maintenance, it is used and recommended Perfect Renewal Process model (PRP) through homogeneous Poisson processes (HPP). For minimal repair, generally it is represented by Nonhomogeneous Poisson Process (NHPP), the most widely used "Power Law" model. However, for the great application of the aforementioned models, there are various situations that are not covered, since the most of cases repair are between the perfect and minimum conditions (imperfect repair). For these situations, it enunciates and develops the "Generalized Renewal Model" (GRP). Moreover, since the probability of finding the values that give the maximum overall function of maximum likelihood is virtually zero by the random search, it is necessary to define a tolerance value for the partial derivatives ($\partial L/\partial B$ and $\partial L/\partial q$) that they are matched to zero, setting this

Table 1 Notation

Notations	
PRP	Perfect renewal process
NHPP	Nonhomogeneous Poisson process
GRP:	Generalized renewal process
$\lambda(t)$	Failure rate of an element in a given time t
t_i	Function time between failure $i - 1$ and i-th
$f(t)$	Probability density function of failure (pdf) of an element with operation time t
$F(t)$	Probability density function of accumulated failure of an element with operation time
$\Gamma(\cdot)$	Gamma function
$f(t; \theta)$	Probability density function of failure of an element with operation time t, with forma and scale parameters given by vector 0
$L(\theta)$	Likelihood function for parameters vector 0 in a pdf given
MTTF	Mean time to failure
MTTR	Mean time to repair
$\hat{a}$	Estimated value of the parameter a (applicable for all parameters)
A_n	Virtual age of system at the immediate moment of the repair of n-th.
T_n	Virtual age (of operation) at the immediate moment of the repair of n-th
q	Parameter which establishes the defect of the repair
TOL	Numeric grade to consider acceptable a distribution adjustment through the likelihood maximum
TQ	Tolerance that corresponds to higher or lower percentage of the possibilities of value q

tolerance value "TOL" as an acceptable range to consider adjusting distribution. For this, the simulation techniques and especially Monte Carlo emerge as a powerful alternative for resolution. It is recommended to check Refs. [5, 6] to study in details some advantages and defects about the models. The notation used in the models is presented in Table 1.

2 Presentation of the Stochastic Models

2.1 Perfect Renewal Process (PRP)

PRP Conceptualization

The PRP model describes the situation in which a repairable system is restored to a state "as good as new" and the times between failures are considered independent and identically distributed. This process assumes that the equipment restores to an identical condition to the original, as if it is replaced. The graph of the failure rate depending on the time elapsed for equipment with growing failure rate, considering the general case of a Weibull distribution, would be the following in Fig. 1.

Fig. 1 Failure rate $\lambda(t)$ in
PRP. *Source* own elaboration

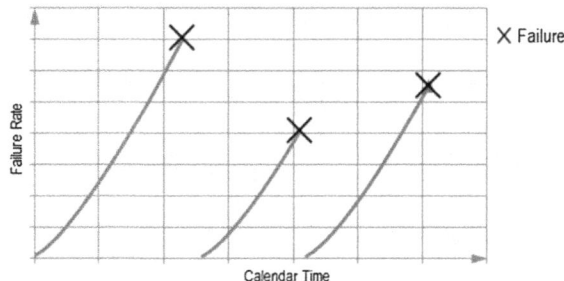

As proved in the former graphic, the evolution of the failure rate is reset after each failure, evidently due to fact that the equipment remains in perfect conditions. The PRP model possesses main application over those equipment that have a complete maintenance over all its components or if it a 100% replacement of the equipment.

In mathematical terms, let t_i be the functioning time between failures $i - 1$ and the i-th. Then, under a PRP model, any time t_i will obey to the same probability distribution with inalterable parameters in time, for example, for the two-parameter Weibull case:

$$f(t_i) = \begin{cases} \frac{\beta}{\alpha}\left(\frac{t}{\alpha}\right)^{\beta-1} e^{-\left(\frac{t}{\alpha}\right)^{\beta}} & t \geq 0 \\ 0 & t < 0 \end{cases},$$

being α and β continuous and inalterable in time. The origin of this type of distribution is in order to consider increasing or decreasing failure rates along the time from the last repair, being α and β the scale and form parameters, respectively.

PRP Parameterization Model—Two-parameter Weibull

From a practical point of view, it is characterized by having two parameters, where β corresponds to the form parameter linked to the well-known bath curve and to the respective phase of life cycle of the asset, and the parameter α known as the scale parameter, which is linked directly with the variability and dispersion of the life data that the asset analyzed has. The probability density function of failure (pdf) corresponds to (1)

$$f(t) = \frac{\beta}{\alpha}\left(\frac{t}{\alpha}\right)^{\beta-1} e^{-\left(\frac{t}{\alpha}\right)^{\beta}}. \tag{1}$$

In this case the failure rate is defined as (2)

$$\lambda(t) = \frac{\beta}{\alpha}\left(\frac{t}{\alpha}\right)^{\beta-1}. \tag{2}$$

The mean time to failure (MTTF) and reliability $R(t)$ are (3)

$$\text{MFTF} = \alpha \cdot \Gamma \left(1 + \frac{1}{\beta}\right), \quad R(t) = e^{-\left(\frac{t}{\alpha}\right)^\beta}. \tag{3}$$

Note that Γ corresponds to gamma function (4):

$$\Gamma(t) = \int_0^\infty x^{t-1} e^{-x} dx. \tag{4}$$

The Weibull function itself is a generalization for the exponential function, knowing that $\beta = 1$, $\alpha = 1/\lambda$, and $\gamma = 0$. With Weibull distribution it is possible to represent the state of the asset for any of three phases in the bathtub curve (life cycle perspective): infant mortality, useful life, or life wear out. For parameterization it is required a maximum likelihood function resolution, and then a natural logarithm is applied partially with respect to each parameter to finally derive it and equals to zero. It is presented in (5) and (6):

$$\alpha = \left(\frac{-\sum_{i=1}^n \left[t_i^\beta\right]}{n}\right)^{1/\beta} \tag{5}$$

$$\frac{n}{\beta} + \sum_{i=1}^n [\ln[t_i]] = \left(\frac{n \sum_{i=1}^n \left[t_i^\beta \ln(t_i)\right]}{\sum_{i=1}^n \left[t_i^\beta\right]}\right). \tag{6}$$

For the resolution of this type of adjustment, there are specialized software applications. One of these is RelPro [7] which disposes of advanced and efficient algorithm for the resolution of this kind of problems. Also, in case of need to clarify and expand the concepts handled here, the following references are recommended [8, 9].

2.2 Nonhomogeneous Poisson Process

NHPP Conceptualization

NHPP is a Poisson process with a parametric model used to represent events with an occurrence of evolutional failure in time and always with the same tendency.

This case applies especially for those equipments that are composed by many components where the replacement of one of them does not affect the global reliability: consider an equipment composed by hundreds of component that work in series; if one of them fails, this component is replaced and the equipment continues working but with a level of waste almost identical to previous one. For this reason the NHPP model applies for the so-called "minimum maintenance".

Next, Fig. 2 presents the graph for the behavior of the failure rate over time, being this completely accumulative between one and other failure. As we appreciate in the former graphic, for the case of NHPP, the failure rate remains dependent on total time elapsed.

In the case of NHPP, the functions of the reliability and failure probability are expressed as follows (Fig. 3).

Having as a base the former graphic, let us consider that one equipment has a failure in a t_1 time. After being repaired, the functioning is restarted and begins to work in that same point. Then, the reliability function from t_1, for a t time that represents the elapsed time beyond t_1, will be given as follows (7).

$$R(t|t > t_1) = \frac{R(t)}{R(t_1)} = 1 - F(t|t > t_1). \tag{7}$$

This is called by various authors [8] as "Mission Time", where t corresponds to elapsed calendar time.

Fig. 2 Failure rate $\lambda(t)$ in NHPP. *Source* own elaboration

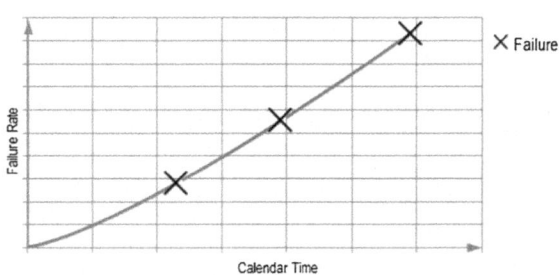

Fig. 3 $R(t)$ and $f(t)$ in NHPP. *Source* own elaboration

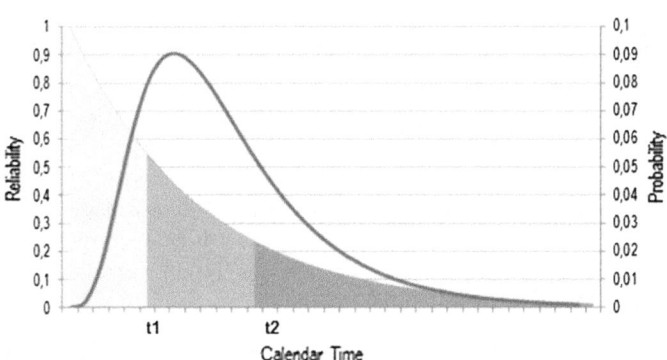

For a Weibull distribution, from the previous equation and since t_{i-1} corresponds to the total time elapsed until the last failure, and t_i the total time (calendar) elapsed after generating the failure i-th, it will be possible to conclude the following probability density function (8):

$$f(t_i|t_i > t_{i-1}) = \frac{\beta}{\alpha}\left(\frac{t_i}{\alpha}\right)^{\beta-1} \exp\left\{\left(\frac{t_{i-1}}{\alpha}\right)^{\beta} - \left(\frac{t_i}{\alpha}\right)^{\beta}\right\}. \tag{8}$$

NHPP Parameterization Model—Weibull 2 parameters

In order to obtain the parameters α and β, the lineal regression is not a choice. It is ideally made an adjustment by maximum likelihood. The likelihood function is expressed as (9)

$$P(x_i \quad \text{en}[x_i, x_i + dx] \forall i \in \{1, \ldots, n\}) = \prod_{i=1}^{n} f(x_i; \theta)$$

$$L(\theta) = \prod_{i=1}^{n} f(x_i; \theta), \tag{9}$$

where θ corresponds to the vector of the parameters of distribution to which it obeys the $f(t)$. Moreover, xi corresponds to the element i-th of the sample. As it is wished to obtain maximum likelihood between the data and one pdf, $f(t; \theta)$, the values of the vector θ are adjusted with the aim to reach that maximum. Conceptually, parameters are searched in order to better fit to a sample $X_i, \ldots, X_n$ in such a way that the probability of the series of values that can be presented in a random sample should be maximal:

Thus, in the present case with the simplified likelihood function and after applying and partial derivatives equal to zero, the results of the estimators for NHPP are (10) and (11):

$$\hat{\alpha} = \frac{t_n}{n^{1/\beta}} \tag{10}$$

$$\hat{\beta} = \frac{n-1}{\sum_{i=1}^{n-1} \ln\left(\frac{t_n}{t_i}\right)}, \tag{11}$$

where t_i corresponds to the elapsed time until the failure i-th and t_n is the elapsed time until the last failure. As in the previous case, if it is needed to clarify and expand the concepts handled here, next references are recommended [8, 9].

2.3 Generalized Renewal Process (GRP)

GRP Conceptualization

The traditional models already shown are only able to model two types of maintenance: the completely perfect and the completely imperfect. GRP model is the generalization for any level of perfection that has the maintenance, including the both mentioned. GRP adds a new parameter, called "virtual age". The parameter A_n represents the age of the system at the immediate instant when the n-th repair is carried out. In this way, if $A_n = y$, the element has a time of functioning associated to a probability distribution conditioned for this age y. That is to say, all the failure times have different probability distributions as the time passes by.

Graphically, the failure rate evolves as shown in Fig. 4.

The way to incorporate this variable is considering that equipment begins to operate with certain waste, which is reflected in the reliability function. In this manner, the accumulated reliability and probability distribution for t_{n+1} are as follows (12):

$$
\begin{aligned}
F(t|A_n = y) &= \frac{F(t+y) - F(y)}{1 - F(y)} \\
R(t|A_n = y) &= \frac{R(t+y)}{R(t)}.
\end{aligned}
\tag{12}
$$

By this way, it is clear that this "virtual age" is the age of waste in which the equipment begins to work again. The reliability function remains similar to the "Mission Time" only that this does not correspond to a real time elapsed, but to an equivalent. X_i is the i-th time of good functioning and T_n the total accumulated time elapsed until failure n-th, as follows (13):

$$
T_n = \sum_{i=1}^{n} x_i.
\tag{13}
$$

Fig. 4 Failure rate $\lambda(t)$ in GRP. *Source* own elaboration

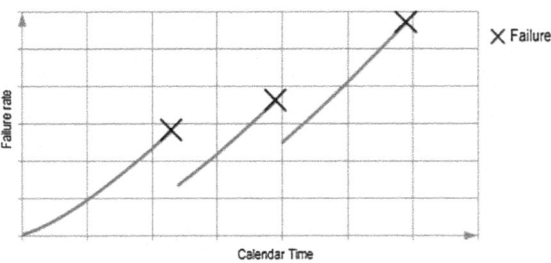

Moreover, the parameter A_n is given by (14)

$$A_n = A_{n-1} + q \cdot x_n. \tag{14}$$

Using (13), then would be (15)

$$A_n = qT_n = q \sum_{i=1}^{n} x_i, \tag{15}$$

where q is the parameter that decides the ineffectiveness of the repair; in this way $q = 0$ implies that $A_n = 0$, that is to say virtual age equals to 0. Therefore, $q = 0$ corresponds to a perfect repair case, that is to say it is completely effective. In the case when it was $q = 1$, it begins to operate in the same part of the reliability function where the equipment failed. This would be as follows (16):

$$0 < q < 1 : \text{GRP}$$
$$q = 0 : \text{PRP(HPP)} \tag{16}$$
$$q = 1 : \text{NHPP}.$$

Plotting the existing relation between real life and virtual age that evolves, it is possible to generate (see Fig. 5) the following comparative graphic for PRP, NHPP, and GRP.

As in NHPP, it is determined the conditioned reliability and the respective pdf. Then, reliability will be modeled according to (17)–(19):

$$R(t_i | t_i > q \cdot t_{i-1}) = \frac{e^{-\left(\frac{t_i}{\alpha}\right)^{\beta}}}{e^{-\left(\frac{q \cdot t_{i-1}}{\alpha}\right)^{\beta}}} \tag{17}$$

$$F(t_i | t_i > q \cdot t_{i-1}) = 1 - e^{\left(\left(\frac{q \cdot t_{i-1}}{\alpha}\right)^{\beta} - \left(\frac{t_i}{\alpha}\right)^{\beta}\right)} \tag{18}$$

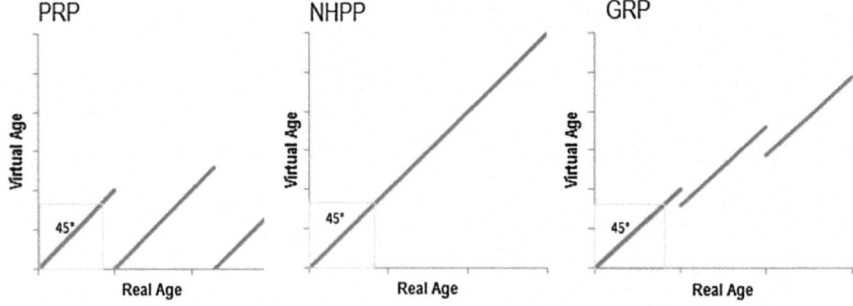

Fig. 5 Virtual age V/S real age in PRP, NHPP and GRP. *Source* own elaboration

$$f(t_i | t_i > q \cdot t_{i-1}) = \frac{\beta}{\alpha} \left(\frac{t_i}{\alpha}\right)^{\beta-1} e^{\left(\left(\frac{q \cdot t_{i-1}}{\alpha}\right)^{\beta} - \left(\frac{t_i}{\alpha}\right)^{\beta}\right)}. \tag{19}$$

Parameterization Model GRP

The adjustment developed is on the basis of a pdf with two-parameter Weibull (α, β), and adding the parameter q, so then we have three parameters to determine. The most common approach for parameters determination, by maximum likelihood, corresponds to the likelihood function. In order to solve this, a partial derivative in each variable is applied, and then it will be obtained as a set of three equations with three unknown quantities; these are α, β, and q.

The parameters α, β, and q are the three values to identify. Searching these parameters is a very exhaustive procedure, as it requires more precision in the procedure, generally is a long process, so it is suggested to use the Monte Carlo simulation. The searching of α, β, and q starts with the simulation of q and β, repeated by uniform distributions (20):

$$\begin{aligned} q &\sim U[0, 1] \\ \beta &\sim U[0, 10] \end{aligned}. \tag{20}$$

Similarly, the estimated parameter $\hat{\alpha}$ for the GRP model yields (21):

$$\hat{\alpha} = \sqrt[\beta]{\frac{\sum_{i=2}^{n} \left[(t_i + t_{i-1}(q - 1))^{\beta} - (q \cdot t_{i-1})^{\beta}\right] + t_1^{\beta}}{n}}. \tag{21}$$

Then, the procedure for the GRP adjustment is plotted through the following diagram of process, see Fig. 6.

As far as the probability to find the values that grant the global maximum of the maximum likelihood function is virtually invalid through random search, it is necessary to define a value of tolerance for the partial derivatives ($\partial L/\partial B$ and $\partial L/\partial q$) equal to 0, being necessary to fix this value of tolerance "TOL" as an acceptable rank to consider that is found in a global maximum and in this way to accept the respective distribution adjustment.

Considering the adding of a new parameter, in this case q, always the adjustment GRP will give a higher likelihood than a PRP or NHPP adjustment. Nevertheless, in order to consider the existence and applicability of these cases, it is necessary to count on selection criteria. This is applied after the adjustment through GRP once obtained the parameter q.

As q value is always a continuous value, the probability to be exactly $q = 1$ or $q = 0$ is practically null; therefore it is considered a new tolerance level, which has been called TQ. This tolerance level corresponds to higher and lower percentage of the possibilities that value q has. The reason of this value (q parameter) is to identify

Fig. 6 Process diagram for
GRP modeling. *Source* own
elaboration

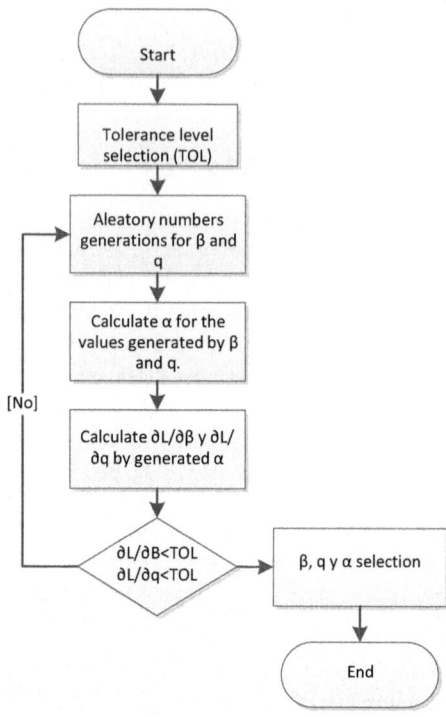

when would be more appropriate to consider a PRP or NHPP model. Therefore, the
practical expression corresponds to (22)

$$
\begin{aligned}
0 + TQ < q < 1 - TQ &: \ \text{GRP} \\
q = \ < TQ &: \ \text{PRP} \\
q > \ = 1 - TQ &: \ \text{NHPP.}
\end{aligned}
\tag{22}
$$

As in previous case, to clarify and expand the concepts handled here, Refs. [8, 9]
are recommended.

3 Numerical Application

According to preliminary conceptual and analytical development, it proceeds to
develop a practical application, which corresponds to the analysis of a slurry pump
(inert material); ID code: P01, it belongs to a process involved in copper mining.
The model to be applied corresponds to GRP, given the flexibility and ability to
generalize and discriminate any of the three models exposed: PRP, NHPP, and
GRP. Table 2 shows the time records of good performance which have been

Table 2 Time between failures for pump P01 (h)

N° of failure	Operating time (h)	N° of failure	Operating time (h)
1	860.05	13	367.41
2	1608.24	14	2757.98
3	1134.24	15	355.50
4	2703.12	16	1084.39
5	645.38	17	855.52
6	95.15	18	280.52
7	1278.48	19	490.48
8	605.34	20	945.55
9	344.33	21	105.32
10	1054.68	22	127.33
11	680.57	23	61.85
12	405.38	24	326.30

Source own elaboration

collected by the entity performing industrial activity, and will be considered as the input data for the simulation. In order to understand and analyze the applicability of the methodology presented and its effects, the selected item is enough. Then, it proceeds to develop the probability distribution adjustment.

Step 1: *Tolerance level*

It must be defined the tolerance level for the partial derivatives as TOL = 0.01 and tolerance for the q value is TQ = 5%.

Step 2: *Distribution parameterization*

Once tolerance level is defined, it proceeds to apply GRP model. For this, it is decided to use a computer tool: RelPro®. It has been developed from the perspective of research and industrial application. Solving the equations with RelPro®, the following parameters are obtained, see Figs. 7 and 8, respectively.

From the parameters obtained ($\alpha = 1986.067$; $\beta = 2.026$; and $q = 0.192$), the partial derivatives are solved and the acceptance of GRP distribution adjustment and level of tolerance are verified properly (23):

$$\left| \frac{\partial[\ln(L)]}{\partial \beta} \right| = 1.76 \times 10^{-7} < \text{TOL} = 0.01$$

$$\left| \frac{\partial[\ln(L)]}{\partial q} \right| = 0.009833 < \text{TOL} = 0.01. \tag{23}$$

Regarding the q value, it has to be (24)

Fig. 7 Parameterization process for GRP, RelPro® software. *Source* own elaboration

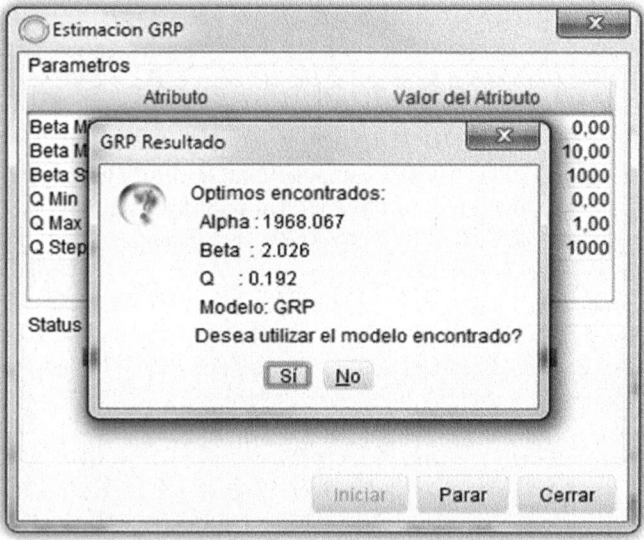

Fig. 8 Parameter estimation adjustment in GRP. RelPro® software. *Source* own elaboration

$$TQ = 0.05 < q = 0.192 < 1 - TQ = 0.95. \tag{24}$$

Therefore, determining an acceptable adjustment solution by maximum likelihood with partial derivatives (quality guarantee of the fit), and a q parameter with a numeric value between the range $0.05 < q < 0.95$, it is possible to affirm that the use of GRP model is suitable for the case. For this, the use of a traditional adjustment would be completely incorrect.

Step 3: *Analysis*

With this result, one possible analysis is to project (correctly) the equipment failures, and indirectly the rate of increase of the frequency of failure and the decreasing of operation times.

The expected time of correct performance, at the previous instant than the operation, is recovery (after failure); it is determined by the difference between the expectancy of probability density function (based on the total elapsed time) and the virtual age of the asset. Graphically, Figs. 9 and 10 represent it.

To estimate the number of accumulated failures over time, it is possible to consider that the first failure occurs at the MTBF, expected value at the beginning of the operation of the equipment. Thus it is possible to project recursively with the expected MTBF and then every occurrence of failure. Obviously, this is a generalization and simplification of the problem. Finally, it is possible to obtain the graph of cumulative number of events for a total time of operation, see Fig. 11 with the fault behavior accumulated v/s total time of operation. Clearly, a rising trend and accelerated failure time less operation are identified. This is understood as active aging in time. Analytically (25)–(30),

$$E[t_i | t_i > q \cdot t_{i-1}] = \int_{q \cdot t_{i-1}}^{\infty} (f(t | t > q \cdot t_{i-1}) \times t) \mathrm{d}t \tag{25}$$

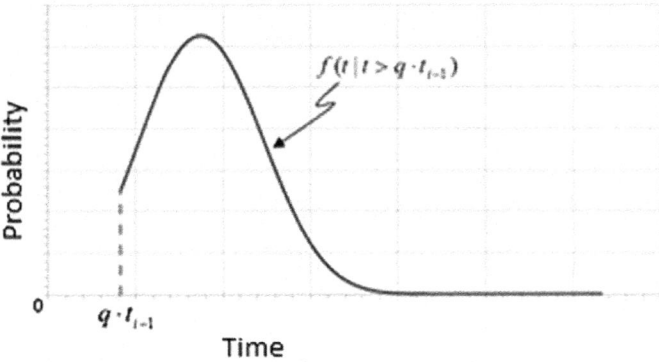

Fig. 9 P.D.F. of GRP model, for elapsed time t_{i-1}. RelPro® software. *Source* own elaboration

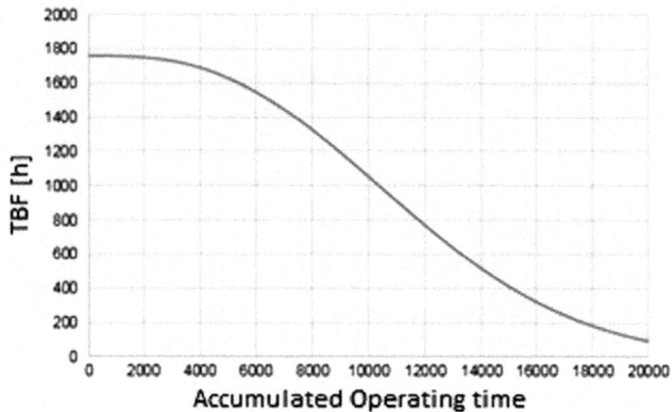

Fig. 10 Expected TBF according to the operating time elapsed until the last intervention. RelPro®
software. *Source* own elaboration

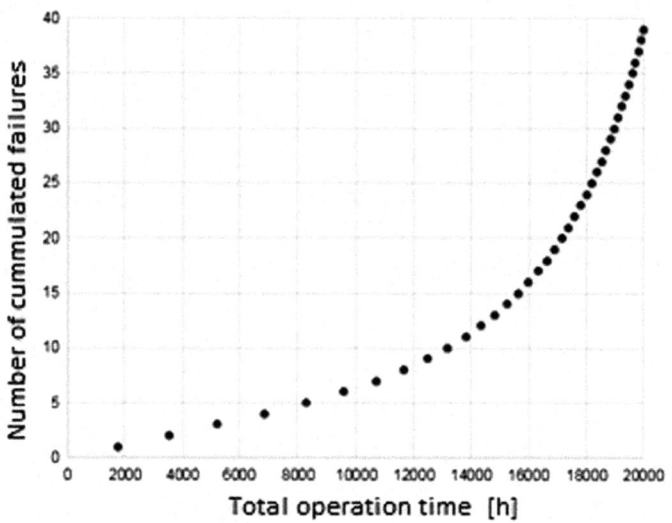

Fig. 11 Forecast for a number of failures. RelPro® software. *Source* own elaboration

$$\text{MTBF}(t_{i-1}) = E[t_i|t_i > q \cdot t_{i-1}] - q \cdot t_{i-1} \tag{26}$$

$$\text{MTBF}(t_{i-1}) = \int_{q \cdot t_{i-1}}^{\infty} (f(t|t > q \cdot t_{i-1}) \times t)\mathrm{d}t - q \cdot t_{i-1} \tag{27}$$

$$\mathrm{MTBF}(t_{i-1}) = \int_{q \cdot t_{i-1}}^{\infty} \left(\frac{\beta}{\alpha}\left(\frac{t}{\alpha}\right)^{\beta-1} e^{\left(\left(\frac{q \cdot t_{i-1}}{\alpha}\right)^{\beta} - \left(\frac{t}{\alpha}\right)^{\beta}\right)} \times t \right) dt - q \cdot t_{i-1} \qquad (28)$$

$$\mathrm{MTBF}(t_{i-1}) = e^{\left(\frac{q \cdot t_{i-1}}{\alpha}\right)^{\beta}} \times \int_{q \cdot t_{i-1}}^{\infty} \left(\frac{\beta}{\alpha}\left(\frac{t}{\alpha}\right)^{\beta-1} e^{\left(-\left(\frac{t}{\alpha}\right)^{\beta}\right)} \times t \right) dt - q \cdot t_{i-1} \qquad (29)$$

$$s - 1 = \frac{1}{\beta} \Rightarrow s = \frac{1}{\beta} + 1$$

$$\mathrm{MTBF}(t_{i-1}) = \alpha \times e^{\left(\frac{q \cdot t_{i-1}}{\alpha}\right)^{\beta}} \times \underbrace{\int_{\left(\frac{q \cdot t_{i-1}}{\alpha}\right)^{\beta}}^{\infty} \left(e^{(-p)} \times p^{s-1} \right) dp}_{\Gamma\left(s, \left(\frac{q \cdot t_{i-1}}{\alpha}\right)^{\beta}\right)} - q \cdot t_{i-1} \qquad (30)$$

$$\mathrm{MTBF}(t_{i-1}) = e^{\left(\frac{q \cdot t_{i-1}}{\alpha}\right)^{\beta}} \alpha \times \Gamma\left(\frac{1}{\beta} + 1, \left(\frac{q \cdot t_{i-1}}{\alpha}\right)^{\beta}\right) - q \cdot t_{i-1}$$

$$\mathrm{MTBF}(t_{i-1}) = \frac{\alpha \times \Gamma\left(\frac{1}{\beta} + 1, \left(\frac{q \cdot t_{i-1}}{\alpha}\right)^{\beta}\right)}{R(t_{i-1})} - q \cdot t_{i-1}$$

In addition, the software tool used (RelPro®) allows diagramming the time evolution curves: probability density function $f(t)$, the cumulative probability function $F(t)$, reliability $R(t)$, and the failure rate $\lambda(t)$, see Fig. 12.

By default, RelPro® graphs the curves which will come under the elements after the next nine events (internal standard). With this it is possible to note how quickly degrades element.

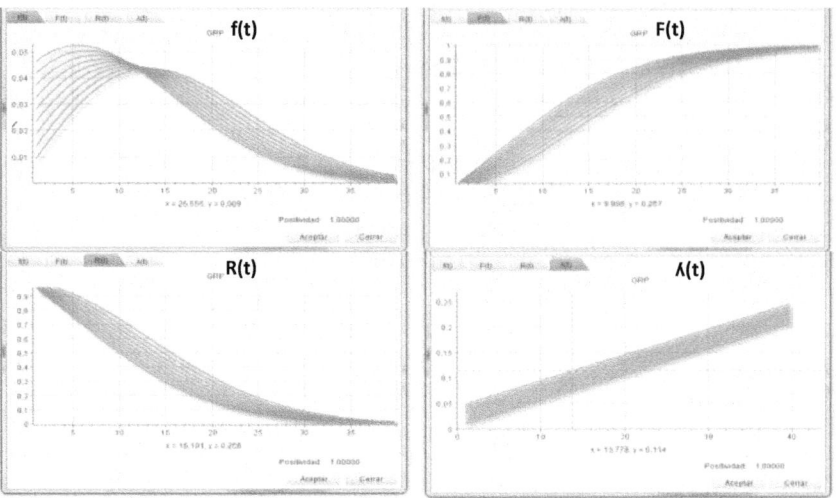

Fig. 12 Curves $f(t)$, $F(t)$, $R(t)$, and $\lambda(t)$, RelPro® software. *Source* own elaboration

Analyzing graphic curves in Fig. 12, the deterioration of reliability curve after the occurrence of a failure is identified. At the same time it is possible to note that the curve of probability density function increases its density, approaching it every time to the origin values. For the failure rate, after each failure it is incremented in equivalent intervals.

4 Future Lines of Application

This methodology is focused on practical application, so further research may focus on different application examples for different types of machines (used in mining or in another sector). In such further applications, it may be interesting to specify the conditions of use of, in which maintenance is carried out, the total operating time, utilization times, environmental or process conditions, etc. With this will be possible to compare the results obtained for both types of machinery according to different boundary conditions.

5 Conclusions

The reliability model is an essential aspect for the management and optimization of physical industrial assets. In order to learn in detail, the step by step of each model is a fundamental task to apply effective and correctly each model. Diverse researches omit the process of resolution and only present final results indicating the use of a model and the use of some computed tool with integrated algorithm. It was identified in different researches, which motivated the research team to develop a specific conceptual pattern and resolution practical for each stochastic parametric model former mentioned. This is fundamental to recognize the value of this work and its contribution for future researchers who wish to learn and apply this knowledge. For this reason, this research becomes an analytic and explicative procedure about the definition, calculation methodology, and criteria that must be considered to parameterize industrial assets under certain degradation level after maintenance, complementing in addition its analysis with a numeric application that allows demonstrating step by step the mathematic development as appropriate. The practical case was developed in mining industry of Chile.

It is worth mentioning here the frequent lack of feedback between users and manufacturers of equipment, resulting in ignorance by manufacturers about the real weaknesses of the machines. The performance of assets under ideal conditions (lab test) is extremely different comparing with real process conditions.

As a second phase of this research for potential publication, the research team is analyzing a presorting to the respective parameter that presents the current article, which corresponds to the identification whether the model is nonparametric. The method is parametric (MP) if the modeling fits a probability distribution function known; on the other hand, if you cannot make this assumption, the method is nonparametric (MNP). There are also models that contain a portion of the parametric function and one not; these are the semi-parametric methods (MSP). The latter classification (MSP) is of great interest for research and application, since in practice generally the assets are subject to many variables that classical models are not included in the modeling and analysis of failure rate, for example, operating temperature, workload, diagnosis of lubricants (parts per million), etc. These variables are not constant and can cause changes in the reliability of a component, which is necessary to analyze and quantify for designing an effective and efficient maintenance policy. In addition, it is expected to incorporate into the analysis the TRP model (Trend-Renewal Process) [10], described and studied in detail by Lindqvist et al. [11], this being a different model of imperfect repair, with similar characteristics to NHPP.

Modern methods allow modeling based on these environmental factors and stresses, but they are bounded to assumptions or restrictions on the number of factors to analyze, which makes more complex application and obviously their systematic use in the reliability analysis.

References

1. Ascher H, Feingold H (1984) Repairable systems reliability: modeling, inference, misconceptions and their causes. Marcel Dekker, New York
2. Veber B, Nagode M, Fajdiga M (2008) Generalized renewal process for repairable systems based on finite Weibull mixture. Reliab Eng Syst Saf 93:1461–1472. doi:10.1016/j.ress.2007.10.003
3. Kijima M, Sumita N (1986) A useful generalization of renewal theory: counting process governed by non-negative markovian increments. J Appl Probab 23:71–88. doi:10.2307/3214117
4. Krivtsov V (2000) Monte Carlo approach to modeling and estimation of the generalized renewal process in repairable system reliability analysis, Ph.D dissertation, University of Maryland
5. Muhammad M, Abd Majid MA, Ibrahim NA (2009) A case study of reliability assessment for centrifugal pumps in a petrochemical plant. In: 4th world congress on engineering asset management, Athens. doi:10.1007/978-0-85729-320-6_44
6. Weckman GR, Shell RL, Marvel JH (2001) Modeling the reliability of repairable systems in aviation industry. Comput Ind Eng 40:51–63. doi:10.1016/S0360-8352(00)00063-2
7. RelPro®, Reliability and Production, Analysis and Simulation (2014) RelPro SpA. Retrieved 10 Apr 2014. http://www.relpro.pro
8. Selivanov AI, Yudkevich E (1972) Fundamentos de la teoría de envejecimiento de la maquinaria
9. Sotskov B (1972) Fundamentos de la teoría y del cálculo de Fiabilidad. Mir
10. Gámiz ML, Lindqvist BH (2015) Nonparametric estimation in trend-renewal processes. Reliab Eng Syst Saf. doi:10.1016/j.ress.2015.08.015
11. Lindqvist BH, Elvebakk G, Heggland K (2003) The trend-renewal process for statistical analysis of repairable systems. Technometrics 45:31–44

Author Biographies

Pablo Viveros Gunckel Researcher and Academic at Technical University Federico Santa María, Chile, has been active in national and international research, both for important journals and conferences. Also he has developed projects in the Chilean industry. He is the consultant specialist in the area of Reliability, Asset Management, System Modeling, and Evaluation of Engineering Projects. He is Industrial Engineer and Master in Asset Management and Maintenance.

Adolfo Crespo Márquez is currently Full Professor at the School of Engineering of the University of Seville, and Head of the Department of Industrial Management. He holds a PhD in Industrial Engineering from this same University. His research works have been published in journals such as the International Journal of Production Research, International Journal of Production Economics, European Journal of Operations Research, Journal of Purchasing and Supply Management, International Journal of Agile Manufacturing, Omega, Journal of Quality in Maintenance Engineering, Decision Support Systems, Computers in Industry, Reliability Engineering and System Safety, and International Journal of Simulation and Process Modeling, among others. Prof. Crespo is the author of seven books, the last four with Springer-Verlag in 2007, 2010, 2012, and 2014 about maintenance, warranty, and supply chain management. Prof. Crespo leads the Spanish Research Network on Dependability Management and the Spanish Committee for Maintenance Standardization (1995–2003). He also leads a research team related to maintenance and dependability management currently with five PhD students and four researchers. He has extensively participated in many engineering and consulting projects for different companies, for the Spanish Departments of Defense, Science and Education as well as for the European Commission (IPTS). He is the President of INGEMAN (a National Association for the Development of Maintenance Engineering in Spain) since 2002.

René Tapia Peñaloza is an Industrial Engineer graduated from the Technical University Federico Santa Maria, Chile; he serves leading the Reliability Engineering projects for HighService Corp, a major maintenance services provider for mining; he has several scientific publications in the area, participating actively in international conferences. He is a designer and expert modeler in RelPro Software. Also he has experience in Software Development Projects, with over 15 years of programming experience. His specialties include Analysis of Production Systems in terms of Reliability, Availability, Maintenance and Production, Sizing of Production Lines; Implementation of Reliability Engineering and Performance Metrics; Simulation of Production Processes; Development of models for optimal maintenance policies; Technical and financial studies for optimal replacement of equipment and components; Simulation and design of transport systems; and Simulation and financial evaluation of projects in early stages.

Fredy Kristjanpoller Rodríguez is an Industrial Engineer and Master in Asset Management and Maintenance of Federico Santa Maria University (USM, Chile), and doctoral candidate in Industrial Engineering at the University of Seville. He is a researcher, academic, and master program coordinator linked to asset management. He has important scientific papers on indexed journals and international proceedings congress on the following areas: Reliability engineering and Maintenance Strategies. He has developed consultant activities in the main Chilean companies.

Vicente González-Prida Díaz holds PhD (Summa Cum Laude) in Industrial Engineering by the University of Seville, and Executive MBA (First Class Honors) by the Chamber of Commerce. He has been honored with the following awards and recognitions: Extraordinary Prize of Doctorate by the University of Seville;National Award for PhD Thesis on Dependability by the Spanish Association for Quality;National Award for PhD Thesis on Maintenance by the Spanish Association for Maintenance;and Best Nomination from Spain for the Excellence Master Thesis Award bestowed by the EFNSM (European Federation of National Maintenance Societies). He is member of the Club of Rome (Spanish Chapter) and has written multitude of articles for international conferences and publications. His main interest is related to industrial asset management, specifically the reliability, maintenance, and after-sales organization. He currently works as Program Manager in the company General Dynamics—European Land Systems and shares his professional performance with the development of research projects in the Department of Industrial Organization and Management at the University of Seville.

Economic Impact of a Failure Using Life-Cycle Cost Analysis

**Carlos Parra Márquez, Adolfo Crespo Márquez,
Vicente González-Prida Díaz, Juan Francisco Gómez Fernández,
Fredy Kristjanpoller Rodríguez and Pablo Viveros Gunckel**

Abstract This chapter aims to investigate technical and economic factors related to failure costs (non reliability costs) within the life-cycle cost analysis (LCCA) of a production asset. Life-cycle costing is a well-established method used to evaluate alternative asset options. It is a structured approach that addresses all the elements of this cost and can be used to produce a spend profile of the assets over its anticipated life span. The results of an LCC analysis can be used to assist management in the decision-making process where there is a choice of options. The main costs can be classified as the 'capital expenditure' (CAPEX) incurred when the asset is purchased, and the 'operating expenditure' (OPEX) incurred throughout the asset's life. This chapter will explore different aspects related to the "failure costs" within the LCCA, and will describe the most important aspects of the stochastic model called: Non-homogeneous Poisson Process (NHPP). This model will be used to estimate the frequency of failures and the impact that could cause diverse failures in the total costs of a production asset. This paper also contains a case study for the Rail Freight Industry (Chile) and in the Oil Industry

C. Parra Márquez (✉) · A. Crespo Márquez · V. González-Prida Díaz ·
J.F. Gómez Fernández
Department of Industrial Management, School of Engineering,
University of Seville, Seville, Spain
e-mail: parrac@ingecon.net.in

A. Crespo Márquez
e-mail: adolfo@us.es

V. González-Prida Díaz
e-mail: vicente.gonzalezprida@gdels.com

J.F. Gómez Fernández
e-mail: juan.gomez@iies.es

F. Kristjanpoller Rodríguez · P. Viveros Gunckel
Department of Industrial Engineering, Universidad Técnica Federico Santa María,
Valparaíso, Chile
e-mail: fredy.kristjanpoller@usm.cl

P. Viveros Gunckel
e-mail: pablo.viveros@usm.cl

© Springer International Publishing AG 2018
A. Crespo Márquez et al. (eds.), *Advanced Maintenance Modelling
for Asset Management*, DOI 10.1007/978-3-319-58045-6_9

(PETRONOX, Venezuela) where the above-mentioned model and concepts will be applied, and respectively compared in terms of results. Finally, the model presented provides maintenance managers with a decision tool that optimizes the LCCA of an asset and will increase the efficiency of the decision-making process related to the control of failures.

Keywords Asset analysis · Failures · Life-cycle cost analysis (LCCA) · Non-homogeneous poisson process (NHPP) · Maintenance managment · Reliability models · Repairable systems · Parameter estimation

1 Introduction

With the purpose of optimizing costs and improving the profitability of the productive processes, the denominated organizations of World Class category [1], dedicate enormous efforts to visualize, analyze, implement, and execute strategies for the solution of problems, which involve decisions in high-impact areas: security, environment, production goals, product quality, operation costs and maintenance. In recent years, specialists in the areas of value engineering and operations direction have decided to focus on asset management field, due to the great opportunities that presents improved. On this scope, one of the most interesting challenges is to improve the quantification process of costs, including the use of techniques that quantify the Reliability factor and the impact of the failure events on the total costs of a production system throughout their life cycle [2]. Taking as reference the maintenance management model (MMM) of the 8 phases proposed in the Fig. 1 [3], this section related to the evaluation of the impact of reliability at the cost of an asset life cycle, is part of phase 7 of the MMM. The concept of Life-Cycle Cost Analysis (LCCA) began to be applied in a structured form from the decade of the 70s, specifically in the Department of Defense of the United States, in the area of military aviation. However, most of the methodologies developed at this stage by the Department of Defense was oriented toward the process of procurement and logistics and did not include the design and production phase. In an attempt to improve the design of assets and reduce changes in time, the so-called concurrent engineering (life cycle engineering) has emerged as an effective technique in the process of optimizing costs [3]. Life cycle engineering believes that the initial phase of development of an asset begins with the identification of the need for it and other phases will be subsequently generated such as design (conceptual, preliminary and detailed), production (manufacturing), use (operations, maintenance), support (logistics), and disincorporation (replacement).

As a life-cycle cost analysis (LCCA) and vision, these improvements have diminished the uncertainty in the process of decision-making in vitally important areas such as design, development, maintenance, substitution, and acquisition of

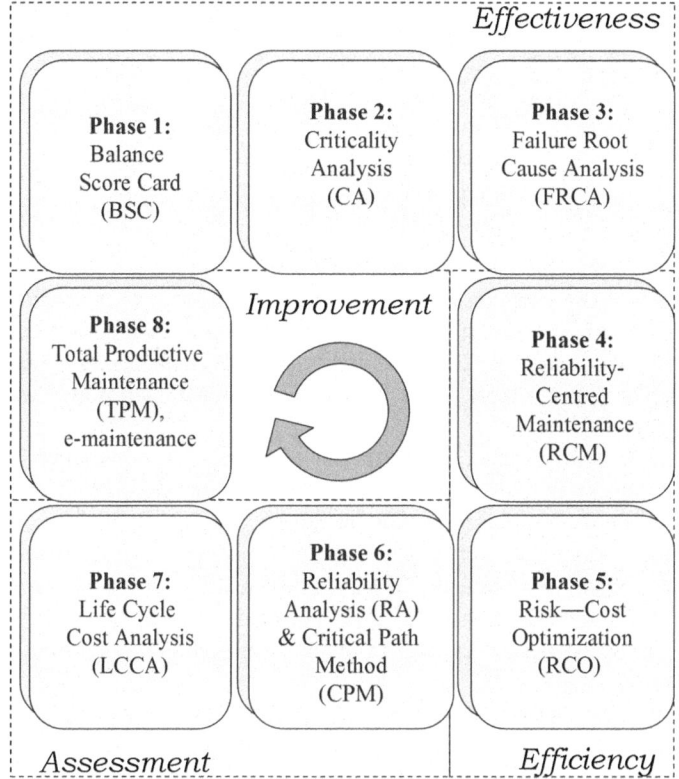

Fig. 1 Sample of techniques within the maintenance management framework (adapted from [3])

production assets. It is important to clear that up; in this whole process, many decisions and actions exist, technical as well as nontechnical, that should be adopted through the whole use period of an industrial asset. Product support and maintenance needs of systems are more or less decided during the design and manufacturing phase [4], but they have a great impact over all the asset life based on LCCA.

Outlines of most of these actions, particularly those that correspond to the design phase of the production system, have a high impact on the total life cycle of the asset, being of particular interest, those decisions related to the improvement process of the "Reliability" factor (quality of the design, used technology, technical complexity, frequency of failures, costs of preventive/corrective maintenance, maintainability levels and accessibility), since these aspects have a great influence on the total cost of the asset's life cycle, and they influence in great measure the possible expectations to extend the useful life of the production systems to reasonable costs (e.g., [4–9]). The traditional methodologies estimate the LCCA with average behaviour, due to it is necessary to search an application that can model the degradation effect, as NHPP, to have a most realistic analysis to define the right Asset Management strategy.

2 Basics Aspects of the LCCA

During recent years, the investigation area related to the Life-cycle Costs Analysis, has continued its development, as much on the academic level as to the industrial level. It is important to mention the existence of other methodologies that have emerged in the area of LCCA, such as: Life-cycle Costs Analysis and Environmental Impact, Total Costs Analysis of Production Assets, among others [10].

These methodologies have their particular characteristics, although regarding the estimation process of the costs for failure events impact; they usually propose Reliability analysis based constant failure rates.

The early implementation of the cost analysis techniques allows for early evaluation in advance of potential design problems and to quantify the potential impact in the costs along the life cycle of the industrial assets [10]. For this, procedures exist that group together in the denominated: Techniques of Life-cycle Costs Analysis.

Life-cycle cost analysis is defined [11] as an economic calculation technique which supports the optimal making decisions linked to design process, selection, development, and substitution of the assets in a production system. It, ideally, evaluates the costs associated to the economical period of expected useful life in a quantitative way, expressed in yearly equivalent monetary units (Dollars/year, Euros/year, Pesos/year).

Another definition [12] states that LCCA is a systematic process of technical–economical evaluation, applied in the selection and replacement process of production systems that allows it to consider Economics and Reliability aspects in simultaneous way, with the purpose of quantifying the real impact of all costs along the life cycle of the assets ($/year), and in this way, be able to select the asset that contributes the largest benefits to the productive system.

The great quantity of variables that directly and indirectly affect the real costs (inflation, rise/decrease of the costs, reduction/increase of the purchasing power, budget limitations, increase of the competition, and other similar characteristics), must be managed for estimating the real costs of an asset along its useful life. Those characteristics of the model generate a scenario of high uncertainty [10], and at the same time have generated a restlessness and interest about the total cost of the assets. Often, the total cost of the production system is not visible, in particular those costs associated with: operation, maintenance, installation tests, personnel's training, among others. Additionally, the dynamics of the economic scenario generate problems related to the real determination of the asset's cost. Some of them [13] are:

- The factors of costs are usually applied incorrectly. The individual costs are inadequately identified and, many times, they are included in the wrong category: the variable costs are treated as fixed (and vice versa); the indirect costs are treated as direct, etc.
- The countable procedures do not always allow a realistic and timely evaluation of the total cost. Besides, it is often difficult (if not impossible) to determine the costs, according to a functional base.
- Many times the budgetary practices are inflexible with regard to the change of funds from a category to another, or, from one year to another.

2.1 Characteristics of the Costs in a Production Asset

The cost of a life cycle is determined identifying the applicable functions in each one of its phases, calculating the cost of these functions and applying the appropriate costs during the whole extension of the life cycle. So that it is complete, the cost of the life cycle should include all the costs of design, fabrication, and production [14]. In the following paragraphs the characteristics of the costs in the different phases of an asset's life cycle are summarized [15]:

- Investigation, design, and development costs: initial planning, market analysis, product investigation, design and engineering requirements, etc.
- Production, acquisition and construction costs: industrial engineering and analysis of operations, production (manufacturing, assembly and tests), construction of facilities, process development, production operations, quality control, and initial requirements of logistics support.
- Operation and support costs: operation inputs of the production system, planned maintenance, corrective maintenance (depending on the Reliability Factor), and costs of logistical support during the system's life cycle.
- Removal and elimination costs: elimination of non-repairable elements along the life cycle, retirement of the system, and recycling material.

From the financial point of view, the costs generated along the life cycle of the asset are classified in two types of costs:

- CAPEX: Capital costs (design, development, acquisition, installation, staff training, manuals, documentation, tools and facilities for maintenance, replacement parts for assurance, withdrawal).
- OPEX: Operational costs: (manpower, operations, planned maintenance, storage, recruiting, and corrective maintenance—penalizations for failure events/low Reliability).

2.2   Impact of the Reliability in the LCCA

Woodhouse [12] outlines that to be able to design an efficient and competitive productive system in the modern industrial environment, it is necessary to evaluate and to quantify in a detailed way the following two aspects:

- Costs: aspect that is related with all the costs associated to the expected total life cycle of the production system. Including: design costs, production, logistics, development, construction, operation, preventative/corrective maintenance, withdrawal.
- Reliability: factor that allows to predict the form in which the production processes can lose their operational continuity due to events of accidental failures and to evaluate the impact on the costs that the failures cause in security, environment, operations, and production.

The key aspect of the term "Reliability" is related to the operational continuity. In other words, is possible to affirm that a production system is "Reliable" when it is able to accomplish its function in a secure and efficient way along its life cycle. Now, when the production process begins to be affected by a great quantity of accidental failure events (low Reliability), this scenario causes high costs, associated mainly with the recovery of the function (direct costs) and with growing impact in the production process (penalization costs). See Fig. 2.

The totals costs of Non-Reliability are described next in Table 1 [2, 16, 17].

The economic impact (cost terms) that generates an asset with low reliability will be associated directly with the behaviour of the following index:

$$\Lambda(t) = \text{Expected number of failures in a time interval } [0, t]. \tag{1}$$

According to [12], the increase of costs is caused, in great majority, by the lack of forecast in the case of unexpected failures appearances, a scenario provoked by ignorance and lack of analysis in the design phase of the aspects related with Reliability. As a result, this situation causes an increase in the operation costs (costs that were not considered in a beginning) affecting the profitability of the production process.

This paper aims to investigate technical and economic factors related to failure costs (non reliability costs) within the LCCA of a production asset. So, to avoid the uncertainty in cost analysis, the studies of economic viability should approach all

Fig. 2 Economic impact of the reliability

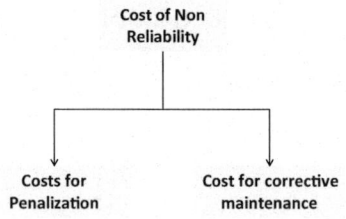

Table 1 Description of costs of non-reliability

Cost for penalization, due to downtimes	(a) Opportunity looses/deferred production (b) Production looses (unavailability) (c) Operational looses (d) Impact in the quality (e) Impact in security and environment
Cost for corrective maintenance	(a) Manpower (own or hired) associated to solve non planned event (b) Material and replacement parts direct costs related with the consumable parts and the replacements used in the event of an unplanned action

the aspects of the life cycle cost. The tendency of variability in the main economic factors, together with the additional problems already defined, have driven to erroneous estimates, causing designs and developments of production systems that are not suitable from the point of view of cost–benefit [13]. It can be anticipated that these conditions will worsen unless the design engineers assume a bigger grade of consideration in the costs. Inside the dynamic process of change, the acquisition costs associated with the new systems are not the only ones to increase, but rather the operation and maintenance costs of the systems already in use also do it in a quick way. This is due mainly to a combination of such factors as [13]:

- Inaccuracies in the estimates, predictions, and forecasts of the events of failures (Reliability), ignorance of the probability of occurrence of the different failure events inside the production systems in evaluation.
- Ignorance of the deterioration processes behaviour.
- Lack of forecast in the maintenance processes and ignorance of the modern techniques of maintenance management.
- Engineering changes during the design and development.
- Changes in its own construction of the system.
- Changes in expected production patterns.
- Changes during the acquisition of system components.
- Setbacks and unexpected problems.

It is important to mention that the results obtained from the LCCA, reach their maximum effectiveness during the phases of: initial development, visualization, and conceptual, basic and details engineering.

Once the design has been completed, it is substantially difficult to modify the economic results. Also, the economic considerations related to the life cycle should be specifically outlined during the phases previously mentioned, if the goal is to take advantage of effective economic engineering. It is necessary to keep in mind that almost two-thirds of the life cycle cost of an asset or system are already determined in the preliminary conceptual and design phase (70-85% of value creation and costs reduction opportunities) [18].

3 Main Stochastic Model Considered for Analysis of Reliability

For this research, the application of life cycle cost analysis must consider technical aspect related with the failure behaviour, performance function, or some information related to the asset deterioration. According to this framework, will be required the use of some stochastic technique, model that will support the assessment and calculation in the life cycle cost.

In the next section we will review the most important stochastic modeling, associated with the condition of repairable or non-repairable system. The technical characteristic is another important aspect, considered in the study for modeling and LCCA calculation.

3.1 General Review of Stochastic Modeling

A non-repairable system is defined as: when it fails, it is discarded (the repair is physically infeasible or noneconomical) [18] so, in this case the most important concept will be survival probability. The times between failures of a non-repairable system are independent and identically distributed (iid). [20]. This is the most common assumption made when analyzing time-to-failure data, but in some situations, it might be unrealistic. The usual non-repairable methodologies to analyze this scenario will be the statistical distribution fitting such as Weibull analysis [21].

On the other hand, repairable systems are those that can be restored up to their fully operational capabilities, not considering the replacement of the entire system as a solution [22]. In this sense, reliability is interpreted as the probability of not failing for a particular period t. In this case, this analysis does not assume that the times between failures are independent or identically distributed, and to model the reliability using stochastic point processes is necessary. The number of failures in an interval of time can be represented through a stochastic point process. It could be interpreted as a counting process, and what it counts is the number of events (failures) in a certain period of time.

Focusing on the repairable systems, five main stochastic models exist [22]:

- The renewal process (RP).
- The homogeneous Poisson process (HPP).
- The branching Poisson process (BPP).
- The superposed renewal process (SRP)
- The nonhomogeneous Poisson process (NHPP).

The RP model assumes that the system is returned to an "as new" condition every time it is repaired, so it is conceptually similar to a "non-repairable system approach," at which time failure can be modeled by a statistical distribution and the iid assumption is valid (see Fig. 3). The HPP is a special case of the RP, which assumes that times between failures are independent and identically exponentially distributed, so the iid assumption is also valid, and the time to failure is described by an exponential distribution (constant hazard rate) [19]. Variations of the RP can also be defined. The modified renewal process, where the first interarrival time differs from the others, and the superimposed renewal process (union of many independent RPs) are examples of these possible variations [22].

The BPP will be implemented to represent time-to-failure data that can be assumed to be identically distributed, but not independent. As one author [22] mentions, this process is applicable when a primary failure (or a sequence of primary failures having iid times to failure) can trigger one or more subsidiary Failures; thus there is dependence between the subsidiary failures and the occurrence of the primary, triggering failure. Very few practical applications of this model are found in the literature. More details about this process can be found in [23].

The SRP is a process derived from the combination of various independent RPs (parts components of the main asset), and in general it is not an RP. Each part component can be modeled as an RP, and then the system would be modeled using an SRP [24]. In addition, the superposition of independent RPs converges to a Poisson process (possibly nonhomogeneous), when the number of superimposed processes grows [25].

When the repair or substitution of the failed part in a complex system does not involve a significant modification of the reliability of the equipment as a result of the repair action, the NHPP is able to correctly describe the failure–repair process. Then, the NHPP can be interpreted as a minimal repair model [26], and it assumes that the unit returns to an "as bad as old" (ABAO) condition after a repair. The NHPP differs from the HPP in that the rate of occurrence of failures varies with time rather the being constant [22]. Unlike the previous model, in this process the interarrival times are neither independent nor identically distributed.

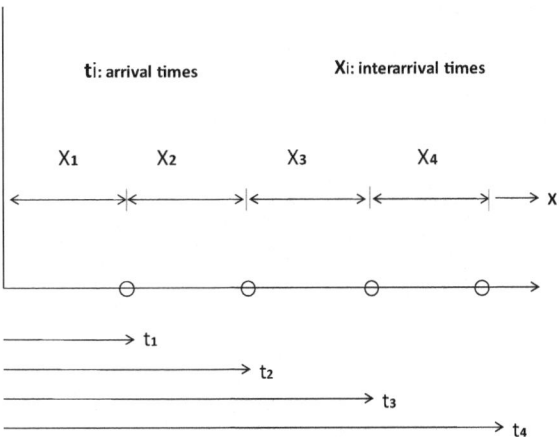

Fig. 3 Basic notation for a stochastic point process

Another common classification for repairable systems is considering that it may end up in one of the five possible states after a repair:

- a. As good as new
- b. As bad as old
- c. Better than old, but worse than new
- d. Better than new
- e. Worse than old

The RP and NHPP described before, account for the first two states respectively. However, the last three repair states have received less attention since they involve more complex mathematical models. Kijima and Sumita [27] proposed a probabilistic model for all the after-repair states called Generalized Renewal Process (GRP).

In order to extend this general (imperfect) repair model, several papers have been published in recent years [28–32]. This GRP approach has been published also by [33] and they have offered a Monte Carlo (MC) based approximate solution for certain application areas [34].

According to this approach, the RP and the NHPP are considered specific cases of the generalized model.

The GRP theory of repairable items introduces the concept of virtual age (An). This value represents the calculated age of the element immediately after the nth repair occurs. For $A_n = y$ the system has a time to the $(n + 1)$th failure, x_{n+1}, which is distributed according to the following cumulative distribution function (*cdf*):

$$F(x|A_n = y) = \frac{F(x+y) - F(y)}{1 - F(y)},$$
(2)

where $F(x)$ is the *cdf* of the time to the first failure (TTFF) distribution of a new component or system.

The summation:

$$S_n = \sum_{i=1}^{n} x_i.$$
(3)

With $S_0 = 0$, is called the real age of the element. The model assumes that the nth repair only compensates for the damage accumulated during the time between the $(n - 1)$th and the nth failure. With this assumption, the virtual age of the component or system after the nth repair is:

$$A_n = A_{(n-1)} + qx_n = qS_n,$$
(4)

where q is the repair effectiveness (or rejuvenation) parameter and $A_0 = 0$. According to this model, the result of assuming a value of $q = 0$ leads to a RP (as good as new), while the assumption of $q = 1$ corresponds to a NHPP (as bad as old). The values of q that fall in the interval $0 < q < 1$ represent the after-repair states in which the condition of the element is better than old but worse than new, whereas the cases where $q > 1$ correspond to a condition worse than old. Similarly, cases with $q < 0$ would suggest a component or system restored to a state better than new. Therefore, physically speaking, q can be seen as an index for representing the effectiveness and quality of repairs [35]. Even though the q-value of the GRP model constitutes a realistic approach to simulate the quality of maintenance, it is important to point out that the model assumes an identical q for every repair in the item's life. A constant q may not be the case for some equipment and maintenance processes, but it is a reasonable approach for most repairable components and systems.

The three models described above have advantages and limitations. In general, the more realistic model will be that which is more complex and complete as the mathematical expression involved.

The NHPP model has been proved to provide good results even for realistic situations with better-than-old but worse-than-new repairs [35].

Based on all those mentioned, and given their conservative nature and manageable mathematical expressions, the methodology of the case study that will be presented next concentrates on the NHPP because of its simplicity and for the following reasons [36]:

(a) It is generally suitable for the purpose of modeling data with a trend, due to the fact that the accepted formats of the NHPP are monotonously increasing/decreasing functions.
(b) NHPP models are mathematically straight forward and their theoretical base is well developed.
(c) Models have been tested fairly well, and several examples are available in the literature for their application.

3.2 Nonhomogeneous Poisson Process Analytical Modeling

The NHPP is a stochastic point process in which the probability of occurrence of n failures in any interval $[t1, t2]$ has a Poisson distribution with the mean:

$$\bar{\lambda} = \int_{t2}^{t1} \lambda(t)\mathrm{d}t, \tag{5}$$

where $\lambda(t)$ is defined as the Rate of Occurrence of Failures (ROCOF) in time interval $[t1, t2]$ [22, 37].

Therefore, according to the Poisson process:

$$\Pr[N(t_2) - N(t_1) = n] = \frac{\left[\int_{t_1}^{t_2} \lambda(t)\mathrm{d}t\right]^n \exp\left[-\int_{t_1}^{t_2} \lambda(t)\mathrm{d}t\right]}{n!}, \tag{6}$$

where $n = 0, 1, 2, \ldots$ are the total expected number of failures in the time interval $[t1, t2]$. The total expected number of failures is given by the cumulative intensity function:

$$\Lambda(t) = \int_0^t \lambda(t)\mathrm{d}t. \tag{7}$$

One of the most common forms of ROCOF used in reliability analysis of repairable systems is the Power Law Model (Weibull Intensity) [22, 37]:

$$\lambda(t) = \frac{\beta}{\alpha}\left(\frac{t}{\alpha}\right)^{\beta-1}. \tag{8}$$

This form comes from the assumption that the interarrival times between successive failures follow a conditional Weibull probability density function, with parameters α and β. The Weibull distribution is typically used in maintenance areas due to its flexibility and applicability to various failure processes. However, solutions to Gamma and Log-normal distributions are also possible. This model implies that the arrival of the ith failure is conditional on the cumulative operating time up to the $(i-1)$th failure. Figure 4 shows a schematic of this conditionality [35]. This conditionality also arises from the fact that the system retains the condition of as bad as old after the $(i-1)$th repair. Thus, the repair process does not restore any added life to the component or system.

In order to obtain the maximum likelihood (ML) estimators of the parameters of the power law model, consider the following definition of conditional probability:

$$P(T \leq t|T > t_1) = \frac{F(t) - F(t_1)}{R(t_1)} = \frac{1 - R(t) - 1 + R(t)}{R(t_1)} = 1 - \frac{R(t)}{R(t_1)}, \tag{9}$$

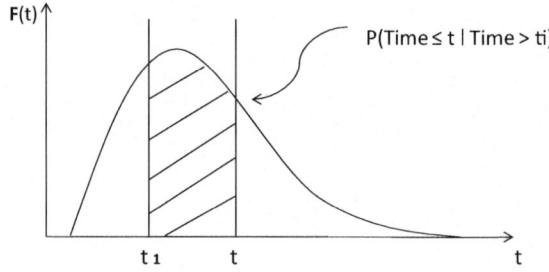

Fig. 4 Conditional probability of occurrence of failure

where $F(\bullet)$ and $R(\bullet)$ are the probability of component failure and the reliability at the respective times. Assuming a Weibull distribution, Eq. (9) yields:

$$F(t_i) = 1 - \exp\left[\left(\frac{t_{i-1}}{\alpha}\right)^{\beta} - \left(\frac{t_i}{\alpha}\right)^{\beta}\right]. \tag{10}$$

Therefore, the conditional Weibull density function is:

$$f(t_i) = \frac{\beta}{\alpha}\left(\frac{t_i}{\alpha}\right)^{\beta-1} \cdot \exp\left[\left(\frac{t_{i-1}}{\alpha}\right)^{\beta} - \left(\frac{t_i}{\alpha}\right)^{\beta}\right]. \tag{11}$$

For the case of the NHPP, different expressions for the likelihood function may be obtained.

For case study next presented the expression based on estimation at a time t after the occurrence of the last failure and before the occurrence of the next failure will be used. See details on these expressions in [38].

3.3 Time Terminated NHPP Maximum Likelihood Estimators

In the case of time terminated repairable components, the maximum likelihood function L can be expressed as:

$$L = \prod_{i=1}^{n} f(t_i) = f(t_1) \prod_{i=2}^{n} f(t_i)R(t_n|t). \tag{12}$$

Therefore,

$$L = \left\{\frac{\beta}{\alpha}\left(\frac{t_1}{\alpha}\right)^{\beta-1}\exp\left[-\left(\frac{t_1}{\alpha}\right)^{\beta}\right]\right\}$$
$$\times \left\{\left(\frac{\beta}{\alpha}\right)^{n-1}\prod_{i=2}^{n}\left(\frac{t_1}{\alpha}\right)^{\beta-1}\exp\left(\sum_{i=2}^{n}\left[\left(\frac{t_{i-1}}{\alpha}\right)^{\beta} - \left(\frac{t_i}{\alpha}\right)^{\beta}\right]\right)\right\} \tag{13}$$
$$\times \left\{\exp\left[\left(\frac{t_n}{\alpha}\right)^{\beta} - \left(\frac{t}{\alpha}\right)^{\beta}\right]\right\}(13)$$

Again, the ML estimators for the parameters are calculated. The results are [21, 36]:

$$\hat{\alpha} = \frac{t_n}{n^{\frac{1}{\beta}}} \tag{14}$$

$$\hat{\beta} = \frac{n}{\sum_{i=1}^{n} \ln\left(\frac{t_n}{t_i}\right)}, \qquad (15)$$

where t_i is the time at which the ith failure occurs, t_n is the total time where the last failure occurred, and n is the total number of failures. The total expected number of failures in the time interval $[t_n, t_{n+s}]$ by the Weibull cumulative intensity function is [38]:

$$\Lambda(t_n, t_{n+s}) = \frac{1}{\alpha^{\beta}}\left[(t_n + t_s)^{\beta} - (t_n)^{\beta}\right], \qquad (16)$$

where t_s is the time after the last failure occurred in the one which needs to be considered the number of failures and t_n is:

$$t_n = \sum_{i=1}^{n} t_i. \qquad (17)$$

The Non-Homogeneous Poisson Process (NHPP) is widely used to model the failure process of repairable systems.

4   NHPP Model Proposed for the Evaluation of the Costs Per Failure

Most of the methodologies proposed in recent years include basic analysis that allow for quantifying the economic impact that the failures inside a production system generate (LCCA review [39]). In relation to the quantification of the costs for non Reliability in the LCCA, in this article the use of NHPP model is recommended. This model proposes to evaluate the impact of the main failures on the costs structure of a production system, starting from a simple process, which is summarized as: first, the most important types of failures are determined; then, assigned to each failure type is a constant value of occurrence frequency per year (this value will not change along the expected useful life); later on, the impact in costs per year is estimated, generated by the failures in the production, operations, environment and security; and finally, the total impact on costs of failures for the years of expected useful life is considered in the present value to a specific discount rate. The Following details the steps to estimate the costs for failures according to NHPP model:

1. Identify for each alternative to evaluate the main types of failures. This way for certain equipment there will be $f = 1\ldots F$ types of failures.

2. Determine for the n (total of failures), the times to failures t_f. This information will be gathered by the designer based on records of failures, databases and/or experience of maintenance and operations personnel.

3. Calculate the Costs for failures C_f ($/failure). These costs include: costs of penalization for production loss and operational impact C_p ($/h), costs of corrective maintenance C_c ($/h) and the mean time to repair MTTR (h). The expression used to estimate the C_f is shown next:

$$C_f = (C_p + C_c) \times \text{MTTR}. \tag{18}$$

4. Define the expected frequency of failures per year $\Lambda(t_n, t_{n+s})$. This frequency is assumed as a constant value per year for the expected cycle of useful life. The $\Lambda(t_n, t_{n+s})$ is calculated starting from the expression (16). This process is carried out starting from the times to failures registered t_f by failure type (step 2). The parameters α and β are set starting from the following expressions (14) and (15). In the expression (16), t_s, it will be a year (1 year) or equivalent units (8760 h, 365 days, 12 months, etc.). This time t_s represents the value to estimate the frequency of failures per year.

5. Calculate the total costs per failures per year TCP_f, generated by the different events of stops in the production, operations, environment and security, with the following expression:

$$\text{TCP}_f = \sum_f^F \Lambda(t_n, t_{n+s}) \times C_f. \tag{19}$$

The obtained equivalent annual total cost, represents the probable value of the money that will be needed every year to pay the problems of reliability caused by the event of failure, during the years of expected useful life.

6. Calculate the total costs per failures in present value PTCP_f. Given a yearly value TCP_f, the quantity of money in the present (today) that needs to be saved, to be able to pay this annuity for the expected number of years of useful life (T), for a discount rate (i). The expression used to estimate the PTCP_f is shown next:

$$\text{PTCP}_f = \text{TCP}_f \times \frac{(1+i)^T - 1}{i \times (1+i)^T}. \tag{20}$$

Later on, the rest of the evaluated costs (investment, planned maintenance, operations, etc.) are added to the costs calculated by non-reliability, the total cost is calculated in the present value for the selected discount rate and the expected years of useful life, and then the obtained results of both the options will be economically compared.

5 Development of Case Studies and Comparison

The following case studies propose the evaluation of the economic impact of the failures using the method NHPP. The analysis was developed for the oil company PETRONOX–SHELL (Venezuela) and for the Rail Freight Industry (Chile).

The maintenance strategy for both assets is time based—preventive maintenance: Case 1: Semiannual10079. Case 2: Weekly. The MTTR was calculated including the corrective and preventive times for repair.

5.1 Case Study 1: Oil Company

This analysis was developed for the oil company PETRONOX–SHELL (contractor of Petróleos of Venezuela), located in the field of gas and petroleum Naricual II, in Monagas, Venezuela. In general terms, it is required to install a compression system to manage an average flow of 20 million cubic feet of gas per day. The organization PETRONOX, evaluates the information of two compressor suppliers. Next shows the data costs of: initial investment, operation, and maintenance for the two options to evaluate (value estimated by the suppliers):

Option A: Reciprocant Compressor, 2900–3200 hp, caudal: 20 millions of feet cubic per day.
Option B: Reciprocant Compressor, 2810–3130 hp, caudal: 20 millions of feet cubic per day (Table 2).

With this information the organization PETRONOX carries out a first economic analysis of the life cycles with which a comparison was made between the two alternatives, in this first evaluation, the economic impact of the failures was not considered. Next, the results of this analysis are presented.

In the results presented in Table 3, the oil company does not consider the possible costs in event of failure, obtaining option B as the best economic alternative of the two evaluated (more economic alternatives for the period of life span of 15 years), with a difference of approximately: 224.917,133 $ (this quantity would

Table 2 Economical data

Data	Option A	Option B
I: investment	1,100,000 $	900,000 $
OPC: operationals costs	100,000 $/year	120,000 $/year
PRC: preventive costs	60,000 $/year	40,000 $/year
OVC: overhauls costs	100,000 $ every 5 years	80,000 $ every 5 years
i: interest	10%	10%
T: expected useful life	15 years	15 years

Table 3 Economical results without to evaluate the costs per failures

Results	Option A ($)	Option B ($)
1. I: invesment	1,100,000	900,000
2. OPC(P): operational? costs in present value	760,607,951	912,729,541
3. PRC(P): preventive costs in present value	45,636,477	30,424,318
4. OVC(P): overhauls costs in present value (t = 5 years)	620,921,323	496,737,058
5. OVC(P): overhauls costs in present value, (t = 10 years)	385,543,289	308,434,632
6. OVC(P): overhauls costs in present value, (t = 15 years)	239,392,049	191,513,639
TLCC(P): total life cycle costs in present value, i: 10%, T: 15 years (Sum 1...6)	2,441,558,387	2,216,641,254

be the potential saving to select option B, without considering the possible costs for failures).

Later on, a proposal was made to the oil organization, before it made some decision with regard to the acquisition of the two evaluated options. This proposal consists of evaluating the possible costs by events of failures through the NHPP model. For this evaluation, the total expected number of failures in the interval of time $[t_n, t_{n+s}]$, is estimated by the NHPP stochastic model (Weibull cumulative intensity function), see [33]. Next, are shown the data of costs and times of failures to be used inside the NHPP model (the data of times to failures t_f were gathered by PETRONOX of two similar compression systems that operate under very similar conditions in those that will work the compressor to be selected).

With the information of Table 4, Eq. (16) was used to calculate the frequency of failures per year $\Lambda(t_n, t_{n+s})$. The parameters α and β of the Distribution of Weibull contained in Eq. (16) were calculated from the Eqs. (14) and (15). The total costs for failures per year TCP_f were calculated from the Eqs. (18) and (19); these costs are converted to present value $PTCP_f$ with Eq. (20). Next, are shown the results of the frequency of failures and the total costs for failures for year obtained starting from the NHPP model, for the two evaluated options (Table 5).

Table 4 Failure costs and maintainability/reliability data

Data	Option A	Option B
Cp ($/h)	6.000	6.000
Cc ($/h)	700	400
MTTR (h)	9	8
t_f (months)	5, 7, 3, 7, 2, 4, 3, 5, 8, 9, 2, 4, 6, 3, 4, 2, 4, 3, 8, 9	2, 3, 3, 5, 6, 6, 5, 6, 5, 6, 4, 3, 2, 2, 2, 2, 3, 2, 2, 3, 2, 2, 3, 3
t_n (total of months)	98	82
n (total of failures)	20	24

Table 5 Results from NHPP model

Results	Option A	Option B
α	6.98	6.14
β	1.13	1.23
Λ (t_n, t_{n+s}) = failure/year	2.7987 = 2.8	4.3751 = 4.38
TCP_f = \$/year	168.840	224.256
$PTCP_f$ = \$ (i = 10%, T = 15 years)	1,284,210	1,705,709

Later on, a second economic evaluation was carried out including the results of costs for failures obtained starting from the NHPP model. Next, the results are presented in Table 6.

Analyzing the results obtained in this second economic evaluation (see, Table 6), in which the total costs for failures are included in present value $PTCP_f$, option A turns out to be the best economic alternative compared to option B, with a difference of approximately: 196.581,368 \$ (this quantity would be the potential savings to select option A).

5.2   Case Study 2: Rail Freight Industry

This case study was developed to assess the impact of failures of diesel locomotives Railway System in Chile. This country is characterized by its length, resulting in extensive travel between stations to meet specific tasks. In this case locomotives that comply with the main routes will be evaluated: transfer of concentrate from the mine ANDINA to the smelter VENTANAS and the Port of VALPARAÍSO; and

Table 6 Economical results with the costs per failures

Results	Option A	Option B
1. I: invesment	1,100,000	900,000
2. OPC(P): operationals costs in present value (\$)	760,607,951	912,729,541
3. PRC(P): preventives costs in present value (\$)	45,636,477	30,424,318
4. OVC(P): overhauls costs in present value (t = 5 years) (\$)	620,921,323	496,737,058
5. OVC(P): overhauls costs in present value (t = 10 years) (\$)	385,543,289	30.843,4632
6. OVC(P): overhauls costs in present value (t = 15 years) (\$)	239,392,049	191,513,639
7. $PTCP_f$: total costs per failures in present value	128,421,046	170,570,897
TLCC(P): total life cycle costs in present value, i: 10%, T: 15 years (Sum 1.7)	3,725,768,851	392,235,022
$PTCP_f$/TLCC(P) = % (total costs per failures/total life	34.46%	43.48%

Table 7 Economical data

Data	Option A	Option B
I: Investment	1,625,333 $	1,272,000 $
OPC: operationals costs	502,448 $/year	467,984 $/year
PRC: preventive costs	68,264 $/year	91,582 $/year
OVC: overhauls costs	433,333 $ every 10 years	406,819 $ every 10 years
i: interest	11%	11%
T: expected useful life	40 years	40 years

Table 8 Economical results without to evaluate the costs per failures

Results	Option A	Option B
1. I: invesment ($)	1,625,333	1,272,000
2. OPC(P): operationals costs in present value ($)	4,497,438	4,188,949
3. PRC(P): preventive costs in present value ($)	611,033 $	819,757
4. OVC(P): overhauls costs in present value ($t = 10$ years) ($)	152,613	143,275
5. OVC(P): overhauls costs in present value, ($t = 20$ years) ($)	53,748	50,459
6. OVC(P): overhauls costs in present value, ($t = 30$ years) ($)	18,929	17,771
7. OVC(P): overhauls costs in present value, ($t = 40$ years) ($)	6667	2204
TLCC(P): total life cycle costs in present value, i: 11%, T: 40 years (Sum 1...7) ($)	7,118,375	6,637,690

also the transport of industrial waste removal in the south of Chile. Rail operations cover an area of over 2.500 km. The main equipment of each rail is its engine, whose main component is the diesel engine. The company is evaluating two different models of diesel engines (Option A and Option B) with a total capacity of 2.300 HP. The economic data of each option are detailed in Table 7.

Similar to case study 1, with this information it is possible to carry out a first economic analysis of life cycle, where the economic impact of the failures was not considered. Next, the results are shown in Table 8.

In the results presented in the Table 8, option B is chosen, considering the period of lifespan of 40 years and a difference of approximately: 480.684 $.

Now, as was done for case study 1, the cost of events of failure through the NHPP model will be considered.

Next, Table 9 show the data of costs and times of failures to be used in the model.

With this information, it is possible to economically evaluate both alternatives, taking into account the results of the Table 10, which has the NHPP parameters. The data was calculated with the same steps described before for case study 1.

Next, the results are presented in Table 11.

Analyzing the results obtained in this second economic evaluation (see Table 11), option A turns out to be the best economic alternative compared to

Table 9 Failure costs and maintainability/reliability data

Data	Option A	Option B
C_p ($/h)	200	200
C_c ($/h)	20	19
MTTR (h)	59.1	55.2
t_f (h)	383, 386, 362, 346, 349, 348, 333, 407, 250, 500, 345, 424, 654,124, 234, 412, 352, 253, 165, 456, 779, 394, 298, 148, 336, 376, 548, 618, 80, 703, 22, 448, 134, 381, 54, 594, 358, 279, 634, 773, 620, 123, 231, 174, 281, 174, 270, 542, 162, 562, 334, 70, 635, 123, 372, 373, 126, 306, 466, 711, 509, 316, 80, 762, 145, 57, 262, 501, 304, 558	125, 429, 323, 284, 415, 618, 327, 251, 201, 630, 367, 578, 345, 589, 317, 232, 184, 211, 348, 515, 217, 356, 506, 512, 273, 593, 234, 92, 42, 151, 23, 225, 35, 304, 600, 465, 423, 173, 188, 148, 55, 230, 347, 62, 406, 591, 26, 259, 239, 638, 166, 284, 485, 61, 528, 299, 342, 302, 134, 379, 415, 87, 71, 525, 247
t_n (total of hours)	25,189	20,027
n (total of failures)	70	65

Table 10 Results from NHPP model

Results	Option A	Option B
α	457.99	478.65
β	1.06	1.12
Λ (t_n, t_{n+s}) = failure/year	26.05 = 26	32.52 = 33
TCP_f = \$/year	338,754	393,104
$PTCP_f$ = \$ (i = 11%, T = 40 years)	3,032,205	3,518,690

Table 11 Economical results with the costs per failures

Results	Option A	Option B
1. I: invesment (\$)	1,625,333	1,272,000
2. OPC(P): operationals costs in present value (\$)	4,497,438	4,188,949
3. PRC(P): preventives costs in present value (\$)	611,033	819,757
4. OVC(P): overhauls costs in present value (t = 10 years) (\$)	152,613	143,275
5. OVC(P): overhauls costs in present value (t = 20 years) (\$)	53,748	50,459
6. OVC(P): overhauls costs in present value (t = 30 years) (\$)	18,929	17,771
6. OVC(P): overhauls costs in present value (t = 40 years) (\$)	6667	2204
8. $PTCP_f$: total costs per failures in present value (\$)	3,032,205	3,518,690
TLCC(P): total life cycle costs in present value, i: 11%, T: 40 years (Sum 1…8) (\$)	10,150,580	10,156,381
$PTCP_f$/TLCC(P) = % (total costs per failures/total life cycle costs)	29.9%	34.7%

option B, with a difference of approximately: 5.801 \$ (this quantity would be the potential savings to select option A).

5.3 Sensibility Analysis

An important assessment for the cases studies is the sensibility analysis considering the variation of the main variables (economical and/or technical data). For this particular analysis, as an example, it is considered the variation of the Costs of penalization for production loss and operational impact (\$/h).

The sensibility analysis was developed for the Rail Freight Industry—Option A and B (Table 12).

Graphically (Fig. 5).

With this analysis we can visualize how the costs of penalization for production loss and operational impact, which is part of the cost of failure, could lead to large variations in the Total life cycle cost of the asset. Analyzing the graph shown in Fig. 5, we identify a point of intersection for both options, where Cp belongs to the range of 200–180 [\$/h], leading to indifference in the decision.

Table 12 Sensibility analysis for the Rail Freight Industry—Option A and B

C_p ($/h)	TLCC (P) [$] Option A ($)	TLCC (P) [$] Option B ($)
300	115,288,550	117,630,884
280	112,532,000	114,417,469
260	109,775,449	111,204,053
240	107,018,899	107,990,637
220	104,262,349	104,777,221
200	101,505,799	101,563,805
180	98,749,249	98,350,390
160	95,992,698	95,136,974
140	93,236,148	91,923,558
120	90,479,598	88,710,142
100	87,723,048	85,496,727
80	84,966,497	82,283,311
60	82,209,947	79,069,895
40	79,453,397	75,856,479
20	76.696,847	72,643,063

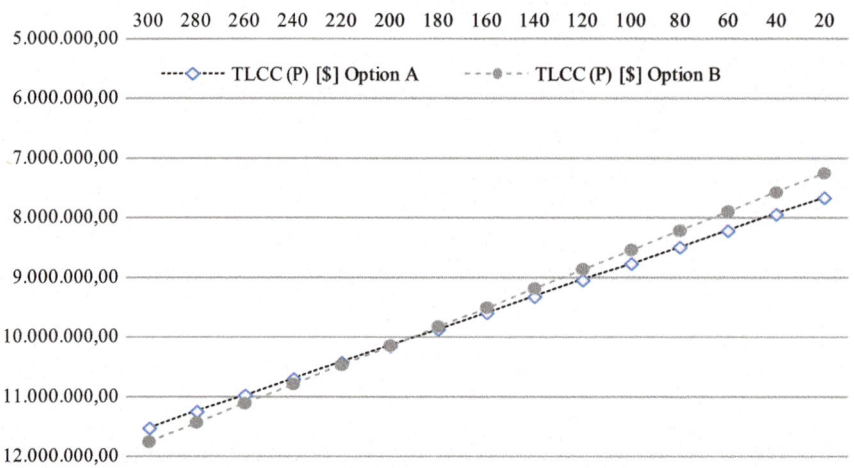

Fig. 5 Sensibility analysis of TLCC (P) [$] varying Cp

Another interesting analysis will be to identify how maintenance policy as defined (i.e., constant age) affects variables n (number of failures), maintenance budget (PRC), and the total life cycle cost. According to the last propose, it is developed as an approximation considering also the impact of: Preventive Cost factor/(CC + Cp).

The approaches were made by specialists in the company, integrating variables of maintenance, operations, and finance (Table 13).

Table 13 Sensibility analysis for the Rail Freight Industry—Option A

Preventive cost ($/h)/[$C_c + C_p$] ($/h)	No. of failures	PRC $/ year	No. of preventive intervention	TLCC (P) ($)	TLCC (P) normalized	PTCPf/TLCC(P) (%)
0.1	26	68,264	58	101,505,799	1000	29.872
0.2	24	155,743	66	10.9,563,573	1079	27.883
0.3	21	252,340	71	11.6,603,425	1149	24.822
0.4	19	359,621	76	124,708,795	1229	22.008
0.5	17	473,524	80	134,157,246	1322	19.901
0.6	15	591,639	83	14.3,664,042	1415	17.842
0.7	14	703,612	85	153,117,908	1508	16.369
0.8	13	815,056	86	162,878,435	1605	15.256
0.9	12	923,634	87	172,164,578	1696	14.182
1	11	1,029,219	87	179,973,191	1773	12.654

The challenge for future work is to analyze in detail how the maintenance policies decision by the organization (technically and economically) affects the asset's life cycle, evidently considering the impact of failures.

5.4 General Analysis of Results of the Case Studies

An important aspect to be considered in this analysis is that when introducing the potential costs for consequences in the events of failures, this category of costs $PTCP_f$ transforms into the economic factor of more weight inside the process of comparison of the two alternatives to select, specifically, this category of costs represents:

- **Oil Company**: For option B it represents the 43.48% and for option A the 34.46% of the total of costs to the prospective life cycle for these two assets (with an interest rate of 10% and a prospective cycle of life of 15 years).
- **Rail Company**: For option B it represents the 34.7% and for option A the 29.9% of the total of costs to the prospective life cycle for these two assets (with an interest rate of 11% and a prospective life cycle of 40 years).

Finally,

- It is recommended to the Oil organization PETRONOX takes into account the importance of the costs estimated by consequences of failures inside the process of selection of the two previously evaluated alternatives and selects option A like the asset of economic better opportunity in the prospective lifespan cycle.
- If the amount of money 5.801 $ is significant for the Rail Industry, the decision should be similar to the Oil industry (select option A), which takes into account the economic consequences of the future failures. In case that the difference of money for both alternatives is not important, the decision will be taken based on qualitative factors which are described below:

  - Professional relationship with the suppliers.
  - Experience with the suppliers.
  - Level of risk of the organization, e.g., some organization will choose the alternative which has the lower percentage % (total costs per failures/total life cycle costs).
  - Corporative contracts with specifics suppliers.

Additionally, both organizations should design an internal procedure that allows evaluating the costs of opportunity of reliability, this procedure will be used in a continuous and obligatory way every time that different options are analyzed inside the processes of design, selection, substitution and/or purchase of assets.

5.5 General Evaluation of the Model

The main strengths and weakness of this model are summarized next:
Strengths:

- The NHPP is one of the most popular models for repairable systems, so many models have been developed based on this process, such as the Crow-AMSAA model [37] and bounded intensity process (BIP) model [40].
- It is a useful and quite simple model to represent equipment under aging (deterioration).
- Involves relatively simple mathematical expressions.
- It is a conservative approach and in most cases provides results very similar to those of more complex models like GRP [41].

Weakness:

- Is not adequate to simulate repair actions that restore the unit to conditions better than new or worse than old.
- This model does not consider the Time to Repair as an independent indicator (TTR) to calculate the number of failures.

6 Conclusions and Future Directions

The specific orientation of this work toward the analysis of the Reliability factor and its impact in the costs, is due to, that great part of the increment of the total costs during the expected cycle of useful life of a production system, is caused in its majority, by the lack of prevision in the face of unexpected appearance of failure events, a scenario basically provoked by ignorance and by the absence of a technical evaluation in the design phase of the aspects related to Reliability. This situation brings as a result an increment in the total costs of operation (costs that were not considered in the beginning) in this way affecting the profitability of the production process.

In the process analysis of the costs along the life cycle of an asset, many decisions and actions exist that should be taken, being of particular interest for this work, those aspects related to the improvement process of Reliability (quality of the design, used technology, technical complexity, frequency of failures, costs of preventive/corrective maintenance, maintainability levels and accessibility), since these have a great impact on the total cost of the life cycle of the asset, and they influence in great measure the possible expectations to extend the useful life of the assets to reasonable costs. For these reasons, it is of supreme importance inside the process to estimate the life cycle of the assets, to evaluate and to analyze in detail the aspects related with the failure rate. According to [21], the following points should be considered in failure rate trend analyses:

- Failure of a component may be partial, and repair work done on a failed component may be imperfect. Therefore, the time periods between successive failures are not necessarily independent. This is a major source of trend in the failure rate.
- Imperfect repairs performed following failures do not renew the system, i.e., the component will not be as good as new; only then can the statistical inference methods using a Rate Of Occurrence Of Failures (ROCOF) assumption be used.
- Repairs made by adjusting, lubricating, or otherwise treating component parts that are wearing out provide only a small additional capability for further operation, and do not renew the component or system. These types of repair may result in a trend of an increasing ROCOF.
- A component may fail more frequently due to aging and wearing out.

It is important to mention that inside the LCCA techniques a potential area of optimization related to the evaluation of the Reliability impact exists. In the near future the new proposals of evaluation of the costs generated by aspects of low Reliability will use advanced mathematical methods such as:

- Stochastic methods see [23, 35, 38, 41–43].
- Advanced maintenance optimization using genetic algorithms see [44, 45].
- Monte Carlo simulation techniques see [46–48].
- Advanced Reliability distribution analyses see [17, 49–55].
- Markov simulation methods see [27, 55, 56] (Table 14).

Table 14 Stochastic processes used in reliability analysis of repairable systems

Stochastic process	Can be used	Background/Difficulty
Renewal process	Spare parts provisioning in the case of arbitrary failure rates and negligible replacement or repair time (Poisson process)	Renewal theory/Medium
Alternating renewal process	One-item repairable (renewable) structure with arbitrary failure and repair rates	Renewal theory/medium
Markov process (MP)	Systems of arbitrary structure whose elements have constant failure and repair rates *during the stay time (sojourn time) in every state* (not necessarily at a state change, e.g., because of load sharing)	Differential equations or integral equations/low
Semi-Markov process (SMP)	Some systems whose elements have constant or Erlanggan failure rates (Erlang distributed failure-free times) and arbitrary repair rates	Integral equations/medium
Semi-regenerative process	Systems with only one repair crew, arbitrary structure, and whose elements have constant failure rates and arbitrary repair rates	Integral equations/high
Non-regenerative process	Systems of arbitrary structure whose elements have arbitrary failure and repair rates	Partial diff. eq.; case by base sol/High to very high

These methods will have their particular characteristics and their main objective diminish the uncertainty inside the estimation process of the total costs of an asset along the expected useful life cycle.

Finally, it is not feasible to develop a unique LCCA model, which suits all of the requirements. However, it is possible to develop more elaborate models to address specific needs such as a reliability cost-effective asset development.

Finally, is important to mention that the results of the developed real case studies help us to understand and confirm the importance of the procedures that consider the economic impact of failures in the life cycle analysis. This consideration will support making the decision.

Acknowledgements The research leading to these results has received funding from the European Community's Seventh Framework Programme (FP7/2007-2013 under grant agreement n° PIRSES-GA-2008-230814).

References

1. Mackenzie J (1997) Turn your company's strategy into reality. Manuf Manage 6–8
2. Woodhouse J (1993) Managing industrial risk. Chapman Hill Inc., London
3. Parra C, Crespo A (2015) Ingeniería de Mantenimiento y Fiabilidad aplicada en la Gestión de Activos, INGEMAN, 2da Edición, Sevilla, España
4. Markeset T, Kumar U (2001) R&M and risk analysis tools in product design to reduce life-cycle cost and improve product attractiveness. In: Proceedings of the annual reliability and maintainability symposium, Philadelphia, pp 116–122, 22–25 Jan 2001
5. Blanchard BS (2001) Maintenance and support: a critical element in the system life cycle. In: Proceedings of the International Conference of Maintenance Societies, paper 003, Melbourne, May 2001
6. Blanchard BS, Fabrycky WJ (1998) Systems engineering and analysis, 3rd edn. Prentice-Hall, Upper Saddle River
7. Goffin K (2000) Design for supportability: essential component of new product development. Res Technol Manage 43(2):40–47
8. Smith C, Knezevic J (1996) Achieving quality through supportability: part 1: concepts and principles. J Q Maintenance Eng 2(2):21–29
9. Woodward DG (1997) Life cycle costing—theory, information acquisition and application. Int J Project Manage 15(6):335–344
10. Durairaj S, Ong S (2002) Evaluation of life cycle cost analysis methodologies. Corp Environ Strategy 9(1):30–39
11. Kirk S, Dellisola A (1996) Life cycle costing for design professionals. McGraw Hill, New York, pp 6–57
12. Woodhouse J (1991) Turning engineers into businessmen. In: 14th National Maintenance conference, London
13. Fabrycky WJ, Blanchard BS (1993) Life cycle costing and economic analysis. Prentice Hall, Inc., Englewood Cliff
14. Ahmed NU (1995) A design and implementation model for life cycle cost management system. Inf Manage 28:261–269
15. Levy H, Sarnat M (1990) Capital investment and financial decisions, 4th edn. Prentice Hall, New York

16. Ruff DN, Paasch RK (1993) Consideration of failure diagnosis in conceptual design of mechanical systems. In: Design theory and methodology, ASME, New York, pp 175–187
17. Barlow RE, Clarotti CA, Spizzichino F (1993) Reliability and decision making. Chapman & Hall, London
18. Dowlatshahi S (1992) Product design in a concurrent engineering environment: an optimization approach. J Prod Res 30(8):1803–1818
19. Louit DM, Pascual R, Jardine A (2009) A practical procedure for the selection of time to failure models on the assessment of trends in maintenance data. Reliab Eng Syst Saf 94: 1618–1628
20. Saldanha PLC, de Simone EA, Frutoso e Melo PF (2001) An application of non- homogeneus Poisson point processes to the reliability analysis of service water pumps. Nucl Eng Des 210:125–133
21. Weckman GR, Shell RL, Marvel JH (2001) Modeling the reliability of repairable systems in the aviation industry. Comput Ind Eng 40:51–63
22. Ascher H, Feingold H (1984) Repairable system reliability: modeling, inference, misconceptions and their causes. Marcel Dekker, New York
23. Rigdon SE, Basu AP (2000) Statistical methods for the reliability of repairable systems. Wiley, New York
24. Ansell J, Phillips MJ (1989) Practical problems in the statistical analysis of reliability data (with discussion). Appl Stat 38:205–231
25. Thompson WA (1981) On the foundations of reliability. Technometrics 23:1–13
26. Calabria R, Pulcini G (2000) Inference and test in modeling the failure/repair process of repairable mechanical equipments. Reliab Eng Syst Saf 67:41–53
27. Kijima M, Sumita N (1987) A useful generalization of renewal theory: counting process governed by non-negative Markovian increments. J Appl Prob 23:71–88
28. Whitaker LR, Samaniego FJ (1989) Estimating the reliability of systems subject to imperfect repair. J Am Statist Assoc 84:301–309
29. Syamsundar A, Naikan VNA (2008) A proportional intensity segmented model for maintained systems. Proc Inst Mech Eng Part O J Risk Reliab 222(4):643–654. doi:10.1243/1748006XJRR158
30. Baxter LA, Kijima M, Tortorella M (1996) A point process model for the reliability of a maintained system subject to general repair. Comm Statist Stochast Models 12:37–65
31. Dorado C, Hollander M, Sethuraman J (1997) Nonparametric estimation for a general repair model. Ann Statist 25:1140–1160
32. Finkelstein M (2000) Modeling a process of non-ideal repair. In: Limnios N, Nikulin M (eds) Recent advances in reliability theory. Birkhauser, Boston, pp 41–53
33. Kaminskiy M, Krivtsov V (1998) A Monte Carlo approach to repairable system relaibility analysis. In: Probabilistic safety assessment and management. Springer, New York, pp 1063–1068
34. Veber B, Nagode M, Fajdima M (2008) Generalized renewal process for repairable systems based on finite Weibull mixture. Reliab Eng Syst Saf 93:1461–1472
35. Yañez M, Joglar F, Mohammad M (2002) Generalized renewal process for analysis of repairable systems with limited failure experience. Reliab Eng Syst Saf 77:167–180
36. Coetzee J (1997) The role of NHPP models in the practical analysis of maintenance failure data. Reliab Eng Syst Saf 56:161–168
37. Crow LH (1974) Reliability analysis for complex repairable systems. In: Proschan F, Serfling RJ (eds) Reliability and biometry. SIAM, Philadelphia, pp 379–410
38. Modarres M, Kaminskiy M, Krivtsov V (1999) Reliability engineering and risk analysis. Marcel Dekker Inc., New York
39. Asiedu Y, Gu P (1998) Product lifecycle cost analysis: state of art review. Int J Prod Res 36(4):883–908
40. Pulcini G (2001) Abounded intensity process for the reliability of repairable equipment. J Q Technol 33(4):480–492

41. Hurtado JL, Joglar F, Modarres M (2005) Generalized renewal process: models, parameter estimation and applications to maintenance problems. Int J Perform Eng 1(1):37–50, paper 3
42. Zhao J, Chan AHC, Roberts C, Stirling AB (2006) Assessing the economic life of rail using a stochastic analysis of failures. Proc Inst Mech Eng Part F J Rail Rapid Transit 220(2): 103–111. doi:10.1243/09544097JRRT30
43. Vasiliy V (2007) Recent advances in theory and applications of stochastic point process model in reliability engineering. Reliab Eng Syst Safety 92(5):549–551
44. Martorell S, Carlos S, Sánchez A, Serradell V (2000) Constrained optimization of test intervals using a steady-state genetic algorithm. Reliab Eng Syst Saf 67:215–232
45. Martorell S, Villanueva JF, Nebot Y, Carlos S, Sánchez A, Pitarch JL, Serradell V (2005) RAMS + C informed decision-making with application to multi-objective optimization of technical specifications and maintenance using genetic algorithms. Reliab Eng Syst Saf 87:65–75
46. Paul BH, Weber DP (1996) Life cycle cost tutorial. In: Fifth international conference on process plant reliability. Gulf Publishing Company, Houston
47. Paul BH, Weber DP (1997) Life cycle cost & reliability for process equipment. In: 8th annual ENERGY WEEK conference & exhibition. George R. Brown Convention Center, Houston, Texas, Organized by American Petroleum Institute
48. Kaminskiy M, Krivtsov V (1998) A Monte Carlo approach to repairable system reliability analysis. In: Probabilistic safety assessment and management. Springer, Berlin, pp 1063–1068
49. Elsayed EA (1982) Reliability analysis of a container spreader. Microlelectron Reliab 22 (4):723–734
50. Grant IW, Coombs CF Jr, Moss RY (1996) Handbook of reliability engineering and management, 2nd edn. McGraw-Hill, New York
51. Elsayed EA (1996) Reliability engineering. Addison Wesley Longman INC, New York
52. Zio E, Di Maio F, Martorell S (2008) Fusion of artificial neural networks and genetic algorithms for multi-objective system reliability design optimization. Proc Inst Mech Eng Part O J Risk Reliab 222(2):115–126. doi:10.1243/1748006XJRR126
53. Syamsundar A, Naikan VNA (2008) A proportional intensity segmented model for maintained systems. Proc Inst Mech Eng Part O J Risk Reliab 222(4):643–654. doi:10.1243/1748006XJRR158
54. Dhillon BS (1999) Engineering maintainability: how to design for reliability and easy maintenance. Gulf, Houston
55. [Markov]Welte TM (2009) A rule-based approach for establishing states in a Markov process applied to maintenance modeling. Proc Inst Mech En Part O J Risk Reliab 223(1):1–12. doi:10.1243/1748006XJRR194
56. Bloch-Mercier S (2000) Stationary availability of a semi-Markov system with random maintenance. Appl Stoch Models Bus Ind 16:219–234

Author Biographies

Carlos Parra Márquez Rewarded Titles:—Naval Engineering, Polytechnic Institute of the National Armed Forces (IUPFAN), 1986–1991, Caracas, Venezuela.—Master in Maintenance Engineering, Universidad de los Andes, School of Mechanical Engineering, Master Program in Maintenance Engineering, 1994–1996, Merida, Venezuela.—Reliability Specialist Engineering, PDVSA Convention—University Maryland—ASME, 2002–2003, United States.—Specialist in Industrial Organization Engineering, School of Industrial Engineering, University of Seville, 2004–2006, Seville, Spain.—Diploma of Advanced Studies, Industrial Engineering area Organization, Ph.D. in Industrial Engineering, School of Industrial Engineering, University of Seville, 2006–2008, Seville, Spain.—Doctor (Ph.D.) In Industrial Organization Engineering,

University of Sevilla, School of Industrial Engineering, Industrial Engineering Department Organization, 2004-2009, Sevilla, Spain. AWARDS/HONORS: Gran Mariscal de Ayacucho Scholarship to study: Master Maintenance Engineering at the University of the Andes, Merida, Venezuela 1994. OEA (Organization of American States) Scholarshipto to study: Doctorate in Industrial Organization Engineering, University of Seville, Spain, 2004. Award for best technical work in the 1st. World Congress of Maintenance Engineering, Bahia/Brazil, September 2005. Presentation/Publication: "Optimizing Maintenance Management process in the Venezuelan oil industry from the use of Reliability Engineering Methodologies."

Adolfo Crespo Márquez is currently Full Professor at the School of Engineering of the University of Seville, and Head of the Department of Industrial Management. He holds a Ph.D. in Industrial Engineering from this same University. His research works have been published in journals such as the International Journal of Production Research, International Journal of Production Economics, European Journal of Operations Research, Journal of Purchasing and Supply Management, International Journal of Agile Manufacturing, Omega, Journal of Quality in Maintenance Engineering, Decision Support Systems, Computers in Industry, Reliability Engineering and System Safety, and International Journal of Simulation and Process Modeling, among others. Professor Crespo is the author of seven books, the last four with Springer-Verlag in 2007, 2010, 2012, and 2014 about maintenance, warranty, and supply chain management. Professor Crespo leads the Spanish Research Network on Dependability Management and the Spanish Committee for Maintenance Standardization (1995–2003). He also leads a research team related to maintenance and dependability management currently with five Ph.D. students and four researchers. He has extensively participated in many engineering and consulting projects for different companies, for the Spanish Departments of Defense, Science and Education as well as for the European Commission (IPTS). He is the President of INGEMAN (a National Association for the Development of Maintenance Engineering in Spain) since 2002.

Vicente González-Prida Díaz is Ph.D. (Summa Cum Laude) in Industrial Engineering by the University of Seville, and Executive MBA (First Class Honors) by the Chamber of Commerce. He has been honoured with the following awards and recognitions:—Extraordinary Prize of Doctorate by the University of Seville;—National Award for Ph.D. Thesis on Dependability by the Spanish Association for Quality;—National Award for Ph.D. Thesis on Maintenance by the Spanish Association for Maintenance;—Best Nomination from Spain for the Excellence Master Thesis Award bestowed by the EFNSM (European Federation of National Maintenance Societies). Dr. Gonzalez-Prida is member of the Club of Rome (Spanish Chapter) and has written multitude of articles for international conferences and publications. His main interest is related to industrial asset management, specifically the reliability, maintenance, and aftersales organization. He currently works as Program Manager in the company General Dynamics—European Land Systems and shares his professional performance with the development of research projects in the Department of Industrial Organization and Management at the University of Seville.

Juan Francisco Gómez Fernández is Ph.D. in Industrial Management and Executive MBA. He is currently part of the Spanish Research & Development Group in Industrial Management of the Seville University and a member in knowledge sharing networks about Dependability and Service Quality. He has authored publications and collaborations in journals, books and conferences, nationally and internationally. In relation to the practical application and experience, he has managed network maintenance and deployment departments in various national distribution network companies, both from private and public sector. He has conduced and participated in engineering and consulting projects for different international companies, related to Information and Communications Technologies, Maintenance and Asset Management, Reliability Assessment, and Outsourcing services in Utilities companies. He has combined his business activity with academic life as an associate professor (PSI) in Seville University, being awarded as Best Thesis and Master Thesis on Dependability by National and International Associations such as EFNSM (European Federation of National Maintenance Societies) and Spanish Association for Quality.

Fredy Kristjanpoller Rodríguez is an Industrial Engineer and Master in Asset Management and Maintenance of Federico Santa Maria University (USM, Chile), and doctoral candidate in Industrial Engineering at the University of Seville. Researcher, academic and master program coordinator linked to asset management. He has important scientific papers on indexed journals and international proceedings congress on the following areas: Reliability engineering and Maintenance Strategies. He has developed consultant activities in the main Chilean companies.

Pablo Viveros Gunckel Researcher and Academic at Technical University Federico Santa María, Chile, has been active in national and international research, both for important journals and conferences. Also he has developed projects in the Chilean industry. Consultant specialist in the area of Reliability, Asset Management, System Modeling and Evaluation of Engineering Projects. He is Industrial Engineer and Master in Asset Management and Maintenance.

Online Reliability and Risk to Schedule the Preventive Maintenance in Network Utilities

Adolfo Crespo Márquez, Juan Francisco Gómez Fernández,
Pedro Moreu de León and Antonio Sola Rosique

Abstract This paper presents a methodology and a case study where the proportional hazard model is used to determine reliability and risk of existing repairable systems in a distribution network utility. After introducing different issues relating to conditioning maintenance management in these companies, we discuss the modeling possibilities to obtain online values for systems reliability and risk. In this case we try to model reliability considering current operating time of the equipment, the number of maintenance interventions, and the value of specific monitored parameters. Reliability is then used to calculate equipment risk for a given failure mode during a certain period, and in order to do so, failure mode affection to the network is estimated from historical data. Finally, the paper presents a possible business process to schedule preventive maintenance activities according to previous findings and a case study.

Keywords Risk management · Proportional hazard model · Reliability estimation · Monitoring · Network utilities · Maintenance

1 Introduction

Network Utilities are companies providing certain services to their clients, supporting and distributing these services through a network infrastructure linked directly to these clients or to their residences (e.g., communication services, electricity, gas, etc.). Utilities are normally capital-intensive industries [1], allowing for

A. Crespo Márquez (✉) · J.F. Gómez Fernández · P.M. de León
Department of Industrial Management, School of Engineering, University of Seville,
Seville, Spain
e-mail: adolfo@us.es

A. Sola Rosique
Technical Services, Iberdrola Generación S.A. Madrid, Madrid, Spain

© Springer International Publishing AG 2018
A. Crespo Márquez et al. (eds.), *Advanced Maintenance Modelling for Asset Management*, DOI 10.1007/978-3-319-58045-6_10

decades of payback on investments, and maintenance contributes to extend asset life cycle and to reduce assets depreciation. In order to study network assets life cycle and their depreciation, functional factors related to changes in the use of the assets and in their operating conditions have to be considered since they can modify importantly assets durability in these types of companies [2].

This paper is concerned with network repairable assets and we will focus on physical causes of failures as a consequence of: their operating time; resulting states after maintenance interventions that could modify—we are concerned about reductions—the useful life of the assets; and the operating environment, since environment and geographical characteristics of the territory could have a great influence on the equipment deterioration.

The aim of this paper is to deduce a practical method, for computational purposes, offering real-time reliability estimations according to changes in environmental factors, and in cumulative failures/repairs impact on the asset, over time. Also we offer a practical vision of how this can be linked to risk when approaching the maintenance scheduling decision-making process. This is done obtaining quantitative results from probabilistic risk assessments (PRAs) under different maintenance strategy scenarios and can be used to convince utilities that certain kinds of maintenance can be performed ensuring goals for the dependability of systems and components, thus giving freedom to the owner utility to create the organizational processes that are best suited to satisfying these goals.

For PRA we use, as most of the quantitative techniques, a variation of a concept known as the "probability/risk number" (PRN) [3]. A PRN is derived by attaching a numerical value to the probability of failure of an asset for a certain period (the higher probability, the higher the value), and attaching another value to the severity of the different categories of failure consequences (the more serious consequences for each category, the higher the value). The two numbers are multiplied to give a third which is the PRN.

To illustrate this, in the sequel we have organized the paper considering the reliability function building process first, and then the methodology to apply that function to estimate the risk of a failure mode and to schedule specific PM activities later. Finally, some conclusions will be presented.

2 Modeling Reliability as a Function of Operating Time and Cumulative Failure Impact (or Repair Effectiveness)

Different techniques could be utilized to obtain a failure pdf (probability distribution function) and the corresponding impact of maintenance effectiveness on that pdf (see [4]), but in order to simplify, in this paper we will initially consider exponential reliability distributions.

$$R(t) = e^{-\lambda t}; \quad h(t) = \lambda, \quad \text{with } \lambda > 0.$$

We do believe this is suitable according to the type of electrical equipment considered, and to the data provided by the network utility, that was basically including number of failures, for a given failure mode and for a given time frame, without capturing operating times to failure of the asset. We will consider that the maintenance interventions will leave the equipment in a situation between "as good as new" and "as bad as before the failure" as in [5], but never "worse than before the failure."

To model the relationship representing the reduction of reliability per accumulated maintenance activity on the equipment we will assume in this work that the equipment reliability function decreases with the number of maintenance incidents, although this effect becomes relatively less important when the number of maintenance tasks increases. This effect, of course, can be very different for each failure mode that is modeled and therefore the analyst will play an important role here identifying failure characteristics and maintenance activities implications on reliability of the equipment for the failure mode being studied (for a discussion on measuring performance of maintenance the reader is addressed to Oke et al. [6]).

In this paper we suggest, for instance, to assume that reliability is reduced in a fix percentage (for this we use the parameter q) each time the number of specific maintenance activities increases. With these considerations we can define the virtual operating time after n failures (T_n), for a given failure mode, as in Eq. (1). This type of formulation is a modified version of the one that the reader can find in literature to model the time to repair when the number of repairs increases (see for instance Ritter and Schooler [7]).

$$T_n = t \cdot q^{(-\ln(n+1)/\ln 2)}. \tag{1}$$

In Eq. (1), t represents the real equipment operating time for the failure mode and n the cumulative number of failures per the failure mode (and therefore accumulated repairs of the equipment for the failure mode). Notice that with this representation we assume that reliability reductions follow an exponential behavior. Values for q can vary within the interval (0.5, 1), with $q = 1$ representing no reliability reductions [see an example of the effects of Eq. (1) in Fig. 1, where we take $t = 100$ and we represent increases in operating time T_n and decreases in reliability $R(T_n)$].

Estimated reliability over the number of maintenance activities can then be formulated, considering the two factors mentioned above, as in Eq. (2), and the representation of the variable for different values of the q parameter is in Fig. 1.

$$R(T_n) = e^{-\lambda T_n} = e^{-\lambda t \cdot q^{-(\ln(n+1)/\ln 2)}} \tag{2}$$

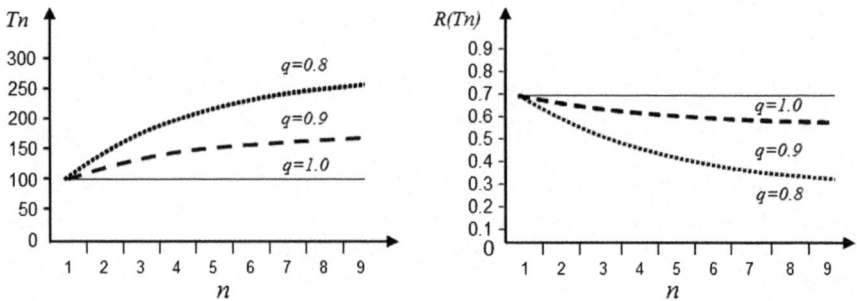

Fig. 1 T_n and $R(T_n)$ for different number of repairs (n) and repair effectiveness factors (q) when $t = 100$

Of course, according to the same reasoning, we are considering the hazard rate to be:

$$h(T_n) = \lambda \cdot q^{-(\ln(n+1)/\ln 2)}. \tag{3}$$

3 Adding Monitoring Parameters Impact on Reliability with a PHM

Conditions of different equipment parameters (for example, temperature or humidity) could accelerate, or could be a consequence of, assets degradation. The consideration of equipment environmental and operating conditions, when evaluating equipment reliability, can be very important for network utilities. Equipment can be placed in very different locations and therefore working under extremely different conditions. Assuming that maintenance procedures are applied similarly in all of geographic areas, we could use the concept of "location or environment related obsolescence" to represent the specific negative effect of the referred environmental factors (see [8]), and their application of this definition to buildings and infrastructure assets).

The influence of these environmental factors may obey different patterns. In this paper we will consider that the unique effect of a unit increase in an environmental factor (covariate) is multiplicative with respect to the failure rate. To simplify, we will also consider that the effect of an abnormal situation in a given equipment condition (as a consequence of the asset deterioration) will also produce a similar multiplicative effect on reliability (the reader is referred to [9]), for the discussion about a more precise considerations of environmental monitored parameters vs. failure consequence monitored parameters). Therefore, in this paper we may use the Cox's proportional hazard model (PHM, published in [10]) to represent the impact of both type of monitored parameters on reliability, and consequently we can express the hazard rate as follows:

$$h(t,x) = h_0(t) \cdot e^{X(t)} = h_0(t) \cdot e^{\sum\limits_{j=1}^{k} \gamma_j \cdot x_j}, \tag{4}$$

where $h_0(t)$ will now be $h_0(T_n)$ in Eq. (3), $h(t)$ is the failure rate or conditional probability of failure at time t; x_i are the environmental or failure consequence factors (with $i = 1 \dots k$) or covariates contributing to $h(t)$; and γ_i are constant coefficients (with $i = 1 \dots k$) representing each x_i contribution.

$$g(t,x) = R_0(t)^{\exp[X(t)]} = R_0(t)^{\exp\left[\sum\limits_{j=1}^{k} \gamma_j \cdot x_j\right]}. \tag{5}$$

Consider the hazard rate as in Eq. (4), consider now $g(t,x)$ in Eq. (4), under certain conditions we can demonstrate that $g(t,x)$ can be a proper representation of the corresponding reliability function. Now instead of $R_0(t)$ we will consider $R_0(T_n)$ as in Eq. (6).

$$R_0(T_n) = e^{-\lambda T_n} = e^{-\lambda t \cdot q^{(-\ln(n+1)/\ln 2)}}. \tag{6}$$

We now present a demonstration in Eqs. (7)–(14). Of course, we know that the hazard rate can be obtained as per the expression in Eq. (7).

$$h(t) = -\frac{R'(t)}{R(t)}, \tag{7}$$

where $R'(t)$ represents the partial derivative of $R(t, x)$ with respect to variable t.

Assuming $g(t,x)$ to be $R(t, x)$ taking logarithms in Eq. (5), and substituting the value of $R_0(t)$ from Eq. (6), we have:

$$L(R) = \exp[X(t)] \cdot \left(-\lambda t \cdot q^{-(\ln(n+1)/\ln 2)}\right) \tag{8}$$

Taking now derivatives in Eq. (7) we obtain

$$\begin{aligned}\frac{R'}{R} &= X'(t)\exp[X(t)] \cdot \left(-\lambda t \cdot q^{-(\ln(n+1)/\ln 2)}\right) \\ &+ \exp[X(t)] \cdot \left(-\lambda \cdot q^{-(\ln(n+1)/\ln 2)}\right).\end{aligned} \tag{9}$$

If we consider Eq. (10),

$$h_0(t) = \lambda \cdot q^{-(\ln(n+1)/\ln 2)}. \tag{10}$$

And if we also consider Eq. (7), then Eq. (9) can be written as

$$-h(t,x) = X'(t) \exp[X(t)] \cdot \left(-\lambda t \cdot q^{-(\ln(n+1)/\ln 2)}\right)$$
$$+ \exp[X(t)] \cdot (-h_0(t)). \tag{11}$$

Then notice that, in case that $X'(t) = 0$, we have

$$h(t) = h_0(t) \cdot \exp[X(t)]. \tag{12}$$

And Eq. (11) is equivalent to Eq. (4) and therefore we can write that

$$R(t,x) = R_0(t)^{\exp[X(t)]} = R_0(t)^{\exp\left[\sum_{j=1}^{k} \gamma_j \cdot x_j\right]}. \tag{13}$$

And if we consider Eq. (6),

$$R(T_n,x) = \left[e^{-\lambda t \cdot q^{-(\ln(n+1)/\ln 2)}}\right]^{\exp\left[\sum_{j=1}^{k} \gamma_j \cdot x_j\right]}. \tag{14}$$

Notice that we can use Eq. (3), for the hazard rate, and 14, for the reliability, to describe the same phenomena when $X'(t) = 0$, that means very precisely that: "when there is no variation of the covariates over time, I can use the expression in Eq. (13) or (14), to represent the reliability function associated to the hazard rate as the one proposed by Cox in Eq. (4)." This important result will be later used in the methodology that we define in the paper to estimate the risk of a failure mode over a certain period. Let us see how we certainly use this result.

Assume that, for a given mode of failure, the equipment has survived an operating time t, consider that we want to estimate the probability of that failure mode to appear during the time interval $(t, t + \Delta t)$, if we can use the expression in Eq. (13) or (14), that probability can be estimated using Eq. (15), of course assuming $X'(t) = 0$:

$$\frac{R(t,x) - R(t+\Delta t, x)}{R(t,x)}. \tag{15}$$

And therefore the risk of the failure mode for that period can be evaluated as

$$\text{Risk}(t, t+\Delta t) = \frac{R(t,x) - R(t+\Delta t, x)}{R(t,x)} \cdot C, \tag{16}$$

where C is the sum of the direct cost—cost of corrective maintenance—and the indirect cost—customer affection—of the failure mode, expressed in economic terms. Calculated in this way, the risk for period $(t, t + \Delta t)$, is obtained again as a probability risk number as mentioned in the introduction of the paper, where Eq. (15) attaches a numerical value to the probability of failure of the asset for this period, and C attaches the severity of the different categories of the mode of failure consequences.

Again, the most important thing to take into consideration is the way we compute $R(t + \Delta t)$, which assumes $X'(t) = 0$, i.e. no changes in the monitored parameter during the time interval $(t, t + \Delta t)$. We pay special attention to the covariates value, instead to their value trend; we do not use historical data and therefore safe easy the computational process of reliability estimation.

Nevertheless, notice that, although we consider that $X'(t) = 0$ to estimate the risk for a certain period $(t, t + \Delta t)$, existing in a given point in time t, $X(t)$ is "really" changing over time. The monitored parameter values will change and the covariates will also change. Therefore, in our case study the risk values are recalculated over time on a permanent basis. The more the covariates (functions of the parameter values) separate from their initial or ideal conditions, the higher the risk changes recalculated for that failure mode.

4 Defining a Methodology

In order to reduce the computing needs for the network risk estimation (for the different failure modes of the different equipment of a large network), and subsequent PM scheduling policy determination, the following methodology was implemented:

1. Estimation of the failure rate (λ) for a failure mode.
2. Determination of the failure mode criticality (priority) assessing consequences of the failure mode for the network (interruption time and number of customers affected). In order to do so. Consider the following notation:

i $1 \dots n$ Failure modes
C_i Failure mode i criticality
Nfe_i Number of failure events of failure mode i
Ta_i Time of affection of failure mode i to a customer
Nnc_i Average number of normal customers affected by failure mode i
Nsc_i Average number of singular customers affected by failure mode i
Mnc Maximum numbers of normal customers affected by a failure mode
Msc Maximum numbers of singular customers affected by a failure mode
wnc Weight of the normal customer
wsc Weight of the singular customer

Then for this particular paper and case study in Sect. 5, the criticality of the failure mode will be obtained according to the following algorithm:

$$C_j = Nfe_i \cdot Ta_i \cdot \left[wnc \left(\frac{Nnc_i}{Mnc} \right) + wsc \left(\frac{Nsc_i}{Msc} \right) \right], \quad (17)$$

where

$$Mnc = \text{Max}_i(Nnc_i) \tag{18}$$

$$Msc = \text{Max}_i(Nsc_i). \tag{19}$$

Of course, the reader should consider this algorithm as an example, which can change according to every particular business situation and suggested criticality assessment criteria.

3. Determination of the corrective maintenance impact as a percentage (q) and for the failure mode considered, agreed with the review team.
4. Selection of the parameters, among those monitored, offering proper information regarding the presence of a potential failure (for the failure mode under analysis), and selection of the covariates (sometimes the covariate can be a combination of the monitored parameters, instead of just only one of them).
5. Elicitation of the covariates coefficients with the review team. A final consideration to understand the case study in the next section is the following: A partial likelihood function (Quasi-Likelihood Model, QLM) can be used, as recommended by Cox [10], Cox and Oakes [11], Hosmer and Lemeshow [12], in order to estimate the γ_i coefficients, taking into account only the samples where a failure event occurred. However, for the initial implementation of the case study presented in this paper, we have adopted a more empirical and practical approach to speed up the coefficient determination process, and to be more consistent with the type of parameters that we may sometimes monitor. This process is defined as a "maintenance supervisor's knowledge elicitation process." This process consists of the following steps:

– Discuss about the impact on reliability of the different covariates with the review team;
– Select the values of the covariates to produce the function loss ($h(t, x) = 1$);
– Agree on threshold values for the covariates (monitored parameters or a combination of them).
– Simulate the impact of the selected values on reliability and risk of the failure mode. This was done using a computer simulation tool developed using Excelsius (® SAP Business Object running over an excel spreadsheet file) to easy knowledge elicitations with the maintenance experts. This tool is not yet a commercial tool but an *ad hoc* tool developed for the purpose of this research with the assumptions and formulation included in this paper. Using this tool the expert may experience changes in reliability and risk [modeled with Eq. (16)] when the different monitored parameters suffer (forced artificially) variations and Cox model coefficients can be tested according to the covariates defined and so on. Several workshops were developed to cover more critical failure modes and all coefficients could be initially defined.
– Register the selected values for each failure mode.

6. Determination of $R(t)$ as in Eq. (14).
7. Determination of the online failure mode risk associated to the asset, as in Eq. (16).
8. Determination of the criteria governing the PM scheduling according to risk.

5   Case Study for a Power Distribution Network

A summary of the vicissitudes experimented and results obtained during the implementation of the methodology explained in Sect. 4 were the followings:

1. Determination of the failure rate. This was done using a database where failures (for each failure mode) were registered over a 5-year time frame and for similar equipment "classes" (for instance 2500 similar transformers, in our example in Table 1), without the consideration of environmental factors influence. Failure rates for the different failure modes were found after a deep screening of the CMMS databases and results are presented in Table 1. In that table we include the number of failure events per the failure modes listed, besides the consequences of the failures. Notice that each different failure mode had very different consequences in terms of customers' affection.
2. The determination of the failure mode criticality was done according to formulation in Eq. (17). Results of the algorithm are presented in Table 1, in the column corresponding to the criticality estimations.
   In this case study results are considering that $wsc = wnc = 1$. This of course could be different in other valuation exercises. Criticality analysis was carried out in early stages of the project to make sure that the most important failure mode were analyzed, and to offer recommendation for future parameters monitoring according to this criteria.
3. The maintenance impact as a percentage (q), and for each one of the failure modes considered was estimated (for instance, for failure mode 1.4.4, was agreed to be $q = 0.9$ with two previous repairs ($n = 2$) according to the historic files presented by the team). This was important to set up the initial reliability function used in the case study [just substituting these values in Eq. (6)].
4. Selection of parameters and covariates. When doing the analysis, the team discovered that some of the parameters were not physically available at that time and therefore they could not be introduced in the expression for the reliability that would be used (these parameters are marked with N/A in Table 3). In Table 3 we present parameters and covariates for the different failure modes. Unless otherwise specified, we initially assumed that $x_i(t) = A_i(t)$ (of course assuming all the covariates to be independent), and the table contains the corresponding γ_i coefficient values of the covariate in Eq. (4). For each one of the covariates the team had to find out the corresponding coefficients (for instance, failure modes 1.6.1, 1.1.3, 1.3.3 and 1.2.1 follow this rule). However, this was not the normal rule since, in many cases, the covariate $x_i(t)$ could result to be

Table 1 Sample list of failure mode criticality, reliability, and condition monitoring data

Failure mode code	Failure mode name	Events in 5 years	Average number of normal customers affected	Average number of singular customers affected	Mean time of affection (h)	Failure mode identification once the failure process has finished	Early failure mode detection	Variables required for the diagnosis	Criticality estimation (dimensionless)	Failure rate ($\times 10^{-6}$ events/year)
1.6.1	Loss of tightness	5	1211.80	33.20	13.14	Non unique	Yes	A4, A9	26.4	0.040
1.1.3	Internal bypass	7	4386.00	97.71	2.91	Non unique	Non unique	A5, A8	25.4	0.056
1.1.1	Short	18	2649.33	50.72	1.51	Yes	Signs, non unique, define inspections	A5, A6, A7, A8	18.4	0.143
1.1.2	Circuit opening	12	9864.25	95.42	0.86	Non unique	Signs, non unique, define inspections	A5, A6, A7, A8	17.2	0.096
1.3.3	Bad connection	2	12505.50	109.00	0.78	Non unique	Non unique	A5, A8	3.1	0.016
1.4.4	Lack of outflow	34	202.71	3.29	1.12	Yes	Yes	A1, A2, A3, A6, A9, A10	1.8	0.271
1.2.1	Dielectric loss	3	1713.67	63.00	0.41	Non unique	Non unique	A4, A5, A8	0.9	0.024
⋯	⋯	⋯	⋯	⋯	⋯	⋯	⋯		⋯	⋯

$x_i(t) = f(A1(t), \ldots, A10(t))$ with $1 \leq i \leq 10$, instead of $x_i(t) = A_j(t)$, and therefore the number of coefficients would not correspond with the number of monitored parameters. For example, see failure mode 1.4.4, this failure mode has two covariates x_1 and x_2 while the monitored parameters are three A1, A2, and A6.

Besides operational reliability and maintainability data, Table 1 includes data related to the possibilities for condition monitoring of the failure modes. The different variables/parameters required for the diagnosis of the failure mode are included, although sometimes a clear early detection of the potential failure is not possible, or maybe, when possible, this detection cannot be unique for this failure mode. In Table 2 we have included the monitored parameters that were considered (A1 … A10) in Table 1.

5. Elicitation of the parameters coefficients with the review team. Coefficient parameters are presented in Table 3. In Table 3 we use the symbols "*" and "º" to refer to the monitored parameters involved in a covariate, and in column "considerations" we make clear the equation of the covariate. Also, the table contains the value of the corresponding covariates coefficients.

6. Determination of $R(t)$ as in Eq. (14). Let us see an example of the algorithm application to a failure mode affected by more than one monitored parameter.

   Failure mode: 1.4.4 (lack of outflow)

   Influential monitored parameters: A1, A2, A3, A6, A9, A10

   Available variables taken into consideration:

   – A3 (Air temperature)
   – A9 (Oil level)
   – A10 (lack of fans' feed)

   Parameters not available (N/A) at the time of the analysis:

   – A1 (Upper oil layer temperature)
   – A2 (Lower oil layer temperature)
   – A6 (load current intensity)

Table 2 List of monitored parameters for the transformers analysis in Table 1

A1	Upper oil layer temperature (°C)
A2	Lower oil layer temperature (°C)
A3	Air temperature (°C)
A4	Humidity (%)
A5	Hydrogen (%)
A6	Load current (intensity)
A7	Service voltage
A8	CO (%)
A9	Oil level
A10	Lack of fans' feed

Table 3 List of parameters coefficients (in Eq. 3) after the elicitation process (N/A parameter finally not available, * or ° meaning that the covariate is a function of referred monitored parameters $x_i(t) = f(A1(t), \ldots, A1(t))$, with $1 \leq i \leq 10$)

Failure mode code	Parameter	Coefficients (γ_i) for the Algorithm in Eq. (3)										Considerations made by the team of experts
		A1	A2	A3	A4	A5	A6	A7	A8	A9	A10	
1.6.1	A4, A9				**5, 5**					N/A		An increase in humidity above 40 ppm oil is a clear indication of this failure mode. The value placed ensures sufficient alert time for inspection and possible subsequent action by the observed condition
1.1.3	A5, A8					**5, 5**			N/A			Given the values of this parameter we should perhaps smooth the signal and then apply the coefficient indicated in this table
1.1.1	A5, A6, A7, A8					5*			N/A			The team decided to use the new variable defined as $x_1(t) = A5(t)/(A6(t) * A7(t))$, as covariate variable, because the reliability depends on the H_2 associated with a decrease of the transmitted energy. The coefficient of x_1 is the one identified (4) in this table row (notice that the load current/intensity and voltage, in isolation, do not inform about reliability in this case)
1.1.2	A5, A6, A7, A8					5*			N/A			Same comments than for failure mode 1.1.1
1.3.3	A5, A8					**5, 5**			N/A			Same comments than for failure mode 1.1.3
1.4.4	A1, A2, A3, A6, A9, A10	1*	1°	N/A			*, °			N/A	N/A	The team decided to use the new variables defined as $x_1(t) = A1(t)/A6(t)$ and $x_2(t) = A2(t)/A6(t)$, because reliability does not depend on intensity or on load but on the ratio temperature vs. load intensity. Now we have two coefficients as indicated in this table row, one (1) for x_1 and another (1) for x_2.
1.2.1	A4, A5, A8				**2, 5**	1			N/A			We assume that the moisture is detrimental to the dielectric strength in a greater extent than hydrogen. The combined effect of the coefficients may generate an alert of this failure mode in time for inspection and subsequent action

All monitored parameters will be used in the algorithm with normalized values [The value oscillates within the interval (0, 1)]. Then the procedure followed was to decide, according to best maintenance technical experience, which covariates were suitable for the analysis of the failure mode potential presence. The team decided to consider two covariates that could inform about the presence of the potential failure.

$$x_1(t) = \frac{A1(t)}{A6(t)}, \quad \text{and}$$
$$x_2(t) = \frac{A2(t)}{A6(t)}.$$

This was simply because the team understood that reliability was not dependent on load current intensity, but on the relationship between temperature and load current. Whenever this ratio increases over time, the failure mode will have a greater risk to show up. The coefficients γ_i are finally two, and it shall be preceded similarly with the function $X(t)$, which now has the form:

$$X(t) = \gamma_1 x_1(t) + \gamma_2 x_2(t) = 1 \cdot \frac{A1(t)}{A6(t)} + 1 \cdot \frac{A2(t)}{A6(t)}. \tag{20}$$

Then resulting graph for the main variables of the failure mode are presented in Fig. 3. In this figure we show a scenario where load current intensity would decrease (A6) while upper and lower layers oil temperatures (A1 and A2) would increase. This would indicate the potential presence of the failure mode 1.4.4. After selecting the coefficients for the two covariates in Eq. (20), we can plot the resulting reliability and hazard rate as in Fig. 2.

7. Determination of the online failure mode risk associated to the asset and failure mode. The last step of the methodology is to obtain an economic expression of risk. In this case study, we do consider a symbolic (nonrealistic) customer affection value of 1€ per customer affected and hour of affection, and an irrelevant direct cost of maintenance. With these assumptions, the risk can be obtained and represented over time as in Fig. 3. Of course we can see that risk curve topology is similar to that of the hazard rate as presented in Fig. 3, since the risk value is that rate times the consequence of the failure.

8. Determination of the criteria governing the PM scheduling according to risk. In this case study, the team decided that the value obtained for the online risk (in Fig. 4) would be compared with the cost of an alternative strategy of preventively maintaining (PM) the equipment for that mode of failure. In order to estimate the cost of this alternative PM strategy we have to:

- Estimate the cost of the PM to restore the equipment to a certain condition (in this case study this condition is reached by updating only covariates values to normal equipment operating condition values).
- Update the new reliability function with new covariates values.

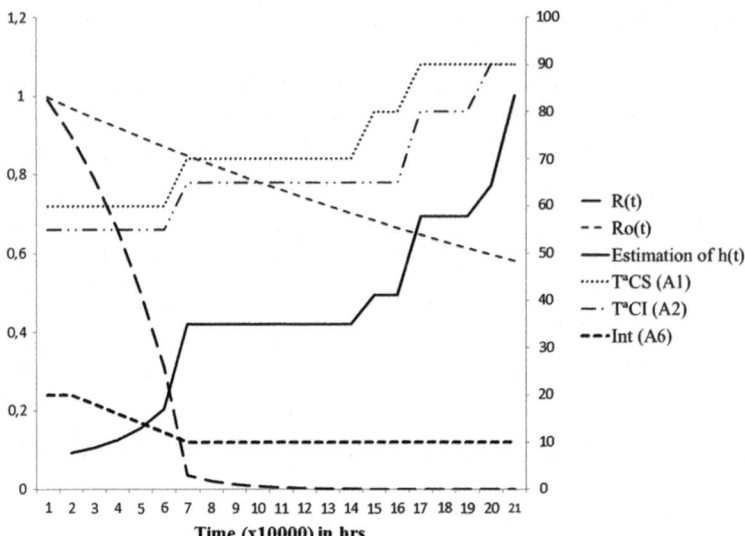

Fig. 2 Evolution of the ideal and actual reliability, failure probability, temperature, and load current intensity for a moment in time (*t*)

Fig. 3 Risk (in euro) for the failure mode over time

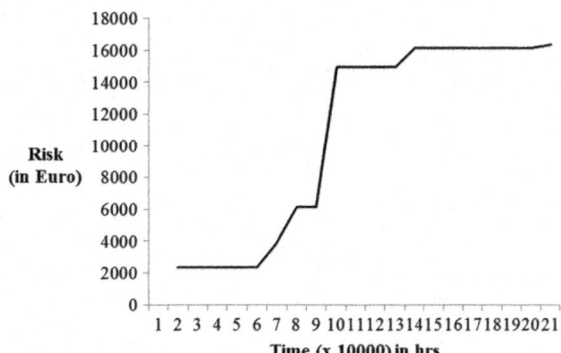

- Calculate the risk for the selected period considering the new reliability function after PM.

When the online risk of doing only corrective exceeds the alternative PM strategy risk to a certain extent, then PM maintenance is scheduled, released, and accomplished. This risk exceeding extent is understood here as a company policy. Figure 4 shows this concept. In this figure both strategies' risks are represented. We have selected two moments in time to show the change in the selected maintenance strategy according to the strategy risk assessment. For instance, T1 would be a moment where "doing PM" strategy would have a higher risk than "CM only" and PM maintenance should not be recommended. T2, however is a moment in time

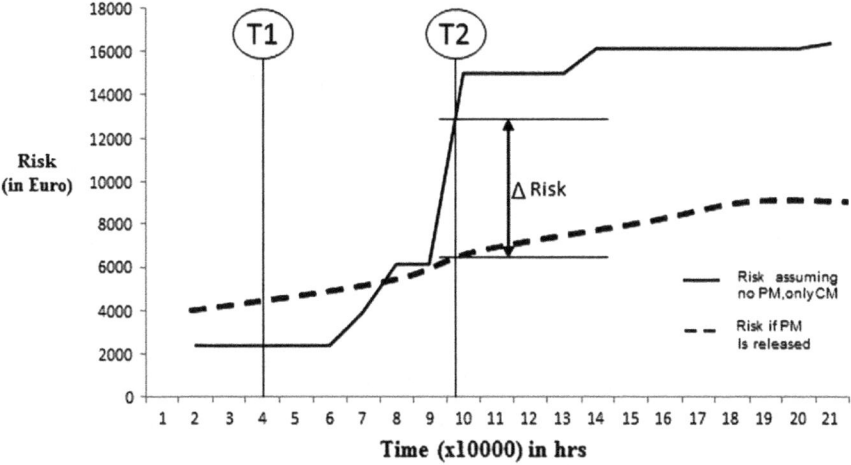

Fig. 4 Risk-based PM scheduling decision

where the "CM only" strategy risk increases more than ΔRisk over the PM strategy risk, PM would be then scheduled and released. Note how decisions are therefore taken based on strategies probability risk numbers and on-line.

6 Conclusions, Implications and Extensions

The above-mentioned methodology can generate suitable estimations of risk per critical failure modes. This can help in the process of releasing PM activities for these assets, minimizing them, besides minimizing also CM activities by increasing reliability. Maintenance activities minimization can increase assets durability and reduce considerably network technical deterioration.

There are different implications of this work from wide variety of perspectives:

- From a logistic point of view, spare parts management can now be improved by considering times to deterioration instead of times to failure. Stock levels can be therefore more precisely estimated with considerable savings in inventory holding cost and working capital requirements.
- From an economic point of view, a clear implication is that assets depreciation could be better defined, according to their reliability and their original purchasing value. Obviously, the equipment reliability affects equipment aging depreciating the equipment purchase value. Consequently, proper maintenance strategies ensure that the asset's books value are accurate, and that repair or replacement decisions are accountable.
- This methodology also offers managers the possibility to control maintenance budgets based on assets risk expectations for a certain period.

We have presented a case study where all these policies where implemented for a network utility and explained the methodology followed exemplifying the results obtained. The pilot project has been implemented and final mass deployment will be accomplished according to the results obtained for a 2-year testing period. Extensions of this work can be related to the study of maintenance logistics and support improvements when using this risk-based PM strategies or to accurately measure above-mentioned economic implications over the 2-year testing period that has been selected.

References

1. Newbery DMG (2002) Privatization, restructuring, and regulation of network utilities. MIT Press
2. Lemer AC (1996) Infrastructure obsolescence and design service life. J Infrastruct Syst 2 (4):153–163
3. Moubray J (1997) Reliability-centred maintenance, 2nd edn. Butterworth-Heinemann, Oxford
4. Wang W (2010) Modelling in industrial maintenance and reliability. http://www.deepdyve. com/browse/journals/ima-journal-of-management-mathematics/2010/v21/i4-1471-678XIMA. J Manag Math 21(4):317–318
5. Kaminskiy M, Krivtsov V (1998) A Monte Carlo approach to repairable system reliability analysis. Probabilistic safety assessment and management. Springer, New York, pp 1063–1068
6. Oke SA, Ayomoh MKO, Oyedokun IO (2007) An approach to measuring the quality of maintenance performance. http://www.deepdyve.com/lp/oxford-university-press/an-approach-to-measuring-the-quality-of-maintenance-performance-myU9INHki3-1471-678XIMA. J Manag Math 18(1):17–32
7. Ritter FE, Schooler LJ (2002) The learning curve. In international encyclopedia of the social and behavioral sciences, Amsterdam, pp 8602–8605
8. Mansfield JR, Pinder JA (2008) Economic and functional obsolescence: their characteristics and impacts on valuation practice. Property Manag 26(3):191–206
9. Wang W (2008) Condition based maintenance modeling. In Kobbacy KAH, Murthy DNP (eds) Complex systems maintenance handbook. Springer, London
10. Cox DR (1972) Regression models and life-tables. J R Stat Soc Ser 34:187–220
11. Cox DR, Oakes D (1984) Analysis of survival data. London Chapman and Hall, London
12. Hosmer DW, Lemeshow S (1999) Regression modeling of time to event data. Wiley, New York

Author Biographies

Adolfo Crespo Márquez is currently Full Professor at the School of Engineering of the University of Seville, and Head of the Department of Industrial Management. He holds a Ph.D. with Honours in Industrial Engineering from this same University. His research works have been published in journals such as Reliability Engineering and System Safety, International Journal of Production Research, International Journal of Production Economics, European Journal of Operations Research, Omega, Decision Support Systems, and Computers in Industry, among others. Prof. Crespo is the author of eight books, the last five with Springer-Verlag (2007, 2010, 2012, 2014) and Aenor (2016) about maintenance, warranty, supply chain and assets management. Prof. Crespo is Fellow of ISEAM (International Society of Engineering Assets Management) and leads the Spanish Research Network on Assets Management and the Spanish Committee for Maintenance Standardization (1995–2003). He also leads the SIM (Sistemas Inteligentes de Mantenimiento) research group related to maintenance and dependability management and has extensively participated in many engineering and consulting projects for different companies, for the Spanish Departments of Defense, Science and Education as well as for the European Commission (IPTS). He is the President of INGEMAN (a National Association for the Development of Maintenance Engineering in Spain) since 2002.

Juan Francisco Gómez Fernández is Ph.D. in Industrial Management and Executive MBA. He is currently part of the SIM research group of the University of Seville and a member in knowledge sharing networks about Dependability and Service Quality. He has authored a book with Springer Verlag about Maintenance Management in Network Utilities (2012) and many other publications in relevant journals, books and conferences, nationally and internationally. In relation to the practical application and experience, he has managed network maintenance and deployment departments in various national distribution network companies, both from private and public sector. He has conduced and participated in engineering and consulting projects for different international companies, related to Information and Communications Technologies, Maintenance and Asset Management, Reliability Assessment, and Outsourcing services in Utilities companies. He has combined his professional activity, in telecommunications networks development and maintenance, with academic life as an associate professor (PSI) in Seville University, and has been awarded as Best Master Thesis on Dependability by National and International Associations such as EFNSM (European Federation of National Maintenance Societies) and Spanish Association for Quality.

Pedro Moreu de León is an expert in maintenance engineering and maintenance management. He holds a Ph.D. in Industrial Management from the University of Seville. He is the author of several books and publications on these topics, is Chairman of the Technical Committee for Standardization of Maintenance CTN AEN-151 "Maintenance" of AENOR (Spanish Association for Standardization) (2002–2014) and Expert Spanish Officer (AENOR) and Secretary Delegation European Committee for Standardization CEN 319 "Maintenance" and Convenor of the Working Group for Maintenance Management standard (WG7, "Maintenance Management") of the Committee. Dr. Moreu is an expert in maintenance of different process plants. He has participated in many engineering and consulting projects for different companies. He is currently member of the Board of Directors of INGEMAN (a National Association for the Development of Maintenance Engineering in Spain) since 2002.

Antonio Sola Rosique is a civil engineer currently acting as Vice President of INGEMAN (a National Association for the Development of Maintenance Engineering in Spain). His professional experience (34 years) is very much related to the field of dependability and maintenance engineering in different types of power generation plants for Iberdrola Generación (nuclear, thermal, hydro, and combined cycles gas steam). Antonio has coordinated various collaborative projects between Iberdrola and the University of Seville, and is active member in the Board of European Safety, Reliability and Data Association (ESREDA), Asociación Española de Mantenimiento (AEM), Asociación Española de la Calidad (AEC), and Asociación Española de Normalización (AENOR). At present Antonio is about to finish a Ph.D. within the field of Maintenance and Risk Management in the University of Seville.

Customer-oriented Risk Assessment in Network Utilities

Juan Francisco Gómez Fernández, Adolfo Crespo Márquez
and Mónica Alejandra López-Campos

Abstract For companies that distribute services such as telecommunications, water, energy, gas, etc., quality perceived by the customers has a strong impact on the fulfillment of financial goals, positively increasing the demand and negatively increasing the risk of customer churn (loss of customers). Failures by these companies may cause customer affection in a massive way, augmenting the intention to leave the company. Therefore, maintenance performance and specifically service reliability has a strong influence on financial goals. This paper proposes a methodology to evaluate the contribution of the maintenance department in economic terms based on service unreliability by network failures. The developed methodology aims to provide an analysis of failures to facilitate decision-making about maintenance (preventive/predictive and corrective) costs versus negative impacts in end customer invoicing based on the probability of losing customers. Survival analysis of recurrent failures with the General Renewal Process distribution is used for this novel purpose with the intention to be applied as a standard procedure to calculate the expected maintenance financial impact, for a given period of time. Also, geographical areas of coverage are distinguished, enabling the comparison of different technical or management alternatives. Two case studies in a telecommunications services company are presented in order to illustrate the applicability of the methodology.

Keywords Network utilities · Customer affection · Reliability analysis · Generalized renewal process

J.F. Gómez Fernández · A. Crespo Márquez (✉)
Department of Industrial Management, School of Engineering,
University of Seville, Camino de los Descubrimientos s/n, 41092 Seville, Spain
e-mail: adolfo@us.es

J.F. Gómez Fernández
e-mail: juan.gomez@iies.es

M.A. López-Campos
Department of Industrial Engineering, Universidad Técnica Federico
Santa María, Av. España, 1680 Valparaíso, Chile
e-mail: monica.lopezc@usm.cl

263

1   Introduction

Within the Services Sector, Network Utilities provide services to clients distributed in an infrastructure network (gas, water, electricity, telecommunications, etc.). Their infrastructures are usually organized and composed by a high number of dispersed elements, supported in hierarchical structures and replicated by distribution areas. These companies are capital intensive [1], meaning decades for payback on investments. Additionally, they have an intense and long-lasting relationship with customers and consequently, quality perceived and demanded by them has a strong impact on the fulfillment of financial goals, through both positive and negative ways, increasing services demand and increasing risk of customer churn respectively. Accordingly, in a competitive market, these companies are always trying to increase their market share and "customer life cycle value." The main strategies by which this is done are retaining actual customers; building customer loyalty and; capturing new potential customers in geographical territories. Therefore, customer opinion is essential and extremely decisive for the consideration of future investments [2].

Customer requirements, attitudes, and behavior are not always the same, even among similar groups or at different times. There are several methods for quality measurement according to the attributes of a service, considering their importance or their contribution for the company to provide value [3]. However, a global measurement of the quality perceived by customers is not an easy task, because of the influence of subjectivity in their opinions ("... *every customer perceives service quality differently,*" [4]). This is why many authors concentrate their efforts in evaluating significant interactions (called *critical incidents*) [5, 6]. Quality measurement in a service must evaluate feasible customer requirements about it, such as those related to the ability to respond to contingencies, the reliability and the security of the service [7]. Subsequently, service quality must be analyzed considering the positive and negative feelings of customers concerning service issues and the supplying company. For this, analysis of historical information on customer behavior, including, where applicable, geographical location of customers, is necessary to correctly define new segmentation criteria for future tailored actions [8].

Moreover, in a very competitive environment a fast response to problems may generate customer retention and loyalty [9, 10]; better availability and cost reduction may allow a decrease in the price of the services; and the sum of these parts is internal motivation, image, and external business reputation [11]. Thus, more reliable services are appreciated in the sector of network utilities where contracts and standards revolve around of service-level agreements (SLAs). Service quality will be accepted by the customers within a tolerance level, but how do we know what this level is? We know that customer perception will be affected by failure occurrence and recurrence. For that reason, maintenance departments should be considered crucial for network utilities [12], pursuing to keep the service delivery reliable, with maximum quality and performance. Focusing on the customer-oriented service quality, a maintenance department contributes [13–15] strongly to:

- Satisfy customer needs and loyalty, fulfilling the service reliability;
- Enhance the business image along with the ability to capture new customers;
- Reduce service costs and avoid unexpected failure costs;
- Improve productivity, increase availability.

Questions at this point arise: How can we measure service quality and the impact that maintenance has on it? How can we estimate maintenance quality (and mainly non-quality) costs? In special cases, as for important clients, or when we may suffer risk of financial loss, it could be useful to launch alerts and alarms about these issues when failures appear due to maintenance performance. It is important to assess the value implications of existing maintenance policies based on maximizing value, instead of minimizing cost of maintenance [16]. The difficulty of measuring the impact of maintenance activities on the quality of service and estimation of non-quality costs complicates decision-making in maintenance departments [17]. Strategic, tactical, and operative decisions will in turn become easier to handle if we could see the trade-offs between gaining in service performance versus increasing maintenance costs [18, 19]. Well-managed proactive maintenance, through proper prevention and inspection, will reduce internal [20] and external non-quality costs in network utilities.

Thus, perceived and demanded service quality could be measured through critical incidents, positive or negative, depending on the maintenance effectiveness solving them. Consequently, the number and type of customer complaints because of service failures can be shown of this. That is, in order to assess the maintenance value, it has to surveillance not only asset reliability but also its financial impact during all the asset life cycle [21]. This research is aimed at raising these companies' awareness in order to define flexible and robust policies based on asset reliability according to its impact on customers.

Many authors on marketing and quality have tried to model customer behaviors through proportionality among quality features as covariates, using qualitative methods as opinion polls, or quantitative methods as parametric/semi-parametric models [22–26]. This analysis allows the evaluation of the customer behavior against reliability service incidents, not only as a direct covariate but also considering the occurrence rate, and quantifying their direct and indirect financial consequences. We develop a methodology that technically and financially describes how the recurrence of failures is correlated with customer abandonment.

Important previous authors on reliability [16, 21, 27–31] have analyzed maintenance contribution linking engineering/reliability concepts to financial concepts in repairable systems. They explore the impact of a system's reliability on its revenue generation capability, considering direct relation between system reliability and the system's performed technical function from a financial standpoint. Thanks to the GRP statistical technique, it is estimated a dynamic model of the duration of customer–company relationship based on the experience on the recurrence of service failures. Non-satisfactory experiences and more recently can reduce this relationship and financial revenues. That is, previous reliability authors model survival system probability, and here it is modelled the customer life survival probability.

They treat recurrent failure modes of the technical system and the effect of partial renewal repairs in system status, now it is considered recurrent service failures (independently the physical system in the network that causes them) and the impact of these events in the mind of the customer orientated to abandon the contract. The aim of this work is to model through GRP, psychological behavior instead of physical behavior, according to recurrent events (failures) and how the dynamism of their occurrence rate influences on trigging the customer abandonment and so on to extend/reduce the customer lifetime. Therefore, this paper deals with indirect behaviors and indirect impacts due to bad reputation of the services in its customers. To illustrate this, this paper provides a basic guidance for maintenance decision-making processes, in evaluating the impact of network reliability on customer satisfaction and on customer retention and loyalty. With this purpose, our article is organized as follows: Sect. 2 reviews maintenance impact on quality in terms of costs in order to orientate the value of maintenance performance on financial goals. In Sect. 3 we develop a methodology to evaluate the impact of network unreliability on customer loyalty when companies deliver services, applying the Generalized Renewal Process (GRP) model from statistical data analysis to predict customer behaviors toward service abandonment (especially when considering recurrent failures). Within Sect. 4, we present a case study from a telecom company that analyzes data sets containing information about the loss of customers and the failure recurrence for a certain period of time. We finish with conclusions and future research challenges.

Notation:

GO	Group of customers without failure experience,
GF	Group of customers with failure experience,
CPV(l)	Customer Present Value (CPV) is the updated benefit per customer (l),
$j = 1, ..., J$	The accounting periods in the customer life,
pr(l, j)	Price paid by a customer (l) during the period j,
c(l, j)	Direct cost of servicing the customer (l) during the period j,
AC(l)	Acquisition costs per customer (l),
r	Discount rate for the company,
t_i	Time when a failure occurs ($t_1, t_2,..., t_n$ with $t_0 = 0$),
x_i	Variables representing the intervals between successive failures ($x_1, x_2, ..., x_n$), $x_i = t_i - t_{i-1}$,
$f(t)$	Probability density function of failures (pdf),
$F(t)$	Cumulative distribution function of failures (cdf),
$R(t)$	Reliability function, $R(t) = 1 - F(t)$,
α, β	Scale and shape parameters of the Weibull distribution of failures,
q	Repair efficiency of an asset failure mode,
q_s	Repair efficiency of service failures,
G	Total number of customers included in the groups to analyze,

v_i Virtual Life represents the calculated age of a customer immediately after the ith repair and occurs taking into account the produced overall damage due to all the preceding failures (with $v_0 = 0$ for $t_0 = 0$),

$R_s(l, j)$ Probability of customer (l) retention/survival in the period (j) modeled it with a Weibull, then $(1 - R_s(l, j))$ = probability of customer abandonment in the period j,

δ_j Periods affected by failures, 1 for the first affected period and followings, 0 otherwise

n_c Number of affected customers per failures,

$cc(l, j)$ Direct cost of corrective activities per customer (l) during the period j,

$Pc(j)$ Probability contribution equals to difference of customer survival probability in each period of the customer life, $Pc(j) = \Delta R_s(j) = R_s^{GO}(j) - R_s^{GF}(j)$,

2 Quality, Customer Experience, and Their Relation to Maintenance

Assuming that the evaluation of maintenance's contribution to business is often based solely on the occurrence of *critical incidents*, it will only be visible to general managers when properly converting preventive and corrective maintenance actions into terms of costs [32]. This refers not only to the maintenance budget, but mainly to related non-quality costs ("... *the knowledge of quality costs helps managers to justify the investment in quality improvement and to assists them in monitoring the effectiveness of the efforts made*" [33]). This is the main objective of the methodology developed in this paper and in order to do this it will be supported by quality costs theory. Most literature related to quality associations recommends classifying quality costs in under Prevention, Appraisal and Failure [34, 35] (the PAF scheme):

- Prevention costs (P) are those associated with actions to ensure the desired quality of products/services;
- Appraisal costs (A) are those incurred due to tracking and controlling quality;
- Failure costs (F) are produced to correct failure or damages in products/services. They can be generated internally or externally. These quality costs are known as "non-quality" costs. External non-quality costs are difficult to measure, directly generate customer dissatisfaction, and are probably the most relevant costs to the business.

Inside the PAF scheme applied to this maintenance evaluation, prevention and appraisal costs are considered tangible and directly quantifiable from one year to the next. However, failure costs cannot be considered from a deterministic point of view, due to the uncertainty of occurrences and their consequences. Therefore, it must be handled from a probabilistic approach [36, 37] based on critical events (incidents in the customer relations). In addition, indicators to support maintenance decisions, for example to invest in preventive or appraisal activities instead of wait for failures (because Prevention and Appraisal costs are inversely proportional to Failure costs), our failure analysis does not only have to focus on a financial year, but also it has to calculate the total impact during the equipment life cycle or customer relationships.

During the customer relationship, failures impact service quality in the following ways [38]:

- Failures decrease customer satisfaction. Each failure may modify customer perception of service quality, transmitting bad propaganda into the market (up to 10 partners). According to Keaveney [39], service failures generate 44% of customer losses, and as a consequence, important changes in "customer life cycle value."
- Recurrent failures decrease customer satisfaction even more. Recurrent service failures (against those established in a contract, or determined by the market standards) accelerates negative perception in the customer. According to Svantesson, the evolution of quality should be measured by comparing it with other companies in the sector, competitors, standards and regulations. Some 80% of customer satisfaction is due to delivering services correctly, and the remaining 20% is due to resolving claims or problems.

Besides previous considerations, customer behavior can fluctuate; the customer may remain with the service waiting a monetary refund, or may pursue legal action. In both situations, the customer could do negative propaganda or could keep the bad experience in mind. Therefore, quantifying this potential behavior according to the failure impact on the service should be beneficial.

In services companies, the most important source of income comes from end customer invoicing, and these customers are the major justification for business sustainability. Then, understanding of customer value, in the entire commercial relationship with the company is important and represented by the Customer Present Value (CPV) [27]. CPV can be formalized including total income from a customer during a period of time minus all the costs required to serve that customer [40], which is updated depending on the lived failures. CPV can be used as a reference benchmark, not only to support the estimation of future profits, but also to improve the customer segmentation process. According to Gupta and Lehmann [40] the equation for CPV could be expressed as follows:

$$\text{CPV}(l) = \sum_{j=1}^{J} \left[\frac{[\text{pr}(l,j) - c(l,j)] \cdot [R_s(l,j)]}{(1+r)^j} \right] - AC(l) \tag{1}$$

Consequently, CPV is the present value benefit per customer; it is the total income from a client obtained during its relationship with the company, less direct costs of sale, acquisition, customer loyalty and advertising, discounted according to the interest rate of reference at the time of the study, and considering the probability of retention in each period of the customer life. The discount rate is employed to calculate the present value of future cash flows. Although many companies utilize their weighted average cost of capital as discount rate in order to compare the CPV with the profitability of an investment activity depending on its actual cost of necessary funds. CPV can be increased if the pricing $pr(l, j)$ is increased, the servicing costs are decreased $c(l, j)$, or the probability of retention or survival $R_s(l, j)$ is improved. Therefore, if maintenance department has a good disposition and performance, the company income can be increased by reducing costs of operation, increasing customer satisfaction and reducing customer loss due to bad services [6, 41, 42]. In addition, if the quality of the service is well accepted in the market, then company reputation will increase, so new customer acquisition will be facilitated. Based on this, we may evaluate the efficiency of maintenance investments through the violation of existing SLAs as well as historical CPVs. Therefore, CPV is very sensible to reliability of future cash flows.

Accordingly, depending on failures such as critical incidents, the customer behavior about the service quality could be represented by a probabilistic approach of the retention, based on failure analysis methodologies [43–45]. In order to quantify the failures consequences, this paper deals with how failures (and the actions to adopt in order to detect, avoid and fix the faults [10, 42]) impact perceived service quality and associated risk of customer abandonment. This calculation is made through a stochastic calculation of the service level. Thus, probability of retention or survival is derived from parametric estimations of the related SLA variables [46, 47], such as MTBF and MTTR (which are characterized with probability distribution functions), accomplishing a "Survival data Analysis." This statistical method demonstrates how a group of individuals will react to a failure after a certain length of time [48–50].

When analyzing failure recurrence with the purpose of making accurate decisions oriented to business results, measurement of *customer survival probability* (R_s) becomes of vital importance in term of costs, and this can be more important than the repair costs (see Eq. (1)). A very interesting opportunity appears when we link this type of analysis to historical business data, per different areas of a company, or against competence and regulatory principles in the sector.

3   Development of the Methodology

The study of *retention probability* and its relationship with the recurrence of failures, considered as a "Survival data Analysis," focuses on surviving customers facing multiple failures (customers who remain in the company even after experiencing one or more faults). There are different techniques to solve this type of analysis [51–56]. The intention of our paper is to develop a methodology considering how the recurrence of failures and the probability of survival are correlated with the customer abandonment.

The Generalized Renewal Process (GRP) is proposed to be applied given its ability to describe the rate about the occurrence of events in a repairable system over time. This method is the most flexible technique for modeling the behaviour of a specific system before failures and the quality of repairs considering all possible states on the system age [28, 57]. There are different formalizations of the GRP process to model equipment evolution over time. The approach is built on the GRP II by Kijima and Sumita [58], to cover complex systems with multiple equipment and repairs, reducing historical data points and processing power required for the analysis. In this case, simplification of the implementation of the GRP method is based on the study by Mettas and Zhao [59] on Weibull probability distribution function. This function is one of the most broadly used in reliability studies allowing representation of different functions such as exponential, Rayleigh or normal), and evaluated with the Maximum likelihood Estimation (MLE) [47, 60]. In short, GRP employs recurrent failures, modeling the probability of system survival with Weibull functions.

GRP method is particularly useful applied to the Weibull distribution function, because it returns the values of α and β: as well as an indicator of repair quality or efficiency (q). To show repair efficiency, a variable denominated Virtual Life (v_i) is introduced, whose concept represents the calculated age of a system immediately after the ith repair occurs and taking into account the produced overall damage due to all the successive failures (with $v_0 = 0$ for $t_0 = 0$). The repair could compensate a proportion of the damage produced during the time between failures i and $i-1$, but also damages produced in previous intervals (see Eq. (2)). Properly performed repairs may improve system virtual states (life), while poorly solved failures could aggravate it (always speaking in terms of reliability). In this equation, the failure and repair data can affect future failures and they are treated as one type of recurrence event data based on the time intervals length. GRP method models the rate of occurrence of failure events over time in a repairable system: the recurrence rate may remain constant, increase or decrease. Thanks to q parameter, the trend is captured showing the effects of the repairs on the age of that system (modeling partial renewal repair). Thus, we could employ the virtual life in comparison to the real age of the system as a measurement of the maintenance contribution to avoid the recurrence of failures over one system (until the last failure $x_i = t_i - t_{i-1}$).

$$v_i = q \cdot \sum_{j=1}^{i} q^{i-j} x_j = q \cdot (x_i + v_{i-1}) \qquad (2)$$

Our methodology reutilizes GRP with recurrent failures over groups of customers, that is, it is not applied in physical systems. Each customer is considered as a system for the GRP application, so if the GRP over systems evaluates the survival probability with a repair efficiency estimation in the same type of failure modes, the GRP over customers pretend to evaluate the customer behaviour for surviving before recurrent critical incidents (service downtime event) in the customer–company relationship (CCR) due to failures in customer service (independently the physical system of the network that has suffer the failure). In short, conventional

GRP searches to estimate the asset reliability, and our adaptation of GRP searches to model the probability of customer survival according to its service failures.

This application of GRP lets us estimate the impact of each single service failure on customer age and also the contribution of maintenance performance avoiding the service failures as a measurement of service quality. In our methodology, Virtual Life (v_i) is the calculated age of a customer immediately after the ith service repair occurs and taking into account the produced overall damage due to all the successive service failures (with $v_0 = 0$ for $t_0 = 0$). Properly performed service repairs may improve perception, while poorly solved service failures could aggravate it (always speaking in terms of reliability). Thus, we could employ the virtual life in comparison to the real age of the customer as a measurement of the maintenance contribution to avoid the recurrence of service failures over one customer.

The service repair efficiency, from a mathematical point of view thanks to the repair influence in the last and previous failure intervals, could be considered as a trend factor of the failure interval length in the time series of them, showing the tendency contribution of successive repairs to the customer status conservation. Well performed service repairs of failures may extend the time between failures and hence the customer relationship with the company reducing its virtual age rejuvenating the customer perception (as in a new captured customer where there is a minor probability of customer abandonment). Meanwhile bad performed service repairs could reduce the time between failures and so making the customer relationship virtually older (with a higher probability of abandonment). Therefore, based on the GRP model there are five possible states of the service repair efficiency (q_s):

- $q_s < 0$, the customer relationship (CR) is rejuvenated better than in the case of a new customer.
- $q_s = 0$, the CR is improved to the case of a new customer.
- $0 < q_s < 1$, the CR is rejuvenated better than old but worse than new.
- $q_s = 1$, the CR is maintained equal to old.
- $q_s > 1$, the CR is deteriorated worse than old.

For this purpose, the following starting points have been defined:

- All service failures causing downtime are detected and, the preventive and corrective times are negligible compared to the times between failures. The service repair could compensate the produced service failure impact during the time between failures i and $i-1$, but also impacts produced in previous intervals. Consequently, depending on the recurrence of service failures (critical incidents), the service repair efficiency (q_s) symbolizes the contribution of maintenance in customer perception (improving or getting worse the intention of abandonment) during the time between service failures i and $i-1$, but also previous damages.
- The relationship between the probability of customer abandonment and recurrent service failures is analyzed considering the survival and abandonment times in two customers groups with similar characteristics. Groups where customers

are characterized by the same demanded service usage features, one based on customers with recurrent failures experience (GF) searching the impact of service failures as critical incident, and other collecting the customers without service failure affection (GO) of downtime.

- For each group, the time series in which the customer requests abandonment (real time) is computed using right censored data (with starting customer age equals to 0), see Fig. 1.

The mathematical format of the model using Weibull function requires the definition of the failure density function $f(t)$ and the failure distribution function F (t) to be expressed in a conditional probability function as follows:

$$F(x_i|v_{i-1}) = \frac{F(x_i + v_{i-1}) - F(v_{i-1})}{1 - F(v_{i-1})} = 1 - e^{-\left[\frac{(x_i + v_{i-1})^\beta - v_{i-1}^\beta}{\alpha^\beta}\right]} \tag{3}$$

$$f(x_i|v_{i-1}) = \frac{\beta}{\alpha}\left(\frac{x_i + v_{i-1}}{\alpha}\right)^{\beta-1} \cdot \exp\left[-\left[\frac{(x_i + v_{i-1})^\beta - v_{i-1}^\beta}{\alpha^\beta}\right]\right] \tag{4}$$

We estimate the $f(t)$, fitting on three parameters α, β and q_s based on the service failure data (n observations t_1, t_2, ..., t_n) for a time truncated estimation (up to time T), using the Maximum Likelihood Estimation (MLE) method, searching the maximum point in three dimensions for the product of conditional $f(t)$:

$$L\{\text{data}|\alpha, \cdot\beta, q_s\} = L\{x_1, x_2, \ldots, x_n|\alpha, \cdot\beta, q_s\} = \prod_{i=1}^{n} f(x_i|\alpha, \cdot\beta, q_s) = L \tag{5}$$

Continuing with the MLE resolution, according to all the observations for G (total number of) customers with similar characteristics in the group, we take logarithms (see Eq. (6)) and equal to 0 partial derivatives with respect to α, β and q_s in order to obtain the best fit on these three parameters. To solve with fast

Fig. 1 Customer survival depending on time

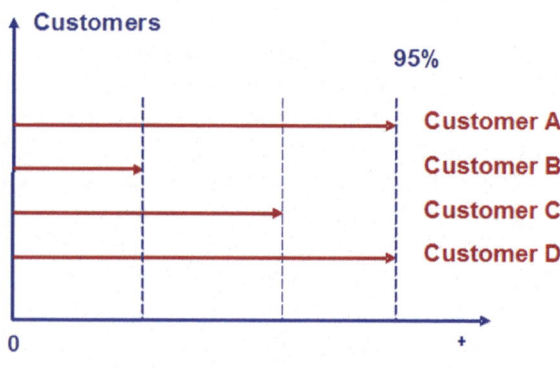

convergence Newton method is recommended and the goodness of results can also be analyzed according to Fisher matrix.

$$\Lambda = \log L\{\text{data}|\alpha, \cdot\beta, \cdot q\} = \sum_{g=1}^{G} G \cdot (-\ln \alpha + \ln \beta) - \sum_{g=1}^{G} \left[\left(\frac{T_g - t_{g,G} + v_G}{\alpha} \right)^{\beta} - \left(\frac{v_G}{\alpha} \right)^{\beta} \right]$$

$$- \sum_{g=1}^{G} \sum_{i=1}^{n} \left[\frac{(x_{g,i} + v_{g,i-1})^{\beta} - v_{g,i-1}^{\beta}}{\alpha^{\beta}} \right] + (\beta - 1) \cdot \sum_{g=1}^{G} \sum_{i=1}^{n} \ln \left(\frac{x_{g,i} + v_{g,i-1}}{\alpha} \right) \qquad (6)$$

The following steps comprise the methodology:

1. Firstly, α, β and q_s parameters have to be calculated in the two customer groups (GO and GF) considering the survival and abandonment times.
2. As a result, the behavior of the two customer groups can be patterned parametrically by the $R(t) = 1 - F(t)$ that could be estimated for each group and discretizing for each period, $R_s(l, j)$. Because of applying GRP in customer groups, reliability distribution function is considered the same for all the members, then $R_s(l, j) = R_s(j)$ inside each group.
3. Now, the service failure impact can be customized as the difference of behavior in the group with recurrent service failures against the group without service failures experience. This is represented by the difference of R_s among both groups depending on time, where there must be a probability increment (Pc, Probability Contribution) per period of losing customers due to perceived service failures. Hence, there is a better probability of retention in the group of customers without service failure experience.

$$Pc(j) = \Delta R_s(j) = R_s^{GO}(j) - R_s^{GF}(j) \qquad (7)$$

4. Economic loss due to recurrent service failures can be determined by the reduction of the above mentioned Customer Present Value (CPV) as a risk, reflecting in the lost monthly payments as a reduction in the probability of survival over an ideal customer relationship without service failures. Thus, in order to assess the maintenance contribution in terms of cost, a residual CPV could be deduced due to service failures because the expected customer life has been reduced by the occurrence of service failures. Consequently, the total Risk is calculated multiplying for all the affected customer per service failures (n_c), the Probability Contribution Pc(j) by the respective proportion of the CPV for the residual life in (J) total periods since the first failure occurrence ($\delta_j = 0$ in each j period except in the period where the first failure occurs and the followings periods in which is equal to 1).

$$\text{Risk} = \sum_{l=1}^{n_c} \sum_{j=1}^{J} \left[\frac{[\text{pr}(l,j) - c(l,j)] \cdot [\delta_j \cdot Pc(j)]}{(1+r)^j} \right] \qquad (8)$$

5. Finally, the Total Failure Costs (F), from a quality point of view, is calculated counting the cost of all the produced/estimated corrective actions (C_F) and plus the total Risk of losing customers for all the affected customers per any service failures (see formula (9), where $cc(l, j)$ is the mean corrective costs per affected customer and period).

$$F = C_F + \text{Risk} = \sum_{l=1}^{n_c} \sum_{j=1}^{J} \frac{cc(l, j)}{(1+r)^j} + \sum_{l=1}^{n_c} \sum_{j=1}^{J} \left[\frac{[pr(l,j) - c(l,j)] \cdot [\delta_j \cdot Pc(j)]}{(1+r)^j} \right]$$

(9)

The calculated Total Failure Cost (directly linked to maintenance performance) can be used to compare alternatives or scenarios, for example against preventive/predictive maintenance actions in order to avoid failures or evaluation activities. Thus, network utilities companies can evaluate, globally or per area, the capital invested and its effect on the customer satisfaction, in order to analyze the use of financial resources in maintenance activities, predicting the financial behavior if there are changes in the reliability of services.

The proposed methodology could be more precise developing the survival reliability per each number of service failures (f) inside the group of affected customers in each period $\left[R_{s1}^{GF}(j), R_{s2}^{GF}(j), R_{s3}^{GF}(j), \ldots, R_{sJ}^{GF}(j) \right]$. Although to simplify the implementation and computation in executive reports, instead of evaluating on real time the risks per segments of customer with the same characteristics per period (discretized yearly or semiyearly) and area of the network, we can also employ mean values evaluated globally as an approximation. Thus, in terms of the reliability function, the mean $C\dot{P}V$ per non-affected customer compared to the reduction of the mean lived time for all the affected customers could be employed as well as the mean corrective cost for all the service failures and periods ($\dot{c}c$) per affected customer (see formula (10)). That is to say, we could use for all the customers the average mean of the differences among the $\left[R_s^{GO}(j) \right]$ and $\left[R_s^{GF}(j) \right]$ for all the periods (i.e., for each semester).

$$F = C_F + \text{Risk} = n_c \cdot \dot{c}c + n_c \cdot C\dot{P}V \cdot \text{mean}_{\forall j} \left\{ \left[R_s^{GO}(j) - R_s^{GF}(j) \right] \right\}$$

(10)

This methodology assesses the impact of reliability on customer satisfaction and retention, streamlining prioritization and decision-making. In order to show the conservation degree of the customer relationship, it can be utilized the survival reliability, the estimated mean virtual life of customers (rejuvenated o degraded from real life) or the service repair efficiency q_s (to eliminate service failures). Therefore, these variables can be useful to obtain a reasonable comparison among:

- Geographical areas,
- Technical groups,
- Procedures,
- Changes in operations or organization,
- Equipment and technologies from different vendors,
- Environmental conditions and operations,
- With competitors, etc.

The same methodology could be utilized to segment the group of affected customers depending on the number of suffered failures in order to obtain different survival reliability functions, distinguishing the contribution of each additional suffered failure per customer.

4 Case Study in a Telecommunication Company

We will apply the methodology to a telecommunications company to show the implementation of the survival method after repeated failures, compared with the abandonment after no failures occurrence; and assuming there are not unreported service failures causing downtime, every service failure in this company is detected by monitoring information systems (80%) or by customer call claims (20%).

We present a case study based on two customer groups, GO and GF:

- Group GO refers to customers that abandoned the telecommunications services (canceling the contract) after a certain time period. They never experienced a service failure incident.
- Group GF is about customers that experienced several recurrent service failures. After a determined number of service failures, they abandoned the telecommunications services.

Each sample group corresponds to 200 customers, and considering the lost and maintained customers in their related time, we would deduce from the Weibull estimation the survival probability and other parameters. To do this, our methodology is developed based on the 3-parameter Weibull GRP method to obtain the probability of abandonment per customer. The solution is calculated by solving the Maximum Likelihood Estimation (MLE) through the Newton–Raphson statistical method to converge quickly. The smoothing of the solution is evaluated through the previously mentioned Fisher Matrix.

4.1 Group GO—Customers Leaving the Company After no Failures

Figure 2 presents the data sample for group GO. The histogram shows the dispersion of the data for the abandonment time of 200 customers. From the histogram, it can be noticed that the majority of the abandonments occur between time 600 and 900 days. The arithmetic mean is 717.37 and the standard deviation is 292.12. The probability density function also is characterized in the histogram as a solid line.

The Weibull parameters are $\beta = 2.564$ and $\alpha = 800.64$. Other important values such as reliability (R), unreliability (Q), MTTF and failure rate of the analyzed sample, are shown in Table 1. All the factors have been calculated when $t = 100$. MTTF and Failure Rate both represent in this case mean life for customer abandonment and abandonment rate.

4.2 Group GF—Customers Leaving the Company After Recurrent Failures

In this case, the group is composed of 200 customers from the telecommunications company who suffered one or more failures before abandoning the services.

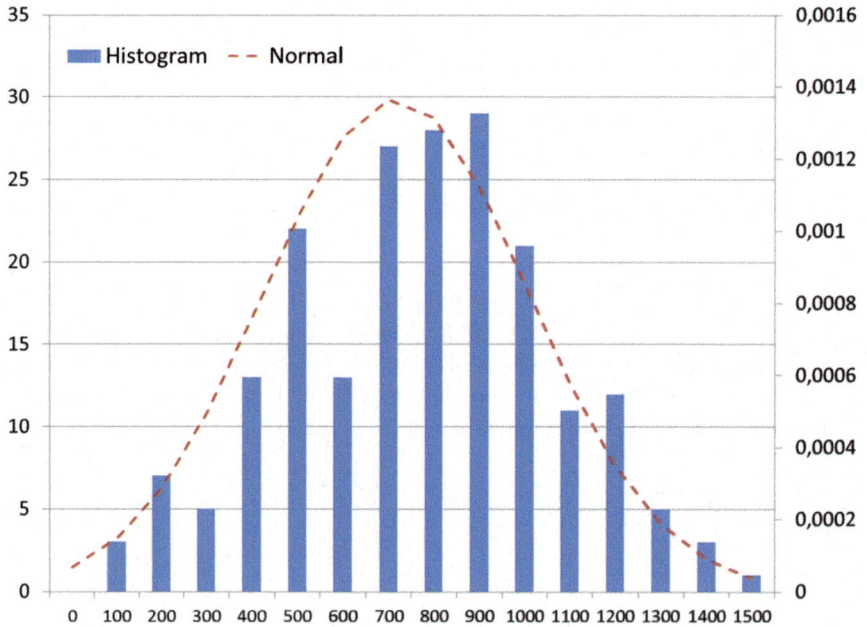

Fig. 2 Histogram of customers leaving the company. Data for case GO

Table 1 Reliability values for Case GO

Index	Value
$R(t = 100)$	0.995183
$Q(t = 100)$	0.004817
MTTF	710.84
Failure rate	0.000124

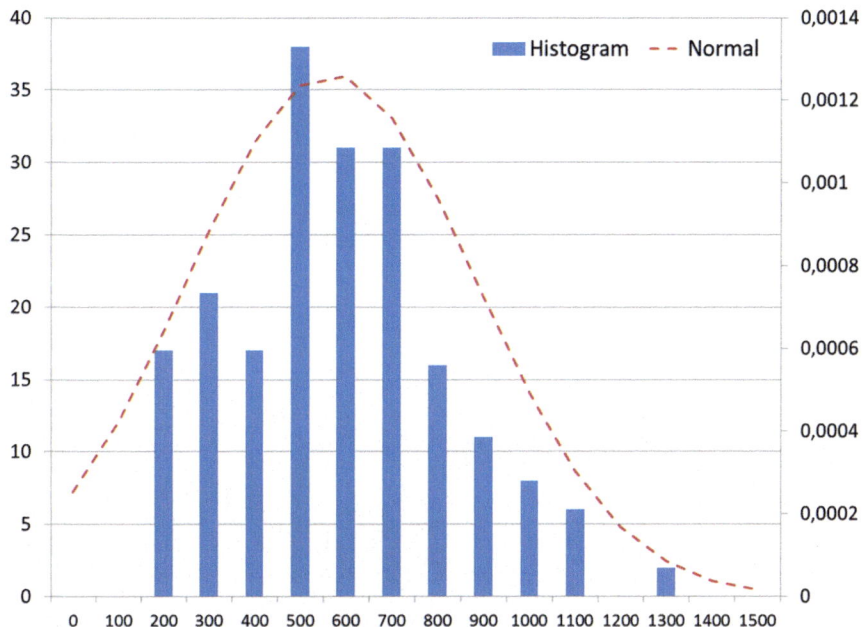

Fig. 3 Histogram of abandonment times for case study GF

Figure 3 presents a histogram of the abandonment times. From the histogram, it can be noticed that the majority of the abandonments occur around 534 days. The arithmetic mean is 566.36 and the standard deviation is 315.60. According to the number of failures, all 200 customers experienced at least one service failure, 77 customers experienced two service failures, and 10 customers experienced three service failures before the abandonment. Figure 4 shows the distribution of the number of service failures experienced by customers, sorted according to the occurrence time. Each line color represents a first, second or third service failure and the abandonment moment per customer number in GF.

We estimate the maximum point in three dimensions (on three parameters α, β, and q_s) for the Maximum Likelihood Estimation (MLE), according to the GRP method for case GF (a parametric recurrent event analysis) are $\beta = 1.3067$ and $\alpha = 442.09$. In the third dimension, repair effectiveness (q_s) between failures is 0.75766 which indicates a rejuvenation of the customer relationship in each successive failure, the appearance of 2nd and 3rd failures are less frequent than the

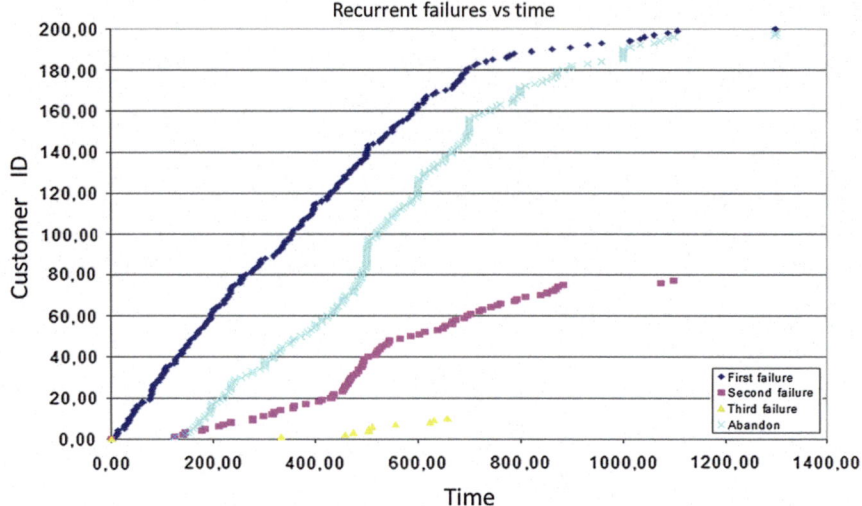

Fig. 4 Recurrent service failures per customer versus. Time for case study GF

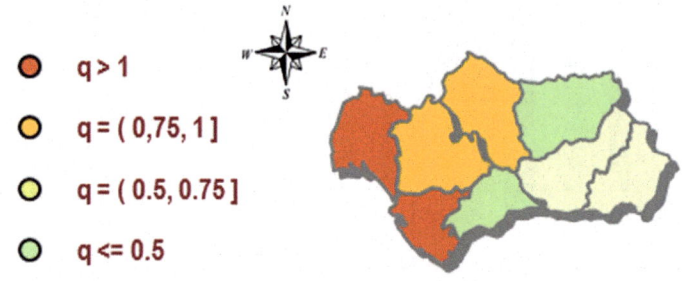

Fig. 5 Internal benchmarking of service repair efficiency among areas

1st failure. That is, according as the recurrence of failure increases, the frequency of appearance of new services failures decreases. Consequently, the virtual life could be representative indicator of the maintenance impact on perceived quality, developing comparisons among geographical areas within the same company (see example in Fig. 5), and guiding loyalty campaigns in order to reinforce company reputation in areas with q next to or greater than 1.

Other important values as reliability (R), unreliability (Q), instantaneous MTBF and instantaneous failure intensity are shown in Table 2. The consistency of the obtained Weibull GRP according to the considered total number of customers (G) in GF, as sensibility analysis, is included in Table 3. Particular attention should be paid in customer selection inside each group (GO and GF), searching consistency in the characterization by the same demanded service usage features, avoiding rare non-failure influenced behaviors about abandonment.

Table 2 Reliability values for case GF

Index	Value
$R(t = 100)$	0.866432
$Q(t = 100)$	0.133568
MTBF	407.88
Inst. failure intensity ($t = 100$)	0.001854

Table 3 Analysis according to the considered number of customers (G)

	$G = 50$	$G = 100$	$G = 150$	$G = 200$
β	1.05	1.15	1.24	1.31
α	489.49	457.36	444.01	442.09
Var(B)	1.10E−02	6.44E−03	4.95E−03	4.15E−03
Cov(Eta, B)	1.09E−04	3.70E−06	1.70E−06	9.00E−06
Var(Eta)	0.000001	2.1232E−07	5.89286E−08	2.17728E−08
LK value	−495.45	−1027.08	−1539.92	−1990

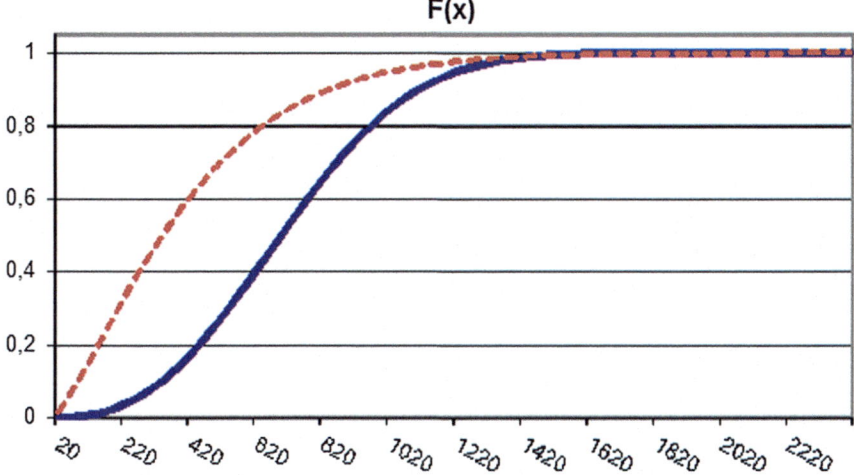

Fig. 6 Comparison unreliability versus time

4.3 Comparative Analysis Case GO Versus Case GF

For analyze the differences between the no failures recurrence case study (solid lines), and the recurrent failure case study (dashed lines), several graphics have been generated to compare the main reliability indicators.

- Fig. 6 shows unreliability behavior for both cases. From this image, it can be observed that the cumulated probability of customer abandonment has a faster growth in case GF (recurrent service failures) than in case GO (without service failures).

- Figure 7 presents the instantaneous failure probability for both cases. This Figure shows that the abandonment probability is higher for case GF, and above all is higher at the beginning of the customer relationship when recurrent failures occur.
- Figure 8 compares the reliability for both cases. The downfall in reliability for case GF is faster than for case GO.
- Finally, Fig. 9 shows the comparison of cumulative failure rates for both cases.

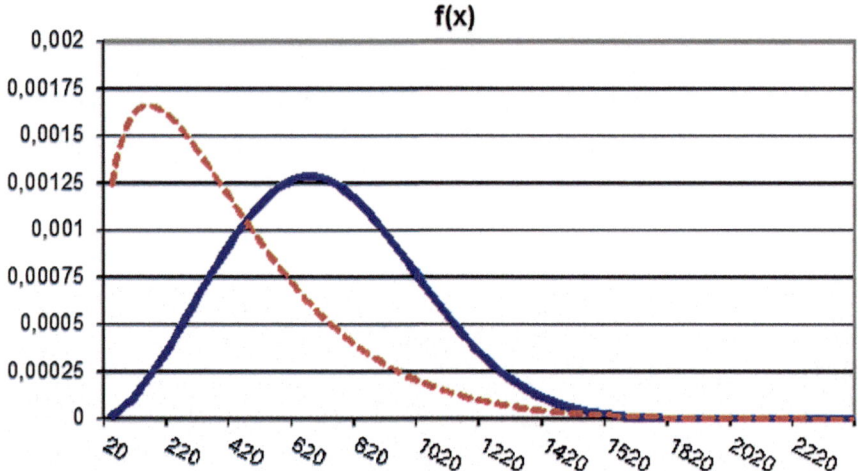

Fig. 7 Comparison instantaneous failure probability

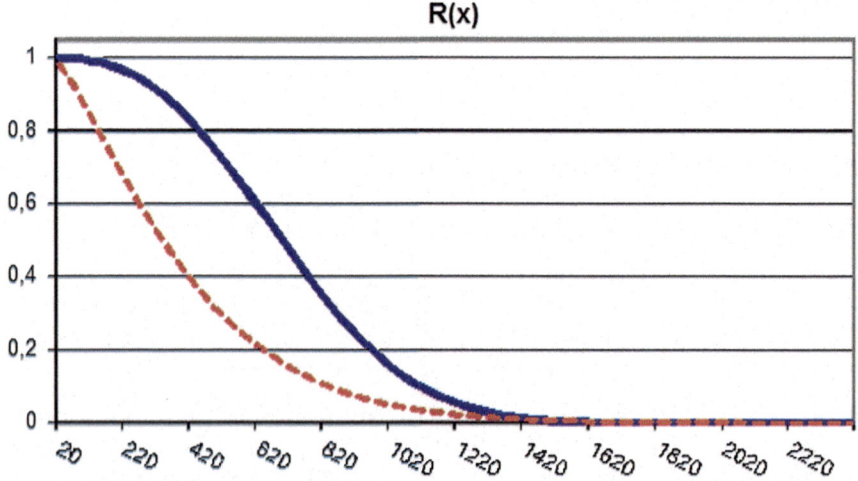

Fig. 8 Comparison reliability versus time

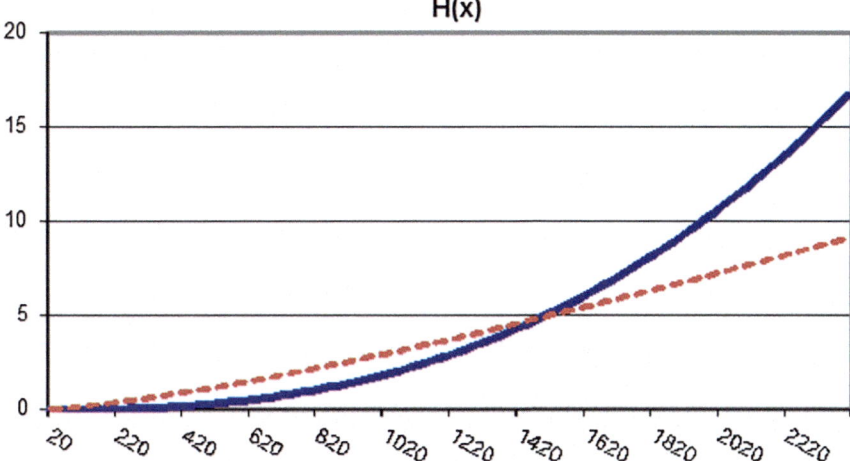

Fig. 9 Comparison cumulative failure rate

These graphs are crucial to understand and characterize the behavior of customers depending on the perceived quality related to service reliability. In addition, this analysis can allow companies to forecast the effect of reliability changes. Consequently this is decisive information for decision-making in maintenance.

4.4 Results and Discussion

The first important data obtained from the previous analysis are the mean times when abandonments occur. For case GO, the MTTF is 710.84 days. This indicates that in cases without service failures, the customer decides to abandon the service at time 710.84. The arithmetic mean time of abandonment is 717.37. For case GF, the instantaneous MTBF is 407.88 and the arithmetic mean time of abandonment is 566.36. This comparison indicates that customer abandonments occur earlier in case GF, when recurrent service failures are experienced. In consequence this reduction in the customer relationship impacts directly on the quantity of end customer invoicing, producing company financial losses via the reduction in CPV. Other reliability parameters shown in Table 4, also indicate the negative impact in the customer relationship due to failures in group GF compared to the group GO. In this table, the reliability factor has been calculated for each semester period of the CPV, where the difference is remarkable between the $\left[R_s^{GO}(j)\right]$ and $\left[R_s^{GF}(j)\right]$, which is not constant, but rather changes from one period to other following the behaviour of the two reliability Weibull functions (although the average mean of the differences for all the periods is 32%).

Table 4 Comparison of some reliability parameters. Case GO versus case GF

	Case study GO (no failures)	Case study GF (recurrent failures)	Difference (%)
Arithmetic mean for abandonment	717.37	566.36	21
MTTF/MTBF	710.843705	407.88	43
$R(j = 1, t = 180)$	0.978444464	0.734133324	24
$R(j = 2, t = 360)$	0.879116396	0.465523471	41
$R(j = 3, t = 540)$	0.694661556	0.272858239	42
$R(j = 4, t = 720)$	0.466854332	0.150842271	32
$R(j = 5, t = 900)$	0.259301801	0.079505553	18

Therefore, if the number of failures or the affected customers by them is high, the impact of maintenance performance in terms of costs is significant.

4.5 Risk Comparison of Losing Customers in Different Areas of GF Group

Finally, this analysis can be particularized in the company Customer Relationship System (CRM) for any customer directly using real-time calculations or simplifying through approximations using mean values. Thus, this methodology calculates the risk of losing customers and the total failure costs (F), and distinguishes areas characterized by different survival reliability $\left[R_s^{GF}(t)\right]$ and service repair effectiveness (q_s). For this example, two groups of customers with affection of recurrent failures are compared from different areas, employing the simplified implementation through mean values of mean corrective cost ċc, mean CṖV, and the mean difference among the survival reliabilities.

Supposing that Evaluation and Prevention costs are the same in both GF areas, 0.45 and 1.15 M€ respectively, then the comparison among areas must be determined by failure costs as sum of corrective and risk costs. To simplify, for the both areas we will employ as $\left[R_s^{GO}(t)\right]$ the Weibull determined in Group GO previously. Therefore, if one area (area1), has arithmetic mean = 566.36, standard deviation = 315.6, $\left[R_s^{GF1}(t)\right]$ (obtained in point 4.2 by Weibull GRP analysis) with parameters $\beta_1 = 1.3067$, $\alpha_1 = 442.09$ and $q_1 = 0.75766$. Another area (area2) has arithmetic mean = 509.77, standard deviation = 214.47, $\left[R_s^{GF2}(t)\right]$ (deduced in the same way by Weibull GRP analysis) with parameters $\beta_2 = 1.52966$, $\alpha_2 = 412.24$ and $q_2 = 0.67318$. Consequently, both areas are compared through Failure Costs, where the average mean of the difference among the $\left[R_s^{GO}(j)\right]$ and $\left[R_s^{GF}(j)\right]$ for all the periods is 32% in the area1 and 35% in the area2 (see Table 5).

Table 5 Risk analysis comparison between the two areas

	Area1	Area2
Total appraisal costs	0.45 M€	0.45 M€
Total prevention costs	1.150 M€	1.150 M€
Mean corrective costs (ċc)	105 €	105 €
n_c (mean number affected customers)	11,240	13,570
Total corrective costs	1.180 M€	1.425 M€
mean{$[R_{GO} - R_{GF}]$}	32%	35%
CPV	1415 €	1415 €
Risk (€)	11,240·1415·0.32 = 5.089 M€	13,570·1415·0,35 = 6.720 M€
Total F cost (€)	6.270 M€	8.145 M€

Table 5 shows failure costs can be different depending on the maintenance performance among areas. The number of service failures and the produced risks due to customer abandonments by failures contribute to the difference of failure costs. The risk can be critical when service failures have a high impact customer abandonments, mainly due to a decrease in retention probability, and consequently reducing the customer lived time. Thus, in the case of area1 the risks are 5.089 M€, much higher than prevention or corrective costs, however in area2 the risks are 6.72 M€ because the number of failures and probability of abandonment are both higher.

As a result, the impact in terms of costs on perceived quality is determined mainly due to the difference in Risk costs (see Fig. 10) for each area. Maintenance contribution can be analyzed by balancing $A + P$ costs versus F costs, considering direct costs as corrective actions and indirect costs as perceived quality, and even so utilizing these evaluations characterized per any failure mode deciding to act with prevention instead of correction activities. Consequently, the maintenance contribution of both areas to the perceived quality and customer retention is unequal mainly due to the failure costs (area2 > area1), directly proportional to the number of failures and indirectly to the potential loss of customers. In addition, the repair efficiency also shows maintenance performance in the area in order to avoid recurrent failures ($q_{s2} = 0.67318 < q_{s1} = 0.75766$). This example indicates the possibility of providing more material resources to preventive/predictive maintenance in order to reduce losses.

Additionally, our methodology classifies groups of affected customers by service failures in several subgroups with the same number of suffered failures. For example, one subgroup with customers with only one failure, other with two, and so on. In this way, by applying the same formulae we specify recurrent increments of $[R_s^{GF}(l,j)]$ for each supplemental suffered failure. In our example the increment of the abandonment probability from customers with 1 failures against 0 is ×1.133, with 2 failures instead of one is the ×1.38 and from 3 failures instead of two

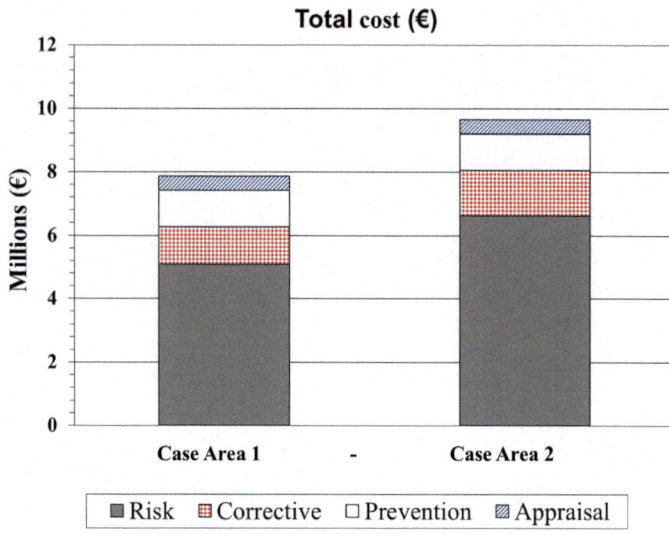

Fig. 10 Costs comparison for case area1 and case area2

		α	β	MTBF	% inc. probability (%)
Table 6 Probability increment of customer abandonment according to number of recurrent failures	GO	800.64	2.56	710.85	
	A1 $\times$ 1	614.99	2.27	544.76	13.30
	A1 $\times$ 2	325.21	1.28	301.17	38.18
	A1 $\times$ 3	267.81	1.44	242.98	43.67

is $\times 1.43$ as average means of the difference among the $\left[R_s^{GO}(j)\right]$ and $\left[R_s^{GF}(j)\right]$ for all the periods (see Table 6 and Fig. 11).

From our analysis, the following inferences can be deduced:

(a) Maintenance contributes to the profits of the company, due to maintenance increases the probability that a customer stays (does not abandon the service).
(b) Not only is it important to prevent failures, but also their recurrence. As failures occur more sporadically, the repair effectiveness is improved. This improvement causes a decrease in customer abandonment. Also, according as the recurrence of failures is increased, the potential lost customers increases, especially if the service disruptions are more than those stipulated in the contract.
(c) Comparison of these two cases reveals that the investment in preventive/predictive maintenance is widely recommended to achieve good levels of reliability, reducing the implicit risk of customer abandonment and the costs of corrective activities.
(d) From an automated Weibull analysis, it is possible to determine the expected reliability of services for a determined time, in order to prevent a possible

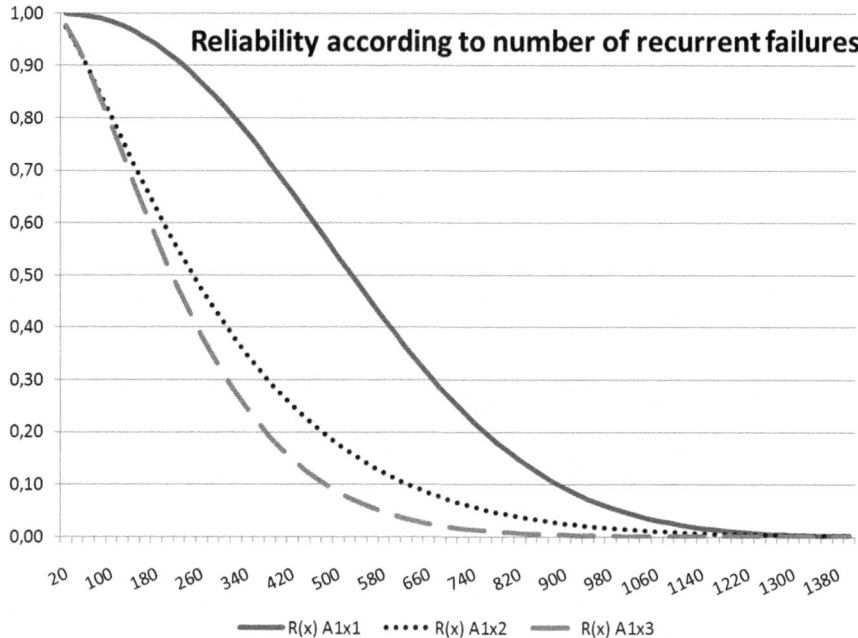

Fig. 11 Reliability according to number of recurrent service failures

impact on the customer, and to design marketing and loyalty campaigns appropriate to each customer history. These campaigns can be targeted specifically at geographical areas or environmental/technological conditions with higher risk of recurrent failures, in order to prevent abandonments and company image loss.

(e) This kind of analysis can be useful also to compare the performance of different subcontractors or technologies, and to compare improvements within the same process; for example by analyzing the repair effectiveness through the recurrence of failures.

(f) Furthermore, the total Failure costs (F) are higher if it is considered that the transmission of bad propaganda up to 10 partners or customers increases the abandonment probability of them and, that the bad propaganda and the failure impact decreases the number of new purchases.

Further research areas of this work could be orientated to deepen understanding customer behaviour, not only about failure times, but also considering repair times or the effect of loyalty/information campaigns in relation to maintenance or the effect of technological service renewals due to advances. The proportionality among these other factors should be considered in order to evaluate additional significant interactions, trying to identify customer perception of service quality. Consequently, Proportional Hazard Model (PHM) and Neural Network techniques could enhance our methodology.

5 Conclusions

Network utilities are complex companies in which failure occurrence impacts on the quality perceived by customers. Use of ICT systems/tools and quality improvement methodologies for maintenance management are essential to stand in the market and to obtain benefits. Maintenance and quality are closely related areas, especially in what corresponds to customer satisfaction. Loyalty of customers depends on service reliability. When the quality perception becomes worse than the conditions offered in the contract, or infringes on the patience of customer (due the number of failures or their duration), the customer abandons the service, meaning financial losses.

Failures analysis for prediction, prevention, correction, and the estimation of their consequences is an issue of vital importance. The methodology proposed in this paper is focused on analyzing the economic implications of maintenance and the occurrence (and reoccurrence) of failures in different scenarios, calculating the economic risk of the failure and comparing it with the costs of prevention.

Maintenance performance affects the offered and the perceived quality by the customers, circumstance that is shown through the proposed methodology in our case studies. Investments in maintenance activities (especially preventive/ predictive) represent economic advantages. The cases analyzed in this paper showed that investment in preventive maintenance actions is economically viable, as that its direct effect is reducing customer's abandonment rates, and even diminishing the costs of corrective activities also. On the contrary, when number of failures or affected customers increases, especially if failures are recurrent, the impact on costs is significant. In this way a direct relation is observed among the execution of preventive maintenance activities, reliability of services and customer satisfaction, producing significant financial consequences.

The proposed methodology, with the intention of calculating the abandonment probability based on the lifetime of each customer, can be used also as an automated routine to determine the expected reliability of the services for a determined time, and geographical zone. Furthermore, it can be utilized to prevent a possible negative impact on the customer, as well as to design marketing and loyalty campaigns appropriate to each customer circumstance. The methodology is useful when comparing different technical or management alternatives, or to determine if the investment in preventive/predictive maintenance has been profitable.

Assessment of maintenance in network distribution companies should not be only done in terms of budget, but also in terms of real profits directly observable and those due to the avoided damages. Therefore, our methodology takes into account the impacts of risks, later reduced, allowing maintenance management to be accountable and comparable with other organizations or reference standards, from different areas and levels of detail, identifying improvements in network reliability and ensuring service quality from customer, business, and society perspectives.

Acknowledgements The authors wish to thank the Institution "Fundación Iberdrola" for providing a research grant during the years 2011 and 2012, making the development of projects possible related to the implementation of advanced maintenance strategies, technologies and services.

This research is funded by the Spanish Ministry of Science and Innovation, Project EMAINSYS (DPI2011-22806) "Sistemas Inteligentes de Mantenimiento. Procesos emergentes de E-maintenance para la Sostenibilidad de los Sistemas de Producción, besides FEDER funds."

References

1. Newbery DMG (2002) Privatization, restructuring, and regulation of network utilities. MIT Press, Cambridge
2. Van Vliet B, Chappells H, Shove E (2005) Infrastructures of consumption: environmental innovation in the utility industries. Earthscan Publications Limited, London. ISBN 1-85383-996-5
3. Johnson RL, Tsiros M, Lancioni RA (1995) Measuring service quality: a system approach. J Serv Mark 9(5):6–19
4. Peters T (1987) Thriving on chaos: handbook for a management revolution. Alfred A. Knopf, New York
5. Bitner M, Booms B, Tetreault M (1990) The service encounter: diagnosing favourable and unfavourable incidents. J Mark 54:71–86
6. Zeithaml VA, Bitner MJ (2003) Services marketing: integrating customer focus across the firm. McGraw-Hill Higher Education, Boston
7. Parasuraman A, Zeithaml VA, Berry LL (1985) A conceptual model of service quality and its implications for future research. J Mark 49:41–50
8. Gellings GW (2009) The retail electricity service business in a competitive environment. In: Andreas Bausch, Burkhard Schwenker (eds) Handbook utility management. Springer, Berlin, pp 545–558
9. De Matos CA, Henrique JL, Vargas Rossi CA (2007) Service recovery paradox: a meta-analysis. J Serv Res 10(1):60–77
10. Maxham JG III (2001) Service recovery's influence on consumer satisfaction, positive word-of-mouth, and purchase intentions. J Bus Res 54:11–24
11. Tschohl J (1996) Achieving excellence through customer service. Best Sellers Publishing, Minnesota
12. Gómez JF, Crespo A (2009) Framework for implementation of maintenance management in distribution network service providers. Reliab Eng Syst Saf 94(10):1639–1649
13. Küssel R, Liestmann V, Spiess M, Stich V (2000) "Teleservice" a customer oriented and efficient service. J Mater Process Technol 107:363–371
14. Murthy DNP, Atrens A, Eccleston JA (2002) Strategic maintenance management. J Qual Maint Eng 8(4):287–305
15. Zhu G, Gelders L, Pintelon L (2002) Object/objective-oriented maintenance management. J Qual Maint Eng 8(4):306–318
16. Marais KB (2013) Value maximizing maintenance policies under general repair. Reliab Eng Syst Saf 119:76–87
17. Dixon JR (1966) Design engineering: inventiveness, analysis, and decision making. McGraw-Hill, Inc., New York
18. Woodhouse J (1993) Managing industrial risk. Chapman Hill, London
19. Wilson RL (1986) Operations and support cost model for new product concept development. In: Proceedings of the 8th annual conference on components and industrial engineering, pp 128–131

20. Remy E, Corset F, Despréaux S, Doyen L, Gaudoin O (2013) An example of integrated approach to technical and economic optimization of maintenance. Reliab Eng Syst Saf 116:8–19
21. Saleh JH, Marais KB (2006) Reliability: how much is it worth? Beyond its estimation or prediction, the (net) present value of reliability. Reliab Eng Syst Saf 91:665–673
22. Janawadea V, Bertranda D, Léo P-Y, Philippea J (2015) Assessing 'meta-services': customer's perceived value and behaviour. Serv Ind J 35(5):275–295
23. Günthera C-C, Tvetea I, Aasa K, Sandnesb G, Borganc Ø (2014) Modelling and predicting customer churn from an insurance company. Scand Actuar J 1:58–71
24. Wang K-Y, Hsu L-C, Chih W-H (2014) Retaining customers after service failure recoveries: a contingency model. Manag Serv Qual 24(49):318–338
25. Knox G, van Oest R (2014) Customer complaints and recovery effectiveness: a customer base approach. J Mark 78(5):42–57
26. Bolton R (1998) A dynamic model of the duration of the customer's relationship with a continuous service provider: the role of satisfaction. Mark Sci 17(1):45–65
27. Marais KB, Saleh JH (2009) Beyond its cost, the value of maintenance: an analytical framework for capturing its net present value. Reliab Eng Syst Saf 94:644–657
28. Martorell S, Sanchez A, Serradell V (1999) Age-dependent reliability model considering effects of maintenance and working conditions. Reliab Eng Syst Saf 64(1):19–31
29. Wua F, Niknamb SA, Kobzac JE (2015) A cost effective degradation-based maintenance strategy under imperfect repair. Reliab Eng Syst Saf 144:234–243
30. Jiang R (2010) A simple approximation for the renewal function with an increasing failure rate. Reliab Eng Syst Saf 95(9):963–969
31. Crespo A, Sánchez A (2002) Models for maintenance optimization: a study for repairable systems and finite time periods. Reliab Eng Syst Saf 75(3):367–377
32. Crosby PB (1979) Quality is free. Mentor Books, New York
33. Dale BG, Plunkett JJ (1991) Quality costing. Gower Publishing, Hampshire
34. ASQC (1970) Quality costs: what and how. American Society for Quality Control, New York
35. BS 4778 (1987) Quality vocabulary. British Standards Institute, London
36. Avizienis A, Laprie JC, Randell B (2001) Fundamental concepts of dependability. LAAS-CNRS; Research report no. 1145. April
37. Gómez Fernández JF, Crespo Márquez A (2012) Maintenance management in network utilities: framework and practical implementation. Springer, London
38. Goodman J (1986) Technical assistance research program (TARP). US Office of Consumer Affairs Study on Complaint Handling in America, USA
39. Keaveney S (1995) Customer switching behavior in service industries: an exploratory study. J Mark 59:71–82
40. Gupta S, Lehman DR (2005) Managing customers as investments: the strategic value of customers in the long run. Wharton School Publishing, New Jersey
41. Wacker G, Tollefson G (1994) Electric power system customer interruption cost assessment. Reliab Eng Syst Saf 46(1):75–81
42. Yanamandram V, White L (2006) Switching barriers in business-to-business services: a qualitative study. Int J Serv Ind Manag 17(2):158–192
43. UNE 200001-3-11 (2003) Gestión de la confiabilidad. Parte 3–11: Guía de aplicación. Mantenimiento centrado en la fiabilidad. UNE
44. Dekker R (1996) Applications of maintenance optimization models: a review and analysis. Reliab Eng Syst Saf 51:229–240
45. Wang W (2008) Condition based maintenance modeling. In: Kobbacy KAH, Murthy DNP (eds) Complex Systems maintenance handbook. Springer, London
46. Greves D, Schreiber B (1993) Engineering costing techniques in ESA. Available online: http://esapub.esriu.esa.it.pointtobullet/greves1.html
47. Harter HL, Moore AH (1965) Point and interval estimators based on order statistics, for the scale parameter of a Weibull population with known shape parameter. Technometrics 7(3):405–422

48. Parmar MKB, Machin D (1996) Survival analysis: a practical approach. Wiley, Chichester
49. Cox DR, Oakes D (1984) Analysis of survival data. London Chapman and Hall, London
50. Klein J, Moeschberguer M (1997) Survival analysis techniques for censored and truncated data. Springer, New York
51. Andersen PK, Borgan O, Gill R, Keilding N (1993) Statistical models based on counting process. Springer, New York
52. Blischke WR, Murthy DNP (2000) Reliability modelling, prediction and optimization. Wiley, New York
53. Hosmer DW, Lemeshow S (1999) Regression modeling of time to event data. Wiley, New York
54. Hougaard P (2000) Analysis of multivariate survival data. Springer, New York
55. Lee ET (1992) Statistical methods for survival data analysis. Wiley, New York
56. Harrell FE (2001) Regression modeling estrategies. Springer, New York
57. Yañez M, Joglar F, Mohammad M (2002) Generalized renewal process for analysis of reparable systems with limited failure experience. Reliab Eng Syst Saf 77:167–180
58. Kijima M, Sumita N (1986) A useful generalization of renewal theory: counting process governed by non-negative Markovian increments. J Appl Probab 23:71–88
59. Mettas A, Zhao W (2005) Modeling and analysis of repairable systems with general repair. In: Proceedings of the reliability and maintainability symposium, pp 176–182. ISBN: 0-7803-8824-0
60. Cohen AC (1965) Maximum likelihood estimation in the Weibull distribution based on complete and on censored samples. Technometrics 7(4):579–588

Author Biographies

Juan Francisco Gómez Fernández is Ph.D. in Industrial Management and Executive MBA. He is currently part of the Spanish Research & Development Group in Industrial Management of the Seville University and a member in knowledge sharing networks about Dependability and Service Quality. He has authored publications and collaborations in journals, books and conferences, nationally and internationally. In relation to the practical application and experience, he has managed network maintenance and deployment departments in various national distribution network companies, both from private and public sector. He has conduced and participated in engineering and consulting projects for different international companies, related to Information and Communications Technologies, Maintenance and Asset Management, Reliability Assessment, and Outsourcing services in Utilities companies. He has combined his business activity with academic life as an associate professor (PSI) in Seville University, being awarded as Best Thesis and Master Thesis on Dependability by National and International Associations such as EFNSM (European Federation of National Maintenance Societies) and Spanish Association for Quality.

Adolfo Crespo Márquez is currently Full Professor at the School of Engineering of the University of Seville, and Head of the Department of Industrial Management. He holds a Ph.D. in Industrial Engineering from this same University. His research works have been published in journals such as the International Journal of Production Research, International Journal of Production Economics, European Journal of Operations Research, Journal of Purchasing and Supply Management, International Journal of Agile Manufacturing, Omega, Journal of Quality in Maintenance Engineering, Decision Support Systems, Computers in Industry, Reliability Engineering and System Safety, and International Journal of Simulation and Process Modeling, among others. Professor Crespo is the author of seven books, the last four with Springer-Verlag in 2007, 2010, 2012, and 2014 about maintenance, warranty, and supply chain management. Professor Crespo leads the Spanish Research Network on Dependability Management and the

Spanish Committee for Maintenance Standardization (1995–2003). He also leads a research team related to maintenance and dependability management currently with five Ph.D. students and four researchers. He has extensively participated in many engineering and consulting projects for different companies, for the Spanish Departments of Defense, Science and Education as well as for the European Commission (IPTS). He is the President of INGEMAN (a National Association for the Development of Maintenance Engineering in Spain) since 2002.

Mónica Alejandra López-Campos is an Industrial Engineer, and Ph.D. in Industrial Organization from the University of Seville (Spain). Currently, she is Professor in the Federico Santa Maria University (Chile), where she teaches the subjects of Maintenance Management, Operations Management and Quality Management. Also, she is member of the National System of Researchers (Mexico). Her research works have been published in journals such as Reliability Engineering and System Safety, International Transactions in Operational Research, Computers in Industry, Quality and Reliability Engineering International, and Journal of Quality in Maintenance Engineering, among others. Her interests in research include maintenance management, modeling of processes, logistics, quality, simulation and educational methodologies for the engineering. She has participated in several research projects in Mexico, Spain and Chile, as well as in international organizations such as the Economic Commission for Latin America and the Caribbean (ECLAC).

Dynamic Reliability Prediction of Asset Failure Modes

Juan Francisco Gómez Fernández, Jesús Ferrero Bermejo,
Fernando Agustín Olivencia Polo, Adolfo Crespo Márquez
and Gonzalo Cerruela García

Abstract In this paper a reliability model based on artificial neural networks and the generalized renewal process is developed. The model is used for failure prediction, and is able to dynamically adapt to changes in the operating and environmental conditions of assets. The model is implemented for a thermal solar power plant, focusing on critical elements of these plants: heat transfer fluid pumps. We affirm that this type of model can be easily automated within the plant's remote monitoring system. Using this model we can dynamically assign reference values for warnings and alarms and provide predictions of asset degradation. These in turn can be used to evaluate the associated economic risk to the system under existing operating conditions and to inform preventive maintenance activities.

Keywords Renewable energy · Maintenance · Dynamic reliability analysis · Artificial neural networks · Generalized renewal process · Thermal solar plant

J. Ferrero Bermejo · F.A. Olivencia Polo
Magtel Systems, Seville, Spain
e-mail: jesus.ferrero@magtel.es

F.A. Olivencia Polo
e-mail: fernando.olivencia@magtel.es

J.F. Gómez Fernández (✉) · A. Crespo Márquez
Department of Industrial Management, School of Engineering,
University of Seville, Seville, Spain
e-mail: juan.gomez@iies.es

A. Crespo Márquez
e-mail: adolfo@us.es

G. Cerruela García
Department of Computer and Numerical Analysis,
University of Cordoba, Córdoba, Spain
e-mail: gcerruela@uco.es

1 Introduction

The consistency of reliability predictions in the renewable energy sector is important because of the high impact of production losses. These predictions are complex to generate due to the fact that assets' operating conditions change seasonally and geographically. In this sector, the optimization of maintenance programs for an asset must consider operational variables (configurations, preventive maintenance, undue handling, etc.) as well as environmental conditions (cleanliness, fastening, temperature, etc.). Then, reliability estimations that take these contributing factors into account may be informative. Besides this, the need to update these estimations over time, and the proper consideration of explanatory variables or covariates, is critical to predict time to system failure.

There are many techniques for survival analysis and estimation [1, 2] that use explanatory variables. These techniques can be parametric when the failure distributions are known, semi-parametric in the case of unknown failure distribution but with defined assumptions of proportionality with time covariates (independent among them), or non-parametric when the failure distributions are not specified [3, 4]. Flexibility and complexity of computational implementation increase from parametric to non-parametric methods [5].

In the search for efficiency, artificial intelligence (AI) can be used to solve, computationally, these prediction problems [6, 7], and well-accepted and recommended methods for these purposes are artificial neural network (ANN) models [8–11]. These networks are composed of multiple, connected units (neurons). The standard ANN architecture is one input layer, one output layer and generally, one or more hidden layers. To achieve autoadjustment, an often used ANN is the backpropagation neural network (BANN) which prevents overtraining. This technique filters the noise and recognizes the most overt and accessible patterns, overcoming in an order of magnitude the linear conventional methods and the polynomial methods [12]. Consequently, BANN provides a strong tolerance to noisy data, due to the storage of redundant information, and adaptation in the presence of explicit knowledge for the resolution of problems [13, 14].

In our proposed prediction model, the neural network model under consideration is a feedforward single layer perceptron [15], which is composed of one input layer with P neurons, an intermediate or hidden layer with M neurons, and an output layer with 1 neuron, Y. The neural network output, Y, is a function based on a linear combination of a set of weights with the outputs from the intermediate layer neurons. All the weights used in the linear combination are learned by a backpropagation algorithm [16], in which the sum-of-squared errors

$$R^2 = \sum_{i=1}^{N} R_i = \sum_{i=1}^{N} \left(Y^{(i)} - f(X^{(i)}) \right)^2 \tag{1}$$

is optimized, reaching a global minimum by a loop with a maximum of S steps, where in each s-step a couple of forward and backward actions are executed in the

N-elements training set $TS = \left\{ \left(X^{(i)}, Y^{(i)} \right) \middle| i \in \{1, \ldots, N\} \right\}$, where $X^{(i)} = \left(X_1^{(i)}, X_2^{(i)}, \ldots, X_P^{(i)} \right)$ and where $Y^{(i)} = \left(Y_1^{(i)}, Y_2^{(i)}, \ldots, Y_P^{(i)} \right)$ are an input and output values respectively from the training set.

This paper is fundamentally concerned with the utilization of a BANN to obtain predictions about failures before they occur. For this purpose, the paper is structured in three main parts: a first part, Sect. 2, which includes the development of our prediction model; then, a second part including Sects. 3 and 4 which include a case study; finally a section that is dedicated to the conclusions of the paper.

2 Proposed Prediction Model

In this Section we assume that there are sufficient experimental data about failures. Due to their experimental nature, these raw data do not follow a formal parametric failure distribution function, and in addition these have nonlinear correlations with the covariates that determine accelerating process in the failure appearance. In order to simplify, we assume these covariates are independent among them and with no time dependency. In these cases, the maintenance decision-making for renewal energy equipments, under different operating environments, can be supported by ANN, with the following benefits:

- Suitable to the managed amount of failure data, the more information you provide the more relevant your result becomes as continuous improvement,
- Implementable in remote monitoring systems, searching automation not only for prediction but also for self-adjusting based on real data,
- Flexible to changes, either for the elimination of any covariate or for the incorporation of new covariates in the input layer or by combining several ANN hierarchically.

Numerous papers [5, 17–21] present the comparison between several existing methods to fit *survival functions* showing relations between the reliability and the covariates; for instance, comparing the widely applied semi-parametric Cox's proportional hazard model (PHM) [22] versus several ANN models. When the model complexity is low, based on a few covariates and with proportional relations with the reliability, there are no significant differences between predictions of Cox regression and ANN models. In the case of complex models with many covariates and with any interaction term, predictions of ANN models have important advantages compared to Cox regression models.

In practice, depending on the way that covariates are considered, different ANN models can be built; for instance:

- ANN can be used instead of the linear combination of weight coefficients in the Cox PHM, as in Faraggi and Simon [18]. In this case it is necessary to solve the PHM using the Partial Maximum Likelihood Estimation (P-MLE).

- Using an input with the *survival status* over disjoint time intervals where the covariate values are replicated. A binary variable can be used with value 0 before the interval of the failure and 1 in the event of failure or later, as in Liestbl et al. [19]. In this case, each time interval is an input with a *survival status*, then a vector of survival status is defined per failure; or

- Employing the Kaplan–Meier (K-M) estimator to define the time intervals as two additional inputs instead of a vector; one is the sequence of the time intervals defined by de K-M, and the other is the survival status at each time of the sequence. This is the case of Ravdin and Clark [20] or Biganzoli et al. [17], models that are known as Proportional Kaplan–Meier.

Since the aim of this research is to improve results of parametric methods, combining them with the self-adaptive property of ANN, the proposed survival ANN model is based on the ideas of Ravdin and Clark [20].

At the same time some mathematical modifications are introduced for simplification and in order to facilitate reliability surveillance (see Table 1). Considering these modifications the application of the proposed model is sequenced in two phases:

1. In a first phase, for easy understanding of results and for applying another reliability analysis over time to failures. The re-utilization of parametric method outcomes is recommended, as a previous estimation and without the consideration of covariates. Because of this, we will have a parametric estimation of the survival curve.

   - Instead of using the Kaplan–Meier estimator, the General Renewal Weibull Process II (GRP-II) method is selected in order to fit the curve better and to reduce the negative effect of a non-monotonically decreasing survival curve. In our study, we also propose to combine ANN with the GRP-II parametric model, which evaluates survival probability in repairable and non-repairable systems, and models the repair efficiency estimation in avoiding the overall damage produced due to all the successive failures (González-Prida et al. [23]. The GRP-II model has more accuracy than the GRP-I for complex systems, or with data from multiple devices of the same type [24]. GRP-II gives three Weibull parameters, α-scale, β-shape and q-repair efficiency. In the q parameter, the recurrence rate of failures is captured showing the effects of the repairs on the age of that system n (modelling partial renewal repair). Properly performed repairs ($0 < q_n < 1$) may improve system virtual states (life), while poorly solved failures ($q_n > 1$) could aggravate it (always speaking in terms of reliability). Then, the virtual age of the system i is updated after a failure j according to the following Eq. (2) (where T_{nj} are the time intervals between successive failures) and the failure probability distribution conditioned to the survival new virtual age is calculate in Eq. (3):

$$V_n^{\text{new}} = q_n \cdot (V_n^{\text{old}} + T_{nj}) \tag{2}$$

$$F(t|V_n^{\text{new}}) = P\left[T_{nj} \leq t | T_{nj} > V_n^{\text{new}}\right] = \frac{F(t) - F(V_n^{\text{new}})}{1 - F(V_n^{\text{new}})} \tag{3}$$

- Only based on the times to failure (T_{nj}), GRP-II is applied over k groups of failures with similar covariate values, for example for each plant. Then, Weibull parameters α, β and q are obtained for each group, and without using the covariates, only based on time to failures.

$$F(t|V_{nk}^{\text{new}}) = \exp\left[\left(\frac{V_n^{\text{new}}}{\alpha_{nk}}\right)^{\beta_{nk}} - \left(\frac{t}{\alpha_{nk}}\right)^{\beta_{nk}}\right] \tag{4}$$

2. After that, in order to adapt the parametric estimation of the survival function according to covariates, the backpropagation ANN is utilized with the following criteria:
A sigmoid (logistic) function

$$f(x) = \frac{e^x}{(1 + e^x)}, \tag{5}$$

- over normalized variables between 0 and 1, is used for hidden layer neurons and a linear combination of the latter, for determining the neural network output. It is not recommendable to use more than two times the number of input neurons in the hidden layer [16].
- Discretized times inside the intervals T_{nkj} over k groups of failures with similar covariate values X_{nk} are defined according to the available and representative information for the selected failure. In order to be upgradeable iteratively and with real data and for all the intervals, the real covariate values will be used as inputs $X_{nk}^{(i)}$, but only up to the time the failure occurs, after which the average covariate value is taken $X_{nk}^{(i)} = \dot{X}_{nk}$. In order to homogenize the selected intervals T_{nkj}, all of them are extended after the failure time in discretized times $t^{(i)}$ of the training set up to the maximum length of the intervals (which corresponds to the maximum time to failure), resulting all $(\dot{T}_{nkj})$ with the same length.

$$TS_{nk} = \left\{\left(X_{nk}^{(i)}, Y_{nk}^{(i)}\right) | i \in \{1, \ldots, N\}\right\} \tag{6}$$

- The additional input of the survival status, in our model, has a gradual increment from 0 to 1 in the failure event or after each specific time to failure. To do this, the Weibull Cumulative Distribution Function (CDF) can be used for all intervals, with the previously obtained β_{nk} per group by GRP-II method, searching to maintain the shape of predictions. Besides, in order to obtain proportionality, the gradual increment from 0 to 1 of the survival status, the Weibull scale α_{nk} in each interval is adapted to the specific time interval T_{nkj} (between successive failures) pondered by the Median Life. Therefore, the output layer contains a single output neuron corresponding to the estimated survival status (probability to failure) that ascends from 0 to 1 until the time to failure and after, see Eq. (7), where for all the intervals: β_{GRP} is obtained by GRP-II for each group of times to failures, and $\alpha_{nkj} = T_{nkj} \cdot Ln(2)^{1/\beta_{nk}}$:

$$
\begin{aligned}
CDF\left(t^{(i)}\right) &= 1 - \left[1/\exp\left(\frac{t^{(i)}}{\alpha_{nkj}}\right)^{\beta_{nk}} \right] \\
&= 1 - \left[1/\exp\left(\frac{t^{(i)}}{\left(T_{nkj} \cdot Ln(2)^{1/\beta_{nk}}\right)}\right)^{\beta_{nk}} \right]
\end{aligned}
\tag{7}
$$

- The training of the network, which gives us the network settings, is carried out based on 75% of available data; and the other 25% is used for the network testing in order to subsequently validate the behaviour pattern. The learning backpropagation algorithm used is a supervised error correction, minimizing the penalized mean square error through the Quasi-Newton method in the free software R.

$$
RMSE = \sqrt{\frac{\sum_{i=1}^{N}\left(Y^{(i)} - f(X^{(i)})\right)^2}{N}}
\tag{8}
$$

As a result, the modification of the Ravdin and Clark type of ANN is shown as an example in Table 1 in comparison with the normal model for two pumps (1 and 2) of the same group (1) of similar covariates and for two homogenized time intervals of failures $\dot{T}_{nkj}$ (one of each pump). In this example, discretized times $t^{(i)}$ are shown in the intervals $\dot{T}_{111}$ and $\dot{T}_{211}$ are shown jointly with the covariate values. Thus, Ravdin and Clark's ANN (R&C ANN) was trained with equal covariate value for all the discretized times in each interval, for example $X_{nk}^{(i)} = \dot{X}_{nk}$, and it was interrupted at each specific time to failure T_{nkj}. While in the modified R&C ANN real data covariates are used $X_{nk}^{(i)}$ and the their average value $\dot{X}_{nk}$ after the T_{nkj} time to failure, but in this case for each homogenized time to failure $\dot{T}_{nkj}$. The original ANN uses a binary system 0 or 1 as survival status and the modified ANN uses estimated probability of failure from a Weibull according to the Eq. (7) (with $\alpha_{111} = 6 \cdot Ln(2)^{1/1.5}$ and $\alpha_{211} = 8 \cdot Ln(2)^{1/1.5}$).

Table 1 Training set of normal and modified R&C ANN

	$t^{(i)}$	Normal R&C ANN			Modified R&C ANN		
		X_{1n1}	X_{1n1}	Survival status	X_{1n1}	X_{1n1}	Survival status, $CDF(t^{(i)})$
Pump$_{11}$ $\dot{T}_{111}$	1	1	1	0	0.7	0.6	0.27
	2	1	1	0	0.8	0.7	0.47
	3	1	1	0	0.7	0.8	0.62
	4	1	1	0	0.9	0.8	0.72
	5	1	1	0	0.8	0.8	0.80
	6	1	1	1	0.8	0.9	0.85
	7	–	–	–	0.8	0.8	0.89
	8	–	–	–	0.8	0.8	0.92
Pump$_{21}$ $\dot{T}_{211}$	1	1	0	0	0.8	0.2	0.21
	2	1	0	0	0.7	0.1	0.38
	3	1	0	0	0.9	0.3	0.51
	4	1	0	0	0.6	0.1	0.62
	5	1	0	0	0.8	0	0.70
	6	1	0	0	0.7	0.1	0.76
	7	1	0	0	0.8	0	0.81
	8	1	0	1	0.9	0.2	0.85

This modification, based on the combination of parametric and AI methods, aims to show how existing information and analysis in the plants, jointly with a monitoring system, may be used to improve decisions, mixing offline statistical models with online real data from remote monitoring systems.

As any ANN, the main weak point in this model is the necessity to adjust the covariate values according to their representative influence in the selected failure in discrete times. That is, depending on their influence in failure degradation with time, it avoids random and bad acquired values but keeps the right data seasonality. Besides this, normalization among different geographical locations is required in order to replicate the analysis. However, once the analysis is accomplished, the developed model can be applied easily in a remote monitoring system, requiring only a model refining each 2 or 3 years, or maybe when operating circumstances change radically. A set of alarms for observed abnormal tendencies may also be implemented (i.e. for q-repair efficiency, warning about tendency of successive repairs to the system status conservation). This could also be used as a warning about the lack of model consistency, i.e. about the need to restart the analysis with new T_{nkj} intervals to capture new repair stages.

In the sequel, the proposed model will now be built and tested in a case study. The idea is also to implement it in a remote monitoring system.

3   Case Study. A Thermal Solar Plant

Thermal Solar plants have been in production for more than 25 years. Current decrease in government incentives for renewable energy sources has forced companies to study useful life extension possibilities. Due to this, potential plant re-investments must also be re-evaluated; incorporating future operating and environmental conditions within equipment reliability analysis.

In these plants the combination of mechanical and thermal stresses makes reliability analysis important. This is not only because the direct costs of failures, but also due to their significant indirect loss of profit, as well as the associated environmental and safety risks [25]. By developing a model for failure prediction we can avoid these risks. This model will be applicable to each critical failure mode, because symptoms and causes may be dissimilar among them and the effect of equipment conditions may apply in a different manner. Understanding the previous point is important; efforts in failure mode analysis will be intense but worthwhile. For instance, defining suitable covariates per failure mode, could add enormous value to protecting our assets and their contribution to the business.

This type of thermal solar plant is usually built modularly; therefore the possibility to replicate the same model for different modules and regions is also considered of great interest. With that in mind, we have tried to develop our ANN model, which is easy to reproduce, and to update it with the most common parameters found in this type of plant.

The solar thermal power plant under consideration has a nominal power of 49.9 MW with an annual production of 180 million KWh and occupies an extension of 2,700,000 square metres. It is located in the southern part of Spain and it will supply energy to more than 100,000 dwellings for an operational time of 25 years. Even in the absence of enough solar radiation, its storage subsystem is able to provide energy for 7.5 h. All the energy produced (180 million KWh/year) is provided by the distributor for 51 million €/year of production (at an initial price of 0.2849 €/KWh, and subsequently reduced due to a legal requirement). The main subsystems of this kind of power plant are, see Fig. 1:

- Solar field. It is composed of 8064 parabolic trough solar collectors with 225,792 mirrors. It heats the high-temperature oil circulated in the HTF loop.
- HTF loop, which conveys high-temperature fluid to the heat exchanger in the steam generator.
- Water loop, where steam flow is condensed and cooled and recirculated as a water flow to the steam generator.
- Steam generator, which transforms water into steam to activate the turbine.
- Turbine, which transforms the mechanical energy of the steam flow into electrical power.

In our case study, we have selected a common system to illustrate the model implementation over real data and in a remote monitoring system: electrical pumps

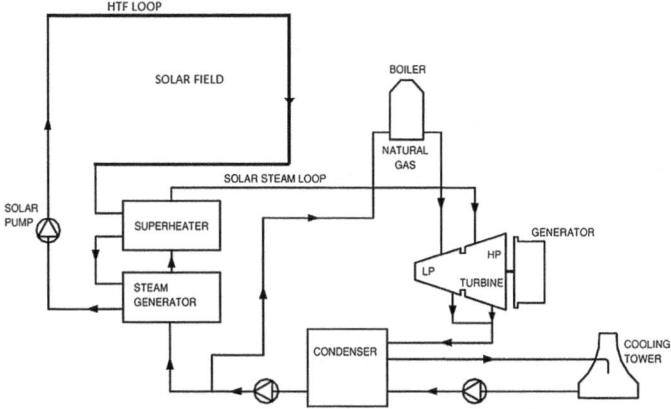

Fig. 1 Thermal solar plant and functional description

in HTF loops. They consist of several large pumps in charge of making the thermal oil (Dowtherm A) to flow throughout the plant. Usually, the tendency is to view HTF pumps in a Thermal Solar Plant as a low potentially hazardous process despite being a heat transfer and under pressure systems that could produce fire and explosion hazards, where leaks can produce a potentially flammable mist or contamination [25]. Moreover, their operation is critical in order to keep the desired availability of the power plant, so all the production is supported by a 2 + 1 HTF pumps configuration (2 of them working in parallel and the other is a spare one). Both active pumps jointly contribute 50% of HTF recirculation oil and a pump failure could reach up to 50% of daily production. Therefore, HTF pumps need surveillance to ensure their efficiency and to control their deterioration. In order to optimize plant efficiency, the remote monitoring system would set the HTF pump speed through changes in the variable frequency drives, according to different temperatures, the direct beam irradiance, pressures, and also the potential fluid density. For example, a bigger difference between inlet–outlet HTF temperatures requires less impelled flow.

For the purpose of this paper, we have selected the failure mode: "damaged mechanical seal", causing significant production losses. This failure mode emerges due to many factors, such as: high seal operating temperature, excessive pump vibration by cavitation, parts misalignment, etc. This problem increases during the summer period when pumps run at full load, at which time production losses are the highest.

Potential mechanical seal failures are predicted using our developed backpropagation neural network, equipped with the last three years pumps' historical data. We focus our attention on failures resulting as a consequence of equipment deterioration due to operational and geographical (environmental) features that could have a great impact on equipment conditions. For instance, we know that extreme fluctuant cycles of inlet–outlet pressure and high temperatures can degrade the oil, producing contamination and corrosion; contraction–expansion may also result in misalignments.

For this case, predicting the problem in real time, using process control variables and with transfer function (thermodynamic approach) was found to be impossible. So, all representative contributions to pump degradation are compound in a single (survival) function which reflects the probability of the failure mode. In this document we show the aptitudes of an ANN to replicate self-adaptive reality by fitting a survival function. This is done in complex and noisy operating conditions.

Our prediction models have innovative features compared to previous works in the literature. The ANN models not only use parametric estimations about the failure times, but also environment variables, such as external humidity, and also assets' condition variables, such as working temperature and different operating times and cycles. In addition, parametric methods are combined with ANN in order to develop a stable model which will be easily and quickly implementable in a remote monitoring system. Through this, an early detection of degradation will be possible before failures affect production, people or the environment, and a quantitative measure of risk can be computed as a percentage.

4 Developing and Implementing the ANN Model

In order to approach the problem of real time condition estimation that could lead to early warnings for the failure mode, the modified ANN architecture is developed based on selected variables from those whose detection is periodically and automatically feasible with our remote monitoring system and they are the most representative showing their effects in the damaged seal of HTF pumps.

Specific information about the developed process is as follows:

- We selected two plants with 8 failures each one.
- The remote monitoring variables for the input layer were:

    - Flow on HTF (l/s).
    - Working Temperature on HTF (°C).
    - Ambient humidity (%).
    - The operation time of the pumps (days).
    - The modelled survival status.
    - The threshold neuron.

- Periods for comparison were selected, to detect the existence of the failure mode and the most representative variables.
- The data was reorganized, eliminating abnormal data that could distort the results. Values were normalized and with the same scale for all the input values to simplify calculations and analysis. Later the normalized values have to be de-normalized before comparison.

- A single hidden layer with nine neurons is used (less than two times the number of the input neurons).

In summary, the implementation of our proposed model is based on the two phases:

1. The estimation, in a first step, of the survival function with a parametric GRP-II Weibull over two groups of the produced time to failures where the covariates are the same, one over the 8 failures of plant 1 and the other over the 8 failures of plant 2. Then, we obtain a characteristic α, β and q for each plant, only based on time to failures (no covariates are used at this level) as shown in Table 2. We take the Weibull Cumulative Distribution Function (CDF), maintaining the β_{nk} in each time interval in each plant (for each group), and taking the $\alpha_{nkj} = T_{nkj} \cdot \mathrm{Ln}(2)^{1/\beta_{nk}}$. As a result for each specific failure, the probabilities to failure ascend from 0 up to 1, next to the maximum failure time. In Table 2, the 16 failures with their time to failure (T_{nkj}) and the modified α_{nkj} with the ponderation are shown for each plant.

2. The modelling, in a second step, of the survival function for each specific failure with adaptation of the parametric estimation according to covariates. For this purpose we have based it on the modified R&C ANN. The discretized time $t^{(i)}$ inside the intervals is selected according to the covariates influence in the degradation of the failure mode. Therefore all the inputs (covariates and CDF) are redefined with this period (notice that this requires the replication of covariates after each specific time to failure with their average value). In our example, for a population of six pumps (three per plant), and sixteen registered failures in three years, after filtering and reorganization, 914 discretized times are trained in the Survival ANN model. Consequently, the data to train and test the GRP-ANN are reorganized (see Eq. (9)) as in Table 3 for plant 1 and failure number 3 ($\alpha_{nkj} = 148.33$, $\beta_{nk} = 3.7$).

Table 2 GRP Weibull parameters and reorganization for parametric estimation of survival function	Plant 1			Plant 2		
	j	T_{nkj}	α_{nkj}	j	T_{nkj}	α_{nkj}
	1	299.53	271.28	1	181.78	167.38
	2	277.76	251.56	2	170.53	157.02
	3	163.78	148.33	3	288.00	265.18
	4	176.22	159.60	4	128.03	117.89
	5	149.21	135.14	5	277.89	255.87
	6	214.71	194.46	6	256.00	235.72
	7	136.59	123.71	7	194.00	178.63
	8	170.90	154.78	8	300.00	276.23
	α_{n1}	220.00		α_{n2}	247.00	
	β_{n1}	3.70		β_{n2}	4.44	

Table 3 Reorganized survival data of failure 3 in plant 1 to train and test the ann

Failure	$t^{(i)}$	$X_{1nk}^{(i)}$, operating hours	$X_{2nk}^{(i)}$, ambient humidity	$X_{3nk}^{(i)}$, working temp	$X_{4nk}^{(i)}$, flow	Normal Ravdin ANN	$X_{5nk}^{(i)}$, modified ANN, $CDF(t^{(i)})$
3	10	10,385	97	291.00	324.42	0	0.00
3	20	10,395	99	301.75	330.27	0	0.00
3	30	10,405	94	322.50	341.57	0	0.00
3	40	10,415	98	301.50	330.14	0	0.01
3	50	10,425	100	306.25	332.72	0	0.02
3	60	10,435	84	307.50	333.40	0	0.03
3	70	10,445	65	299.25	328.91	0	0.06
3	80	10,455	70	318.75	339.53	0	0.10
3	90	10,465	91	307.75	333.54	0	0.15
3	100	10,475	80	321.00	340.75	0	0.21
3	110	10,485	65	327.75	344.43	0	0.28
3	120	10,495	77	330.00	345.65	0	0.37
3	130	10,505	99	304.00	331.50	0	0.46
3	140	10,515	86	326.12	343.54	0	0.55
3	150	10,525	87	324.13	342.46	0	0.65
3	160	10,535	72	326.40	343.69	0	0.73
3	170	10,545	85	313.48	336.66	1	0.81
3	180	10,555	85	313.48	336.66	1	0.87
3	190	10,565	85	313.48	336.66	1	0.92
3	200	10,575	85	313.48	336.66	1	0.95

$$\text{If}\ \left[0 \le t^{(i)} \le T_{nkj}\right] \rightarrow \begin{cases} X_{nk}(i) = \text{real value of vector}\ X_{nk}\ \text{in each}\ t^{(i)} \\ CDF\left(t^{(i)}\right) = 1 - \left[1/\exp\left(\dfrac{t^{(i)}}{\left(T_{nkj}\cdot\text{Ln}(2)^{\frac{1}{\beta_{nk}}}\right)}\right)^{\beta_{nk}}\right] \end{cases}$$

$$\text{If}\ \left[T_{nkj} \le t^{(i)} \le \max[T_{nkj}]\right] \rightarrow \begin{cases} X_{nk}(i) = \dot{X}_{nk}\ \text{of previous}\ t^{(i)}\ \text{to}\ T_{nkj} \\ CDF\left(t^{(i)}\right) = 1 - \left[1/\exp\left(\dfrac{t^{(i)}}{\left(T_{nkj}\cdot\text{Ln}(2)^{1/\beta_{nk}}\right)}\right)^{\beta_{nk}}\right] \end{cases}$$

$$(9)$$

The result, the output of the ANN model is the probability of failure estimation, developed from the GRP-II model, and with covariates affection as roughly proportional to Weibull Survival probability. The ANN analysis done, going through the processes of training, predictions and test, produced the results in Table 4.

The learning algorithm parameters were as follows: (a) maximum number of cycles = 1000, (b) maximum validation failures = 40, (c) min_grad = 1.0e−10,

Table 4 Data set of variables

Variables of vector X_{nk}	Max.	Ref.	Min.	Unit
X_{1nk}, operating hours	20,000	15,000	0	h
X_{2nk}, flow	600	300	55	Kg/s
X_{3nk}, ambient humidity	100	75	25	%
X_{4nk}, temperature	400	300	290	°C
X_{5nk}, survival function	1	0.5	0	

Table 5 Results of training in developed model

Results	Value
MSE training	90.02918
MSE test	335.9361
R^2 training	0.948243
R^2 test	0.8262953

Table 6 Results of training with Ravdin and Clark

Results	Value
MSE training	352.2306
MSE test	492.8387
R^2 training	0.8590198
R^2 test	0.7971156

(d) goal = 0, (e) μ = 0.005, (f) μ_dec = 0.1, (g) μ_inc = 10, (h) λ = 0, (i) min Error = 0.00001833. The results obtained in this case guarantee a good optimization model, as shown in Table 5. MSE (Mean Square Error), in the training and testing, validates the ANN signifying the average distance between the prediction obtained and the real production. Besides that, Table 5 shows the results of the model training process.

Whereas, if we had used the Ravdin and Clark model directly, the results would have been with less accuracy (as Table 6 shows).

In this developed model, R^2 is consistent with this result, explaining 94.8% of the predicted model. Figures 2 and 3 are a representation of deduced predictions. Figure 2 show the training of both Ravdin and Clark ANN and Survival ANN with a dashed line. (2a) in the case of normal R&C ANN, and (2b) in the case of modified R&C ANN. Figure 3 has a straight line to indicate the best approximation for error minimization. For validation purposes, the 25% of historical data is used to estimate the generalization error.

This case study has generated a good prediction of a real failure based on three year of data of it. Using less than three years of data is possible, but to deduce covariate relationships with degradation may be difficult and the seasonal behaviour of some of them would degenerate future predictions. Our recommendation is to employ more than 2 years in the case of environmental influence. However, the number of discretized times in other applications may be less if the covariates are more stable over time.

Fig. 2 **a** Normal R&C ANN training. **b** Modified R&C ANN training. In both graphs *straight lines* are the modelled CDF and *dashed lines* are the predicted CDF by the ANNs

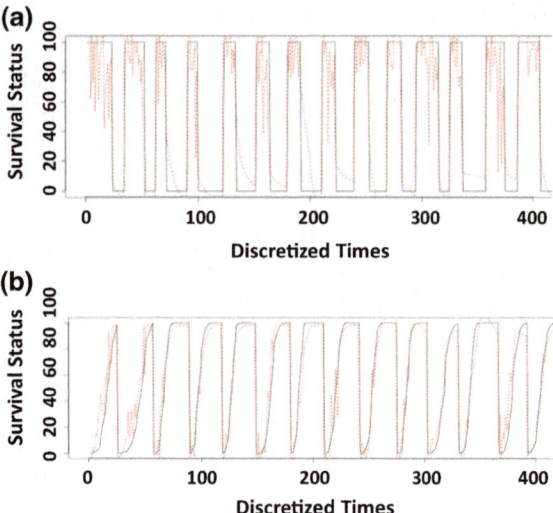

Fig. 3 Modified R&C ANN CDF—$Y'^{(i)}$ predictions versus modelled CDF—$Y^{(i)}$

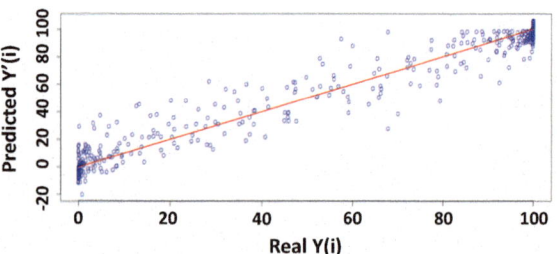

Returning now to the preventive maintenance, our mathematical tool allows one to implement an intelligent preventive maintenance strategy. The strategy is self-adaptive to observed imperfection in repairs and the influence of selected covariates on potential failures. Finally the strategy can to trigger a preventive maintenance action according to two possible business rules:

1. A rule based on a determined level of confidence or failure probability as general reference, Proportion of CDF($t^{(i)}$). In our case study, the level of estimated CDF($t^{(i)}$) which triggers preventive maintenance is 0.6.
2. Another rule based on risk–cost–benefit Analysis. In this case, we can consider not only the failure probability, but also the cost of the possibility to reach its minimum expected value $C\left(t^{(i)}\right) = \dfrac{\text{CDF}(t^{(i)})}{t^{(i)}} \cdot \text{Cost}_{\text{corrective}} + \left[\dfrac{1-\text{CDF}(t^{(i)})}{t^{(i)}}\right] \cdot \text{Cost}_{\text{preventive}}.$

That is, considered the risk of being preventive several steps ahead, and the risk of waiting for the failure (due to corrective unavailability).

For the second business rule, the last step of the methodology is to obtain an online economic estimation of risk, as in Fig. 4. The idea is to determine the optimal interval between preventive actions ($t^{(i)}$) [26] to minimize the total expected cost of the equipment maintenance per unit time. In order to do so, the criteria governing the PM action release is determined by comparing the economic value of risk (for a specific period to be selected) of the following two maintenance strategies:

- Strategy 1: Doing preventive ASAP, This would restore the equipment to a certain condition minimizing the risk of a failure for certain period (in this case study this condition is reached by updating only covariate values to normal equipment operating condition values), but would cost the price of the corresponding preventive maintenance activity (in our case $Cost_{preventive}(t^{(i)}) = 14,500$ €). This calculation is computed online and compared to the risk of:
- Strategy 2: Doing nothing. The economic value of this strategy would be calculated by only computing the online risk of doing only corrective maintenance when failure takes place (in this case with a higher probability than if we follow strategy 1), $Cost_{corrective}(t^{(i)}) = 8000€ + \left(\frac{5822€}{h} \cdot 8h\right) = 90,440€$ (where the average corrective cost of a corrective is considered to be about 8000 € and the indirect cost 82,440 €, estimated as loss of profit 5822 €/h with a $MTTR = 8$ h).

When the online risk of doing nothing (Strategy 2) exceeds the risk of the PM activity (Strategy 1) to a certain extent, then PM maintenance is automatically released and accomplished. This risk exceeding extent is understood here as a company policy. Figure 4 shows this concept. For instance, between $t^{(i)} = 120$ and 140 days would be a moment in time where Strategy 2 risk increases more than the PM Strategy 1 risk, PM would then be scheduled and released. Note how decisions are therefore taken based on strategy probability risk numbers and online.

Finally, the repercussions of the chosen prediction model have to be evaluated with a cost-benefit analysis, prior to their implementation and communication to the entire organization.

Fig. 4 Risk–cost analysis based on expected costs and searching the *right* time to trigger preventive action. **a** *straight line* is the minimum expected cost, **b** *dashed line* is the expected preventive cost, and **c** *dotted line* is the expected corrective cost

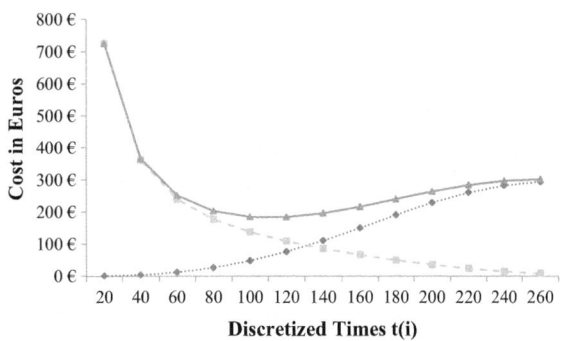

The most vulnerable (and/or sensitive) points of these pumps are mechanical seals. They are responsible for preventing fluid leakage (dangerous fluid at high temperature and pressure). Thanks to this research, the associated risk to "damaged seals" failure mode could be reduced by 247,319 €/plant a year, with an estimated potential impact on the life cycle of the plant (25 years) of 7.24 M€.

5 Conclusions

Thermal Solar Plant managers want to ensure longer profitability periods with more reliable plants. To ensure profitability during the life cycle of the plant we must ensure critical equipment reliability and maximum extension of their life cycle, otherwise failure costs will penalize the expected profit.

Throughout this document, we suggest applying an ANN model per failure mode and we foster a practical implementation in SCADA systems for different plants. This methodology may ease and may improve decision-making and risk modelling, enabling reductions in corrective maintenance direct and indirect costs or allowing the display of residual life until total equipment failure.

In cases when enough data for significant training is available, a better implementation of our methodology will help to reduce the costs and will improve the knowledge of the life cycle of the plant when suffering non-homogeneous operational and environmental conditions.

ANN capacity for self-learning among sources of data (sometimes noised or deprived of communication) thanks to reiterative memory is important. In our case study, we had a vast quantity of data, although sometimes this data was affected by problems of sensor readings or communications. Backpropagation perceptron ANN is recommend for automation developments with real-time utilization. Furthermore, advanced ANN models could be applied when supporting additional variables.

Acknowledgements Part of the funding for this research was provided by the SMARTSOLAR project (OPN—INNPACTO-Ref IPT-2011-1282-920000).

The authors would like to acknowledge the support of the Scientific Chair of MM BinLadin for Operation and Maintenance Technology at Taibah University, Madina, Saudi Arabia.

References

1. Cox DR, Oakes D (1984) Analysis of survival data, vol. 21. CRC Press
2. Smith PJ (2002) Analysis of failure and survival data. CRC Press
3. Hougaard P (2012) Analysis of multivariate survival data. Springer Science and Business Media
4. Lee ET, Wang J (2003) Statistical methods for survival data analysis, vol 476. John Wiley and Sons

5. Ohno-Machado L (2001) Modeling medical prognosis: survival analysis techniques. J Biomed Inform 34(6):428–439
6. Cohen PR, Feigenbaum EA (eds) (2014) The handbook of artificial intelligence, vol 3. Butterworth-Heinemann
7. Kalogirou S (2007) Artificial intelligence in energy and renewable energy systems. Nova Publishers
8. Caner M, Gedik E, Keçebaş A (2011) Investigation on thermal performance calculation of two type solar air collectors using artificial neural network. Expert Syst Appl 38(3): 1668–1674
9. Kuo C (2011) Cost efficiency estimations and the equity returns for the US public solar energy firms in 1990–2008. IMA J Manag Math 22(4):307–321
10. Martín L, Zarzalejo LF, Polo J, Navarro A, Marchante R, Cony M (2010) Prediction of global solar irradiance based on time series analysis: application to solar thermal power plants energy production planning. Sol Energy 84(10):1772–1781
11. Mellit A, Benghanem M, Arab AH, Guessoum A (2005) An adaptive artificial neural network model for sizing stand-alone photovoltaic systems: application for isolated sites in Algeria. Renew Energy 30(10):1501–1524
12. Lapedes A, Farber R (1987) Nonlinear signal processing using neural networks: prediction and system modelling (No. LA-UR-87-2662; CONF-8706130-4)
13. Curry B, Morgan P, Beynon M (2000) Neural networks and flexible approximations. IMA J Manag Math 11(1):19–35
14. Malcolm B, Bruce C, Morgan P (1999) Neural networks and finite-order approximations. IMA J Manag Math 10(3):225–244
15. Rosenblatt F (1958) The perceptron: a probabilistic model for information storage and organization in the brain. Psychol Rev 65(6):386
16. Lawrence S, Giles CL, Tsoi AC (1998) What size neural network gives optimal generalization? Convergence properties of backpropagation
17. Biganzoli E, Boracchi P, Mariani L, Marubini E (1998) Feed forward neural networks for the analysis of censored survival data: a partial logistic regression approach. Stat Med 17(10): 1169–1186
18. Faraggi D, Simon R (1995) A neural network model for survival data. Stat Med 14(1):73–82
19. Liestbl K, Andersen PK, Andersen U (1994) Survival analysis and neural nets. Stat Med 13 (12):1189–1200
20. Ravdin PM, Clark GM (1992) A practical application of neural network analysis for predicting outcome of individual breast cancer patients. Breast Cancer Res Treat 22(3):285–293
21. Xiang A, Lapuerta P, Ryutov A, Buckley J, Azen S (2000) Comparison of the performance of neural network methods and Cox regression for censored survival data. Comput Stat Data Anal 34(2):243–257
22. Cox PR (1972) Life tables. Wiley
23. González-Prida, V, Barberá L, Márquez AC, Fernández JG (2014) Modelling the repair warranty of an industrial asset using a non-homogeneous Poisson process and a general renewal process. IMA J Manag Math dpu002
24. Dagpunar JS (1997) Renewal-type equations for a general repair process. Qual Reliab Eng Int 13(4):235–245
25. Ennis T (2009) Safety in design of thermal fluid heat transfer systems. In: Symposium series, vol 155, pp 162–169
26. Campbell JD, Jardine AK (2001) Maintenance excellence: optimizing equipment life-cycle decisions. CRC Press

Author Biographies

Juan Francisco Gómez Fernández is Ph.D. in Industrial Management and Executive MBA. He is currently part of the Spanish Research & Development Group in Industrial Management of the Seville University and a member in knowledge sharing networks about Dependability and Service Quality. He has authored publications and collaborations in journals, books and conferences, nationally and internationally. In relation to the practical application and experience, he has managed network maintenance and deployment departments in various national distribution network companies, both from private and public sector. He has conduced and participated in engineering and consulting projects for different international companies, related to Information and Communications Technologies, Maintenance and Asset Management, Reliability Assessment, and Outsourcing services in Utilities companies. He has combined his business activity with academic life as a associate professor (PSI) in Seville University, being awarded as Best Thesis and Master Thesis on Dependability by National and International Associations such as EFNSM (European Federation of National Maintenance Societies) and Spanish Association for Quality.

Jesus Ferrero Bermejo is currently Ph.D. candidate in Industrial Engineering, and his main academic education is Statistician and with a Master in Industrial Organization. He is the author of several national and international publications in journals, books, and conferences. He works in Magtel Operations as Outsourcing Manager in the System's Division, leading several projects in relation to TIC sector for public companies. He has participated in different projects of consultancy in national and international companies, with a high technical and managerial specialization in different areas about maintenance and with a wide experience in Photovoltaic Installations. He has collaborated in diverse R+D+i projects for Magtel, being awarded with the Best Record in both of them. He was also awarded with the Best Master Thesis by AEC (Spanish Association for Quality) and AEIPRO (Spanish Engineering and Projects Management Association).

Fernando Agustín Olivencia Polo is Ph.D. in Technology and Engineering, Telecommunications Engineer and International Commerce MBA. He has more than 20 years of experience working for utilities in different sectors: telco, energy, water and railways, and has a sound knowledge of business and operation Information systems for this kind of companies. He has combined his business activity with academic and research life as an associate professor (PSI) for the University of Córdoba, Telefónica I+D and MAGTEL, respectively.

Adolfo Crespo Márquez is currently Full Professor at the School of Engineering of the University of Seville, and Head of the Department of Industrial Management. He holds a Ph.D. in Industrial Engineering from this same University. His research works have been published in journals such as the International Journal of Production Research, International Journal of Production Economics, European Journal of Operations Research, Journal of Purchasing and Supply Management, International Journal of Agile Manufacturing, Omega, Journal of Quality in Maintenance Engineering, Decision Support Systems, Computers in Industry, Reliability Engineering and System Safety, and International Journal of Simulation and Process Modeling, among others. Prof. Crespo is the author of seven books, the last four with Springer-Verlag in 2007, 2010, 2012, and 2014 about maintenance, warranty, and supply chain management. Prof. Crespo leads the Spanish Research Network on Dependability Management and the Spanish Committee for Maintenance Standardization (1995–2003). He also leads a research team related to maintenance and dependability management currently with 5 Ph.D. students and 4 researchers. He has extensively participated in many engineering and consulting projects for different companies, for the Spanish Departments of Defense, Science and Education as well as for the European Commission (IPTS). He is the President of INGEMAN (a National Association for the Development of Maintenance Engineering in Spain) since 2002.

Gonzalo Cerruela García is associate professor in the area of Computer Science and Artificial Intelligence and is attached to the Department of Computing and Numerical Analysis in Córdoba University. Electronic Engineering in 1989 and Ph.D. in Computer Science in 1999, belonging to the research group ISCBD from that same year. He has been developing his teaching and research activities in different universities and research centres, from 1989 to 1995 as author of the Materials and Reagents for Electronics Institute in Havana University, and from 1995 to 1999 as a fellow joined to the Computer Science Faculty in Malaga University. His teaching responsibilities include the disciplines of Databases, Data Structures and Information, Methodology and Programming Technology and Software Engineering. His research interest focus on computational chemistry new solutions, systems support to education, parallel computing, among others.

A Quantitative Graphical Analysis
to Support Maintenance

Luis Barberá Martínez, Adolfo Crespo Márquez,
Pablo Viveros Gunckel and Adolfo Arata Andreani

Abstract This paper proposes a logical support tool for maintenance management decision-making. This tool is called GAMM (Graphical Analysis for Maintenance Management) and it is a method to visualize and analyze equipment dependability data in a graphical form. The method helps for a quick and clear analysis and interpretation of equipment maintenance (corrective and preventive) and operational stoppages. Then, opportunities can be identified to improve both operations and maintenance management (short-medium term) and potential investments (medium-long term). The method allows an easy visualization of parameters like: number of corrective actions between preventive maintenance, accumulation of failures in short periods of time, duration of maintenance activities and sequence of stops of short duration. In addition, this tool allows identifying, a priori, anomalous behavior of equipment, whether derived from its own functioning, maintenance activities, from misuse, or even as a result of equipment designs errors. In this method we use a nonparametric estimator of the reliability function as a basis for the analysis. This estimator takes into account equipment historical data (total or partial) and can provide valuable insights to the analyst even with few available data.

Keywords Management and maintenance optimization · Quantitative graphical method · Graphical data analysis · Efficiency and effectiveness in maintenance

L. Barberá Martínez · A. Crespo Márquez (✉)
Department of Industrial Management, School of Engineering, University of Seville, Camino de los Descubrimientos s/n, 41092 Seville, Spain
e-mail: adolfo@us.es

P. Viveros Gunckel
Department of Industrial Engineering, Universidad Técnica Federico Santa María, Avenida España 1680, Valparaíso, Chile

A. Arata Andreani
Escuela de Ingeniería Industrial, Pontificia Universidad Católica de Valparaíso, Avenida Brasil 2950, Valparaíso, Chile

© Springer International Publishing AG 2018   311
A. Crespo Márquez et al. (eds.), *Advanced Maintenance Modelling for Asset Management*, DOI 10.1007/978-3-319-58045-6_13

1 Introduction

The development and application of graphical tools supporting decision-making in the area of operational reliability [1] is a fundamental task to achieve an accurate and efficient management of assets and resources in an organization, even when there is a large number of devices with functional configuration that is highly complex. To obtain real applications of analytical models, practical, functional, innovative and simple tools must be generated. This will help to make tactical and operational decisions easier.

The management of physical assets of an organization involves processes of innovation and continuous improvement at all levels [2, 3]. Therefore, the need for reliable information to allow appropriate study and analysis of reliability and maintainability is one of the main pillars for decision making at a tactical and operational level. Furthermore, they remain correctly aligned with the vision, strategy and economic indicators of business [4, 5]. Additionally, the development of simple technical tools to facilitate the exercise of analysis and results outcome [6] provides a framework for the control and monitoring of action plans implemented in terms of maintenance activities.

This article proposes a new graphical tool based on data related to the interventions sequence performed to a piece of equipment in during a time horizon. This quantitative graphical analysis method provides easy access to certain variables patterns showing useful information for maintenance management and decision making in the short, medium, and long term.

2 Problem Statement

A typical problem when generating and controlling action plans for improving reliability and maintainability of equipment, is primarily the lack of practical mechanisms to support maintenance management. In particular, there is a need for tools to illustrate, in a clear and simple way, the patterns of deficiencies that can be found in the equipment performance.

Overall indicators of maintenance management [e.g. mean time between failures (MTBF), Mean Time to Repair (MTTR), availability (A), reliability (R), among others] are, besides the consequence or impact of failure, the base to determine the equipment criticality (based on criticality ranking) and to provide system-level guidelines for the allocation of technical and economic resources. However, these indicators do not bring out the performance problems of the actual maintenance, operation, design, etc.

It is therefore necessary to design control tools aimed at improving detection of issues affecting the reliability and maintainability, allowing in turn responding to, for example, the following questions (Table 1).

Table 1 Important questions for maintenance management

1	How many corrective interventions (frequency) are made between planned preventive interventions?
2	Is the execution of maintenance tasks correct?
3	Does the equipment operate properly?
4	What is the trend in the time scale of the interventions?
5	What is the deviation between the optimal execution frequencies of the preventive maintenance?
6	Are the repair times consistent in relation to the amount of work associated with it?
7	Is the number of interventions consistent with the life cycle of the piece of equipment?
8	How do the waiting times of spare parts affect the repair times?
9	Is the maintenance staff training adequate to implement the maintenance strategy for the equipment?
10	What is the impact of outsourcing on the reliability and maintainability of the equipment?

Moreover, usually that the lack of information and its quality difficult the accomplishment of the reliability analysis. For what it was previously commented, the tools design that allows a simple interpretation and analysis, are even more necessary to improve operational performance in the global system, thereby maximizing achievable benefits.

3 Context of the Used Data

For the development of this paper, we use actual data from a process of the extraction and filtration of solids in a wastewater treatment plant located in Chile. In this process, the goal is the elimination of existing solid waste in the fluid using different extraction and filtering systems in function of the size of the solids. Specifically, we used data from a piece of equipment named Grid 1, a specific machine for mid-size residue (Fig. 1).

The grid is powered by three cables, two for moving and one for opening and closing of the rake. In turn, the grid filter is composed of the following elements:

a. Support structure with two side rails that allow the movement of the cleaning rake.
b. Start button which is located a gear motor (which activates the movement cables of the rake), the hydraulic system (which runs the opening and closing thereof), switches the automatic and the rake cleaning system.
c. Mobile cleaning kit.
d. Electric diagram.

A screw conveyor is used to proceed with the removal of waste.

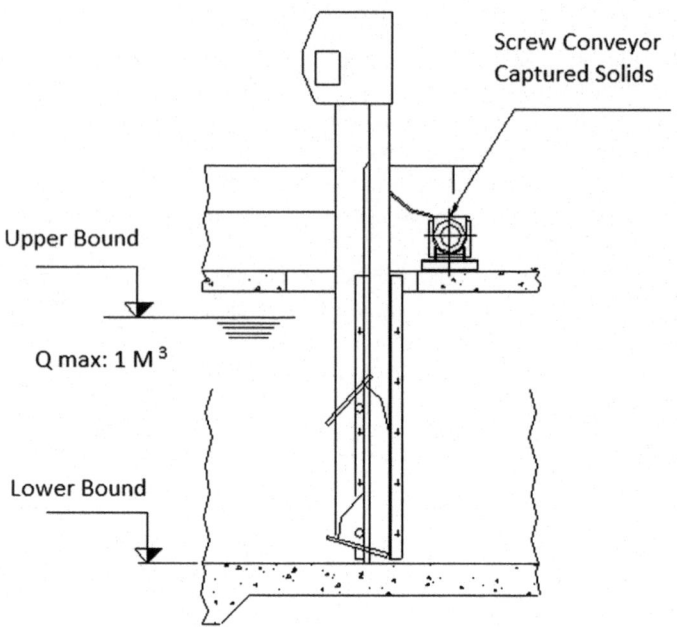

Screw Conveyor
Captured Solids

Upper Bound

Q max: 1 M ^{3}

Lower Bound

Fig. 1 Filtering grid 1 for medium-sized solid residues

4 Dispersion Diagram

The Nelson–Aalen dispersion diagram is a graphical representation of the cumulative number of interventions N versus time t_i. This diagram should be the starting point for any analysis of reliability, as it shows the spatial distribution and trend data versus time, $N(t_i)$ [7]. The construction of the graph may be done on complete or censored a historical data using dispersion diagrams.

Table 2 shows the full real historical data of interventions in a piece of equipment with additional indication of whether the intervention was corrective or preventive, the latter identified with a (*). Preventive interventions are for censored data through the right side, because the equipment is intervened before the failure occurs or corrective intervention, i.e., the range of real-time operation of the equipment is interrupted. Therefore, in the development of this document, all preventive intervention will be considered as a censored data [6].

Considering the data in Table 2 the dispersion diagram can generate the cumulative number of operations $N(t_i)$ versus time t (Fig. 2).

The information provided in the scatter diagram (Fig. 2) refers to the trend of the $N(t_i)$. In particular, with the data used, the diagram shows a function that presents a linear trend, i.e., the time between interventions has the same expected value or is it a stationary system. Additionally, the linear trend of the function allows to assume the possibility of enforcing the i.i.d. hypothesis [7] (time between interventions

Table 2 Historical of interventions 1

N(ti)	Ti [H]	N(ti)	Ti [H]	N(ti)	Ti [H]	N(ti)	Ti [H]	N(ti)	Ti [H]
1*	311	11	4354	21*	6727	31*	8476	41	11,659
2	504	12	4375	22	6790	32*	9031	42*	11,755
3	663	13	4500	23*	6821	33*	9343	43*	12,194
4*	1481	14*	4669	24	7132	34	9646	44*	12,523
5*	2153	15*	4672	25*	7137	35	9816	45*	12,840
6*	2824	16	5052	26*	7499	36	10,029	46*	13,720
7*	3975	17*	5319	27	7735	37*	10,726	47*	14,087
8	4027	18	5689	28	7802	38*	11,207	48	14,103
9	4138	19*	5791	29*	7826	39*	11,211	49*	14,415
10*	4289	20*	6633	30*	8240	40*	11,564	–	–

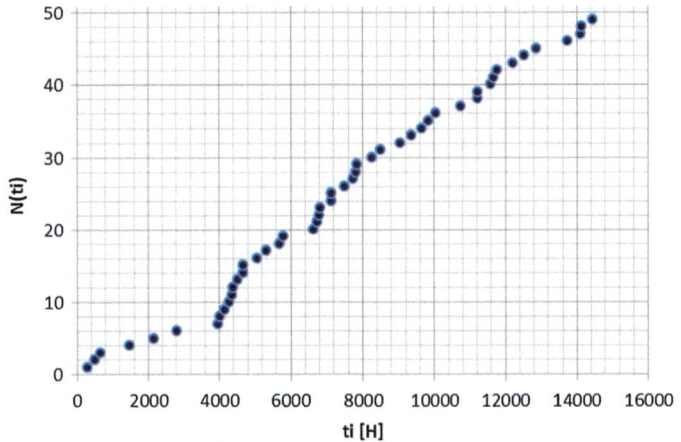

Fig. 2 Dispersion diagram of cumulative number of operations $N(t_i)$

independent and identically distributed), but is not certain. Independent of the result
of the hypothesis, this will determine the use of methods of parametric analysis or
nonparametric analysis.

A parametric-type approach assumes that the function belongs to a particular class of
functions, making it possible to assign a generic probability distribution based on
previous experience or theoretical considerations [7]. By contrast, a nonparametric-type
approach is generally assumed that the function belongs to a group of infinite dimen-
sional function, for example, due to the lack of data. Therefore, the distribution or
generated curve by a nonparametric analysis is particularized to each individual case.

On the other hand, if the curve of the function $N(t_i)$ is concave or convex, the use
of parametric methods based on the performance of the i.i.d. hypothesis would be
inappropriate since, as mentioned, there is a linearity requirement.

In summary, the dispersion diagrams as a starting point of any analysis of reliability provide graphic information only on the trend of the function $N(t_i)$. This determines the linear and nonlinear behavior and finally determines the enforceability of i.i.d. hypothesis. Subsequently, it is not possible to obtain additional relevant information to support and justify decisions at the operational level.

5 GAMM (Graphical Analysis for Maintenance Management)

The Graphical Analysis for Maintenance Management (GAMM) proposed is a quantitative and qualitative analysis method that intends to support decision-making in maintenance management using the logic of a dispersion diagram, stated before, but also integrating the following variables: type of intervention, duration of intervention and state of equipment/system during the intervention. The combination of these new variables with a graphic display of the sequence of interventions generates synergies with regard to the information given by the diagram, thus establishing new sources of analysis.

5.1 Analytical Framework to the GAMM Method

As mentioned and described within the conceptual framework, new variables to consider are added (considering the historical data of interventions illustrated in Table 2): duration of interventions and status of equipment/system at the time to intervene. Consequently, the new history of interventions is shown in Table 3.

Thus, the columns of Table 3 represent.

- $N(t_i)$: cumulative number of interventions. T

  - The data marked with "*" represent preventive interventions.
  - The data without "*" represent the corrective measures.

- t_i: cumulative run time in hours. Is the total time without discounting the interventions times.
- Δt: duration of intervention, expressed in hours.
- Det: state of the system during the intervention.

  - Value 1: the equipment/system does not stop during the intervention.
  - Value 0: The equipment/system is stopped during intervention.

In Table 4, the time of correct functioning of equipment has been calculated (TBF). This parameter is calculated considering the existence or nonexistence of detention of the equipment in its intervention:

Table 3 History of interventions, duration and condition of the equipment intervened

N (ti)	Ti [H]	Δt [H]	Det	N (ti)	Ti [H]	Δt [H]	Det	N (ti)	Ti [H]	Δt [H]	Det	N (ti)	Ti [H]	Δt [H]	Det	N (ti)	Ti [H]	Δt [H]	Det
1*	311	1	1	11	4354	1	0	21*	6727	2	0	31*	8476	1	0	41	11659	1	0
2	504	1	1	12	4375	5	0	22	6790	2	0	32*	9031	1	0	42*	11755	6	0
3	663	1	0	13	4500	5	1	23*	6821	16	1	33*	9343	1	1	43*	12194	1	0
4*	1481	4	1	14*	4669	2	0	24	7132	1	0	34	9646	1	0	44*	12523	1	0
5*	2153	1	1	15*	4672	2	0	25*	7137	4	1	35	9816	25	0	45*	12840	1	0
6*	2824	1	0	16	5052	1	1	26*	7499	1	1	36	10029	5	0	46*	13720	1	0
7*	3975	1	1	17*	5319	4	1	27	7735	1	0	37*	10726	3	0	47*	14087	7	0
8	4027	1	0	18	5689	1	0	28	7802	2	0	38*	11207	1	0	48	14103	1	0
9	4138	20	0	19*	5791	10	0	29*	7826	4	1	39*	11211	2	0	49*	14415	4	0
10*	4289	1	1	20*	6633	1	0	30*	8240	1	0	40*	11564	1	0	-	-	-	-

Table 4 History of interventions with TBF

N (ti)	Ti [H]	TBF [H]	Δt [H]	Det	N (ti)	Ti [H]	TBF [H]	Δt [H]	Det	N (ti)	Ti [H]	TBF [H]	Δt [H]	Det	N (ti)	Ti [H]	TBF [H]	Δt [H]	Det
1*	311	311	1	1	14*	4669	169	2	0	27	7735	236	1	0	40*	11,564	351	1	0
2	504	193	1	1	15*	4672	1	2	0	28	7802	66	2	0	41	11,659	94	1	0
3	663	159	1	0	16	5052	378	1	1	29*	7826	22	4	1	42*	11,755	95	6	0
4*	1481	817		1	17*	5319	267	4	1	30*	8240	414	1	1	43*	12,194	433	1	0
5*	2153	672	1	1	18	5689	370	1	0	31*	8476	235	1	0	44*	12,523	328	1	0
6*	2824	671	1	0	19*	5791	101	10	0	32*	9031	554	1	0	45*	12,840	316	1	0
7*	3975	1150	1	1	20*	6633	832	1	0	33*	9343	311	1	0	46*	13,720	879	1	0
8	4027	52	1	0	21*	6727	93	2	0	34	9646	302	1	0	47*	14,087	366	7	0
9	4138	110	20	0	22	6790	61	2	0	35	9816	169	25	0	48	14,103	9	1	0
10*	4289	131	1	1	23*	6821	29	16	1	36	10,029	188	5	1	49*	14,415	311	4	0
11	4354	65	1	0	24	7132	311	1	0	37*	10,726	692	3	0					
12	4375	20	5	0	25*	7137	4	4	1	38*	11,207	478	1	1					
13	4500	120	5	1	26*	7499	362	1	1	39*	11,211	3	2	1					

- T_i[H] Accumulated time (calendar time in hours).
- TBF$_i$ time of correct functioning (Time between failures).

When the intervention N_{i-1} stops (Det = 0), the calculation is TBF$_i$ = $T_i - T_{i-1} - \Delta t_{i-1}$.

When the intervention N_{i-1} does not stop (Det = 1), the calculation is TBF$_i = T_i - T_{i-1}$.

5.2 Reliability Analysis Based on the GAMM Method

Additionally, GAMM is able to estimate the reliability function, which is calculated using the algorithms based on the nonparametric method of Nelson–Aalen [7] and evaluated in the time t_i, i.e., in the moment before the intervention. The estimation method can consider both the historic full data and censored historical data [6]. The Nelson–Aalen estimator for the cumulative failure rate considers censored data on the right side is expressed in Eq. 1.

$$Z(t) = \sum_v \frac{1}{n - v + 1}, \tag{1}$$

where [7],

- $Z(t)$ is the cumulative failure rate.
- n is the total number of interventions recorded in the history.
- v is equal to the parameter "i" which represents the sequential number of each intervention.

The GAMM estimator for the cumulative failure rate is expressed in Eq. 2

$$Z(ti) = \begin{cases} Z(t_{i-1}) + \frac{1}{n-i+1} & \text{if the event } i \text{ is a failure} \\ Z(t_{i-1}), & \text{otherwise (preventive maintenance)} \end{cases} \tag{2}$$
$$\text{with } Z(t_0) = 0$$

Therefore, the estimator of the reliability function is given by Eq. 3.

$$R(t_i) = e^{-Z(t_i)}, \tag{3}$$

where [7]

- $R(t_i)$ is the reliability function.
- $Z(t_i)$ is the cumulative failure rate.

By organizing the times between interventions TBF, lowest to highest [7], Table 5 shows the estimates calculated using Eqs. 2 and 3.

Table 5 Estimates of cumulative failure rate and reliability function

i	TBF [H]	$1/(n-i+1)$	$Z(t_i)$	$R(t_i)$
1*	1	0.02	0	1
2*	3	0.021	0	1
3*	4	0.021	0	1
4	9	0.022	0.022	0.978
5	20	0.022	0.044	0.957
6*	22	0.023	0.044	0.957
7*	29	0.023	0.044	0.957
8	52	0.024	0.068	0.934
9	61	0.024	0.092	0.912
10	65	0.025	0.117	0.889
11	66	0.026	0.143	0.867
12*	93	0.026	0.143	0.867
13	94	0.027	0.17	0.844
14*	95	0.028	0.17	0.844
15*	101	0.029	0.17	0.844
16	110	0.029	0.199	0.819
17	120	0.03	0.23	0.795
18*	131	0.031	0.23	0.795
19	159	0.032	0.262	0.77
20*	169	0.033	0.262	0.77
21	169	0.034	0.296	0.744
22	188	0.036	0.332	0.717
23	193	0.037	0.369	0.691
24*	235	0.038	0.369	0.691
25	236	0.04	0.409	0.664
26*	267	0.042	0.409	0.664
27	302	0043	0.453	0.636
28	311	0.045	0.498	0.608
29*	311	0.048	0.498	0.608
30*	311	0.05	0.498	0.608
31*	311	0.053	0.498	0.608
32*	316	0.056	0.498	0.608
33*	328	0.059	0.498	0.608
34*	351	0.063	0.498	0.608
35*	362	0.067	0.498	0.608
36*	366	0.071	0.498	0.608
37	370	0.077	0.575	0.563
38	378	0.083	0.581	0.559
39*	414	0.091	0.581	0.559
40*	433	0.1	0.581	0.559
41*	478	0.111	0.581	0.559
42*	554	0.125	0.581	0.559
43*	671	0.143	0.581	0.559
44*	672	0.167	0.581	0.559
45*	692	0.2	0.581	0.559
46*	817	0.25	0.581	0.559
47*	832	0.333	0.581	0.559
48*	879	0.5	0.581	0.559
49*	1150	1	0.581	0.559

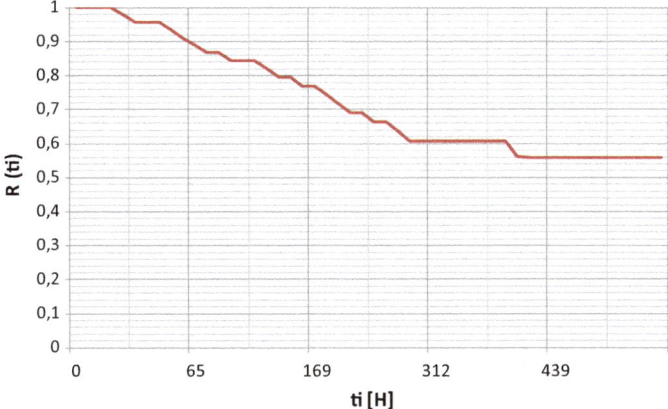

Fig. 3 Reliability function $R(ti)$

It can be assumed that when a unit is subjected to a maintenance operation, it retrieves the highest reliability comparable to that in the time of acquisition, i.e., a reliability $R(t)$ equals 1 [8]. With all this information, the reliability function shown in Fig. 3 is obtained. Thus, from the graph, the reliability can be acquired at the moment immediately before the intervention.

The graphic (Fig. 3) shows a decrease in the reliability of equipment in relation to time of correct functioning, i.e., the reliability decreases as the equipment functioning time increases, reaching very low levels of reliability, close to 55%.

5.3 Graphical Analysis Based on the GAMM Method

Considering the new variables, an algorithm is programmed to enable a quantitative and qualitative graphic analysis, which replaces the dispersion diagram (Fig. 2) for a bubble chart (Fig. 4) with the following considerations:

- The dispersion of the bubbles represents the same function (dot to dot) as the dispersion diagram of the cumulative number of interventions $N(ti)$ proposed in Fig. 2.
- The size (diameter) of the bubbles plotted represents the duration of the intervention.
- The color of the bubbles represents the type of maintenance performed on equipment: white (preventive maintenance) and gray (corrective maintenance).
- The edge or boundary line of each bubble represents the state of the equipment/system during intervention. Thus, if the bubble has a hard shaded

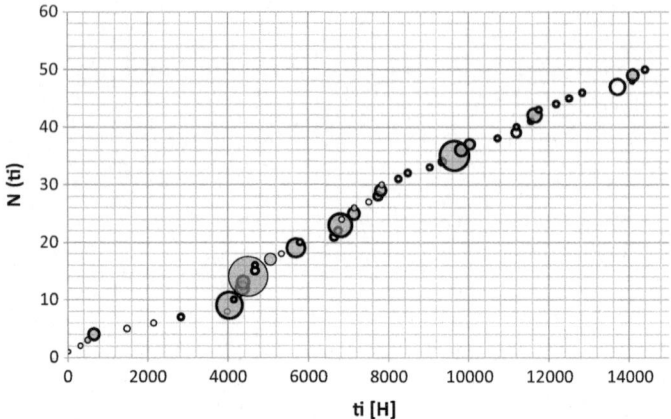

Fig. 4 Graphical analysis for maintenance management (GAMM). Graphic 1

contour line, the value of the variable "Det" is 0, i.e., the intervention involves a stop of the equipment/system. However, if the bubble has a thin contour line, the value of the variable "Det" is 1, i.e., the intervention is performed without stopping the machine.

Finally, we construct the first graphic of the Graphical Analysis for Maintenance Management (GAMM), considering all the features described above. The proposed graphical method is presented in Fig. 4.

Noting the dispersion of the bubbles, which represent the accumulated number of interventions $N(t_i)$, there is a clear upward trend with a sharp increase in the number of interventions depending of the time. Also, there are a large number of corrective interventions (non programmed interventions, in gray color) clearly important, because they are mostly of long duration involving, in most cases, the stop of the system.

The following graphic (Fig. 5), provides further information, from the graph in Fig. 4, to show the reliability of the equipment in the instant before the intervention, either corrective or preventive, from Eq. 2. Thus, we can correlate each of the interventions with the existing level of reliability in the equipment prior to it. This way, through the quantitative and qualitative analysis, based on the lecture and interpretation of the results of both charts, relevant information to aid decision-making in the global maintenance management can be obtained.

In the diagram of Fig. 5 it shows values of reliability (with great variation between them) versus time. In addition, it appears that the high numbers of corrective interventions are made mainly in periods of high reliability of the equipment. These two facts are particularly relevant for further analysis due to the following:

- A correct maintenance management should ensure that the value of the reliability of the system or maintained equipment is close to a value (close to 1) more

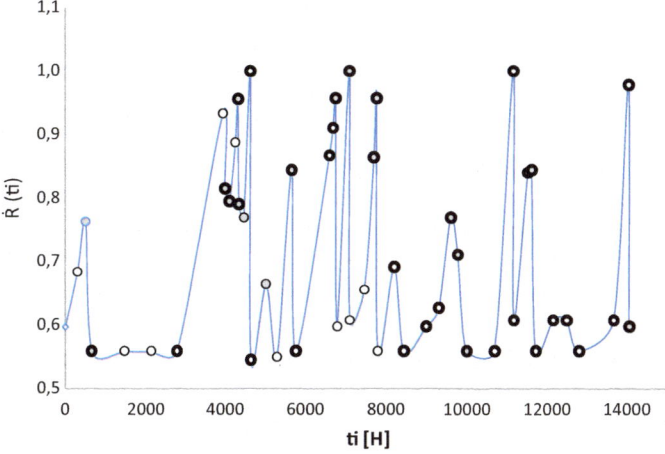

Fig. 5 GAMM, Graphic 2. Estimated reliability diagram of the equipment at the moment of its intervention

or less constant over time. In practice, the reliability values vary with time, but they should not do so in sudden or large fluctuations.

- Preventive maintenance interventions are carried out in periods of high reliability of the equipment, which is why the aim of these is to prevent the value of the reliability to decrease. On the other hand, the corrective interventions (in its case) are usually carried out in periods of low reliability of equipment (achieved by an ineffective preventive maintenance program). This low reliability is the reason for the occurrence of a failure event which forces to perform a corrective maintenance intervention.

One of the possible causes of the observed variability, in the value of reliability, may be due to inadequate planning of a preventive maintenance based on the evolution of the reliability of the equipment itself. Carrying out maintenance that is preventive, unscheduled and not optimized causes the equipment reliability to not be close to an appropriate and constant value in the timescale. This hypothesis is reinforced on the one hand, noting the existence of preventive interventions during periods of low reliability of the equipment (Fig. 5) and, secondly, noting the absence of a consistent temporal pattern of time between preventive interventions, as a result of advanced planning (Fig. 4).

There are also other reasons that add dispersions in the values of the reliability of equipment, for example, the corrective interventions. Such interventions are not programmable, although they can be minimized substantially with an optimized preventive maintenance program.

Moreover, the existence of a large number of corrective interventions, many of them high impact (duration and stop of the equipment, Fig. 4) during periods of high reliability of equipment, may be due to poor operation of equipment by staff responsible for its operation.

The GAMM method [joint analysis of both graphics (Figs. 4 and 5)] offers practical and relevant information about the effectiveness and efficiency of current maintenance performed to equipment or analyzed system. Thus, it is quantitative and qualitative analysis shows important aspects such as: number of procedures performed in the timeline, type of intervention performed (preventive or corrective), duration of it, whether or not stopping the equipment during the intervention, reliability of equipment prior to the intervention or variability of the reliability of equipment in terms of time. Also, it is important to mention that this method requires some level of experience in the maintenance area, because there is some implicit qualitative analysis that will depend directly about the interpretation of the quantitative results.

5.4 Contributions and Improvements Obtained that Support Decision Making for Optimal Maintenance Management

The proposed method (GAMM) allows viewing, from a complete or censored historical record of maintenance interventions performed on the system or equipment analyzed, different patterns of analysis that provide useful information to assist decision-making at the operational level. Thus, the patterns identified are:

- *Trends in the behavior of the interventions*: the GAMM allows viewing the trend of the function of the cumulative number of interventions versus calendar time. The approximate behavior of the function $N(t_i)$: linear, concave or convex, it will determine if the equipment/system is in stationary or non stationary operation stage. This shows the temporal distribution of time between interventions. For a stationary system, the function $N(t_i)$ is linear, the time between interventions are distributed according to a determined expected value and for systems whose function $N(t_i)$ is non linear, concave or convex, the time between interventions tend to increase or decrease depending on ti respectively. Thus, the behavior of the function $N(t_i)$ provides information about the enforceability of the i.i.d. hypothesis, which eventually involves the use of parametric or non-parametric methods of analysis [7].
- *Preventive maintenance deviation*: as a first contribution to the operational maintenance management that can be extracted from GAMM, is the relative information about possible deviations of preventive maintenance. Adding variable analysis as the type of intervention can display the frequency of completion of preventive or programmed maintenance (white bubble) and, therefore, can track and monitor the attainment of the established preventive maintenance frequencies in the program.

- *Quality of operations and/or preventive maintenance*: a graphic display of interventions by type (preventive or corrective) provides the sequence recognition of certain patterns that can define the quality of operation and/or the performing preventive maintenance activities. This way, by observing corrective interventions within preventive ones, sequences of corrective short term stoppages and accumulation of corrective interventions immediately after a preventive maintenance intervention; it is evident the existence of problems in the execution of the duties or how the equipment/systems operates. This provides a knowledge base to identify opportunities for improvement, but does not determine exact causes. Therefore it should be taken into account in the formulation of hypotheses factors such as equipment/system in the continuing stage, low-quality parts, design problems, etc.
- *Reliability Function*: The GAMM shows the reliability function of the equipment/system calculated from the estimates. The Graphic 2 (reliability diagram, Fig. 5) shows the reliability of the equipment/system at pre-intervention, complementing the analysis of the quality of operation and/or preventive maintenance. This demonstrates, for example, the existence of corrective measures in periods of high levels of reliability. Corrective intervention at a high level of reliability indicates abnormal premature failure, which may be motivated by different causes: problems in the quality of duties and/or improper operation of equipment by the staff, poor quality of parts used in preventive maintenance, design issues, among others. Finally, the value of the reliability of the system or maintained equipment should approach a value (close to 1) more or less constant over time, demonstrating adequate frequencies of preventive maintenance with preventive interventions to a predetermined level of reliability and corrective interventions with a low frequency of occurrence.
- *Efficiency and quality in the implementation of interventions*: the duration of each intervention is represented by the size of the bubble in GAMM (Graphic 1). This variable identifies those interventions that are beyond the average times of intervention, questioning if the schedule of the workload is right or the deviation in execution times is due to external factors such as lack of tools, waiting for parts, lack of staff training, etc.
- *Impact on production*: the stop variable (Det) determines the status of equipment/system at the time to intervene. The bubbles with hard shaded border represent interventions in which the equipment/system stops, affecting their individual availability, and possibly affecting the overall system availability. Analyzing the Graphic 1 of the GAMM (Fig. 4) it demonstrates the impact of the intervention on the availability of equipment, i.e., if the equipment/system loses its functionality during the intervention, either preventive or corrective.
- *Opportunistic Maintenance Management*: "Opportunistic maintenance" or "Convenience Maintenance" refers to the situation in which preventive maintenance is carried out at opportunities [9]. For this analysis, a typical example is when one component is out for maintenance and it is decided to take out another

component for maintenance ahead of the maintenance plan, since it is considered to be rational taken in to account the information about the level of reliability, interpreted as the expected remaining life too, provided by the Graphic 2 in Fig. 5.

Finally, the Graphical Analysis for Maintenance Management (GAMM) proposed is a quantitative and qualitative analysis that intends to support the decision-making in the overall maintenance management helps to reduce the high costs of unavailability of the system or the maintained equipment, as well as the costs associated with unforeseen maintenance interventions. The combination of analysis patterns presented, provides valuable information to implement action plans to improve the global maintenance management. This specifically included the performance of equipment/systems and, ultimately, to achieve a management that is efficient, effective and opportunistic, identifying gaps to help determine the root causes thereof.

6 Conclusions

In relation to the stated objectives, it can be concluded that Graphical Analysis for Maintenance Management (GAMM) is a support tool for operational maintenance management of short and medium term. The tool provides useful information regarding the reliability and maintainability of systems or equipment for analysis when considering variables such as type of intervention, duration of intervention or the existence of a stop during the intervention of the equipment/system. This broadens the spectrum display for maintenance management, acquiring new evaluation parameters and graphically determining possible areas of improvement.

The graphics of GAMM (Figs. 4 and 5) illustrates aspects such as: trend of the interventions, deviation in the frequency of preventive interventions, reliability function, efficiency of maintenance operations, impact by unavailability, quality in the intervention performance, quality in use and operation of equipment by staff responsible of it, among others. However, possible variables such as: continuing equipment, poor design, lack of staff training, etc., should be considered in formulating of hypotheses for the search of possible causes.

According to Table 1, important questions for maintenance management can be answered partial or in a complete way within the presented GAMM method. For example:

- Question 1 about "the number of intervention between planned preventive interventions" can be answered analyzing the graphics presented in Fig. 4 and 5. It is a simple counting process.
- Question 2 and 3 about "if the execution of maintenance and property operation were correct" can be preliminary answered by analyzing the time between failure TBF_i and the reliability $R(ti)$ of the asset at the moment of the

intervention. The main idea is to identify if those indicator are normal (close to the average of TBF and with an acceptable reliability level).

- Question 4 about "the existence of trend in the time scale of the intervention" can be identified through a visual analysis of the graphs presented in Figs. 2 and 4.

In general, the others questions could be partially answered by the analysis of the GAMM graphics and some indicators related to reliability and maintainability features. For successful results should be strictly necessary the experience and advanced knowledge about maintenance management process.

Thus, the tool proposed provides information that is useful, clear and easy to interpret. This way it can show quantitative and qualitative maintenance implemented activities and its operational implications in the reliability and maintainability of the equipment within the maintenance management process.

With regard to the applicability of this method, it is necessary to clarify that a relatively simple database it is required for the diagrams elaboration. Also, considering all the variables involved: cumulative time of intervention, duration of intervention and state of the equipment/system during intervention. To process information, common calculus tools can be used, such as VBA programming (spreadsheet) that generates a simple algorithm which allows the elaboration of graphics that are part of the support tool presented.

References

1. Birolini A (2007) Reliability Engineering, 5th edn. Springer, Berlin. ISBN 978-3-540-49388-4
2. Crespo Márquez A, Gupta J (2006) Contemporary maintenance management. Process, framework and supporting pillars. Omega. Int J Manage Sci 34(3):325–338
3. Crespo Márquez A, Moreu P, Gómez J, Parra C, González Díaz V (2009) The maintenance management framework: A practical view to maintenance management. Taylor & Francis Group, London. ISBN 978-0-415-48513-5
4. Arata A (2009) Ingeniería y gestión de la confiabilidad operacional en plantas industriales. 1 Ed. Santiago, Ril editores
5. Crespo Marquez A (2007) The maintenance management framework: models and methods for complex systems maintenance. Springer, London. ISBN 9781846288203
6. Nelson W (1969) Hazard plotting for incomplete failure data. J Qual Technol 1:27–52
7. Rausand M, Hoyland A (2004) System reliability theory: models, statistical methods, and applications. Wiley-Interscience, USA, 644 p
8. Yañez M, Joglar F, Modarres M (2002) Generalized renewal process for analysis of repairable systems with limited failure experience. Reliab Eng Syst Saf 77:167–180
9. Andréasson, N (2004) Optimization of opportunistic replacement activities in deterministic and stochastic multi-component systems, Licentiate thesis Chalmers University of Technology and Göteborg University, Göteborg, Department of Mathematical Sciences, ISSN 0347-2809

Author Biographies

Luis Barberá Martínez is Ph.D. (Summa Cum Laude) in industrial Engineering by the University of Seville, and Mining Engineer by UPC (Spain). He is a researcher at the School of Engineering of the University of Seville and author of more than 50 articles, national and main internationals. He has worked at different international Universities: Politecnico di Milano (Italy), CRAN (France), UTFSM (Chile), C-MORE (Canada), EPFL (Switzerland), University of Salford (UK) or FIR (Germany), among others. His line of research is industrial asset management, maintenance optimization and risk management. Currently, he is Spain Operations Manager at MAXAM Europe. He has three Master's degrees: Master of Industrial Organization and Management (2008/2009) (School of Engineering, University of Seville), Master of Economy (2008/2009) (University of Seville) and Master of Prevention Risk Work (2005/2006). He has been honored with the following awards and recognitions: Extraordinary Prize of Doctorate by the University of Seville; three consecutive awards for Academic Engineering Performance in Spain by AMIC. (2004/2005), (2005/2006) and (2006/2007); honor diploma as a "Ten Competences Student" by the University of Huelva and CEPSA Company (2007); graduated as number one of his class.

Adolfo Crespo Márquez is currently Full Professor at the School of Engineering of the University of Seville, and Head of the Department of Industrial Management. He holds a Ph.D. in Industrial Engineering from this same University. His research works have been published in journals such as the International Journal of Production Research, International Journal of Production Economics, European Journal of Operations Research, Journal of Purchasing and Supply Management, International Journal of Agile Manufacturing, Omega, Journal of Quality in Maintenance Engineering, Decision Support Systems, Computers in Industry, Reliability Engineering and System Safety, and International Journal of Simulation and Process Modeling, among others. Prof. Crespo is the author of seven books, the last four with Springer-Verlag in 2007, 2010, 2012, and 2014 about maintenance, warranty, and supply chain management. Professor Crespo leads the Spanish Research Network on Dependability Management and the Spanish Committee for Maintenance Standardization (1995–2003). He also leads a research team related to maintenance and dependability management currently with five Ph.D. students and four researchers. He has extensively participated in many engineering and consulting projects for different companies, for the Spanish Departments of Defense, Science and Education as well as for the European Commission (IPTS). He is the President of INGEMAN (a National Association for the Development of Maintenance Engineering in Spain) since 2002.

Pablo Viveros Gunckel is Researcher and Academic at Technical University Federico Santa María, Chile, has been active in national and international research, both for important journals and conferences. Also he has developed projects in the Chilean industry. Consultant specialist in the area of Reliability, Asset Management, System Modeling and Evaluation of Engineering Projects. He is Industrial Engineer and Master in Asset Management and Maintenance.

Adolfo Arata Andreani is Professor at the Pontificia Universidad Católica-Valparaíso (Chile) and Visiting Professor in graduate programs at the Universidad de Chile and the Universidad Austral (Argentina). He has been Professor in graduate programs at several universities in Chile and abroad, such as the Politecnico di Milano (Italy), and Full Professor at the Universidad Santa María (Chile), of which he was President, Dean of Engineering and Member of the Board of Directors. He has also been researcher at the Enea Research Center-Ispra (European Union) and at the Politecnico di Milano (Italy). He has participated in research in different programs, like the Cyted (Spain), Università Bocconi (Italy), Altas Corporation (USA) and Conicyt (Chile). He is a member of the board of Millennium Science Initiative (ICM) (Chile). He was Referee de la Agenzia Nazionale di Valutazione dello Sistema Universitario e della Ricerca (ANVUR) (Italy), and member of the National Accreditation Commission of High Education (CNA) (Chile). He is a Director of CGS, applied research center with many activities in large companies worldwide. He was part of a team of consultants in RDA and Segesta (Italy) and was President of the Valparaiso region Chamber of Commerce and Production. He is Mechanical Engineer with a Diploma in Industrial Engineering and Dr.-Ing. (Ph.D.)

Case Study of Graphical Analysis for Maintenance Management

Luis Barberá Martínez, Adolfo Crespo Márquez, Pablo Viveros Gunckel and Raúl Stegmaier

Abstract This paper presents a case for practical application of the GAMM method, which has been developed and published by the authors (Barberá L., Crespo A. and Viveros P). The GAMM method supports decision-making in the overall maintenance management through the visualization and graphical analysis of data. In addition, it allows for the identification of anomalous behavior in the equipment analyzed, whether derived from its own operations, maintenance activities, improper use of equipment or even as a result of design errors in the equipment itself. As a basis for analysis, the GAMM method uses a nonparametric estimator of the reliability function using all historical data or, alternatively, part of the history, allowing it to perform an analysis even with limited available data. In the case study developed, GAMM has been used to analyze two slurry pumps in a mining plant located in Chile. Both pumps are part of the same industrial process, which is described in Sect. 3, and both pumps had a higher failure rate but one more than the other. GAMM identified deficiencies in each of the pumps being studied, thus improving decision-making and problem solving process related to the maintenance of the pumps. Particularly, this work initially provides a description of the GAMM method (Sect. 1), and, afterwards, it is depicted with special attention the approach to the problem (Sect. 2). In Sect. 3, a background of the industrial context is presented. Then, Sect. 4 shows step by step the application of GAMM Method. Finally, results and conclusions are presented in Sect. 5 where the main improvements obtained are summarized.

Keywords Mining industry · Maintenance optimization · Slurry pumps · Graphical data analysis · Efficiency and effectiveness in maintenance

L. Barberá Martínez (✉) · A. Crespo Márquez
Department of Industrial Management, School of Engineering, University of Seville, Camino de los Descubrimientos s/n, 41092 Seville, Spain
e-mail: lubarmar@us.es

P. Viveros Gunckel · R. Stegmaier
Department of Industrial Engineering, Universidad Técnica Federico Santa María, Avenida España 1680, Valparaíso, Chile

© Springer International Publishing AG 2018
A. Crespo Márquez et al. (eds.), *Advanced Maintenance Modelling for Asset Management*, DOI 10.1007/978-3-319-58045-6_14

331

1   Introduction. GAMM Method

The development and application of graphic tools that give support and make decisions in the field of operational reliability [1], is a fundamental task for the proper and efficient management of assets and resources within an organization, even more so with various types of equipment whose functional configuration is highly complex [2, 3].

The GAMM method [4] is a graphical tool based on collected data related to the sequence of technical equipment revisions. This method of graphic analysis uses the logic of a dispersion diagram and adds variables with useful information that allows the management to visualize support patterns and make decisions in the overall maintenance management, such as: number of technical revisions performed, type (corrective or preventative), duration, operational failure during the routine service, reliability of the equipment before the service, or the variability of equipment failure over a period of time. Moreover, GAMM provides a visual representation from a complete or partial historical record of the maintenance work performed, showing different patterns of analysis that provide useful information for decision-making and problem solving [4].

GAMM Method offers practical and relevant information about the effectiveness and efficiency of current maintenance performed to equipment or analyzed system. It is important to mention that this method requires some level of experience in the maintenance area, because there is some implicit qualitative analysis that will depend directly about the interpretation of the quantitative results.

With regard to the applicability of this method, it is necessary to clarify that a relatively simple database it is required for the diagrams elaboration. GAMM could be applied to any system, considering all variables/data involved and required: cumulative time of intervention, duration of intervention and state of the equipment/system during intervention. To process information, common calculus tools can be used, such as VBA programming (spreadsheet) that generates a simple algorithm which allows the elaboration of graphics that are part of the support tool presented.

2   Approach to the Problem

This paper presents a real application of GAMM Method. In the case study developed, GAMM was used to analyze existing equipment in a mining plant located in Chile.

The detected problem involved two slurry pumps, which form part of the industrial process in the mining plant. Both pumps are part of the same industrial process, which is described in Sect. 3, and both pumps had a higher failure rate but one more than the other. The initial objective of this case study was to determine the root cause of that higher failure rate of these two pumps, to improve the current situation. As will be explained later, the application of GAMM has allowed for the detection of the root causes of existing problems and identification of potential improvements, helping to adopt optimal solutions for the problem presented related to the maintenance of the pumps.

To do this, GAMM method was used considering all available historical data from the database of mining plant itself. GAMM facilitated the process of identifying the root causes of problems and subsequently took the necessary measures to solve them.

3   Background of the Industrial Context

In mining industry, comminution is a process that is used to reduce the size of an extracted material; it is performed to separate two minerals from each other, or to achieve the optimal size for manipulation of a material. A mineral extracted from a mine inevitably has a wide particle size distribution, in such a situation, the objective of comminution is to reduce the size of the larger fragments down to a uniform small size. Crushing and grinding are the two primary comminution processes.

Crushing is normally carried out on 'run-of-mine' ore, whereas grinding (normally carried out after crushing) may be conducted on dry or slurried material. For the development of the case study, we will use information from a milling plant copper concentrate located in northern Chile, focusing the analysis in the pumping process flow "Slurry" (water + concentrate copper + inert material) to a continuous process of flotation. Specifically, the mining process under consideration in this paper is the transport of slurry from the buffer to hydrocyclones, which are the previous stage before flotation process. This paper was developed using real information pertaining of this process, which constitutes one of the various operations necessary for the ore purification. In the practical development of this article, two pumps are considered (Fig. 1), specifically these pumps are P1 and P2.

The logic of the processes can be understood by Fig. 1, which shows schematically the main stages and equipments of the selected process.

The first process corresponds to the grinding process, which is developed by the main equipment "SAG Mill", which feeds the accumulation and filtering system "Buffer" with a level of mineral grain size less than ½ in. Subsequently, the "Slurry Pump" process each provides 50% of the expected production, which leads to a corresponding logic configuration of 50%/50% load sharing. The pumped flow "Slurry" is provided by the accumulator system "Buffer". Each "Slurry Pump" feed independently the equipment "Hydrocyclone" which has two functions: the first is to transfer about 90% of the flow to the next process "flotation" and the other 10% it filters and recirculates to the "Ball Mill" which corresponds to the second comminution process, before being sent back to the accumulator "Buffer".

This continuous process has as a security mechanism, an alternative to grinding second recirculation line, which is activated in case of failure or the ball mill

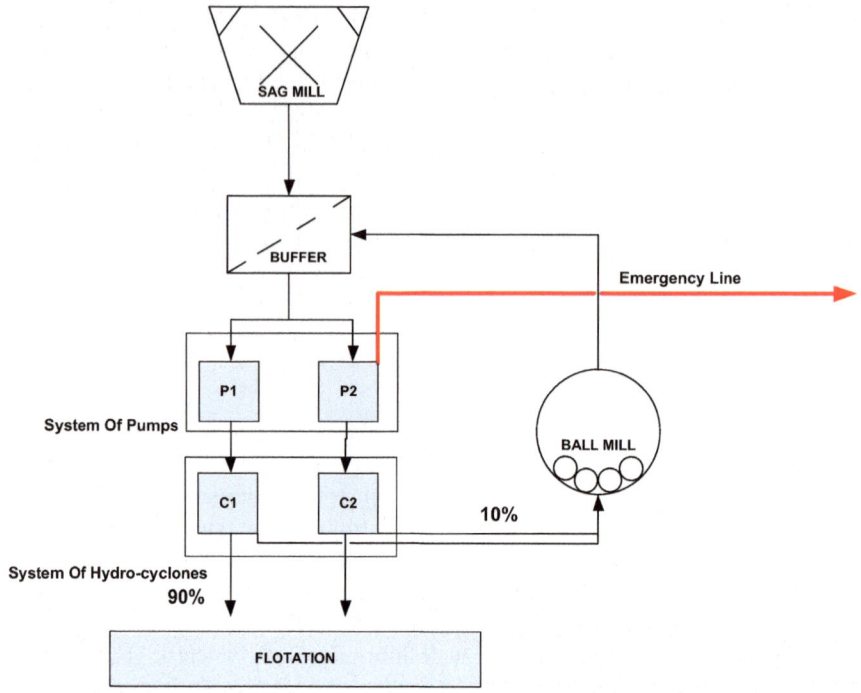

Fig. 1 Process diagram. Slurry pumps location in the mining process

flotation process "down stream". The security system is a special flow line (by pass) from the pump "P2" to another parallel line grinding.

The case study and application of GAMM is focused on the slurry pump process (P1 and P2), which are essential for the continuity of the process (critical), and are exposed to a high stress job, especially by the transfer characteristics of highly abrasive and corrosive flow, and high solids content and strong acidity.

A slurry pump is a rugged heavy duty pump intended for aggressive or abrasive slurry solutions typically found in the mining industry with particles of various sizes. It achieves this by lining the inside of the pump casing as well as the impeller with rubber. Although rubber does eventually wear, the elasticity of its surface allows the hard mineral particles to bounce off thereby reducing what would be otherwise very aggressive erosion. These pumps are used wherever abrasive slurries need to be pumped, especially in the mining industry. The requirement for these types of pumps is typically higher than comparative standard centrifugal pumps. These pumps increase the pressure of liquid and solid particle mixture (aka slurry), through centrifugal force (a rotating impeller) and converts electrical energy into slurry potential and kinetic energy (Fig. 2).

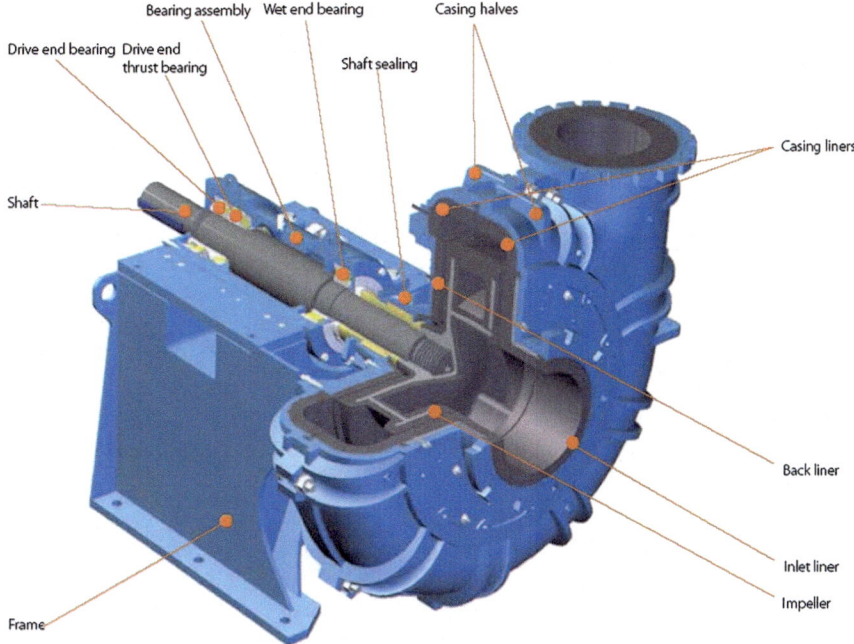

Fig. 2 General scheme of slurry pumps

Fig. 3 Notation of GAMM
method

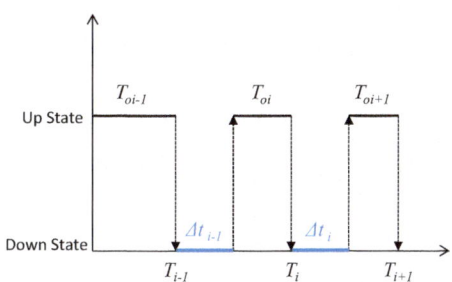

Slurry pumps are designed to allow the passage of abrasive particles which can be extremely large. Therefore, these pumps need much wider and heavier impellers to accommodate the passage of large particles. Also, they are constructed in special materials to withstand the internal wear caused by the solids.

4   Application of GAMM Method

In [4], GAMM Method is explained in detail, however, a summary of steps of this method is provided below (Fig. 3).

1. *Historical data required are*:

- $N(t_i)$: cumulative number of interventions (preventive and corrective interventions).

$$N(t_i) = i \qquad (1)$$

- t_i: cumulative run time in hours. It is the total time without discounting the interventions times.

$$t_i = \sum_{j=1}^{i} \text{To}_j \qquad (2)$$

- Δt: duration of intervention, expressed in hours.
- Det: state of the equipment/system during the intervention (stopped or not).
- T_i: Accumulated time (calendar time in hours).

$$T_i = \sum_{j=1}^{i} \text{To}_j + \Delta t_j \qquad (3)$$

- T_{0i}: Operating time between interventions, expressed in hours.

When the intervention N_{i-1} stops (Det = 0), the calculation is $T_{0i} = T_i - T_i - 1 - \Delta t_{i-1}$.

When the intervention N_{i-1} does not stop (Det = 1), the calculation is $T_{0i} = T_i - T_{i-1}$.

2. *Dispersion diagram*

The dispersion diagram [5] is a graphical representation of the cumulative number of interventions $N(t_i)$ versus time t_i. This diagram should be the starting point for any analysis of reliability, as it shows the spatial distribution and trend data versus time, $N(t_i)$ [6].

3. *Equations*

Additionally, GAMM is able to estimate the reliability function, which is calculated using the algorithms based on the nonparametric method of Nelson–Aalen [5] and evaluated in the time t_i, i.e., at the moment before the intervention. The estimation method can consider both the historic full data and censored historical data. The GAMM estimator for the cumulative failure rate is expressed in Eq. (4).

$$Z(ti) = \begin{cases} Z(t_{i-1}) + \frac{1}{n-i+1} & \text{if the event } i \text{ is a failure} \\ Z(t_{i-1}), & \text{otherwise (preventive maitenance)} \end{cases} \qquad (4)$$
$$\text{with } Z(t_0) = 0,$$

where [4]

- $Z(t_i)$ is the cumulative failure rate.
- n is the total number of interventions recorded in the history.
- i represents the sequential number of each intervention.

Therefore, the estimator of the reliability function $(R(t_i))$ is given by Eq. (5).

$$R(t_i) = e^{-Z(t_i)} \tag{5}$$

4. Graphics generation

Considering all variables, the dispersion diagrams are replaced for bubble charts and the GAMM Graphics 1 are generated with the following considerations: the size (diameter) of the bubbles represent the duration of the technical revisions, the color represents the type of maintenance done to the equipment (white—preventative maintenance, gray—corrective maintenance) the outline of each bubble represents the state of the equipment during the revision. Thus, if the bubble has a bold-shaded outline, the value of the variable "Det" is 0, i.e., the revision requires a shutdown of the equipment/system. However, if the bubble has a thin outline, the value of the variable "Det" is 1, i.e., the revision is performed without stopping the machine.

The Graphics 2 of GAMM, provides further information, from Graphics 1, to show the reliability of the equipment in the instant before the intervention, either corrective or preventive, from Eqs. 4 and 5. Thus, we can correlate each of the interventions with the existing level of reliability in the equipment prior to it.

4.1 Dispersion Diagram

Table 1 show the full real historical data of technical revisions in the slurry pumps with additional indication as to whether the intervention was corrective or preventive, the latter identified with a (*). The notations used in Tables have been explained at the beginning of Sect. 4.

Considering the data in Table 1, generating a dispersion diagram for each pump, which represents the cumulative number of technical revisions $N(t_i)$ versus time t_i (Figs. 4 and 5).

The dispersion diagram, as a starting point for any reliability analysis, offers graphic information relative to the tendency of the function $N(t_i)$, determining its linear or nonlinear behavior and finally determining the possibility of completing the hypothesis i.i.d. Both diagrams show a linear tendency which can be assumed to satisfy the hypothesis i.i.d. [6] (time between technical revisions both independently and identically distributed).

Table 1 History of technical revisions of P1 and P2, respectively

$N(t_i)$	T_i	$N(t_i)$	T_i
1*	602.83	11*	207.451
2	1054.55	12	621.586
3*	421.421	13*	454.33
4	52.2682	14*	249.986
5*	526.72	15*	102.061
6*	126.234	16	942.625
7*	700.97	17*	197.732
8	1089.07		
9*	266.915		
10*	294.097		

$N(t_i)$	T_i	$N(t_i)$	T_i	$N(t_i)$	T_i
1	25	11	2257	21	6201
2*	279	12*	2463	22*	6738
3	619	13*	2809	23*	6986
4*	776	14	2956	24	7355
5*	828	15*	3508	25*	7681
6	1346	16*	3917	26*	7915
7	1488	17	4554		
8*	1543	18*	4899		
9	1768	19	5216		
10*	1870	20*	5720		

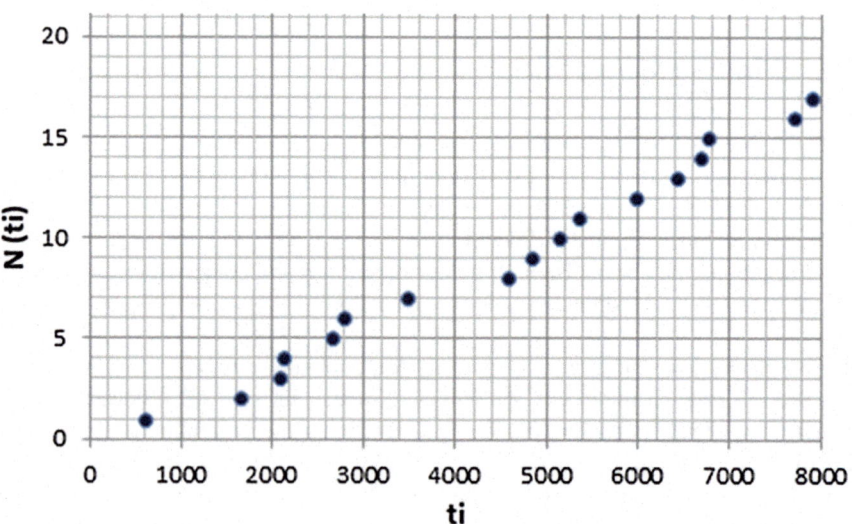

Fig. 4 Dispersion diagrams of the cumulative number of technical revisions $N(t_i)$ for slurry pumps P1 and P2, respectively

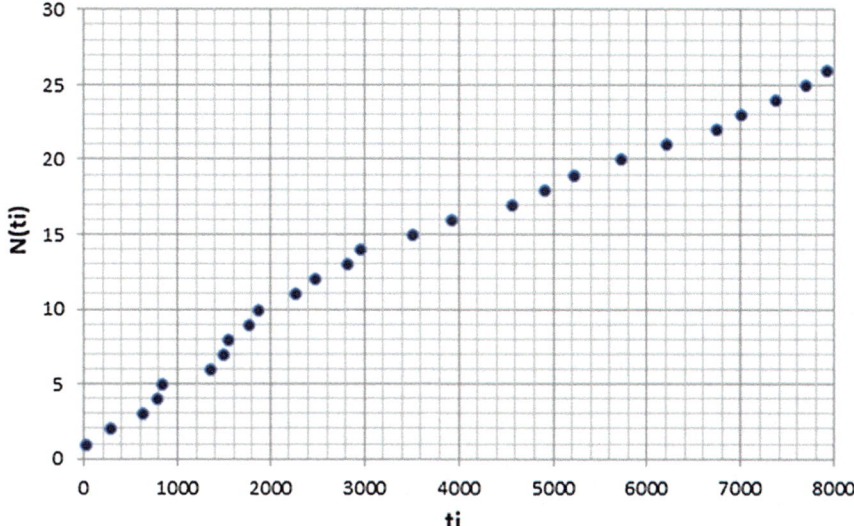

Fig. 5 Dispersion diagrams of the cumulative number of technical revisions $N(t_i)$ for slurry pumps P1 and P2, respectively

4.2 GAMM Method

The application of the GAMM Method integrates a series of variables that are not considered in the rationale of the dispersion diagram previously mentioned [4] such as: type of intervention, duration time and state of the equipment/system at that time. Finally, the time between failure and interventions (TBF) of each slurry pump has been calculated. This parameter is calculated considering whether or not the equipment had to be shut down during the revision.

Based on the nonparametric method of the Nelson–Aalen [5, 6], GAMM is able to estimate the reliability function, the GAMM estimator for the cumulative failure rate and the estimator of the reliability function, as well as how Graphics 1 and 2 (Figs. 6, 7, 8 and 9) are obtained are described in Sect. 4 and developed in [4].

Considering all variables, the dispersion diagrams (Figs. 4 and 5) are replaced for bubble charts and the GAMM Graphics 1 (Figs. 6 and 7) are generated with the same considerations described in Sect. 4.

Finally, the first GAMM graphic is constructed considering all of the features described above (Figs. 6 and 7).

The comparison of both graphics yields the following information (Table 2).

Below, Graphics 2 of GAMM (Figs. 8 and 9) provide further information, from Graphics 1 (Figs. 6 and 7), and show the reliability of the equipment at the moment before the intervention, whether corrective or preventative [4], from Eqs. 4 and 5. In this way, we can correlate each of the revisions with the existing level of the equipment's reliability prior to it.

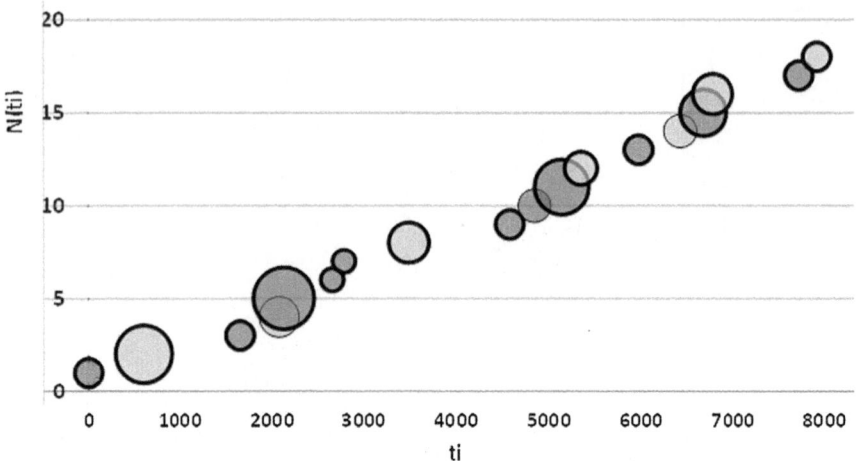

Fig. 6 GAMM, Graphic 1 of the pump P1 ($N(t_i)$ vs. t_i)

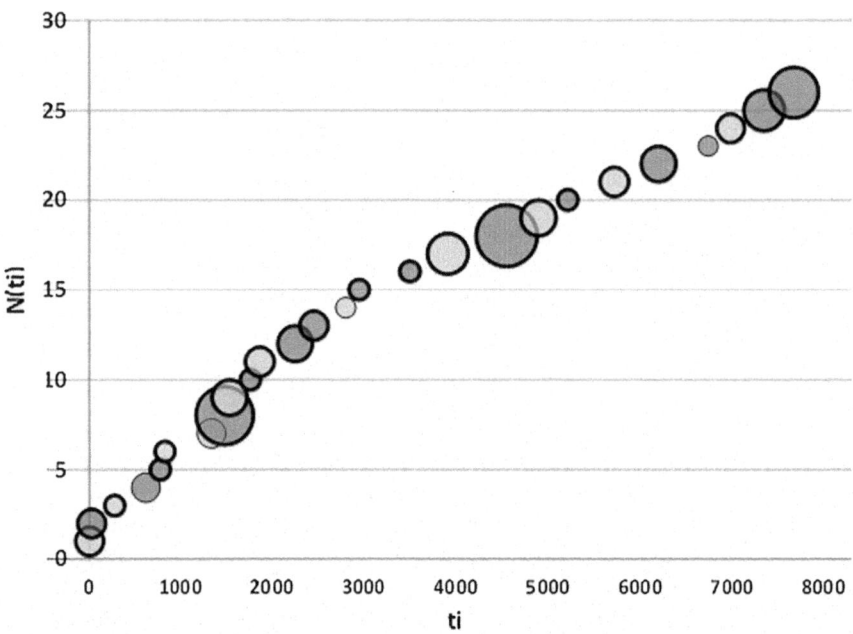

Fig. 7 GAMM, Graphic 1 of the pump P2 ($N(t_i)$ vs. t_i)

In pump P1 and P2, there are not stable reliability values. This means that the data corresponds to equipments which have a deficient preventative maintenance plan and/or program. Both pumps show values of reliability that are very changeable over time. Also, it is noted that a large number of corrective interventions are performed during the equipment's periods of high reliability.

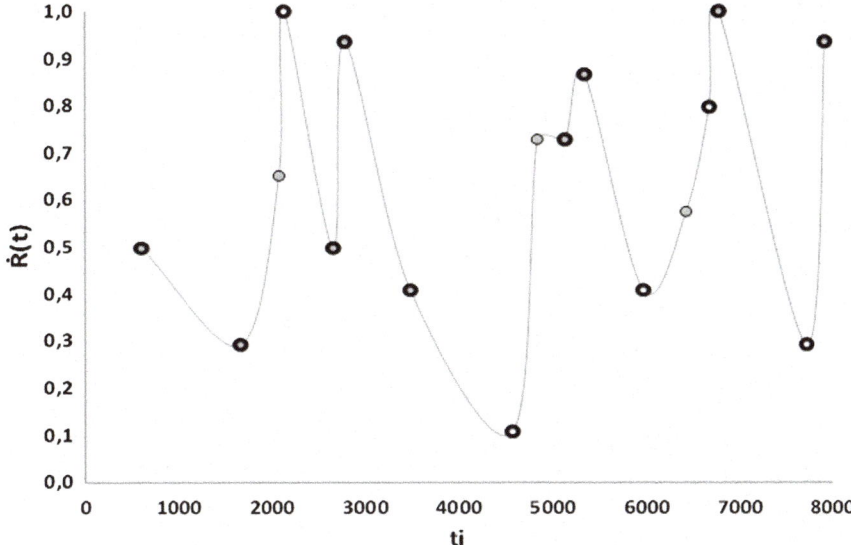

Fig. 8 GAMM, Graphic 2 of pump P1

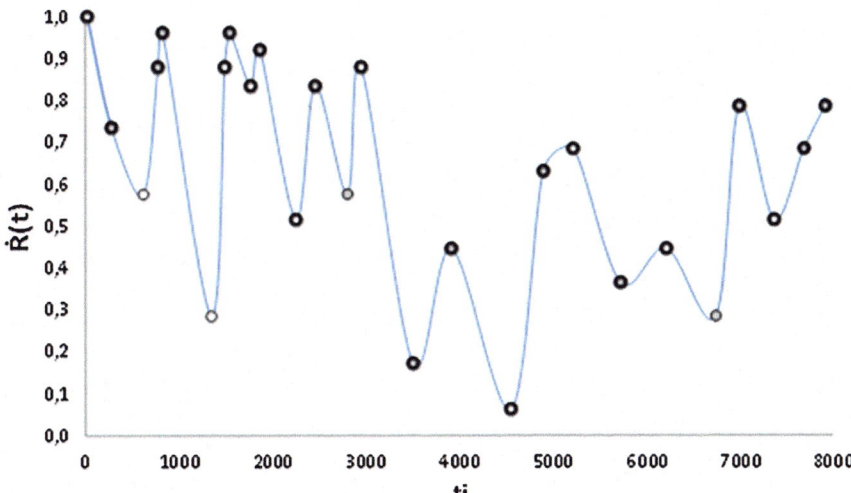

Fig. 9 GAMM, Graphic 2 of pump P2

It is necessary to clarify that Graphics 2 (Figs. 8 and 9) show the reliability of slurry pumps at the instant immediately before the intervention (bubbles). Lines connecting the bubbles are only useful to know certainly which intervention (bubble) comes before the other.

Table 2 Information extracted from GAMM Graphics 1 (Figs. 6 and 7) for each pump

	P1	P2
Steady increase in the cumulative number of interventions	Yes	Yes
High number of corrective interventions (unscheduled)	Yes	Yes
Planning of preventive maintenance	Yes	Yes
High frequency (number) of preventive interventions	No	No
High times of preventive interventions (some interventions are long lasting compared to the rest of preventive interventions included in the graphics)	Yes	No

4.3   Analysis and Decision-Making

GAMM has identified a set of patterns that provide useful information to assist the decision making process at the operational level. The patterns identified in P1 and P2 are as follows:

- A trend in the behavior of the technical revisions: Both in P1 as in P2, the diagrams (Figs. 4 and 5) show a function that presents a linear trend.
- Deviation of preventive maintenance: it is observed that both pumps show a planning for preventative interventions due to the fact that it follows an orderly seasonal pattern, as shown in Graphic 1 (Figs. 6 and 7). However, both pumps show preventative interventions which some of them are highly time-consuming interventions.
- Quality of operation and/or preventative maintenance: in P1 and P2, a large number of corrective interventions is observed, some of which are time-consuming and made shortly after revisions of preventative maintenance, which could be the result of existing problems, the execution of the work itself, or in the operation of the pumps.
- Reliability function: both pumps show corrective interventions during periods of high reliability.
- Efficiency and quality in the execution of interventions: P1 and P2 show several preventative and corrective revisions that exceed the time averages, questioning if the workloads programmed are adequate or if the deviation in execution time is due to external factors, such as lack of tools, waiting for spare parts, lack of trained personnel, etc.
- Impact on production: Graphic 1 (Figs. 6 and 7) shows how the majority of both preventative and corrective revisions force the loss of functionality for the pumps, affecting their individual availability, and due to the configuration of the process, affecting the entire system's overall availability.

4.3.1 Working Hypothesis

One possible cause for the variability observed in the reliability value (Graphic 2, Figs. 8 and 9) could be inadequate planning of preventative maintenance or its partial execution, making it so that the equipment's reliability does not approach a proper constant in the time scale value. This hypothesis is reinforced by observing the existence of preventative revisions in periods of low reliability of the equipment (Figs. 8 and 9).

There are also other reasons that may affect the reliability of the equipment, for example, corrective revisions which can be substantially minimized with a program of optimized preventative maintenance, i.e., many of the observed corrective interventions can be a direct result of a poor preventive maintenance program or its poor application.

For its part, in P2 a high number of interventions is observed, compared to P1 (Graphic 1, Figs. 6 and 7). Keeping in mind that the maintenance program was designed for both slurry pumps, one of the possible causes that can explain this phenomenon would be a deficient design of the maintenance program itself (frequency and scope of the revisions) or specific operating conditions which were not considered in the maintenance plan. At the same time, the time consumption of preventative interventions could be due to low trained operators, the inexistence of explanatory data sheets for each of the revisions to be performed or the existence of accessibility problems to spaces and/or necessary tools for execution.

For the analysis, it should be considered the type of equipment analyzed and its operating context. The slurry pumps (P1 and P2) are exposed to a high stress job, especially by the transfer characteristics of highly abrasive and corrosive flow, and high solids content and strong acidity. Moreover, the continuous process under study (Fig. 1) has as a security mechanism, an alternative to grinding second recirculation line, which is activated in case of failure or the ball mill flotation process "down stream". The security system is a special flow line (by pass) from only the slurry pump P2 to another parallel line grinding.

Finally, the GAMM graphics suggest the possibility that malpractice in the pumps' operation and/or in the interventions could be another possible cause for the deficiencies listed above.

4.3.2 Analysis

From these assumptions, it is appropriate to perform a study of the existing maintenance program for each pump which turned out to be the same for both pumps. Learning the maintenance program, analyzing and later identifying possible malpractices carried out during the operation and maintenance revisions of the slurry pumps (P1 and P2) by those same operators. The existence of poor operational practices is another cause that could provoke the emergence of equipment failures, and those in turn shorten the life span and proper functioning of the pumps. For this reason and attending to the hypothesis based on the GAMM method, a qualitative analysis was performed using the information collected on site in different information exchanges with the executive operators of the maintenance and operational tasks. In this way, a collection of malpractices was identified (Table 3).

4.3.3 Decision-Making and Problem Solving

Once the information provided by GAMM was compared to the technical information of the pump, the existing maintenance program, and the identified malpractices, we then proceeded by redefining the maintenance program based on the methodology or RCM [2, 7] for each slurry pump (P1 and P2), considering the existing differences in the work conditions of each pump with the following objectives: define and justify the maintenance actions, redefine the programmed

Table 3 Existing operational and maintenance malpractices of P1 and P2

Operation and maintenance malpractices identified	
Type	Description
Operation	The existence of overdemanding routines for the pumps, using different mechanisms to avoid security and control measures (without considering the control system signals. example: temperature, output pressure, and seal water pressure system)
Maintenance	The existence of time periods in which the equipment remains broken for more than 2 days, which generates a saturation and internal solidification of the sludge. This generates problems immediately following corrective maintenance of the pump, due to the internal inertia that the rotor must overcome once restarting. This overloads the internal components, reducing its useful lifespan
Operation	Improper use of the pumps' safety accessories: casings cover the motor's propeller, bolts, and instruments, among others
Maintenance	There are no defined procedures for cleaning the equipment, which directly affects their operation and maintenance
Maintenance	The existing maintenance program for P1 and P2 is the same, not considering the different working conditions of each pump
Operation	The process under study has as a security mechanism, an alternative to grinding second recirculation line, which is activated in emergency case. The security system is a special flow line (by pass) from P2 to another parallel line grinding. This means that only P2 is subjected to a high load during certain time periods
Operation	The design of the buffer placed at the pumps inlet could be improved. Oscillation on the slurry intake pressure is high on a regular basis due to small slurry surface of the buffer. Slurry recirculation and SAG mill outputs generate this pressure change leading to abnormal pumps operating conditions and deviation from process standards
Maintenance	Many basic elements of the existing maintenance program for P1 and P2 are the same as for the other pumps (water pumps) located in the plant. It is not considered that the slurry pumps are exposed to a high stress job, especially by the transfer characteristics of highly abrasive and corrosive flow, and high solids content and strong acidity

actions, secure and increase the efficiency of each pump so as to provide a set of recommendations regarding the pumps.

First of all, an analysis of the different failure modes was made, identifying and prioritizing them according to the frequency or severity of the defect. To do this, we performed a functional analysis of the slurry pumps, a failure mode analysis, and finally, a hierarchical structuring depending on the criticality of its components. In the functional analysis, the equipment was broken down into subsystems, establishing functional relationships between them (functional diagram) and indicating the functions and interactions with operational surroundings. Like the FMECA result, critical subsystems were determined: mechanical seal, inlet and casing liners, PLC control system and shaft coupling, the other subsystems being considered noncritical. On the basis of this ranking for the components of each pump, a new maintenance plan was redefined for each of them.

5 Results and Conclusions

In relation to the proposed objectives, it can be concluded that the application of the GAMM method has determined and graphically characterized the deficiencies of the slurry pump which presented an elevated failure rate, facilitating decision-making and problem solving that has allowed for maximum availability and efficiency in both slurry pumps analyzed.

The comparative analysis of the GAMM graphics corresponding to each of the pumps has also determined aspects like: tendency of revisions, deviation in the frequency of preventative revisions, function of reliability, efficiency of the maintenance operations, impact of unavailability, quality of the revisions performed or quality of use and operation of the equipment on the part of the personnel responsible for it. Based on this information, there were a series of hypotheses to search for possible causes of the inefficiency (also other variables were considered such as: equipment in rotation, low quality spare parts or design problems, but all of them were discarded). The final result has been the identification of a combination of existing deficiencies in the operation and maintenance of the slurry pumps (P1 and P2, Sect. 4.3). It can be summarized as follows:

– P1: constant increase in the cumulative number of revisions, an elevated number of corrective revisions (unscheduled) and lower number of interventions than P2. Preventative interventions are also quite time-consuming. In addition, the existence of very changeable values of reliability based on time as well as a large number of corrective revisions in periods of high reliability.
– P2: constant increase in the cumulative number of revisions, higher number of interventions than its counterpart, elevated number of corrective revisions (unscheduled) and existence of very changeable values of reliability based on time as well as a high number of corrective revisions carried out in periods of high reliability.

Once the information provided by GAMM is compared to the technical information of the pump, the existing maintenance program and malpractices identified (prior qualitative analysis of the information captured in the field and in different information exchanges with the executive operators of the maintenance and operation tasks), were able to determine the causes of the problems previously identified:

- Design deficiencies in the maintenance program were detected (frequency and scope). In fact, the same maintenance program was applied to water pumps and slurry pumps, without considering the differences in their operating conditions (input stream).
- A lack of analysis for the components most critical to the pumps was detected, meaning that there was a maintenance plan for preventive revisions for non-critical parts, unnecessarily affecting the availability of the pumps.
- One pump (P2) was subjected to a high load during certain time periods, because the security system was a special flow line (by pass) from P2 to another parallel line grinding. This was not considered, the maintenance plan was the same for P1 and P2.
- The study detected a lack of technical training for personnel responsible for the operation and maintenance of the pumps, and the existence of problems due to inaccessibility of space and necessary tools needed for the execution of revisions (logistical problems). All of these factors wasted excessive time dedicated to maintenance tasks.
- A combination of malpractices was detected, affecting the quality the operation and maintenance work of the pumps.

Finally, several measures were adopted to eliminate the causes of the problems previously identified. First, the FMECA analysis (Failure Mode, Effects and Criticality Analysis) and based on the criticality of each subsystem [2], a new maintenance plan based on the RCM methodology [7] was drafted, for each slurry pumps (P1 and P2), considering the differences in the work conditions. All of this allowed:

- Hierarchization of the components of each slurry pump based on its criticality, i.e., preventive maintenance of critical components, improving the availability of each pump and eliminating unnecessary revisions and thus, the time necessary for them.
- Full compliance with the new corresponding maintenance program designed for P1 and P2, by those responsible for maintenance, through the development of a specific training program aimed at personnel responsible for operation and maintenance.
- Development of a new maintenance plan for each pump, P1 and P2 respectively, for eliminating the deficiencies in the previous maintenance plan, considering all differences between P1 and P2 (security system) as well as the specific requirements of slurry pumps.

- Development of a technical formation program for personnel responsible for the maintenance and operation of the pumps, correcting the overall malpractices detected in both the operation and maintenance of the pumps.
- Elaboration of explanatory technical sheets for each of the preventive maintenance revisions under the maintenance program and implementation of a registry for revisions already performed.
- Referring to the slurry pumps buffer design, a larger surface inverted cone-type buffer recipient was designed to avoid high intake pressure oscillation.

In this case study, GAMM method allowed for the identification of existing deficiencies in each pump under study, supporting the decision-making process and implementation of a new maintenance plan.

Acknowledgements This research is funded by the Spanish Ministry of Science and Innovation, Project EMAINSYS (DPI2011-22806) "Sistemas Inteligentes de Mantenimiento." Procesos emergentes de E-maintenance para la Sostenibilidad de los Sistemas de Producción, besides FEDER funds. The research work was performed within the context of iMaPla (Integrated Maintenance Planning), an EU-sponsored project by the Marie Curie Action for International research Staff Exchange Scheme (project acronym PIRSES-GA-2008-230814 iMaPla).

References

1. Birolini A (2007) Reliability engineering, 5th edn. Springer, Berlin. ISBN 978-3-540-49388-4
2. Barberá L, Crespo A, Viveros P, Stegmaier R (2012) Advanced model for maintenance management in a continuous improvement cycle: integration into the business strategy. Int J Syst Assur Eng Manag (Jan–Mar 2012) 3(1):47–63. doi:10.1007/s13198-012-0092-y
3. Crespo Márquez A, Gupta J (2006) Contemporary maintenance management. Process, framework and supporting pillars. Omega. Int J Manag Sci 34(3):325–338. doi:10.1016/j.omega.2004.11.003
4. Barberá L, Crespo A, Viveros P, Arata A (2012) The graphical analysis for maintenance management method: a quantitative graphical analysis to support maintenance management decision making. J Qual Reliab Eng Int, Copyright © 2012 Wiley, USA (wileyonlinelibrary.com) doi:10.1002/qre.1296
5. Nelson W (1969) Hazard plotting for incomplete failure data. J Qual Technol 1:27–52
6. Rausand M, Hoyland A (2004) System reliability theory: models, statistical methods, and applications. Wiley-Interscience, USA, 644 p
7. Moubray J (1997) Reliability centred maintenance: RCM II. 2nd Ed. Industrial Press, USA

Author Biographies

Luis Barberá Martínez is Ph.D. (Summa Cum Laude) in industrial Engineering by the University of Seville, and Mining Engineer by UPC (Spain). He is a researcher at the School of Engineering of the University of Seville and author of more than 50 articles, national and main internationals. He has worked at different international universities: Politecnico di Milano (Italy), CRAN (France), UTFSM (Chile), C-MORE (Canada), EPFL (Switzerland), University of Salford (UK) or

FIR (Germany), among others. His line of research is industrial asset management, maintenance optimization and risk management. Currently, he is Spain Operations Manager at MAXAM Europe. He has three Master's degrees: Master of Industrial Organization and Management (2008/2009) (School of Engineering, University of Seville), Master of Economy (2008/2009) (University of Seville) and Master of Prevention Risk Work (2005/2006). He has been honored with the following awards and recognitions: Extraordinary Prize of Doctorate by the University of Seville; three consecutive awards for Academic Engineering Performance in Spain by AMIC (2004/2005), (2005/2006) and (2006/2007); honor diploma as a "Ten Competences Student" by the University of Huelva and CEPSA Company (2007); graduated as number one of his class.

Adolfo Crespo Márquez is currently Full Professor at the School of Engineering of the University of Seville, and Head of the Department of Industrial Management. He holds a Ph.D. in Industrial Engineering from this same University. His research works have been published in journals such as the International Journal of Production Research, International Journal of Production Economics, European Journal of Operations Research, Journal of Purchasing and Supply Management, International Journal of Agile Manufacturing, Omega, Journal of Quality in Maintenance Engineering, Decision Support Systems, Computers in Industry, Reliability Engineering and System Safety, and International Journal of Simulation and Process Modeling, among others. Prof. Crespo is the author of seven books, the last four with Springer-Verlag in 2007, 2010, 2012, and 2014 about maintenance, warranty, and supply chain management. Professor Crespo leads the Spanish Research Network on Dependability Management and the Spanish Committee for Maintenance Standardization (1995–2003). He also leads a research team related to maintenance and dependability management currently with five Ph.D. students and four researchers. He has extensively participated in many engineering and consulting projects for different companies, for the Spanish Departments of Defense, Science and Education as well as for the European Commission (IPTS). He is the President of INGEMAN (a National Association for the Development of Maintenance Engineering in Spain) since 2002.

Pablo Viveros Gunckel is Researcher and Academic at Technical University Federico Santa María, Chile, has been active in national and international research, both for important journals and conferences. Also he has developed projects in the Chilean industry. Consultant specialist in the area of Reliability, Asset Management, System Modeling and Evaluation of Engineering Projects. He is Industrial Engineer and Master in Asset Management and Maintenance.

Raúl Stegmaier is Full-time Professor at the Department of Industrial Engineering USM, Valparaíso, Chile. Industrial Engineer at the Universidad Técnica Federico Santa María (USM) and Master of Science in Industrial Engineering at the Universidad de Chile. His teaching is in the field of Operation Management with emphasis in Physical Asset Management and Maintenance and Reliability Engineering in undergraduate and graduate programs. His research work has been published in several journals and books. Additionally, he has directed various industrial and consulting projects in System Reliability Analysis, Reliability Engineering deployment and in Lean Manufacturing Principles Implementation in sectors from mining to services. He has been member of the TC 251—ISO 55000 series, Chilean chapter. Regarding university management, he has held diverse positions at the USM such as Director of Academic Affairs, Director of Planning and Development and Head of the Department of Industrial Engineering. He is currently director of the Master in Asset Management and Maintenance and he is a member of the board of directors of the Universidad Técnica Federico Santa María.

A Graphical Method to Support Operation Performance Assessment

Pablo Viveros Gunckel, Adolfo Crespo Márquez,
Luis Barberá Martínez and Juan Pablo González

Abstract This paper proposes a graphical method to easy decision-making in industrial plants operations. The proposed tool "Graphical Analysis for Operation Management Method" (GAOM) allows to visualize, and to analyze, production related parameters, integrating assets/systems maintenance aspects. This integration is based on the TPM model, using its quantitative management techniques for optimal decision-making in day-to-day operations. On the one hand, GAOM monitors possible production target deviations, and on the other, the tool illustrates different aspects to gain control on the production process, such as availability (A), repair time, cumulative production or overall equipment effectiveness. Through appropriate information filtering, individual analysis by class of intervention (corrective maintenance, preventive maintenance or operational intervention) and production level can be developed. GAOM integrates maintenance information (number of intervention, type of intervention, required/not required stoppage) with production information (cumulative production, cumulative defective products, and cumulative production target) during a certain timeframe (cumulative calendar time, duration of intervention). Then the tool computes basic performance indicators supporting operational decision-making. GAOM provides interesting graphical outputs using scatter diagrams integrating indicators on the same graph. GAOM is inspired in the GAMM (Graphical Analysis for Maintenance Management) method, published by the authors (LB, AC and PV) in 2012.

Keywords Operation management · Graphical analysis · Decision making

P. Viveros Gunckel · J.P. González
Department of Industrial Engineering, Universidad Técnica Federico Santa María,
Avenida España, 1680 Valparaíso, Chile

P. Viveros Gunckel · A. Crespo Márquez (✉) · L. Barberá Martínez
Department of Industrial Management, School of Engineering, University of Seville,
Camino de los Descubrimientos s/n, 41092 Seville, Spain
e-mail: adolfo@us.es

© Springer International Publishing AG 2018
A. Crespo Márquez et al. (eds.), *Advanced Maintenance Modelling for Asset Management*, DOI 10.1007/978-3-319-58045-6_15

1   Introduction

The industry has undergone significant changes over the past three decades, concerning management approaches, technologies, production processes, customer expectations, supplier management, and behavior of competition [1]. Because the intense global competition, companies have worked to improve and optimize productivity in order to remain competitive. The need to improve productivity necessitated further integration of maintenance management to production management, with the ultimate goal of producing at minimum cost.

Efficiency management in production lines and manufacturing environments is gaining more importance, not only because it is a quality neutral way to reduce production cost but also because of its role in ensuring the facility use with the best practices [2]. To achieve these goals, practical and easily implemented tools are needed allowing the identification of potential improvements in the production process are needed. Literature reviewed reveals that industry still lack approaches and tools to better understand the inefficiencies of machines [3], particularly with a focus on production management decisions (quality, maintenance, production planning, among others).

Current business practice is characterized by intense international competition, rapid product innovation, increased use of automation, and significant organizational changes in response to new manufacturing and information technologies. Correspondingly, also the operating function is undergoing many significant changes using new technologically advanced equipment and new forms of organization [4]. The above arguments suggest that the changes in the production environment are relevant for management control system.

According to Fleischer [5], the competitiveness of manufacturing firms mainly depends on variables such as the availability and productivity of their production equipment. The development and application of graphic tools that give support and make decisions in the field of operational reliability [6] is a fundamental task for the proper and efficient management of assets and resources within an organization, even more so with various types of equipment whose functional configuration is highly complex [7, 8]. The emerging trend in current research is the integration of Decision Making (DM) techniques in constructing an effective decision model to address practical and complex production problems [9].

Typically, maintenance management has been analyzed independently to the management of production. This has hindered the comprehensive analysis of cause and effect between the two. It is therefore necessary design tools and methods that facilitate the analysis and identification of losses in the production process, which allows giving answers to common questions in the area of production and maintenance (Table 1). On the other hand, there is generally a lack of quality of information mainly due to collection processes and low operating complex data [10, 11]. It is necessary procedures for the collection and management of simpler data and adjusted to real needs, facilitating the interpretation and analysis [12, 13]. Here

Table 1 Some important questions to be answered by the tool

	Operations/production area
1.	Which is the operating/running time between interventions, preventive intervention (PI) and corrective intervention (CI)?
2.	Which is the non-operating time for operational intervention?
3.	Which is the real production rate for equipment and overall system?
4.	Which is the difference between "real production" and "planned production"?
5.	Which is the difference between "real production" and "nominal production"?

	Maintenance area
1.	Which is the frequency of corrective interventions between preventive interventions?
2.	How variables are the "repair times" for preventive interventions and corrective intervention?
3.	How variables are the "functioning times" after a preventive interventions and a corrective intervention?
4.	Which is the effect of maintenance management decisions over performance indicators as: Intervention time and operation/running time?

	Integrated areas
1.	Is there a deviation in equipment performance? Is it possible to identify the responsible area?
2.	Are there accelerated wear patterns in the equipment?
3.	Is the capacity of the plant adjusted to market demand?

are some questions generally required by the area of production and maintenance, presented in an integrated way.

By definition, the performance measurement process corresponds to the quantification of the action, where the measurement is the quantization process and the action performance leads to [14]. Moreover, there is a relationship between the control and: objectives formulation, setting standards, action programs, budgets, rational use of resources, measurement and verification of results, deviations and performance correction or improvement.

Specifically, 26 DM techniques have been reviewed [9]. Nevertheless, the main revised methods taken into account to develop GAOM have been: Balanced Scorecard (BSC) [15, 16], Total Productive Maintenance (TPM) [1, 5, 10–17], Overall Equipment Effectiveness (OEE) [18] and the GAMM (Graphical Analysis for Maintenance Management) Method [19].

2 GAOM Method

2.1 Introduction

The GAOM method is constructed from a database that integrates information from the production process through three classes of information: intervention (number of intervention, type of intervention, required/not required stoppage), production (cumulative production, cumulative defective products, and cumulative production target) and time (cumulative calendar time, duration of intervention).

GAOM integrates this information through calculation of basic performance indicators related to timing information, production and quality, evaluating production yields and maintenance. To facilitate analysis, GAOM presents information graphically, using scatter diagrams, integrating performance indicators on the same graph. The software used by GAOM to calculate indicators and construct graphs, Microsoft Excel (VBA Language), has been widely used in companies of many sectors.

2.2   Input Information

Input information is necessary for the success of GAOM method. The classification for this input is presented in Tables 2, 3 and 4.

It is necessary to emphasise that the measurement of throughput is taken at the output stage of the production process. Also, it works under the assumption that the actual production process does not exceed the settled nominal production capacity of it. First, is recommended to analyze a system from the overall point of view, identifying production deviations and the most important loss factors. Then, the equipment analysis will be interesting to determine the main causes of deviation or to refine the analysis. Figure 1 represents both possible analysis scenarios

Table 2 Intervention information

Class	Description
N° intervention [N_i]	Assigns the number of occurrence of the intervention i-th. It is a natural number under consecutive order
Stop/do not stop [DET_i]	Reports whether the intervention stopped (0) or not stopped (1) the operation. It is a Binary variable
Type of intervention [CI_i]	Corrective intervention (CM), associated with the value 1. It is a maintenance action after the occurrence of a failure [$CI_i = 1$]
	Preventive intervention (PM), associated with the value 2. It is a maintenance action prior to a failure [$CI_i = 2$]
	Operational intervention (OI), associated with the value 3. Do not execute a maintenance action. It generates a detention or reducing capacity performance [$CI_i = 3$]

Table 3 Time information

Class	Description
Cumulative calendar time [CCT_i]	It is the cumulative time when the i-th intervention occurs. Counted from the beginning of the evaluation horizon ($T = 0$)
Time of Intervention [TI_i]	Time duration requested for the i-th intervention. Related to the time to repair

Table 4 Production information

Class	Description
Cumulative production [CP_i]	It is the cumulative production when the i-th intervention occurs. Counted from the beginning of the evaluation horizon ($T = 0$)
Cumulative defective production [CDP_i]	It is the cumulative defective production (out of quality standard) when the i-th intervention occurs. Counted from the beginning of the evaluation horizon ($T = 0$)
Nominal Production Capacity [NCP]	Nominal capacity of the process. It must be settled by operation

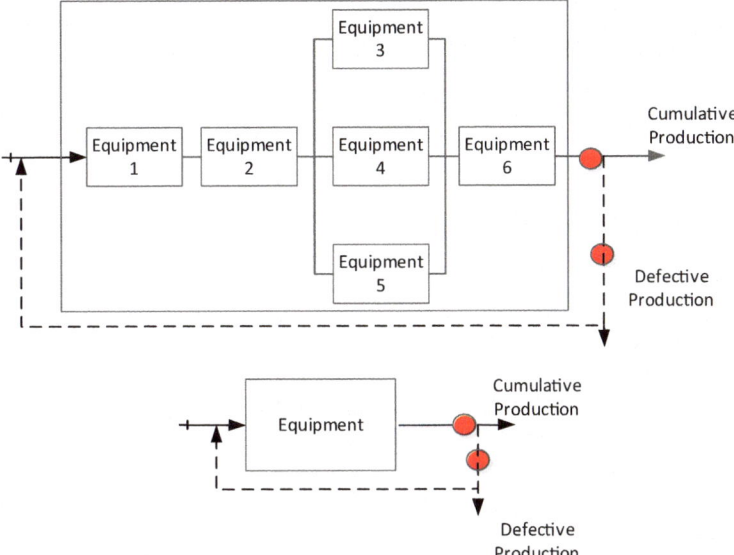

Fig. 1 Analysis from process or equipment point of view

(equipment and system, respectively), showing, at the same time (red points) where measurement of the production input could be captured. This flexibility, understood as an advantage of GAOM, will be applied regarding users' interests and data availability.

2.3 Scatter Diagram and Basic Performance Indicators

To represent information in a clear and simple manner, GAOM uses a scatter diagram of cumulative production (CP_i) based on the cumulative number of interventions (N_i) for the i-th intervention occurred along the time horizon.

The data presented in Table 5 refers to equipment used in a piping line, transporting inert material in a mining company located in Chile. Considering this, it is possible to generate the scatter diagram of the cumulative production (CP_i), based on the cumulative number of interventions (N_i) for the "i-esima" intervention (Fig. 2). This diagram allows the analyst to evaluate the distribution and trend of cumulative production.

According to the information related to "input information" (Tables 2, 3 and 4), it is recommended to collect the historical data considering the format shown in section "Standard format for data collection" in Appendix 1 (Table 8).

From input information, GAOM requires to calculate some basic performance indicators (BPI), which are classified based on time or production (Tables 6 and 7).

To calculate the total cycle time TCT_i indicated in Table 6, there are four possible scenarios of detailed calculation shown in Fig. 3. In this figure, the gray blocks represent interventions that produce equipment shutdown, while the white colored blocks represent interventions that do not require a shutdown of equipment.

Table 5 Historical data of interventions and cumulative production (tons)

N_i	CP_i	N_i	CP_i	N_i	CP_i	N_i	CP_i	N_i	CP_i	N_i	CP_i
1	2165	10	20,192	19	47,243	28	66,217	37	76,751	46	81,002
2	3143	11	25,311	20	53,699	29	68,847	38	77,045	47	81,478
3	3777	12	27,989	21	54,792	30	70,279	39	77,301	48	81,963
4	7668	13	31,248	22	56,873	31	71,517	40	77,646	49	82,516
5	9972	14	32,024	23	57,894	32	72,452	41	78,241		
6	12,386	15	37,540	24	60,228	33	73,062	42	78,715		
7	15,965	16	41,590	25	61,817	34	74,375	43	79,238		
8	17,095	17	43,490	26	62,939	35	74,992	44	79,850		
9	17,992	18	45,236	27	64,174	36	76,264	45	80,457		

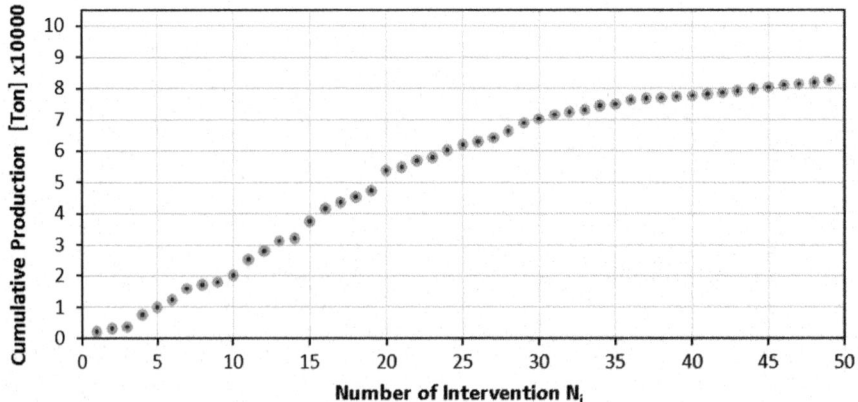

Fig. 2 Scatter plot of cumulative production

Table 6 BPI based on time—GAOM method

No	BPI	Description and formulation
1	Time between intervention [TBI$_i$]	Total time between intervention i and intervention $i-1$ TBI$_i$ = CCT$_i$ − CCT$_{i-1}$
2	Downtime of intervention [DT$_i$]	Time during the equipment does not operate DT$_i$ = TI$_i$ × (1 − DET$_i$)
3	Operating time [OT$_i$]	Time during the equipment is in operation OT$_i$ = TBI$_i$ − TI$_{i-1}$ × (1 − DET$_{i-1}$)
4	Cumulative operating time [COT$_i$]	Cumulated operating time/working time when i th intervention occur. COT$_i$ = ΣOT$_i$
5	Total cycle time [TCT$_i$]	This includes operation and intervention time TCT$_i$ = TBI$_i$ + DT$_i$
6	Planned working time [PWT$_i$]	Time the organization has planned to operate PWT$_i$ = TBI$_i$ − DT$_i$ (to preventive intervention) TTP$_i$ = TBI$_i$ (to corrective or operational intervention)
7	Cumulated planned working time [CPWT$_i$]	Cumulative sum of planned working time. CPWT$_i$ = ΣPWT$_i$

Table 7 BPI based on production—GAOM method

No	BPI	Description and formulation
1	Production between intervention [P$_i$]	Production processed during the time between interventions. P$_i$ = CP$_i$ − CP$_{i-1}$
2	Nominal cumulative production [NCP$_i$]	Cumulated production in chronological time (no intervention of any kind). NCP$_i$ = CCT$_i$ × NCP
3	Cumulative planned production [CPP$_i$]	Production in the planned working time; i.e., the planned production to nominal capacity, excluding the scheduled time of preventive interventions. CPP$_i$ = CPWT$_i$ × NCP
4	Cumulative expected production [CEP$_i$]	Expected production during functioning time, i.e., the expected nominal production capacity, excluding the downtimes. CEP$_i$ = COT$_i$ × NCP

2.4 Analysis and Graphs Construction

For subsequent analysis GAOM requires to develop two new integrated databases, linking the above data and basic indicators. For more details, see section "Integrated databases" in Appendix 2 (Tables 9 and 10).

The GAOM method is constructed from a main production accumulated chart v/s number of intervention, this being a scatter diagram of bubbles (cumulative production) in which filters for the individual analysis of each type of intervention may be used, representing variables of interest including repair time or time between failures.

The bubble's dispersion function represents the cumulative production (CP$_i$) based on the accumulated number of intervention N_i for the i-th intervention. The

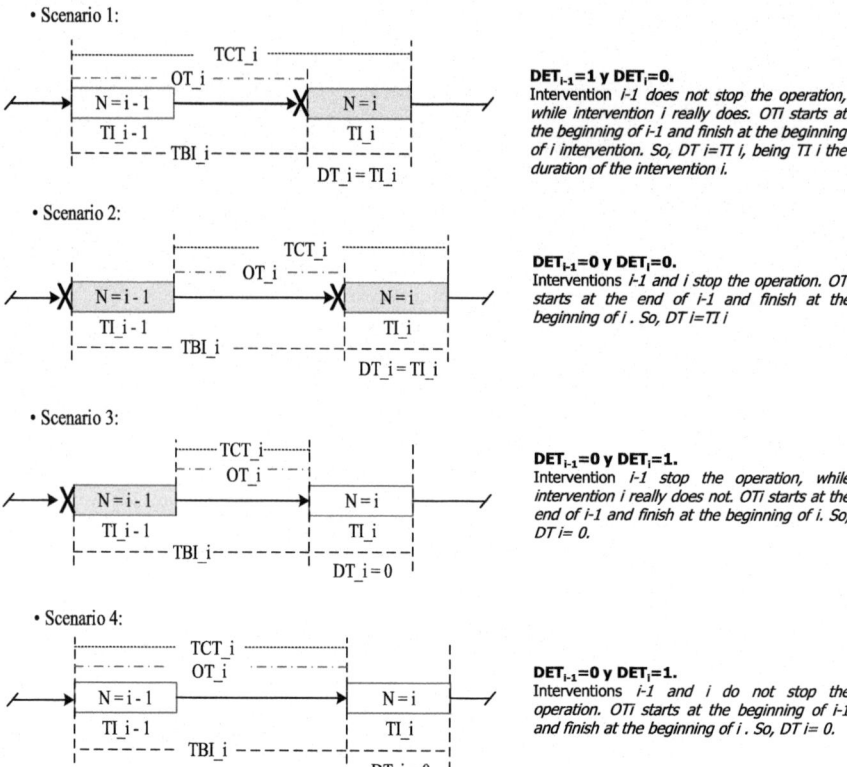

Fig. 3 Interventions timeline

size (diameter) of the bubbles represent, relatively, the indicator results such as the time between intervention (TBI$_i$) shown in Fig. 4, or the time of intervention (TI$_i$) shown in Fig. 5 for the i-th intervention. The bubble's color represents the type of intervention: dark gray (corrective maintenance), light gray (preventive maintenance) and white (operating procedure).

GAOM is capable of generating a total of eight graphics, namely:

- Graph 1: "*Analysis of Time Between Failures/Interventions*". It displays the cumulative production in each intervention. Furthermore, the time between failures or interventions for each N_i is represented by the size of the bubble. This graph allows to analyze the effect of interventions on production. Furthermore, by comparing the size of the bubbles it is possible to identify aging (systematic bubble size decrease) and the type of intervention phenomena, auditing the compliance of the established maintenance policy.
- Graph 2: "*Analysis of intervention time*". Displays the cumulative production in each intervention. Furthermore, the repair time for each intervention N_i is represented by the size of the bubble. This new graphic complement the analysis in

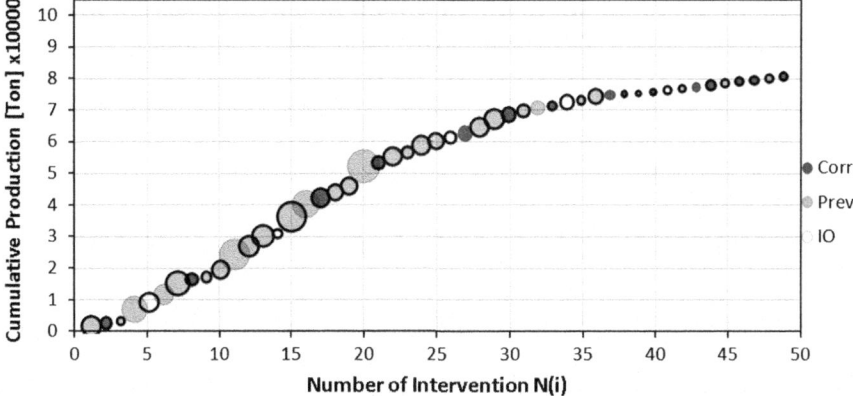

Fig. 4 Graph 1 "analysis of time between failures/intervention". GAOM method

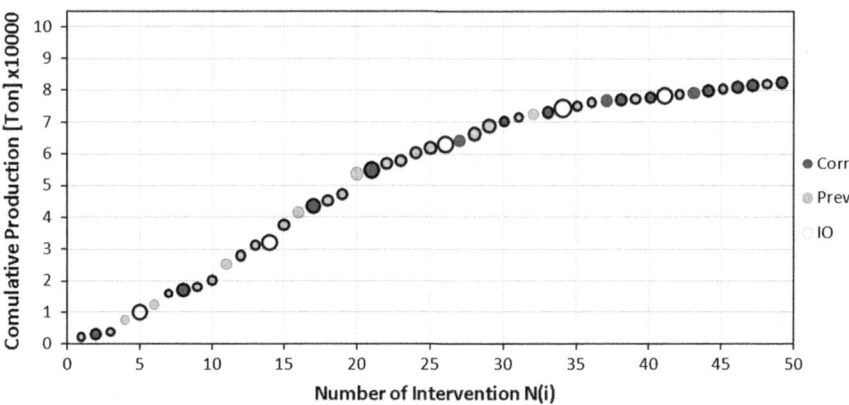

Fig. 5 Graph 2 "analysis of intervention time". GAOM method

graph 1, discriminating whether effect of loss production is cause of high frequency of failure (Chronic failure phenomena) or long times repair. Furthermore, by comparing the size of the bubbles can be detected labor behaviors related to: lack of discipline and/or work procedures for the development of interventions (high variability in the size of the bubbles), important differences between the repair times linked to corrective or preventive interventions, atypical repair (size of bubble too big or too small), among others.

To perform an individual analysis of each type of intervention, GAOM is able to filter based on this variable. Thus, it is possible to analyze independently the behavior of each type of intervention, allowing the user to track variables patterns, analyzing and concluding about it. See graph 3.

- Graph 3: *"Analysis of TBI, filtering corrective maintenance interventions"*. This graph is equivalent to the structure presented in Graph 1, except that only allows corrective maintenance analysis. This representation focuses the analysis on the phenomena of failure by detecting patterns such as representation in the package of interventions (scatter graph), often as cumulative production (differences in vertical axis), imperfect repair resulting in premature failures (size small bubbles), and aging, among others.
- Graph 4: *"Analysis of TI, filtering preventive maintenance interventions"*. This graphic is equivalent to graph 2 in structure, with the difference that allows analysis of preventive maintenance. This graph facilitates the analysis of planned maintenance and its effect on real output, considering: number of interventions, dispersion and frequency, times to repair's variability linked to maintainability (bubble size), among others.

The analysis of corrective maintenance interventions in a production process identifies the effectiveness of the maintenance policy. When equipment reaches a certain age, the failure rate starts to grow rapidly due to wear out [20]. To avoid this, it is necessary to optimize preventive maintenance, increasing, if any, the frequency of planned interventions. GAOM identifies the existence of an increasing failure rate (corrective maintenance) between preventive maintenance interventions. Figure 8 exemplified a production process where a policy of preventive maintenance is applied based on the amount of processed product, defined of course to prevent the occurrence of failures. It is seen as corrective maintenance interventions occur systematically and with similar amounts of processed product. This suggests that it is possible to identify the product processing factor generated by equipment failure, improving the policy of preventive maintenance and avoiding or reducing the occurrence of intermediate failures (dark gray bubble).

- Graph 5: *"Visualizing problems in the maintenance policy"*. This graphic outlines the presence of a classical maintenance phenomenon: the incorrect definition of the frequency of preventive interventions. This phenomenon can be analyzed by an obvious pattern (Fig. 8), as it is the permanent presence of corrective interventions from preventive interventions. This graph allows preventive procedure to correlate with the level of cumulative production. The arrows drawn graphically explain and represent the production post preventive intervention, which is equivalent in time. There is clearly a correlation between production and maintenance, and it is clear too that corrective interventions are not considered to redefine the preventive frequency (scheduling).

Another use of filter type of intervention is the ability to manage operational interventions (OI) (white bubbles). This is possible by analyzing intervention times (linked to the repair time for operational response filter). GAOM is able to perform this type of analysis by displaying the bubble diameter as shown in Fig. 9. Thus, GAOM allows us to focus on the analysis of those OI longer stop operation, helping identify specific causes [18].

- Graph 6: *"Analysis of TI, filtering operational interventions"*. This graphic is equivalent to Graph 2 in structure, allowing only analyze operational interventions. This graph allows you to focus on the presence of equipment operation phenomena, related to frequency and impact on production.

Meanwhile, losses of efficiency in processes generate deviations between actual and expected or planned production. It is necessary to control these deviations identifying the expected production target and the existing deviations. GAOM facilitates the analysis of these deviations, relating the nominal production, planned and expected production:

- Nominal/ideal production: production accumulated over time ((ideal scenario without any kind of intervention) is given by the nominal process capability. Displays how the process should produce under OAE (Overall Effectiveness of the Asset) with value 100% [18].
- Planned Production: corresponds to the expected production (at nominal capacity) during the work planned time. This time horizon does not include planned intervention as a possible time for production.
- Expected production: production quantity in the operating time, using the nominal capacity as the reference.

- Graph 7: *"Analysis of TBI and production targets"*. This chart shows the set of possible variables in a single graph. The dotted lines represent the nominal, planned and expected production. Integrally, this chart allows us to analyze the impact of different phenomena intervention on production, compared to an ideal and/or planned scenario. For the latter, the analysis is complemented with Graph 8.

It is also possible to identify any deviation between the actual cumulative production and expected cumulative production by analyzing the behavior of the vertical distances between the center of the respective bubbles and the curve of cumulative production target (Fig. 11).

- Graph 8: *"Deviation between the actual cumulative production and cumulative production target"*. This graph is an exemplification of a possible analysis from Graph 7. Allows clearly to identify any deviation to the production target, considering one or more interventions. The arrows drawn graphically explain and represent the difference of production between both curves at any i-th intervention, clarifying some kind of patron between the type of intervention and the production delta. Also it is possible to clarify some increase/decrease of production steadily over time.

Regarding presented graphics, it is important to mention how must be integrated each particular analysis and graphical into the GAOM proposal, identifying problems and phenomena to improve and/or research, reaching correct conclusions. For this, will be mandatory to follow some basic rules (3), particular recommendations (2) and the following paragraph 3 with analysis and results.

Basic rules:

- The procedure for GAOM proposal construction must follow this sequence: Data source identification ensuring quantity and quality, relevant input filter, design and calculation of basic performance indicators, and finally, graphics construction and analysis.
- The procedure for GAOM analysis depend of the level of knowledge and particular interest of the analysts, then the starting point for inquiring could be:
  - Having already identified some particular phenomena (e.g. overload, accelerated wear, etc.) using graphs to explain and measure the effect, supporting KPI's results.
  - General search for phenomena that have the effect on the breach of goals (low productivity, availability and/or utilization), then GAOM analysis will support, preliminary, to identify the cause of unproductivity.
- The graphics lecture and analysis must be always from global (graphs 1, 2 and 7) to particular (3, 4, 5, 6 and 8) point of view.

Recommendations:

- Multidisciplinary teamwork for graphs' interpretation.
- To identify trends we recommend using at least 3 months of historical data.
- To analyze phenomena affecting production, we recommend filtering in short periods, ideally every week.
- Integrate GAOM into the ERP (production and maintenance) and standardize sequence analysis.

3 Analysis and Results

GAOM allows answering the questions initially posed in Table 1:

- **Area of operation/production answers**

GAOM allows us to select between the BPIs time between intervention/failures (TBI) and intervention time (TI), making it possible to analyze trends and outliers over time between failures of computer data, as well as intervention times. Moreover, for the specific time display of operational interventions, GAOM allows filtering by type of intervention (Figs. 6, 7 and 8), allowing work only with data type required intervention (Fig. 9).

The production rate is shown by comparing the curves in Fig. 10, which shows the relationship between the ideal scenarios, planned and expected, even if the center of each bubble (actual production) differs considerably from the rest of the dotted curves. This indicator is particularly important for analyzing equipment load.

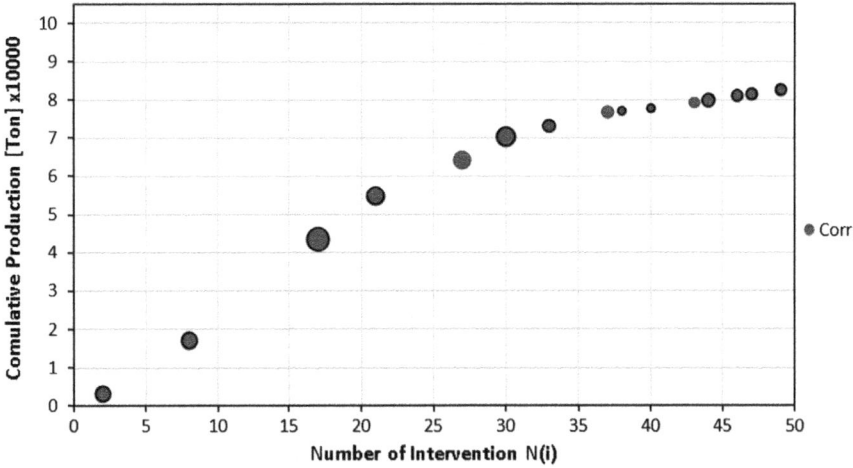

Fig. 6 Graph 3 "analysis of TBF, filtering corrective maintenance interventions". GAOM

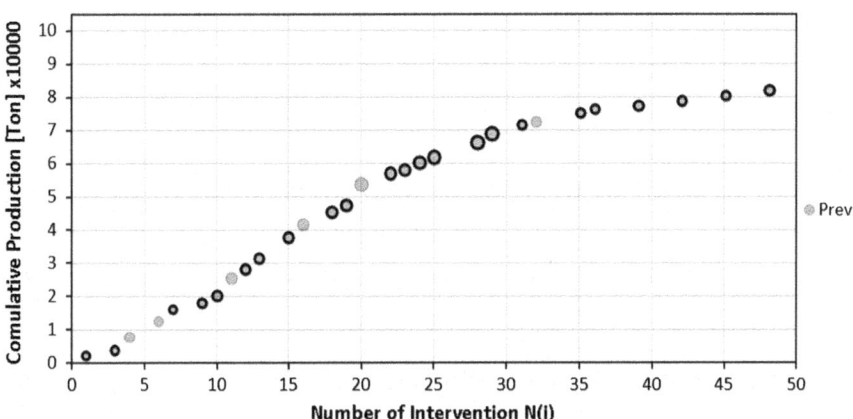

Fig. 7 Graph 4 "analysis of TI filtering preventive maintenance interventions". GAOM

It is possible to quantify the deviation between the actual output and production targets (planned and nominal) by analyzing the difference in the vertical axis (Graph 8, Fig. 11), either for current production status between one or various interventions, or to analyze the effect that an intervention has had on the production system.

- **Answers to maintenance area**

The frequency of corrective interventions "CM" between preventive interventions "PM" can be show by graphic analysis of Figs. 6 and 7, applying the filter of

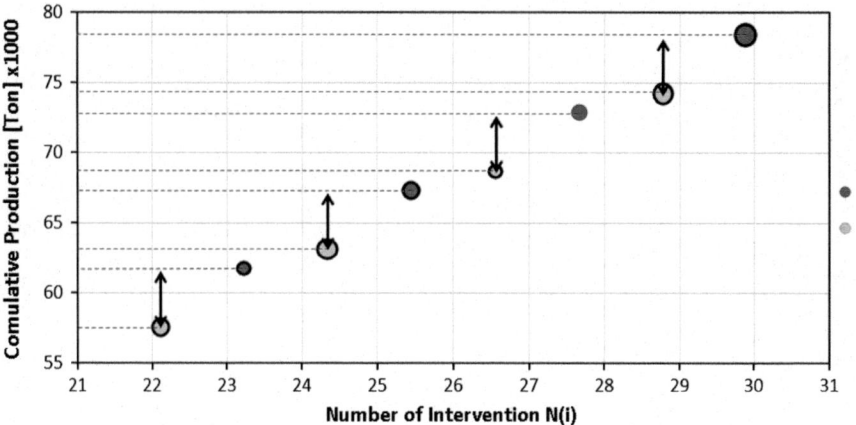

Fig. 8 Graph 5 "display of problems in the maintenance policy". GAOM method

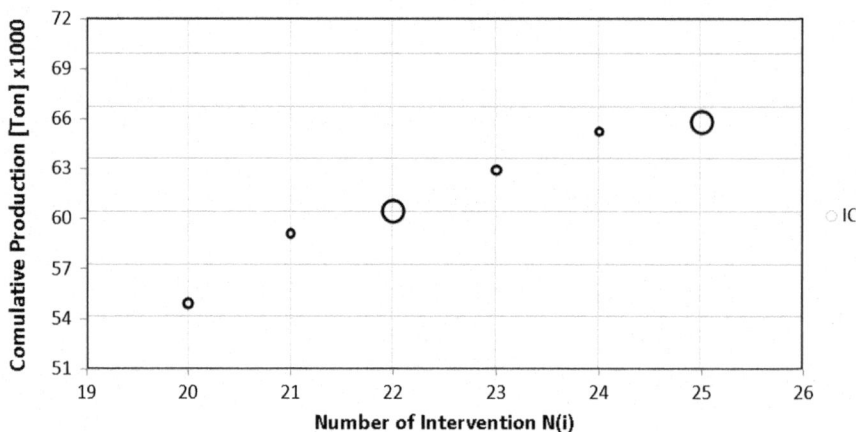

Fig. 9 Graph 6 "analysis of TI with filter of operational interventions". GAOM method

intervention that seeks to simplify the analysis for each type of maintenance. After applying the filter, a simple counting process is executed.

Using graph 2 (Fig. 5) analysis of intervention time, applying preventive and corrective maintenance filters respectively, it is possible to analyze the behavior of the execution times of preventive maintenance, and assess the adequacy of response from the maintenance area to corrective interventions, in terms of time is concerned. This is useful to improving the maintainability of the assets analyzed, for example by a stock management more efficient spare parts, investment in technology, design work areas, staff training, among other.

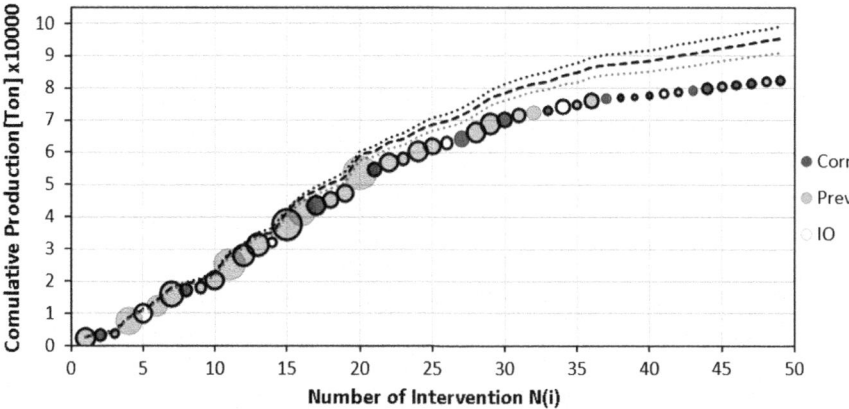

Fig. 10 Graph 7 "analysis of TBI and production targets". GAOM method

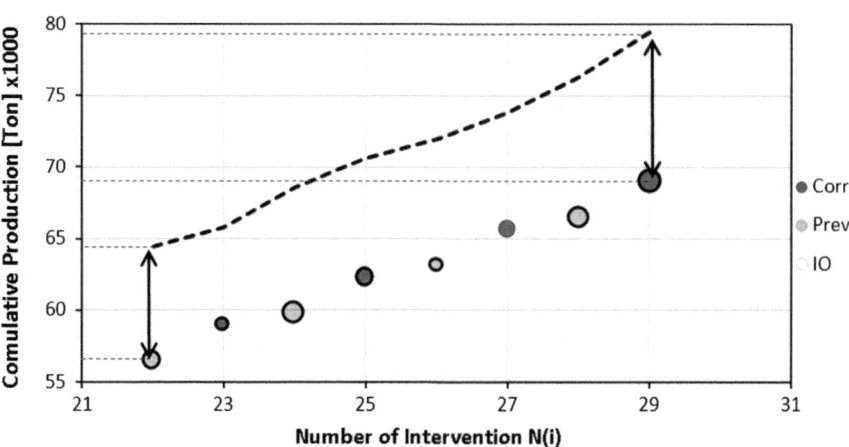

Fig. 11 Graph 8 "Deviation between the actual cumulative production and cumulative production target". GAOM method

It is possible to analyze the evolution in time of the selected indicator using both filters, either in a first phase, time between failures or repair time (bubble diameter) and, secondly, the type of intervention to be analyzed.

- **Answers to integration areas**

It is possible to know if there is a shift in the equipment performance and identify the responsible area by vertical comparison between actual production (bubbles) and the dotted lines of the most ideal scenarios. Furthermore, by comparing the dotted lines (ideal scenario, planned and expected), the distance between them also provides information on possible causes of production losses. Two possible cases occur:

- Case 1: considerable distance between the actual curve (bubble center) and expected curve, indicating that the loss of production is due to a variation on the workload of the equipment or system.
- Case 2: considerable distance between expected and planned curve, indicating that the loss of production is mostly caused by unplanned events.

Common symptoms associated with accelerated wear out of equipment can be identified by analyzing the evolution of changing times of good performance (tendency to decreased bubble radius, Fig. 4) and partly by analyzing increases of intervention time (trend increasing bubble radius, Fig. 5). Analyze the accelerated wear out of the equipment allows to identify potential investments for equipment replacement.

Analyzing the impact and trends of operational interventions (Fig. 5) due to falls in demand can be estimated if the plant capacity is adequate. This information may be useful in evaluating possible decisions regarding plant capacity.

4 Conclusions

This paper presents the GAOM method, which integrates the main indicators of maintenance and production, mainly: cumulative production, intervention times (corrective maintenance, preventive and operational interventions) and time between failures/intervention. With GAOM it is possible to identify individual or systemic phenomenon, measuring, evaluating or auditing the impact of decision related to operational procedures or maintenance policies or strategies; as well as incorporating the tool as a "big picture" of process performance, supporting to control operation and maintenance management. GAOM summarizes and complements the decision process to reduce production losses and thus maximizing business benefits.

The GAOM method consists of seven input information and one optional production data (production target). The information is classified as "intervention" (number of intervention, type of intervention, and stopped/not stopped equipment), "production" (cumulative production and defective products) and "time" (calendar accumulated time and time intervention).

GAOM, through a preliminary analysis, manages to answer all original questions (Table 1), demonstrating its effectiveness to support in a joint and independent manner, maintenance and production areas. For its construction, has sought an innovative design and easy to interpret, using graphics with high synergy in the information provided. It is complemented with simple indicators supporting making decisions. The integration between variables related to maintenance and operations impacting yields indicators for the overall management of the process.

GAOM supports analyst's measurement, computing and validating indicators. However, it should be noted the importance of quality and quantity of historical data. It is recommended the integration with other tools that center their analysis on complementary indicators, such as the GAMM [1] tool, focused maintenance management analysis. Thus, integration of both tools (GAMM and GAOM) would integrate more information about operations management, production and maintenance. This would improve the optimization of maintenance policies.

Another important feature GAOM is its flexibility to control processes at the level of equipment or system. This feature can be used as a mechanism to search for opportunities for improvement, starting from an overview of the system for, afterwards, subsystems and equipment analysis.

Regarding applicability of this method, the design of the tool allows to process information of any production process. To process the information, common calculation tools can be used as VBA (Visual Basic Programming) programming spreadsheet, or generation of algorithms for processing graphs that are part of the tool. As future lines of research, there is great development potential of the tool on automation, using more advanced programming languages that enable its implementation in enterprise resource planning systems (ERP), such as ERP SAP-PM. Another area of development is the application of GAOM in different production scenarios, validating, complemented and/or improving the current tool from the conceptual and practical point of view.

Appendices

Appendix 1

Standard format for data collection.

Regarding authors' experiences, it is proposed a standard format to collect historical data, specifically: Intervention Data, Production Data and Time Data (Table 8).

Appendix 2

Integrated databases.

GAOM requires the development of two integrated databases, which relate the information in Tables 9 and 10. These databases are necessary for graphics construction.

Table 8 Standard format

Intervention data			Production data		Time data	
Number of intervention	Class of intervention	Stop/no stop	Cumulative production	Cumulative defective production	Cumulative calendar time	Time of intervention
$\{0, 1, 2, \ldots, N\}$	$\{1, 2, 3\}$	$\{0, 1\}$	Productive unit	Productive unit	Time unit	Time unit
N_i	CI_i	DET_i	CP_i	CDP_i	CCT_i	TI_i
0	–	–	0	0	0	0
1	2	0	CP_1	CDP_1	CCT_1	TI_1
2	1	1	CP_2	CDP_2	CCT_2	TI_2
⋮	⋮	⋮	⋮	⋮	⋮	⋮
n	3	0	CP_n	CDP_n	CCT_n	TI_n

Table 9 Historical cumulative production data—GAOM

N_i	CP_i	TI_i	DET_i	OT_i	N_i	CP_i	TI_i	DET_i	OT_i
0	0	0	1	0	25	61,817	11	0	93
1	2165	6	0	132	26	62,939	18	0	58
2	3143	9	0	55	27	64,174	9	1	72
3	3777	6	0	32	28	66,217	12	0	126
4	7668	6	1	206	29	68,847	12	0	143
5	9972	16	0	127	30	70,279	8	0	80
6	12,386	6	1	127	31	71,517	7	0	72
7	15,965	6	0	192	32	72,452	7	1	64
8	17,095	13	0	62	33	73,062	9	0	36
9	17,992	7	0	47	34	74,375	21	0	77
10	20,192	8	0	116	35	74,992	7	0	33
11	25,311	8	1	280	36	76,264	7	0	79
12	27,989	8	0	147	37	76,751	9	1	35
13	31,248	8	0	169	38	77,045	9	0	19
14	32,024	16	0	41	39	77,301	8	0	17
15	37,540	9	0	288	40	77,646	9	0	19
16	41,590	9	1	217	41	78,241	17	0	33
17	43,490	15	0	117	42	78,715	7	0	27
18	45,236	9	0	93	43	79,238	9	1	29
19	47,243	9	0	105	44	79,850	10	0	41
20	53,699	10	1	325	45	80,457	7	0	35
21	54,792	17	0	70	46	81,002	9	0	32
22	56,873	10	0	117	47	81,478	10	0	32
23	57,894	10	0	58	48	81,963	7	0	33
24	60,228	10	0	128	49	82,516	10	0	32

Table 10 Nominal, planned and expected cumulative production

N_i	PN_{Acum_i}	PP_{Acum_i}	PE_{Acum_i}	N_i	PN_{Acum_i}	PP_{Acum_i}	PE_{Acum_i}
1**	–	–	–	26	72,923	69,603	69,003
2*	7783	7583	7583	27*	74,723	71,043	70,443
3**	8603	8403	8263	28**	77,233	73,553	72,953
4**	12,833	12,433	12,293	29**	80,333	76,493	75,893
5	15,363	14,963	14,823	30*	82,163	78,163	77,563
6**	18,223	17,463	17,323	31**	83,753	79,753	78,993
7**	22,053	21,293	21,153	32**	85,163	81,023	80,263
8*	23,413	22,453	22,313	33*	85,883	81,743	80,983
9**	24,613	23,653	23,373	34	87,603	83,463	82,523
10**	27,073	25,933	25,653	35**	88,673	84,193	83,253
11**	32,823	31,483	31,203	36**	90,393	85,773	84,833
12**	35,763	34,423	34,143	37*	91,223	86,463	85,523
13**	39,293	37,753	37,473	38*	91,603	86,843	85,903
14	40,263	38,543	38,263	39**	91,923	87,163	86,043
15**	46,333	44,233	43,953	40*	92,263	87,343	86,223
16**	50,853	48,573	48,293	41	93,103	88,183	86,883
17*	51,993	49,713	49,433	42**	93,973	88,713	87,413
18**	54,153	51,873	51,433	43*	94,683	89,283	87,983
19**	56,423	53,963	53,523	44*	95,503	90,103	88,803
20**	63,343	60,703	60,263	45**	96,403	91,003	89,503
21*	63,693	61,053	60,613	46*	97,183	91,643	90,143
22**	66,373	63,733	63,133	47*	98,003	92,463	90,783
23**	67,733	64,913	64,313	48**	98,863	93,323	91,443
24**	70,493	67,513	66,913	49*	99,643	93,963	92,083
25**	72,543	69,383	68,783				

In Table 10, specifically for the first column (N_i), the data is supplemented with a sign (*) or (**), the first one related to corrective interventions and the second one to preventive interventions. The data without a sign will correspond to an operational intervention

References

1. Ahuja IPS, Khamba JS (2008) Total productive maintenance: literature review and directions. Int J Qual Reliab Manag 25(7):709–756
2. Zhu Q, Lujia F, Mayyas A, Omar MA, Al-Hammadi Y, Al Saleh S (2015) Production energy optimization using low dynamic programming, a decision support tool for sustainable manufacturing. J Clean Prod 105(2015):178–183. doi:10.1016/j.jclepro.2014.02.066
3. May G, Barletta I, Stahl B, Taisch M (2015) Energy management in production: a novel method to develop key performance indicators for improving energy efficiency. Appl Energy 149(2015):46–61. doi:10.1016/j.apenergy.2015.03.065
4. van Veen-Dirks Paula (2005) Management control and the production environment: a review. Int J Prod Econ 93–94(2005):263–272. doi:10.1016/j.ijpe.2004.06.026
5. Fleischer J, Weismann U, Niggeschmidt S (2006) Calculation and optimisation model for costs and effects of availability relevant service elements. In: Proceedings of LCE, pp 675–680

6. Birolini A (2007) Reliability engineering, 5th edn. Springer, Berlin. ISBN: 978-3-540-49388-4
7. Barberá L, Crespo A, Viveros P, Stegmaier R (2012) Advanced model for maintenance management in a continuous improvement cycle: integration into the business strategy. Int J Syst Assur Eng Manag 3(January–March (1):47–63. doi:10.1007/s13198-0120092-y
8. Crespo Márquez A, Gupta J (2006) Contemporary maintenance management. Process, framework and supporting pillars. Omega. Int J Manag Sci 34(3):325–338. doi:10.1016/j.omega.2004.11.003
9. Chai J, Liu JNK, Ngai EWT (2013) Application of decision-making techniques in supplier selection: a systematic review of literature. Expert Syst Appl 40(2013):3872–3885. doi:10.1016/j.eswa.2012.12.040
10. Watson H, Goodhue D, Wixom B (2002) The benefits of data warehousing: why some organizations realize exceptional payoffs. Inf Manag 39(6):491–502
11. Park Y (2006) An empirical investigation of the effects of data warehousing on decision performance. Inf Manag 43:51–61
12. March S, Hevner A (2007) Integrated decision support systems: a data warehousing perspective. Decis Support Syst 43:1031–1043
13. Pomponio L, Le Goc M (2014) Reducing the gap be-tween experts' knowledge and data: the TOM4D methodology. Data Knowl Eng 94:1–37
14. Watson H, Goodhue D, Wixom B (2002) The benefits of data warehousing why some organizations realize exceptional payoffs. Inf Manag 39:491–502
15. Neely A, Gregory M, Platts K (1995) Performance measurement system design: a literature review and research agenda. Int J Oper Prod Manag 15(4):80–116
16. Kaplan RS, Norton DP (1992) The balanced scorecard: measures that drive performance. Harvard Bus Rev, pp 71–79
17. Ireland F, Dale BG (2001) A study of total productive maintenance implementation. J Qual Maint Eng 7(3):183–192
18. Biasotto E, Dias A, Ogliari A (2012) Balanced score-card for TPM maintenance management. Santa Catarina Federal University, vol 8(2)
19. Barberá L, Crespo A, Viveros P, Arata A (2013) The graphical analysis for maintenance management method: a quantitative graphical analysis to support maintenance management decision making. Qual Reliab Eng Int 29(1):77–87
20. Crespo A (2007) The maintenance management frame-work. Springer, London

Author Biographies

Pablo Viveros Gunckel Researcher and Academic at Technical University Federico Santa María, Chile, has been active in national and international research, both for important journals and conferences. Also he has developed projects in the Chilean industry. Consultant specialist in the area of Reliability, Asset Management, System Modelling and Evaluación of Engineering Projects. He is Industrial Engineer and Master in Asset Management and Maintenance.

Adolfo Crespo Márquez is currently Full Professor at the School of Engineering of the University of Seville, and Head of the Department of Industrial Management. He holds a PhD in Industrial Engineering from this same University. His research works have been published in journals such as the International Journal of Production Research, International Journal of Production Economics, European Journal of Operations Research, Journal of Purchasing and Supply Management, International Journal of Agile Manufacturing, Omega, Journal of Quality in Maintenance Engineering, Decision Support Systems, Computers in Industry, Reliability Engineering and System Safety, and International Journal of Simulation and Process Modeling, among others. Prof. Crespo is the author of seven books, the last four with Springer-Verlag in 2007, 2010, 2012, and 2014 about maintenance, warranty, and supply chain management. Prof. Crespo leads the Spanish Research Network on Dependability Management and the Spanish Committee for Maintenance Standardization (1995–2003). He also leads a research team related to maintenance and dependability management currently with 5 PhD students and 4 researchers. He has extensively participated in many engineering and consulting projects for different companies, for the Spanish Departments of Defense, Science and Education as well as for the European Commission (IPTS). He is the President of INGEMAN (a National Association for the Development of Maintenance Engineering in Spain) since 2002.

Luis Barberá Martínez is PhD (Summa Cum Laude) in industrial Engineering by the University of Seville, and Mining Engineer by UPC (Spain). He is a researcher at the School of Engineering of the University of Seville and author of more than fifty articles, national and main internationals. He has worked at different international Universities: Politecnico di Milano (Italy), CRAN (France), UTFSM (Chile), C-MORE (Canada), EPFL (Switzerland), University of Salford (UK) or FIR (Germany), among others. His line of research is industrial asset management, maintenance optimization and risk management. Currently, he is Spain Operations Manager at MAXAM Europe. He has three Master's degrees: Master of Industrial Organization and Management (2008/2009) (School of Engineering, University of Seville), Master of Economy (2008/2009) (University of Seville) and Master of Prevention Risk Work (2005/2006). He has been honoured with the following awards and recognitions: Extraordinary Prize of Doctorate by the University of Seville; three consecutive awards for Academic Engineering Performance in Spain by AMIC. (2004/2005), (2005/2006) and (2006/2007); honour diploma as a "Ten Competences Student" by the University of Huelva and CEPSA Company (2007); graduated as number one of his class.

Juan Pablo González business developer at Megacenter Group, USA, has been active in national and international entrepreneurship. Also he has developed projects in the Chilean logistics and transport industry and is director of Chargello Chile. Industrial Engineer graduated of the Technical University Federico Santa Maria.

Value Assessment of e-Maintenance Platforms

Marco Macchi, Luis Barberá Martínez, Adolfo Crespo Márquez,
Luca Fumagalli and María Holgado Granados

Abstract The present paper proposes a methodology to assess the values created through the development of an E-maintenance platform. This is based on the postulation that the platform is providing a set of services developed as an application of a set of E-technologies to support targeted maintenance processes within a given company. Henceforth, the applied service is the building block used in order to assess the value creation for the company. The methodology is firstly presented, discussing on the applied services as initiators of new operational practices in maintenance processes and, subsequently, on the values created for the business of the company deciding to invest in an E-maintenance platform. The methodology is then tested in the case of a manufacturing company: the test proves the potentials of the methodology in order to support the investment appraisal of E-maintenance solutions; the specific case concerns solutions developed in the framework of a SCADA system originally implemented for production purposes.

Keywords E-maintenance · Services · Service development · Value assessment

M. Macchi (✉) · L. Fumagalli
Department of Management, Economics and Industrial Engineering,
Politecnico di Milano, Piazza Leonardo Da Vinci 32, 20133 Milan, Italy
e-mail: marco.macchi@polimi.it

L. Fumagalli
e-mail: luca1.fumagalli@polimi.it

L. Barberá Martínez · A. Crespo Márquez
Department of Industrial Management, School of Engineering,
University of Seville, Camino de los Descubrimientos s/n, 41092 Seville, Spain
e-mail: lubarmar@us.es

A. Crespo Márquez
e-mail: adolfo@us.es

M. Holgado Granados
Department of Engineering, Institute for Manufacturing, University of Cambridge,
17 Charles Babbage Road, CB3 0FS Cambridge Cambridge, UK
e-mail: mh769@cam.ac.uk

© Springer International Publishing AG 2018
A. Crespo Márquez et al. (eds.), *Advanced Maintenance Modelling for Asset Management*, DOI 10.1007/978-3-319-58045-6_16

1   Introduction

This paper is concerned with innovation in maintenance management driven by Information and Communication Technologies (ICTs). E-maintenance can be considered as an avant-garde strategy to this regard, in the spectrum of alternative ways [7, 9, 14]. Muller et al. [23] gave birth to the E-maintenance definition, keeping in mind the European standard [12]: *"maintenance support which includes the resources, services and management necessary to enable proactive decision process execution. This support includes e- technologies (ICT, web-based, tether-free, wireless, infotronics technologies) but, also, e-maintenance activities (operations or process) such as e-monitoring, e- diagnosis, e-prognosis, etc.".* Moreover, different authors consider E-maintenance under different perspectives: (i) as a maintenance strategy (a method of management using real-time electronic data sourced from computers through digital technologies), (ii) as a maintenance plan (a set of tasks with an interdisciplinary approach that includes monitoring, diagnosis, prognosis, decision and process control), (iii) as a type of maintenance (condition-based maintenance, proactive and predictive maintenance) and, in general, (iv) as a maintenance support (consisting of resources and services necessary to carry out the maintenance activities). Further on, an important concern, partially analyzed in literature, is the industrial adoption of E-maintenance platforms. To this end, it is interesting to be capable to assess how maintenance interact with newly introduced technologies and, in particular, how much maintenance processes can get benefits, when changing from an AS-IS to a TO-BE situation, after introducing the technologies provided by E-maintenance. Therefore, besides an in-depth on technical development of the E-Maintenance platform, additional questions, for better business understanding, have to be answered, such as, e.g.: how much money would a company save, or gain, by adopting this kind of technologies? how much is the investment rewarding for the invested capital? Value assessment is an important issue in order to answer to these questions. The paper is developed in this mind set. Henceforth, it studies E-maintenance as a maintenance support, by focusing on the value creation that can be achieved by means of E-technologies (the resources) and their applied services in targeted maintenance processes of a given company. The paper is a further development of previous researches of some authors of the paper, focused on the cost-benefit analysis of E-maintenance (e.g. Fumagalli et al. [13]); it is then a contribution to the still open issue of economic analysis of E-maintenance as a technological platform for maintenance process/system improvement, developed as a derivation of the approach presented in Macchi et al. [22] for value-driven engineering.

Considering such a focus, the processes of a maintenance management system are initially discussed through a brief literature analysis; a quick review is also provided on the most promising E-technologies in order to innovate the maintenance management system (paragraph 2). Keeping in mind this process—technological context, the concept of applied services is introduced (paragraph 3) and the methodology for value assessment of applied services is then presented (paragraph 4). The paper finally discusses a case study taken from the textile industry (paragraph 5), with the objective of studying the use of the methodology to support the investment appraisal of an E-maintenance platform.

2 Literature Review

2.1 Short Review on Maintenance Processes

Maintenance management is a multi-disciplinary area consisting of a wide set of many interrelated approaches [33], TPM and RCM are just a few examples. According to Pintelon and Van Wassenhove [26] maintenance management depends on the internal and external factors of the organization; indeed, the most desirable situation is the full integration of maintenance management in the organization itself [36]. Numerous proposals of integral maintenance management models are shown in literature. Each proposal identifies a list of processes, activities or actions, part of the integral management system. Table 1 shows the main relevant innovations proposed by each maintenance management models through the years.

Table 1 Innovations of maintenance management models in chronological order

Innovation	References
Maintenance performance measurement	[26]
Necessity to link maintenance with other organizational functions + Importance of using quantitative techniques for management	[27]
Focus on effectiveness and efficiency of maintenance + Importance of the managerial leadership in maintenance	[36]
Integrated modelling approach based on the concepts of situational management theory	[31]
Use of Japanese concepts and tools for the statistical control of maintenance processes	[10]
Orientation to the computer use, adopting a standard for information exchange	[16]
Use of e-maintenance + analysis of the outsourcing convenience + incorporation of tacit and explicit knowledge in a computer database + knowledge management	[34, 37]
Qualify function deployment and total productive maintenance into the model	[28]
Maintenance as contributor to the fulfilment of "external stakeholders" requirements + model oriented to the improvement of the operational reliability besides the life cycle cost of assets	[8, 33]
Design of a new maintenance management model aligned to quality management standard	[21]
Design of preventive maintenance plans and needed resources + selection of critical spare parts + involvement of e-technologies	[2]

Adapted from Barberá et al. [2]

Let's refer now to Barberá et al. [2] as a reference. This paper proposes a model consisting of 7 managerial actions to ensure efficiency, effectiveness and continuous improvement of the maintenance management system: (i) definition of objectives, strategies and maintenance responsibilities; (ii) ranking of the equipments; (iii) analysis of weaknesses in high-impact equipments; (iv) design of maintenance and resources plans; (v) maintenance scheduling and resources management; (vi) control, evaluation and data analysis; (vii) life cycle analysis. The 7 actions can be further defined according to the maintenance tasks needed for their implementation. For example, the analysis of weaknesses in high-impact equipments—one managerial action out of the whole set proposed—requires many tasks such as, e.g. technical inspections, analysis of repetitive and chronic failures and Root Cause Analysis implementation. Maintenance scheduling and resources management—another action—is concerned with other tasks such as, e.g. work order/task management, allocation of human and material resources to work orders/tasks, etc.

These are just few examples for reminding of the fact that, once a maintenance management model is selected (the one from Barberá et al. [2] or others), therein one can find a taxonomy of processes/actions/activities to be carried out as well as a set of needed tasks (and eventual subtasks at a detailed level) for implementation. In our understanding, when we get to a sufficient level of understanding—which means also the tasks and subtasks, we can derive the requirements for support, achievable from the enabling technologies.

2.2 Short Review on E-Technologies

The most promising technological components of an E-maintenance platform are herein shortly presented. Each component (hardware, firmware or software) is discussed, by focusing on its main functional features, accordingly with the functions it is capable to support.

A Smart Sensor is a transducer that provides functions in order to generate the digital representation of a sensed/controlled quantity (e.g. temperature, pressure, vibration, strain, flow, pH, etc.). Indeed, it supports the following functional features: the real-time data acquisition from a physical environment; the data processing based on predefined algorithms; the digital data transmission; the connection into applications in a networked environment made of transducers and actuators. Smart Sensors can be improved thanks to the use of MEMS, a technology enabling to integrate mechanical and electronic components, transducers and actuators, on a silicon wafer substrate. The advantage of using MEMS to build Smart Sensors is to achieve their functional features along with miniaturization.

RFID is a technology that can be used to identify an entity—being it either a maintenance operator, a physical asset, a tool or a spare part—and to keep its information. The RFID tag is in fact attached to the entity and enables to exchange data locally with a RFID reader, this leads to the following functional features for the RFID system: the capability to provide the identification of the tagged entity;

the capability to store, read and write entity's data; the connection to a back office system for further data processing (e.g. a CMMS, storing more data for the entity, or a smart sensor, performing some data analysis).

Powerful handheld computing devices, such as PDAs, smart-phones and tablet PCs, can be considered an aid for the maintenance personnel during daily activities. Indeed, they are devices with many functional features useful for supporting the activities done during the personnel's mobility, such as, e.g.: the on field support to office activities (e.g. to take notes, keep records, use spreadsheets while being close to the equipment); the support to read bar codes or RFID tags; the wireless connection to internet or other local area networks, to download and upload data from/to remote computers, equipments and databases; the provision of multi-media tools, to make photos/videos, register audios; the geo-localisation (when the device is equipped with GPS).

A SCADA system enables controlling and monitoring continuously in time processes and equipments in a plant. This is possible thanks to different functional features: the real-time data acquisition from a physical environment, through remote devices (such as PLCs—Programmable Logic Controllers—or RTUs—Remote Terminal Units); the synchronized communication between remote devices in the system; the closed-loop process control of the equipments (through PLCs/RTUs); the capability to process huge data volumes collected from field; the support for performing analysis of different process or equipment variables (online and offline analysis) and for report generation.

The diagnostics and prognostics toolbox is cited as the constituting element of an Intelligent Maintenance System (IMS) platform. It automatically integrates the readings from multiple sensors, in order to assess the degradation of an equipment through various models/algorithms. Indeed, the functional features of such a toolbox are intended for diagnosis, prognosis and further decision-making for carrying out maintenance interventions based on the condition of the equipment. More precisely, the toolbox is typically endowed with: a capability to provide diagnosis, through memorizing significant signatures/patterns in order to recognize a degradation; a capability to provide the prognosis of the degradation, thanks to a set of models/algorithms for the prediction of performances; a synchronized communication with other remote devices/systems (e.g. the CMMS where scheduling of maintenance resources, required on condition, is managed).

A CMMS (Computerized Maintenance Management System) software package maintains information on maintenance activities. Therefore, the CMMS enables to manage the execution of the maintenance activities as well as to monitor their performances. In particular, the CMMS is endowed with a number of functional features in order to support: maintenance planning/scheduling, work order management, performance monitoring, budgeting/cost controlling, spare parts management.

RMESs (Reliability and Maintenance Engineering Systems) are software tools for modelling, analysis and simulation, to support maintenance engineers in designing or redesigning the maintenance system. To this end, a RMES is generally endowed with a number of functional features in order to provide: support for reliability analysis/prediction at the component level, for failure mode, effects and criticality analysis, for reliability and availability analysis at the system level (e.g. through a Reliability Block Diagram). It is worth mentioning that RMESs, available on the market, differ a lot, ranging from pure statistical tools to more dedicated tools specifically supporting maintenance engineering tasks.

The reader should refer to many literature references for more details regarding the above mentioned technological components. In this paper we provide just a very small sample: [3, 29, 39] for smart sensors and MEMS; [5, 6, 11] for RFID systems; [4, 11, 38] for PDAs, smart-phones and, in general, mobile devices; [1] for SCADA system; [18, 20, 32] for diagnostics and prognostics toolboxes; [24] for CMMSs [35] for RMESs.

3 From Processes and Technologies to Applied Services

We introduce the terms service/applied service, to put a bridge among functional features of E-technologies and requirements of maintenance processes.

The term service is largely adopted: several definitions exist in different domains (such as industrial, economic, etc.) and sometimes more than one definition appears in the same domain [17]. Considering the software domain, Huang and Mason [15] present services as particular entities with some unique features that differentiate from other entities, named as objects or components in the Object-Oriented or Component-Based development paradigms. Similarly to objects and components, a service is an entity consisting of data and operations over the data. On the other hand, some specific features apply to services: a service is an entity with its own processes, that can interact with other services; the interaction may have different meanings for objects and components, because objects and components may not necessarily have own stand-alone processes for themselves. Thus, according to the literature, *a service is promising as building block for the development in a process oriented environment* and this stands as a first remark for this research work. Indeed, in our opinion, the features of a service can be more easily conceived in a business environment, as business people are certainly well aware of the concept of process. More precisely, business people are aware of the concept of processes at an enterprise level: since the value assessment deals with understanding of values generated for some processes at the enterprise level, our first remark seems an adequate choice.

This business-oriented assumption can be also supported by technical consid-erations in literature, see, e.g. [40]: according to these authors, service creation is based on a service composition; this allows creating enhanced services from already existing services, using the application of new technologies to existing processes (i.e. we intend this as existing enterprise processes). As a second remark, we can then deduce that *the services can be interpreted as operational processes having different granularities with respect to the enterprise processes they support*, being the operational processes more limited for what concern their organizational scope, if compared to the enterprise processes. This is a concept leading us to think a third,

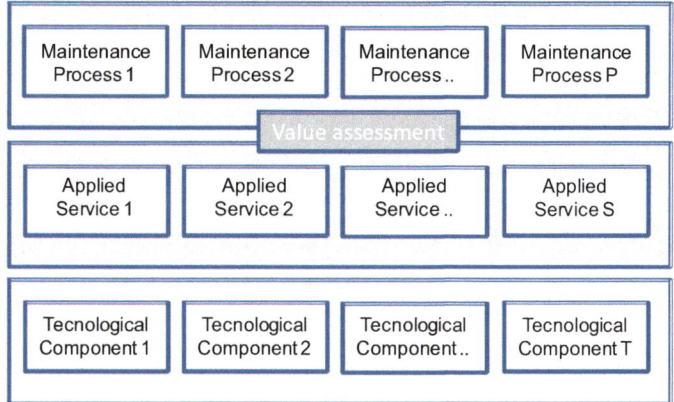

Fig. 1 A multi-layered view of an E-maintenance platform and its link to value assessment

and related, remark: *the development of an E-maintenance platform can be seen as a composition of sets of operational processes (i.e. the services) to finally support sets of enterprise processes in the maintenance area of a company.* A similar concern—i.e. modelling at different levels of process definition, either at operational or enterprise level—comes out from many relevant papers in the E-maintenance literature (e.g. Levrat et al. [19]).

Next Fig. 1 summarizes the whole concept in a 3-layered architecture of an E-maintenance platform: a set of E-technologies/technological components (bottom layer) is applied to develop services (mid layer) to support a set of enterprise processes part of a maintenance management system (top layer). The services' and processes' layer are under concern for value assessment; in particular, services are the building blocks from which reasoning for value assessment can be developed. The methodology proposed in this paper is based on this general assumption.

4 Methodology for Value Assessment

4.1 Process Innovation

The paper postulates that the maintenance processes are driving the identification of the applied services to be developed from one or more enabling E-technologies. Indeed, the innovation is considered process driven. In particular, each process may require one or more applied services as support. At the same time, an applied service may need one or more functional features to be developed, as they are available in the generic capabilities of each technological component of the E-maintenance platform.

The values generated through the applied services should then be assessed to justify the technological development. In particular, we consider the following basic assumptions as background of the value assessment:

the technological development normally leads to the emergence of novel operational practices at process level;

values generated with the technological development can be better grasped and estimated when one think how a process is changing thanks to the introduction of new practices.

Henceforth, in order to help sharing the vision between the different stakeholders involved in the innovation (i.e. the technical personnel, maintenance manager, investor, …), it seems worthy providing a narrative of changes in the operational practices, as an intermediate step before assessing the values generated through the technological development. Since our focus is on the applied services resulting from the technological development, we can be more precise by asserting another remark for this research work: *the value assessment is done through estimating the values generated by the applied services as they are the initiators of new operational practices in the maintenance processes*. Therefore, the new operational practices can be interpreted as "concrete materialization" of the process innovation through applied services, and it is worthy to analyze them for complete understanding of innovation itself.

4.2 Value Assessment

After the identification of the applied services and once new operational practices are characterized (as assumed in previous par. 4.1), in our proposal value assessment is done in three further steps.

Firstly, we identify the value proposition, to motivate the development of the applied services. We use this term as it can be derived from the business literature. Keeping in mind general definitions from many authors, we remind that the value proposition is an aggregation, or bundle, of products and/or services that a firm offers to customers [25] and represents the value for being competitive with respect to other firms in the market [30]. Applying to our case, we can re-interpret the definitions in such an assertion: the *value proposition is the bundle of applied services that the developer of the E-maintenance platform is offering to the maintenance personnel (i.e. the customer) for enabling a more competitive maintenance business*.

As a second step, the identification of values associated to the applied services under concern is done. To this end, a generic value tree is presented in Fig. 2, as a reference model. This model can be considered as a guide to help identifying specific values created by each applied service resulting from a given technological development. In the remainder of this paper, we focus only tangible values. Third

Fig. 2 Structure of tangible and intangible values (adapted from (RFId-IPO 2007))

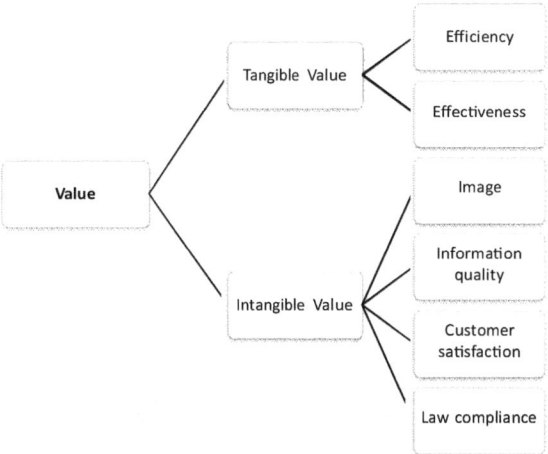

and last step is then concerned with the estimation of the values previously identified; this can be achieved using different information sources (as, e.g. mathematical models, benchmarks, experts' opinions).

5 The Case Study in Manufacturing

5.1 Context

The case study is taken from the textile industry. The plant under concern is part of a multinational group producing high quality textiles for the fashion sector. The strategic objective of high quality requires the accurate selection of raw material, the utilization of control technologies in different production phases, as well as the adoption of advanced machining systems. Considering responsibilities in maintenance organization, the structure of the plant is divided in two functional areas: the former corresponds to the production line; the latter comprises all the utilities for the distribution of electric energy, waters and steams, for processing and draining water outflows, for processing chemicals used by the production process. The utilities are the scope of the case study: herein, a SCADA system is the enabling technology on which the group has invested driven by production purposes. Due to its openness as technological platform and thanks to the competences available in the company for software development, the SCADA system has also become a promising architecture for continuous development of applied services to ease the tasks of the maintenance personnel.

5.2   Process Innovation

Table 2 Summarizes the Technological Development in the Case Study (Read the Table from Left to Right)

We start from the functional features of the technological component (the SCADA system) used in order to develop the applied services supporting the execution of some targeted process (considering just the third column of the table, herein we also intend a task/subtask within a target process/activity/action—see what said in par. 2.1, since now on, for simplicity we will use just the process term).

It is worthy observing that, if one considers the business view of the investor, the approach is exactly the opposite way (from right to left): innovation starts from processes, going backward in order to identify the applied services and functional features that have to be developed through the investment. In this paper we are assuming this view.

Process execution, enriched with new practices initiated thanks to applied services, can then be envisioned through a narrative of the novelties.

Alarm handling—When a fault is detected from real-time data acquisition, a synchronized communication is triggered in order to alert the maintenance manager and the technical personnel. The alert is carried out through different means, e.g. an e-mail sent to the PC in the control room and an SMS/automatic voice sent to a mobile device, such as a smart-phone.

Lubricants' reintegration—The operating time and conditions of interest of the plant (i.e. the machining speeds and temperatures) can be monitored thanks to the real-time data acquisition and capability to process huge data volumes. This allows to build—based on the technical specifications provided by the lubricants' producers—a data model for predicting the remaining useful life. The prediction is shown by means of a visual aid in a PC of the control room to the technical personnel, who may decide to make the lubricants' reintegration. It is worthy noticing that, by using the visual aid in the control room, the technical personnel can reduce walk around inspections through the plant at an acceptable level.

Continuous condition monitoring—Online decision-making is supported thanks to performance analysis of process/equipment variables in the time domain (e.g. power absorption of electric motors …). This can be continuous, thanks to real-time data acquisition, and effective, thanks to the capability to process huge data

Table 2 Functional features, applied services, target processes

Functional features	Applied service	Target process
Real-time data acquisition + Synchronized communication	Maintenance work order request	Alarm handling
Real-time data acquisition + Capability to process huge data volumes	Remaining Useful Life Prediction	Lubricants' reintegration
Real-time data acquisition + Capability to process huge data volumes + Support for performing analysis of process/equipment variables	Online performance analysis	Continuous condition monitoring

Table 3 Value propositions and values

Applied service	Value proposition	Value
Maintenance work order request	Plant surveillance 24 h and quick intervention	Effectiveness due to reduced down times for intervention
Remaining Useful Life Prediction	Prevention from over-lubrication and subsequent risk of drifts of oils, spoiling the product quality	Effectiveness due to higher production quality
	Prevention from under-lubrication and subsequent risk of unexpected downtimes	Effectiveness due to reduced down times for intervention
	Avoidance of useless walk around inspections for periodic control of lubricants	Efficiency due to reduced wasted times
Online performance analysis	Anticipated detection of failures + Reduction of unscheduled plant stoppages	Effectiveness due to reduced down times for intervention

volumes. Besides, the support for performing analysis allows to develop graphical utilities to help technical personnel, based on own expertise, easily analyzing performances of processes/equipments, so to finally advice the plant stoppage, when required, for making a condition-based maintenance.

5.3 Value Assessment

Table 3 reports the value proposition of each applied service and the subsequently created value. Each value is identified as a specification of the value proposition and thanks to the support of the generic value tree of Fig. 2.

The estimates of achieved values—herein not presented—were defined correspondingly. Indeed, according to the maintenance manager, improvement of effectiveness and efficiency were measured as percentages of reduction/increase of operational performances (mean downtimes of interventions, mean quality rate of the production line and workloads of maintenance personnel). This was needed to appraise the investment during the years of technological development in the framework of the SCADA system, in order to finally calculate related financial indicators (e.g. payback time).

6 Conclusions

This research work is a starting point to further develop the theory of value assessment of E-maintenance. Missing issues are the analysis of intangible values as well as the complete coverage of technologies and processes target of interest for the development of applied services. In future works, the research will go through the full review of E-technologies and their functional features together with understanding of main potentials of applied services in different maintenance processes.

Acknowledgements This research work has been originated through empirical research done by the Observatory TeSeM (Technologies and Services for Maintenance, www.tesem.net) of the School of Management of Politecnico di Milano. Its theoretical development has been finalized within the context of iMaPla (Integrated Maintenance Planning), an EU-sponsored project by the Marie Curie Action for International Research Staff Exchange Scheme (project acronym PIRSES-GA-2008-230814—iMaPla).

References

1. Bangemann T, Rebeuf X, Reboul D, Schulze A, Szymanski J, Thomesse JP, Thron M, Zerhouni N (2006) PROTEUS—creating distributed maintenance systems through an integration platform. Comput Ind 57(6):539–551
2. Barberá L, Crespo A, Arata A, Stegmaier R, Viveros P (2012) Advanced model for maintenance management in a continuous improvement cycle: integration into the business strategy. Int J Syst Assur Eng Manage. doi:10.1007/s13198-012-0092-y (Springer)
3. Bult K, Burstein A, Chang D, Dong M, Fielding M (1996) A distributed, wireless MEMS technology for condition based maintenance, US DTIC (Defense Technical Information Center)
4. Campos J, Erkki J, Prakash O (2009) A web and mobile device architecture for mobile e-maintenance. Int J Adv Manuf Technol 45(1–2):71–80
5. Chang YS, Oh CH, Whang YS, Lee JJ, Kwon JA, Kang MS, Park JS, Park U (2006) Development of RFID enabled aircraft maintenance system. In: IEEE international conference on industrial informatics, pp 224–229
6. Ko C-H (2009) RFID-based building maintenance system. Autom Constr 18:275–284
7. Crespo Márquez A, Blanchar C (2006) A decision support system for evaluating operations investments in high-technology business. Decis Support Syst 41:472–487
8. Crespo Márquez A (2007) The maintenance management framework. Models and methods for complex systems maintenance. Springer, London
9. Campbell DA, Sarker S, Valacich JS (2006) Collaboration using mobile technologies: when is it essential?. In: Proceedings of the international conference on mobile business (ICMB'06), IEEE, Copenhagen
10. Duffuaa S, Raouf A, Dixon Campbell J (2000) Maintenance system. Planning and control. Limusa, México
11. Emmanouilidis C, Liyanage JP, Jantunen E (2009) Mobile solutions for engineering asset and maintenance management. J Qual Maintenance Eng 15(1):92–105
12. EN 13306:2001 (2001) Maintenance terminology. European standard. CEN (European Committee for Standardization), Brussels
13. Fumagalli L, Di Leone F, Jantunen E, Macchi M (2010) Economic value of technologies in an e-maintenance platform. In: Preprints of 1st IFAC workshop, A-MEST'10, advanced maintenance engineering, services and technology, Lisboa, Portugal, 1–2 July 2010
14. González-Prida V, Barberá L, Crespo A (2012) E-maintenance applied to the warranty management. Int J E-Bus Dev 2(1):16–22 World Academic Publishing

15. Huang H, Mason RA (2006) Model checking technologies for web services. In: Proceeding of the fourth IEEE workshop on software technology for future embedded and ubiquitous System SEUS 2006 and the second international workshop on collaborative computer integrated quality assurance WCCIA 2006, art. no. 1611738, pp 217–222

16. Hassanain MA, Froese TM, Vanier DJ (2001) Development of a maintenance management model based on IAI standards. Artif Intell Eng 15(1):177–193

17. Grida Ben Yahia I, Bertin E, Pierre Deschrevel J, Crespi N (2006) Service definition for next generation networks. In: Proceedings of the international conference on networking, international conference on systems and international conference on mobile communications and learning technologies, ICN/ICONS/MCL'06 2006, art. no. 1628268

18. Jardine A, Lin D, Banjevic D (2006) A review on machinery diagnostics and prognostics implementing condition-based maintenance. Mech Syst Signal Process 20(7):1483–1510

19. Levrat E, Iung B, Crespo Marquez A (2008) E-maintenance: review and conceptual framework. Prod Plan Control 408–429

20. Lee J, Ni J, Djurdjanovic D, Qiu H, Liao H (2006) Intelligent prognostics tools and e-maintenance. Comput Ind 57:476–489

21. López M, Gómez JF, González V, Crespo A (2010) A new maintenance management model expressed in UML. Reliability, risk and safety: theory and applications. Taylor & Francis Group, London, ISBN 978-0-415-55509-8

22. Macchi M, Crespo Márquez A, Holgado M, Fumagalli L, Barberá Martínez L (2014) Value-driven engineering of E-maintenance platforms. J Manufact Technol Manage 25(4): 568–598

23. Muller A, Crespo Marquez A, Iung B (2005) On the concept of e-maintenance: review and current research. Reliability engineering and system safety, 2008: 1165–1187

24. O'Hanlon T (2005) Computerized maintenance management and enterprise asset management best practices. Available at http://www.cmmscity.com

25. Osterwalder A, Pigneur Y (2010) Business model generation—a handbook for visionaries, game changers, and challengers. Wiley, Hoboken, New Jersey

26. Pintelon L, Van Wassenhove L (1990) A maintenance management tool. Omega 18(1):59–70

27. Pintelon LM, Gelders LF (1992) Maintenance management decision making. Eur J Oper Res 58(3):301–317

28. Pramod VR, Devadasan SR, Muthu S, Jagathyraj VP, Dhakshina Moorthy G (2006) Integrating TPM and QFD for improving quality in maintenance engineering. J Qual Maintenance Eng 12(2):1355–2511

29. Ramamurthy H, Prabhu BS, Gadh R, Madni AM (2007) Wireless industrial monitoring and control using a smart sensor platform. IEEE Sens J 7(5)

30. Richardson J (2008) The business model: an integrative framework for strategy execution. Strateg Change 17:133–144

31. Riis J, Luxhoj J, Thorsteinsson U (1997) A situational maintenance model. Int J Qual Reliab Manag 14(4):349–366

32. Siddique A, Yadava GS, Singh B (2003) Applications of artificial intelligence techniques for induction machine stator fault diagnostics: Review. Proceedings of the IEEE international symposium on diagnostics for electric machines, power electronics and drives, New York 2003:29–34

33. Söderholm P, Holmgren M, Klefsjö B (2007) A process view of maintenance and its stakeholders. J Qual Maintenance Eng 13(1):19–32

34. Tsang A (2002) Strategic dimensions of maintenance management. J Qual Maintenance Eng 8(1):7–39

35. Tucci M, Bettini G (2006) Methods and tools for the reliability engineering: a plant maintenance perspective. In: proceedings of maintenance management '06, Sorrento

36. Vanneste SG, Van Wassenhove LN (1995) An integrated and structured approach to improve maintenance. Eur J Oper Res 82:241–257

37. Waeyenbergh G, Pintelon L (2002) A framework for maintenance concept development. Int J Prod Econ 77(1):299–313

38. Wang W, Tse PW, Lee J (2007) Remote machine maintenance system through Internet and mobile communication. Int J Adv Manuf Technol 31(7–8):783–789
39. Zhang Y, Gu Y, Vlatkovic V, Wang X (2004) Progress of smart sensor and smart sensor networks. IEEE Publications. pp 3600–3606
40. Zhao Z, Laga N, Crespi N (2009) A survey of user generated service. In: Proceedings of 2009 IEEE international conference on network infrastructure and digital content, IEEE IC-NIDC2009, art. no. 5360953, pp 241–246

Author Biographies

Marco Macchi is an Associate Professor at Politecnico di Milano, Department of Management, Economics and Industrial Engineering. He is the chair of the IFAC Working Group on AMEST (Advanced Maintenance Engineering, Services and Technology), the vice-chair of the IFAC Technical Committee 5.1 Manufacturing Plant Control and Book Reviews Editor and Editorial Board Member of the International Journal of Production Planning & Control: The Management of Operations. He is also a member of the IFIP WG 5.7 Advances in Production Management Systems and a Fellow of the International Society of Engineering Asset Management. In his research activity at Politecnico di Milano, he is responsible for several European funded projects in the fields of smart manufacturing, maintenance and industrial sustainability. Furthermore, he is responsible for the Observatory on Technologies and Services for Maintenance and is Research Co-director of the Observatory on Industria 4.0 of the School of Management at Politecnico di Milano. His research interests are advanced production systems, maintenance management and asset life cycle management.

Luis Barberá Martínez is Ph.D. (Summa Cum Laude) in industrial Engineering by the University of Seville, and Mining Engineer by UPC (Spain). He is a researcher at the School of Engineering of the University of Seville and author of more than fifty articles, national and main internationals. He has worked at different international Universities: Politecnico di Milano (Italy), CRAN (France), UTFSM (Chile), C-MORE (Canada), EPFL (Switzerland), University of Salford (UK) or FIR (Germany), among others. His line of research is industrial asset management, maintenance optimization and risk management. Currently, he is Spain Operations Manager at MAXAM Europe. He has three Master's degrees: Master of Industrial Organization and Management (2008/2009) (School of Engineering, University of Seville), Master of Economy (2008/2009) (University of Seville) and Master of Prevention Risk Work (2005/2006). He has been honoured with the following awards and recognitions: Extraordinary Prize of Doctorate by the University of Seville; three consecutive awards for Academic Engineering Performance in Spain by AMIC. (2004/2005), (2005/2006) and (2006/2007); honour diploma as a "Ten Competences Student" by the University of Huelva and CEPSA Company (2007); graduated as number one of his class.

Adolfo Crespo Márquez is currently Full Professor at the School of Engineering of the University of Seville, and Head of the Department of Industrial Management. He holds a Ph.D. in Industrial Engineering from this same University. His research works have been published in journals such as the International Journal of Production Research, International Journal of Production Economics, European Journal of Operations Research, Journal of Purchasing and Supply Management, International Journal of Agile Manufacturing, Omega, Journal of Quality in Maintenance Engineering, Decision Support Systems, Computers in Industry, Reliability Engineering and System Safety, and International Journal of Simulation and Process Modeling, among others. Prof. Crespo is the author of seven books, the last four with Springer-Verlag in 2007, 2010, 2012, and 2014 about maintenance, warranty, and supply chain management. Prof. Crespo leads the Spanish Research Network on Dependability Management and the Spanish Committee for Maintenance Standardization (1995–2003). He also leads a research team related to maintenance and dependability management currently with 5 Ph.D. students and 4 researchers. He has extensively participated in many engineering and consulting projects for different companies, for the Spanish Departments of Defense, Science and Education as well as for the European Commission (IPTS). He is the President of INGEMAN (a National Association for the Development of Maintenance Engineering in Spain) since 2002.

Luca Fumagalli is Assistant Professor at Politecnico di Milano. He is Mechanical Engineer, graduated at Politecnico di Milano in 2006, and obtained Ph.D. in Industrial Engineering at Politecnico di Milano in 2010. He works on different research topics about production management, industrial services and in particular maintenance management related topics, with a specific concern on new technological solutions. His research activity has been related also with European research funded projects. Luca Fumagalli is also presently responsible of the research of the observatory TeSeM (www.tesem.net) and vice-director of Master Megmi (Master Executive in Gestione della Manutenzione Industriale—Executive Master on Industrial Maintenance Management) delivered by MIP—School of Management—Politecnico di Milano and SdM—School of Management Università degli Studi di Bergamo (http://www.mip.polimi.it/megmi).

María Holgado Granados is a Research Associate at Centre for Industrial Sustainability, Institute for Manufacturing, University of Cambridge. She is an Industrial Engineer, graduated at University of Seville, Spain, and obtained a Ph.D. in Management, Economics and Industrial Engineering at Politecnico di Milano, Italy. Her Ph.D. research examined how industrial maintenance can serve as an enabler for practicing more sustainable manufacturing operations through provision of industrial services. Her Ph.D. thesis was granted the mention of Thesis of Excellence by the Italian Maintenance Association (AIMAN) and the Best Ph.D. Thesis Award in Maintenance by the European Federation of National Maintenance Societies (EFNMS) in 2016. She has been a Marie-Curie Research Fellow within the European project iMaPla "Integrated Maintenance Planning" in University Federico Santa Maria in Valparaíso, Chile, and also a Visiting Scholar at the Intelligent Maintenance Systems Center in University of Cincinnati, US, and at the Institute for Manufacturing in University of Cambridge, UK. Her research interests include energy and resource efficiency, industrial symbiosis, sustainable value creation, sustainable business models, product service systems, maintenance management, maintenance services and maintenance contribution to industrial sustainability.

Assistance to Dynamic Maintenance Tasks by Ann-Based Models

Fernando Agustín Olivencia Polo, Jesús Ferrero Bermejo,
Juan Francisco Gómez Fernández and Adolfo Crespo Marquez

Abstract Reliability requirements are increasingly demanded in all economic sectors, practical applications of the reliability theory are pursued often as a tailor-made suit for each case, in order to manage the assets effectively according to specific operating and environmental conditions. In sectors like renewable energy, these conditions can be changing importantly over time and reliability analysis is periodically required. At the same time, adapting a unique model to similar systems placed in different plants has proven to be troublesome. This paper adapts reliability models in order to incorporate monitored assets operating and environmental conditions. This paper introduces a logic decision tool which is based on two artificial neural networks models; allowing updating assets reliability analysis in relation to changes in operating and/or environmental conditions, and even more, this model could be easily automated within a SCADA system. Thus, by using the model, reference values and the corresponding warnings and alarms can be dynamically generated, serving as an online diagnostic or prediction of a potential degradation of the asset. The models are developed according to the available amount of failure data and they are used to detect early degradation in the energy production due to power inverter and solar tracker failures, and to evaluate the associated economic risk to the system under existing conditions. This information can then trigger preventive maintenance activities.

F.A. Olivencia Polo (✉) · J. Ferrero Bermejo
Magtel Systems, Seville, Spain
e-mail: fernando.olivencia@magtel.es

J. Ferrero Bermejo
e-mail: jesus.ferrero@magtel.es

J.F. Gómez Fernández · A. Crespo Marquez
Department of Industrial Management, School of Engineering,
University of Seville, Seville, Spain
e-mail: juan.gomez@iies.es

A. Crespo Marquez
e-mail: adolfo@us.es

© Springer International Publishing AG 2018
A. Crespo Márquez et al. (eds.), *Advanced Maintenance Modelling for Asset Management*, DOI 10.1007/978-3-319-58045-6_17

Keywords Renewable energy · Maintenance · Condition-based maintenance · Artificial neural network · Proportional Weibull reliability

1 Introduction

Renewable energies present a high dependency on the random condition of climatological phenomena. This variability has a great impact in chances of production commitment fulfilment as established by the legislation in many countries (to reach a reliable electrical network).

In recent years, the interest of a great part of the scientific community has been focused on predicting possible failures to minimize the impact on transport networks. Regular performance checks on the functions of grid-connected PV systems need an advanced processes monitoring through a sensor distributed network, and proper acquired time series statistical analysis. As a result, companies face a costly maintenance program, only affordable for large sites and companies with the advantage of certain economies of scale.

Concerning environmental conditions forecasting, there exist many ways of performing this analysis:

- Considering climatological forecasting: physical models, with solar irradiation-generated power curves, for predicting production in terms of irradiation.
- Without climatological forecasting: statistical models based on historical data (monthly averaged solar irradiation). For example, the Ministry of Industry of Spain provides a prediction tool without climatological calculations, including a coefficient table considering as variables: months, time of the day and climatological zones.

Nowadays, automatic failure detection in photovoltaic systems needs logging a great number of electrical variables, such as currents flowing through solar panels and voltages on batteries, together with environmental data such as irradiance and temperature [1, 2]. Some proposals try to reduce the complexity in the failure detection using few variables and more complex statistical analyses [3, 4]. This paper focuses on applying artificial neural networks with this purpose.

Artificial neural networks (ANNs) are mathematical tools with intensive utilization in the resolution of many real-world complex problems, especially in classification and prediction ones. ANNs (Artificial Neural Networks) pretend to emulate biological human neural networks learning from the experience and generalizing previous behaviours as characteristics time series. To do this, the simple unit is the neuron whose mission is to process the received data as an activating function that could be the entry of other neuron, combining neurons as a directed graph that can carry out information processing by means of its state response to continuous or initial input [5]. The ANN architecture consists on an input layer, an output layer and generally, one or more hidden layers. Their main characteristic is its ability to process information features in nonlinear, high-parallelism, fault and noise environments with learning and generalization capabilities [6, 7]. In comparison to traditional model-based methods, ANNs are data-driven self-adaptive methods well implemented in computers on real time, learning from examples and capturing subtle and hidden functional relationships that are unknown or hard to describe. In addition, ANNs provide strong tolerance before noised data because store information redundantly. Thus, ANNs are well suited for solving problems where explicit knowledge is difficult to specify or define, but where there are enough data. [8–10]. In this sense, [11] have shown that backpropagation neural network exceeds by an order of magnitude to the conventional lineal and polynomial methods dealing with chaotic time series of data. Consequently, our interest is using ANNs to analyze data and dismiss predictions errors concerning failure appearance.

The application of these techniques to the renewable energies field, and more specifically to power generation of photovoltaic (PV) systems, has been in continuous development during the last years, including:

- Meteorological data forecasting [12, 13].
- PV systems sizing [14–17].
- PV systems modelling, simulation and control [18].

There are previous works on forecasting PV systems electricity output using ANNs [19, 20], however this paper describes two new algorithms for early detection of failures in PV systems according to the available amount of failure data, with the intention to be included when describing predictive maintenance task resulting from RCM programs implementation (Reliability-Centred Maintenance).

2 Models Evaluation and Decision-Making Process

Reliability centred maintenance (RCM) is the most widespread methodology to study the required maintenance program for an asset in a given operational context [21], quantifying the risks [22] and evaluating the remedial measures to detect, avoid or prevent the functional failures [23]. When considering variable operation and environmental conditions, the study of failures may be complex. Attributable to non-optimal operating conditions, failures often occur in assets suffering of changing environmental (cleanliness, fastening, temperature, etc.) and operational (configurations, preventive maintenance, undue handling, etc.) conditions. Moreover, non-evident defects in the assets (design imperfection, implementation errors, quality of materials, etc.) may also lead to failures [24, 25].

Reliability analysis of Renewal Energy Equipment, in line with the RCM method, is a very complex task depending on operating and environmental conditions. This analysis considers the effects, in the equipment function, of the different failure modes degrading the equipment functionality through deviations from standard operating conditions [26]. Based on real data as historic events, this degradation can be observed or predicted following a failure curve. Due to its own complexity, this analysis is associated to quantitative tools and so it have to be mainly implemented in depth in critical equipment or equipment in which failure consequences are not admissible (due to environment, health and safety, etc.).

An example of this is the "Survival Data Analysis", focused on a group of individuals and how they react to failure after certain length of time [27–29]. Data and information about these contributing factors could be decisive to obtain, and even to update over time, reliability estimations about the contribution of some events, represented through explanatory variables or covariates, in order to obtain the time until the failure (Survival Time). There are several techniques to solve survival estimations [30, 31], in which typical failure distribution functions are asymmetrical (censored to the right). The influence of these explanatory factors may obey different patterns that could be then used to work out the real risk of an asset. These techniques based on explanatory variables could be parametric when the hazard distributions are known, semi-parametric in the case of unknown hazard distribution but with defined assumptions of hazard proportionality with the time and independence between the constant through time covariates, or non-parametric when these are not necessary to be specified [32–34].

In maintenance, the decision-making is usually characterized by conditions of uncertainty, anticipation in order to handle non-controlled variables is frequently required and this is done by studying their historical evolution individually, or on their relation to other variables. In practice, with limited knowledge, maintenance technicians often feel more confident with their experience, and this would influence their decision that could be conservatively based on levels of satisfaction instead of being optimal [35]. Therefore, it is recommended to improve decision capacity using formalized frameworks which are suitable to the level of information required and to the data which is available. Quantitative tools are preferred to seek greater precision in the choice of strategies, but this is the choice of what is "better", among what is "possible" [36]. Also, the decision process is interactive, not only to predict something, but to replicate reality; it should be upgradeable as improvement continues to obtain and share knowledge.

Parametric methods, as Weibull actuarial and graphical models (EM), are usually employed when people have enough information about failures with a regular pattern, so they can be developed to model failures resulting, most of the times, in a tailor-made suit per equipment. On the other hand, as previously it has already mentioned before the utilization of semi-parametric methods, as the widely applied Proportional Hazard Model (PHM) of Cox [37], based on a log-lineal–polynomial expression of the covariates under the assumptions of independency among them and constant with the time. While, in non-parametric methods stand out ANN methods thanks to be a self-adaptive and empirical process even with noised and non-lineal information and/or time-dependency in covariates.

Parametric, semi-parametric and nonparametric techniques are employed to estimate the reliability function mainly depending on the knowledge about the failure time distribution (from major to minor respectively). However, concerning to the flexibility against to above-mentioned covariates assumptions (independency and time-independency), from EM models to ANN models, the flexibility and efficiency showing relationships among the life cycles and other variables are increased, but also the complexity of implementation and the computational load are increased at the same time [38]. Additionally, in numerous papers [38, 39], the PHM and ANN are compared to fit *survival functions* showing no significant differences between predictions of Cox regression and ANN models when complexity in models is low. In case of complex models, with many covariates and any interaction terms the differences in terms of advantages are important, showing the following results:

- ANN predictions were better than Cox PHM predictions with high rates of censoring (censoring rate of 60% and higher [40]), reducing significant biases.
- ANN predictions provide better predictions to detect complex nonlinear relationships between independent and dependent variables.
- ANN predictions can incorporate quantified potential prognostic factors that may have been overlooked in the past.

As a result, the maintenance decision-making in Renewal Energy Equipments under different operating environments can be supported by ANN fulfilling the requirements of:

- Suitable to level of failure information,
- Implementable in SCADA systems,
- Upgradeable iteratively and with reality,
- Flexible and integrated hierarchically.

According to previous paragraphs, this work main contribution is a logic decision tool doing PV systems electricity forecasting, which, at the same time, may serve as predictive maintenance instrument, that can be linked to proper RCM programs outputs to control critical failure modes. In the sequel, this work focuses on applying artificial neural networks (ANNs) to model PV systems failures.

3 Practical Case Materials

To support this practical research, the ANN models over case studies are now presented. The idea is, not only building the models, but also implementing them in a SCADA system.

PV plants have been in production for more than 25 years. Current decrease in government incentives to renewal energy sources has forced companies to study useful life extension possibilities. Due to this, potential plant reinvestments must be also re-evaluated; incorporating future operating and environmental conditions within equipment reliability analysis is considered to be crucial to avoid future production disruptions.

This type of photovoltaic plan was usually built modularly, each 100 KW may represent over 600.000 € of investment, therefore the possibility to replicate the same model for different modules and regions is also considered of great interest. With that in mind, this work tries to develop ANN models that are easy to reproduce, and to update, when the most common parameters found in a photovoltaic plant, determining production, suffer changes.

Our prediction models have innovative features compared to previous works in the literature. The ANN models use, not only environment variables as external temperature or radiation, but also assets' conditions variables as internal temperature for the different operating times. Through this, an early detection of degradation will be possible before failures affect production, and a quantitative measure of risk can be computed. It is important to acknowledge how risk of failures could even reach ten times the purchase equipment cost [41], therefore it has to be classified, and modelled properly, the different non reliability related cost along equipment lifecycle, such as warranties, indemnities, reparations, penalties, etc.

Additionally to be exhaustive in failure predictions per equipment, the analysis of failures has to be accomplished per each critical failure mode because symptoms and causes could be dissimilar among them and the effect of equipment conditions could apply in a different manner.

In our case study, functional analysis and failure mode analysis (Failure Mode Effect and Criticality Analysis—FMECA) was carried out in advance for critical equipment of the photovoltaic plant, understanding that these efforts in failure mode analysis could add enormous value for protecting a production of 6,258,966 €/year in our plant. This effort was completed identifying, at the same time, parameters required to predict failure modes (when that was feasible).

Two common systems are selected to illustrate the model implementation over real data and in a SCADA system: a power inverter and a solar tracker. Both of them are from a 6.1 MW photovoltaic plan opened up in September of 2008 (49,640 operation hours), compound by 37,180 photovoltaic panels in groups of 100 Kw for each inverter. The solar trackers orient photovoltaic panels towards the sun to maximize collected solar energy, while the power inverter transforms direct current (DC) to alternating current (AC) form strings of panels, aggregating its own energy (210,000 KWh/year) each five inverters jointly into a transformer (see configuration of the selected transformer in Table 1) through which energy is

Table 1 Standard configuration of one transformer with inverters and panels

CT	ID	KWn	Ref. module	No strings	No panels strings
CT15	A8-1	100	IS-220	528	12
	A8-2	100	IS-220	528	12
	A8-3	100	IS-220	528	12
	A8-4	100	IS-220	528	12
	A8-5	100	IS-220	528	12

provided to the distributor at an initial price of 0.4886 €/KWh (513,030 €/year of production), and subsequently reduced due to a legal requirement. Consequently reliability aspects are important, not only to consider the direct costs of failures, but also the indirect loss of profit. Anticipation to avoid this loss of profit will be pursued by the monitoring system.

The standard configuration of one of transformer is described in Table 1.

Possible future failures are predicted using a backpropagation neural network that is trained with inverters' historical data of the last 5 years. This paper focuses on failures resulting as a consequence of equipment deterioration and useful life reduction due to operational and geographical (environmental) features that could have a great influence on the equipment. In the following paragraphs of this Section, the paper first describes the backpropagation training process of the network, and then it concentrates on presenting the overall prediction methodology presented in the paper, applying it to the case study.

Backpropagation is a popular learning mechanism for solving predictions in multilayer perceptron networks [42], where differentiability is required in the activation function (as in the case of sigmoid function or the hyperbolic tangent function) in the output layer Z, in which its values may vary between 0 and 1, when using normalized variables in the input of the network. The sigmoid or logistic function that is used is presented in (1).

$$Z = f(x) = \frac{e^x}{(1+e^x)} \tag{1}$$

The backpropagation training consists in the following two steps (see Fig. 1 for a better understanding):

- Forward Steps:

  - Select an input value from the training set $(x_1, x_2, ..., x_{Q-1})$. The number of neurons in the entry layer is (Q), including the inputs variables $(Q-1)$ and another variable (b) which characterizes a threshold used internally by the ANN model and facilitates the convergence properties. An ANN model is composed by an entry layer, one or several hidden layers and the output layer.
  - Apply this entry set to the network and calculates the output $(\hat{y}_d)$.
    In the hidden layers, the value in the nucleus of each neuron is n_j, which is calculated using weights for each input $(w_{j,i})$, applied for each neuron (j) of the hidden layer and the correspondent input (a_i) until the number of neurons of the previous layer (Q).

$$n_j = \sum_{i=1}^{Q} \left(w_{j,i} \cdot a_i \right) + b \tag{2}$$

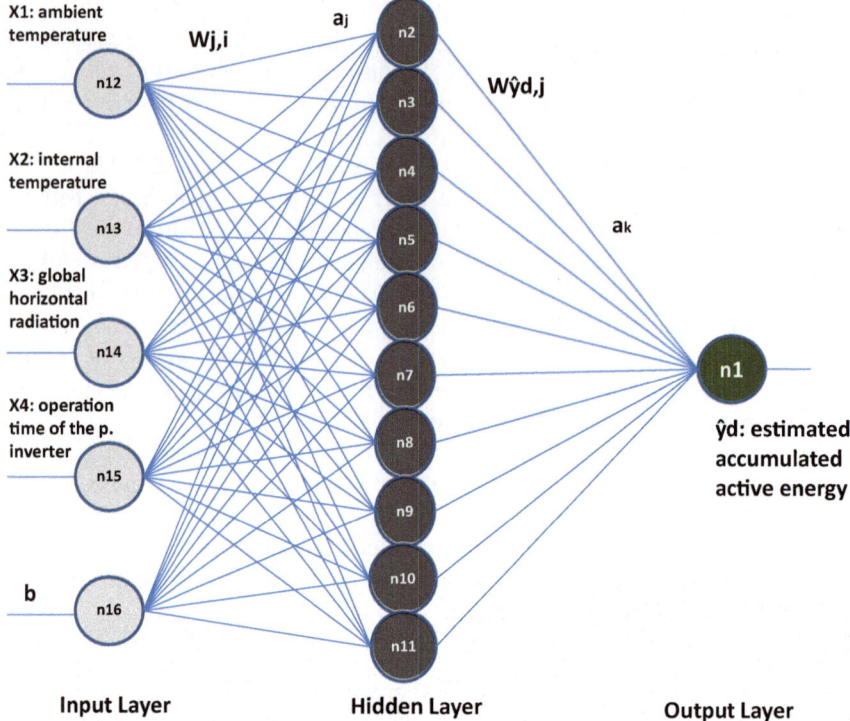

Fig. 1 Developed backpropagation percepton multi-layer ANN

In the first hidden layer the inputs of neurons are $a_i = x_i$. The output of each neuron of the first hidden layer (with J neurons) after the application of the activation function is a_j:

$$a_j = f(n_j) \tag{3}$$

In the following hidden networks, the value in the nucleus of each neuron is now based on the past outputs a_j using other weights ($w_{k,j}$), and producing the output a_k applying the activation function on the nucleus value; and at this way for successive hidden layers.

Finally, the output of the ANN (the output layer in the case of one neuron) is $\hat{y}_d$:

$$\hat{y}_d = f(n_{\hat{y}d}) = f\left(\sum_{j=1}^{J} (w_{\hat{y}d,k} \cdot a_k) \right) \tag{4}$$

- Backward Steps:
  - Calculate the errors between the obtained output ($\hat{y}_d$) and the real output (y_d).
  - Adjust the weights in order to decrease the error in reverse way. For this stage, this work has employed a learning coefficient equals (μ) to 1 and so is included in the second term of the sum in Eqs. (5) and (6).

$$w_{\hat{y}d,k}^{new} = w_{\hat{y}d,k}^{old} + a_k \cdot [\hat{y}_d \cdot (1 - \hat{y}_d) \cdot (y_d - \hat{y}_d)] \tag{5}$$

Equation (3) is the weights adjustment for the output neuron of the output layer, where $[\hat{y}_d \cdot (1 - \hat{y}_d) \cdot (y_d - \hat{y}_d) = \delta \hat{y}_d]$. An Eq. (4) is the adjustment on any neuron of the hidden layers.

$$w_{j,i}^{new} = w_{j,i}^{old} + w_{j,i}^{old} \cdot a_i \cdot \left[a_j \cdot (1 - a_j) \cdot (w_{k,j}^{new} \cdot \delta_j) \right] \tag{6}$$

- Repeat forward and backward steps about all the training set until the global error is acceptably low, for this work based on minizing the root mean square error for the number of observations (n).

$$RMSE = \sqrt{\frac{1}{n} \sum_{d=1}^{n} (y_d - \hat{y}_d)^2} \tag{7}$$

Because of nonlinearity of Z, the learning mechanism of multilayer perceptron networks requires a resolution heuristic algorithm that guarantees the best solution or the global minimum (this is done using the Quasi-Newton resolution method in the free software R or the Levenberg–Marquardt Method in Matlab). To avoid overadjustment of the network repeating the same employ this time MSE (Mean Square Error) with penalty characterized by λ, mainly employed with a bare quantity of historical data (otherwise λ trend to 0):

$$MSE + \lambda \cdot \sum_{j,i}^{Q,J,K} w_{j,i} = \frac{1}{n} \cdot \sum_{d=1}^{n} (y_d - \hat{y}_d)^2 + \lambda \cdot \sum_{j,i}^{Q,J,K} w_{j,i}^2 \tag{8}$$

After describing the process of the backpropagation training of the network, let's now concentrate on the Logic Decision Tool based on ANN models that this paper proposes.

4 Logic Decision Tool and ANN Models

RCM present a generic process for the logic selection of the maintenance actions to correct or prevent the occurrence of failure modes [21], as extension of this for the specific on-condition maintenance actions, the process of decision-making is developed addressing before mentioned requirements, see Fig. 2, which includes the following steps:

- The work flow starts with the inspection and failure data collection of external and internal relationships considering the differences in the operational and environmental conditions.
- Then it continues, evaluating if the symptoms of a gradual function loss can be detected effectively.
- Hereafter, the failure modes analysis is developed, determining their effects in the gradual function loss through a set of variables.
- Next, a logic decision tree analysis (LTA) is employed to select among the different prediction models (based on referenced authors):

  - If there are enough formal statistical training to develop or with lineal covariates, parametric models are recommendable.
  - If there are enough data about failures but not as formal statistical training, fulfil the covariates assumptions (that is when the relationship among the hazards of two similar assets with different operating environment factors is clearly proportional), and censoring rate less of 60%, it is suggested PHM.
  - If there are not enough data about failures (formal statistical or not) and not censored, but with enough data of process control variables (generally noised), it is recommendable to employ direct ANN (even to reproduce complex physic or chemical functions).
  - When there are enough data about failures but not as formal statistical training, and complex interaction with noise and time-dependency among covariates (that is, where the covariates assumptions are not satisfied), or satisfying the covariates assumptions the censoring rate is equals to 60% or higher, then ANN is recommended. Although in this case, ANN has its own limitations, because the data set has to be reorganized to replicate the Survival Function.

- After, the detection mechanism has to be defined.
- Finally, the repercussions of the chosen prediction models have to be evaluated with a cost–benefit analysis, previously to their implementation and communication to the entire organization.
- As a result, the implementation of the approved prediction models is realized.
- Updating with in-service data collection has to be maintained as continuous improvement.

Consequently, two new mathematical ANN models are developed in this document showing the aptitudes of ANN to replicate reality self-adaptively in complex and noised operating conditions: Case (A) of direct ANN in absence of failure data, reproducing energy production of the power inverter which is has a physic complex equation; and Case (B) of Survival ANN with enough non-formal failure data and complex covariates interaction, trying to fit the Survival Function of solar trackers.

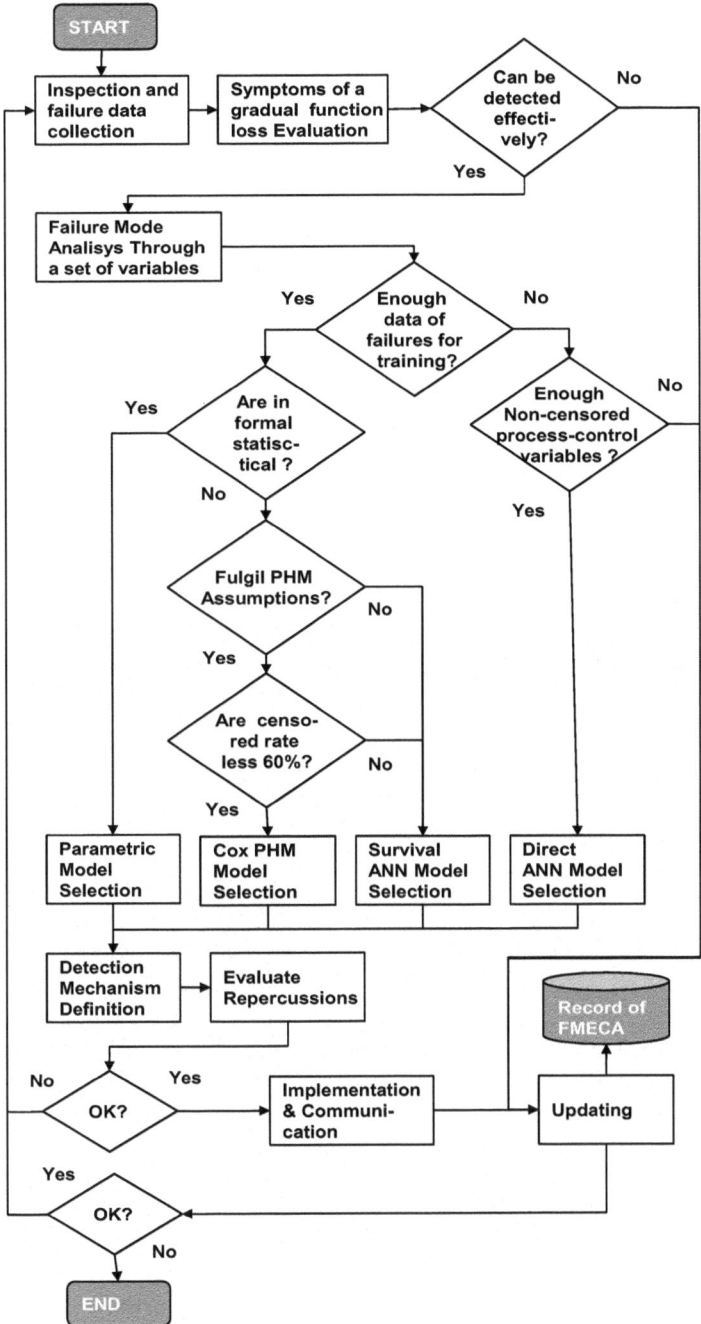

Fig. 2 Work-flow logic decision tree about on-conditions predictions

In both cases, for real-time estimation the variables have to be selected from those whose detection is periodically and automatically feasible. All the representative contributions selected have to be compound in a single function which reflects the degradation of the failure model but with two different methodologies. This is done to facilitate the failure discrimination and analysis. The data normalization is undone to the original range of values in the output layer. The ANN architecture has to be developed according to the number of input variables and the final estimation function, see it in Fig. 1 (presented previously), a multi-layer Perceptron using a linear hyperplane as function type base and with a single hidden layer with ten neurons. The activation function of each hidden neuron is a logistic function. The initial weights are randomly selected. The learning backpropagation algorithm used is an error correction supervised minimizing the penalized mean square error through the Quasi-Newton method in the free software R. The training of the network is realized depending on the architecture and available historical data of variables, where for this work there are the followings:

- Historical Data of variables and production output per hour and days during five years.
- Periods for comparison, to detect the existence of the failure mode when selected for each case.
- In case of a failure mode defect is corrected, the ideal model considers the equipment as new.
- Then, 75% of available data is considered for the network training and the 25% for the network testing.
- In the training the ANN behaviour pattern gives us the network settings such as the number of hidden nodes, which will be validated subsequently with the testing.

(Case A) Failure Mode Prediction: Lack of Insulation

The selected failure mode of the power inverter is the "lack of insulation" failure, which due to the fact that production losses are significant; this failure mode is considered in SCADA with priority. This failure mode emerges due to corrosion and, the environmental conditions could be determinants in different areas and besides the inverter operating time. The most representative variables of operation, external environment and internal conditions have to be selected and tested to show their effects in the failure mode. The available variables in our SCADA in the case of power inverters are: ambient temperature (°C), the internal temperature of the power inverter (°C), the global horizontal radiation (W/m^2), the operation time of the power inverter (h), and the active energy accumulated of the inverter (kWh).

This case, in our PV plant is characterized by the absence of enough failure data and with non-statistical form. Then, prediction can be realized over process control variables as the accumulated active energy of the inverter, where physical models for the different components makes the characterization in a transfer function for an accurate estimate of the state of the generator (voltage, current and power) in real time impossible. For all these reasons, researchers have developed several proposals to model these systems [1, 43]. By this reason, the physical model is reproduced with the ANN model.

This case will estimate the accumulated active energy ($\hat{y}_d$) of the inverter in absence of the failure mode, in order to compare this with the real accumulated active energy (y_d). For this, an ANN will be trained in absence of failures seeking an ideal production model that will be confronted with the real production to distinguish significant changes that denote the failure mode. Deduced by the FMECA analysis of the photovoltaic plant, this ideal production model could be used for detecting at an early stage other failure models even in other equipments, simply comparing it with other internal equipments variables in each case.

The ANN model has in the input layer with five input neurons, corresponding to the ambient temperature (°C), the internal temperature of the power inverter (°C), the global horizontal radiation (W/m^2), the operation time of the power inverter (h), the accumulated active energy of the inverter (kWh), see Fig. 1 and Table 2, and the threshold neuron. The output layer contains a single output neuron corresponding to the estimated active energy accumulated of the module (kWh). The ANN analysis, done going through the processes of training, predictions and test produces the following results (17,700 measures of two years are processed, 3540 measures per variable from 08:00 am to 17:00 pm, four inputs and one output).

The learning algorithm parameters are as follows: (a) maximum number of cycles = 980, (b) maximum validation failures = 40, (c) min_grad = 1.0e−10, (d) goal = 0, (e) μ = 0.005, (f) μ_dec = 0.1, (g) μ_inc = 10, (h) λ = 0, (i) min Error = 19.47. The obtained results in this case guarantee a good optimization model, as shown in Table 3.

MSE (Mean Square Error), in the training and testing, validates the ANN signifying the average distance between the obtained prediction and the real

Table 2 Data set of variables case A

Variable	Max.	Ref.	Min.	Unit
Ambient temperature	61	37	1	°C
Internal temperature	57	40	17	°C
Global horizontal radiation	1291	644	10	W/m^2
Operation time		49,640		h
Accumulated active energy	99	60	1	KWh

Table 3 Results of training case A

Results	Value
MSE training	72.47686
MSE test	83.41932
R^2 training	0.910275
R^2 test	0.8912438

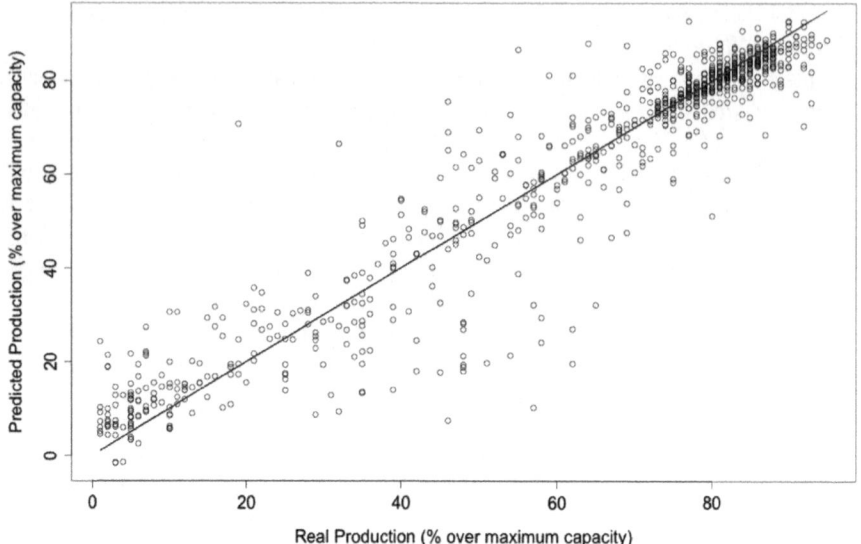

Fig. 3 ANN predictions case A

production. Besides, R^2 is consistent with this result, explained 89.12% of the predicted model versus actual production. Figure 3 is a representation of deduced predictions, remarking a straight line to indicate the best approximation for error minimization. For validation purposes, the 25% of historical data is used to estimate the generalization error.

Once the ANN model of accumulated active energy is validated, the detection mechanism has to be defined. After training five years of data, the ideal production model pretends to be an approximation to real production in absence of failures, through the transformation of combined experience in abstract conceptualization which is sustained by a nonlinear function of a weighted sum of its inputs (modifying the weights of the links that connect the neurons). The ideal production model has to be compared with the real production one, trying to define early prediction. In our "lack of isolation" failure mode, an early warning could easily be set up (see Fig. 4) at least 48 h before failure. Notice that the difference between ideal and real production in that case is 40%.

To protect the model against spurious alarms, the 40% difference generating the warning has to be maintained at least during 4 consecutive hours for alarm generation (circumstance modelled in SCADA incrementing a counter by 1 each hour), and requesting to schedule immediately an corrective action.

Hereafter, the case will show the comparison between ideal and real production values (Table 4).

Therefore, based on SCADA systems the model could be implemented easily and replicated for all power inverters in the plant, or in other plants, and the knowledge about this failure mode may increase comparing results among others

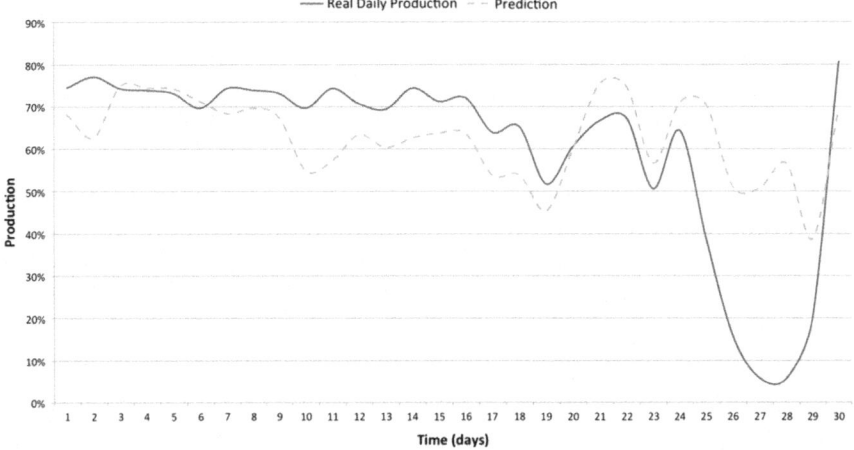

Fig. 4 Detection from ideal and real production comparison

Table 4 Data comparison of ideal and real production from SCADA

Date	Time	Real production	Ideal production	Alarm counter
25-9-12	10	85.03	73.9	0
25-9-12	11	52.82	81.4	0
25-9-12	12	35.88	83.5	1
25-9-12	13	47.99	83.7	2
25-9-12	14	43.59	81.1	3
25-9-12	15	26.37	74.1	4
25-9-12	16	21.06	77.6	5
25-9-12	17	33.69	83	6

inverters and redefining the model, or incorporating new or modified variables as the difference among external–internal temperature; or establishing the early alarm to gain more than 48 h.

Thanks to this research the "lack of isolation" failure mode associated indirect cost, as loss of profit, was reduced by 68,591 € per year and plant (575 KW/day with MTBF = 3 per year and for 61 inverters). Furthermore, extrapolating the potential advantage to the life cycle of the plant (5 years) the profits may have reached to the total production of one inverter during five years.

With the aim of extend the ANN model to other PV plants, the failure mode behaviour has been analyzed in two PV plants in different geographical provinces of Spain, Toledo and Zamora, where the operating environments are different. In both of them, the ANN model predicts the lack of insolation of the inverter but in a soften way versus in Cordoba (southern than Toledo and Zamora), due to the difference of meteorological variables, see Fig. 5.

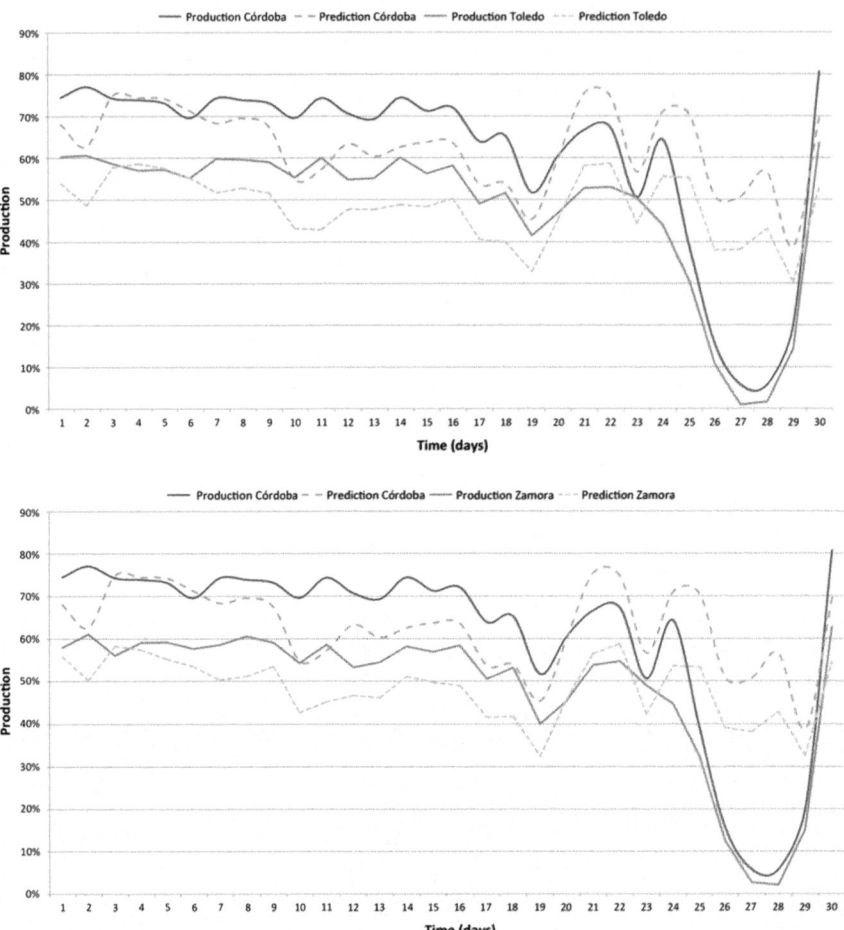

Fig. 5 Detection from ideal and real production comparison with Zamora and Toledo

(Case B) Failure Mode Prediction: Solar Tracker Blocking

The selected failure mode of the solar tracker is the "blocking" failure, which is repetitive in field due to the huge volume of installed units. This failure mode emerges due to corrosion and, the environmental conditions could be determinants in different areas and besides the operating time. This case, in our PV plant is characterized by enough failure data with non-statistical form varying among plants and with high censured rate. Then, prediction can be realized as Survival ANN. The most representative variables of operation and external environment conditions have to be selected and tested to show their effects in the failure mode. The selected variables in our SCADA in the case of solar trackers are relative to diary average: ambient humidity (%), wind speed (m/s), the global horizontal radiation (W/m^2),

Table 5 Semi-parametric Weibull parameters for reorganization of survival data

Pv1 Fn	TTFi	Pondered α_i	Pv2 Fn	TTFi	Pondered α_i
1	105.82	73.34	1	305.58	211.77
2	88.59	61.39	2	119.36	82.71
3	84.06	58.25	3	277.89	192.57
4	128.03	88.73	4	110.34	76.47
5	88.28	61.18	5	99.94	69.26
6	167.21	115.88	6	134.14	92.96
7	188.90	130.91	7	170.53	118.18
8	181.78	125.97	8	375.92	260.51
α	144.15		α	226.58	
β	3.47		β	2.19	

the operation time of the solar tracker (days) (see Table 5). However, to develop Survival Function this case has to reorganize the available data, because the training means adjusts iteratively the weight coefficients given the condition in order to approximate the output to the target, which is an input of the ANN. In practice, survival events have to be included and depending on the way to include them, different ANN models are produced [38, 39] in two manners, for example,

- Using ANN instead of the lineal combination of weights coefficients in the Cox PHM, as Farragi and Simon [44], being necessary solve the PHM with Partial Maximum Likelihood Estimation (P-MLE).
- Using an input with the Survival Status over disjoint time intervals where the covariates values are replicated, with a binary variable 0 before the interval of the failure and 1 in the event or after, as Liestol et al. [45] and Brown et al. [46] where each time interval is an input with Survival Status, then a vector of survival status is defined per failure; or
- Employing the Kaplan–Meier (K-M) estimator to define the time intervals as two additional inputs instead of vector, one is the sequence of the time intervals defined by de K-M survival status, and the other is the survival status in each time of the sequence. This is the case of Ravdin and Clark [47] or Biganzoli et al. [48] models which are known as Proportional Kaplan–Meier.

For similarity with the previous case and the intention to utilize the same ANN architecture, now this case has oriented our proposed Survival ANN model based on the ideas of Ravdin and Clark, but with some mathematical modifications:

- With periodic disjoint intervals (for all the failures) of the maximum time to failure, to be suitable to level of failure information, instead of employing Kaplan–Meier estimator intervals.
- With covariates using real data (the average) in each disjoint interval to be upgradeable iteratively and with reality property, instead of repeating the value in each interval.

- With a semi-parametric Weibull estimation of the Survival Status, instead of employing Kaplan–Meier estimator, in order to fitting the curve better and reduce the negative effect of non-monotonically decreasing survival curve.

Thus, the ANN model would have in the input layer with five input neurons, corresponding to ambient humidity (%), wind speed (m/s), the global horizontal radiation (W/m^2), the operation time of the solar tracker (days), the modelled semi-parametric survival status, and the threshold neuron. The output layer contains a single output neuron corresponding to the estimated survival function with values from 0 to 1.

The developed semi-parametric Weibull model consists into create time intervals of the maximum time to failure with an increment of the survival function instead of to maintain binary (0 or 1). Therefore, our propose resides in:

1. To estimate in a first step, the survival function with a parametric Weibull over groups of the produced time to failures where the covariates are the same. For example, if the case has two PV plants with 8 failures each one, it has to be realized the Weibull model in two groups, one over the 8 failures of PV plant 1 and other over the 8 failures of PV plant 2. Then, it will obtain a characteristic α and β in each plant, and without using the covariates, only based on time to failures as shows Table 5. Due to this, an estimation of the survival curve shape is obtained.

2. After that, maintaining the β in each plant (which represents the slope of the line), in order to model an estimation of the survival function for each specific failure with a gradual increment from 0 to 1, it is taken the β and the specific time to failure to replace in the Weibull Cumulative Distribution Function (CDF) in each time interval. Consequently, the two additional inputs are developed, one with the time intervals and other with the Weibull CDF with an increment discretized in the time intervals. Although, to match up the CDF curve with a gradual increment from 0 at the beginning of the time to 1 in the exact time of the failure and later, the CDF uses the previous β but α pondered by 0693, similar to the Median Life (Median Life = $\alpha \cdot$ Ln(2)^ β = 0.693 but $\beta = 1$). As a result for each specific failure, the probability to failure ascends unto reach 1 at the time of the failure and after, using this semi-parametric model, see Eq. (9) with Fn as number of failure in its plant, TTFi as specifics time to failure, ti as the time interval value, and α_i as pondered α. In Table 5, the 16 failures (Fn) and their time to failure (TTF) are presented for each plant (PV1 and PV2) with the initial α and β, and the modified α_i with the ponderation.

$$
\text{CDF}(t) = \begin{cases} 1 - \left(1/e^{\left(\frac{ti}{0.693 \cdot \text{TTFi}}\right)^\beta}\right) = 1 - \left(1/e^{\left(\frac{ti}{\alpha i}\right)^\beta}\right) & \text{if } ti < \text{TTFi} \\ 1 & \text{if } ti > = \text{TTFi} \end{cases} \tag{9}
$$

Consequently, the data to train and test the ANN are reorganized as in Table 6.

Table 6 Reorganized survival data to train and test the ANN

Failure number (Fn)	1	1	1	1	1	1	1	1	1	1	1
Time interval	10	20	30	40	50	60	70	80	90	100	110
Ambient humidity (%)	95	93	99	91	95	92	84	62	40	53	89
Wind speed (m/s)	11	8.8	6.7	11.3	6.7	11.6	10.9	12.1	8.2	7.3	4.3
G.H. radiation (W/m²)	35.4	53.4	43.5	31.9	38.7	51.7	80.1	68.1	54.7	68	86
Weibull CDF	0.001	0.011	0.044	0.115	0.232	0.392	0.573	0.741	0.869	0.947	1

For failure estimation, the output of the ANN model offers an estimation of the CDF or probability of failure, learning from semi-parametric estimation of a Weibull with covariates affection, as roughly proportional to Weibull Survival probability. The ANN analysis, done going through the processes of training, predictions and test produces the following results (3200 measures of two years are processed, 640 measures per variable diary, four inputs and one output) (Table 7).

Now, the learning algorithm parameters are as follows: (a) maximum number of cycles = 1000, (b) maximum validation failures = 40, (c) min_grad = 1.0e−10, (d) goal = 0, (e) μ = 0.005, (f) μ_dec = 0.1, (g) μ_inc = 10, (h) λ = 0, (i) min Error = 0.00001833. The obtained results in this case guarantee a good optimization model, as shown in Table 8.

While, if the Ravdin and Clark model had been employed directly, the results had been with less accuracy (as Table 9 shows).

In this developed model, R^2 explained 85.4% of the survival data. Figure 6 is the representation of deduced predictions, remarking a straight line to indicate the best approximation for error minimization.

As a result, for quick convergence and fitting of the curve, the initial values to train the ANN this case has utilized the semi-parametric estimation of Weibull CDF as an input to obtain the output as close as possible. Then, this case is researching a proportional semi-Weibull ANN model.

These two developed ANN models pretend to explore the capacity of ANN to obtain knowledge about covariates updating it based on experience with new valid

Table 7 Data set of variables case B

Variable	Max.	Ref.	Min.	Unit
Tiempo	400	205	10	h
Humedad relativa	100	74.5	27.3	%
Velocidad media viento	17.2	4.59	0.6	m/s
Radiación global	379.5	106.64	1.4	W/m^2
Supervivencia	1	0.5	0	

Table 8 Results of training case B in developed model

Results	Value
MSE training	0.01551932
MSE test	0.01641588
R^2 training	0.8681797
R^2 test	0.8540106

Table 9 Results of training case B with Ravdin and Clark

Results	Value
MSE training	0.08595152
MSE test	0.08493271
R^2 training	0.6371432
R^2 test	0.6520446

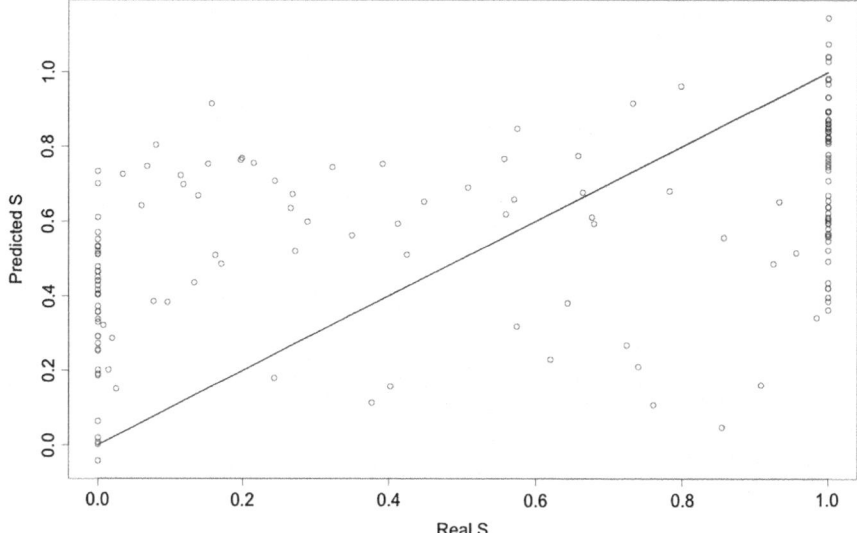

Fig. 6 ANN predictions case B

data. Although, weighted sum of the inputs of the ANN nodes could not be directly interpreted as the coefficients of the covariates. The aim is to estimate failures with one ANN architecture, either as first approximation to the covariates coefficients, or to be employed as input of other model, or to update the obtained coefficients with other techniques, or to incorporating new inputs, or to compare the quality versus failures in different PV plants or different equipments.

PV Plants managers want to ensure longer profitability periods with more reliable plants. To ensure profitability along the life cycle of the plant maintenance departments must ensure critical equipment reliability and maximum extension of their life cycle, otherwise failure costs will penalize the expected profit.

Throughout this document, this paper suggests to apply an ANN model per failure mode and foster a practical implementation in SCADA systems for different plants. This methodology may ease and may improve decision-making processed in condition-based maintenance and risk modelling, enabling reductions of corrective maintenance direct and indirect costs or allowing to show residual life until total equipment failure.

In cases when enough data for significant training is available, a better implementation of our methodology will help to reduce the costs and will improve the knowledge of the life cycle of the plant when suffering non-homogeneous operational and environmental conditions.

ANN capacity of autolearning among sources of data (sometimes noised or deprived of communication) thanks to reiterative memory is important. In our case study, a vast quantity of data from different remote plants was available, although sometimes this data was affected by problems of sensors readings or communications. Backpropagation perceptron ANN is recommend for automation developments with real-time utilization. Furthermore, advanced ANN models could be applied supporting additional variables.

It is important to know the failure mode behaviour in order to pretreatment historical data, eliminating abnormal data that may distort the results.

Values have to be normalized if it is used differentiability activation functions, and with the same scale for all the input values to simplify calculations and analysis. After the normalized values have to be des-normalized before comparison.

Acknowledgements Part of the funding for this research was provided by the SMARTSOLAR project (OPN—INNPACTO-Ref IPT-2011-1282-920000).

References

1. Orioli A, Di Gangi A (2013) A procedure to calculate the five-parameter model of crystalline silicon photovoltaic modules on the basis of the tabular performance data. Appl Energy 102:1160–1177
2. Kostylev V, Pavlovski A (2011) Solar power forecasting performance–towards industry standards. In: 1st international workshop on the integration of solar power into power systems, Aarhus, Denmark
3. Guasch D, Silvestre S, Calatayud R (2003) Automatic failure detection in photovoltaic systems. In: Proceedings of 3rd world conference on photovoltaic energy conversion, vols A–C, pp 2269–2271
4. Olivencia Polo F, Alonso del Rosario J, Cerruela García G (2010) Supervisory control and automatic failure detection in grid-connected photovoltaic systems. Trends Appl Intell Syst, pp 458–467
5. Miller WT III, Glanz FH, Kraft LG III (1990) CMAS: an associative neural network alternative to backpropagation. Proc IEEE 78:1561–1567
6. Basheer IA, Hajmeer M (2000) Artificial neural networks: fundamentals, computing, design, and application. J Microbiol Methods 43:3–31
7. Zhang GQ, Patuwo BE, Hu MY (1998) Forecasting with artificial neural networks: the state of the art. Int J Forecast 14:35–62
8. Curry B, Morgan P, Beynon M (2000) Neural networks and flexible approximations. IMA J Manage Math 11:19–35
9. Malcolm B, Bruce C, Morgan P (1999) Neural networks and finite-order approximations. IMA J Manage Math 10:225–244
10. Kuo C (2011) Cost efficiency estimations and the equity returns for the US public solar energy firms in 1990–2008. IMA J Manage Math 22:307–321
11. Lapedes A, Farber R (1987) Nonlinear signal processing using neural networks
12. Mellit A, Benghanem M, Bendekhis M, IEEE (2005) Artificial neural network model for prediction solar radiation data: application for sizing, stand-alone photovoltaic power system. In: 2005 IEEE power engineering society general meeting, vols 1–3, pp 40–44
13. Yacef R, Benghanem M, Mellit A (2012) Prediction of daily global solar irradiation data using Bayesian neural network: a comparative study. Renew Energy 48:146–154
14. Benghanem M, Mellit A (2010) Radial basis function network-based prediction of global solar radiation data: application for sizing of a stand-alone photovoltaic system at Al-Madinah, Saudi Arabia. Energy 35:3751–3762
15. Mellit A, Benghanem M, Arab AH, Guessoum A, IEEE (2003) Modelling of sizing the photovoltaic system parameters using artificial neural network

16. Mellit A, Benghanem M, Arab AH, Guessoum A, Moulai K (2004) Neural network adaptive wavelets for sizing of stand-alone photovoltaic systems
17. Mellit A, Benghanem M, Arab AH, Guessoum A (2005) An adaptive artificial neural network model for sizing stand-alone photovoltaic systems: application for isolated sites in Algeria. Renew Energy 30:1501–1524
18. Hiyama T (1997) Neural network based estimation of maximum power generation from PV module using environmental information—discussion. IEEE Trans Energy Convers 12:247
19. Ashraf I, Chandra A (2004) Artificial neural network based models for forecasting electricity generation of grid connected solar PV power plant. Int J Global Energy Issues 21:119–130
20. Mellit A, Shaari S (2009) Recurrent neural network-based forecasting of the daily electricity generation of a Photovoltaic power system. In: Ecological vehicle and renewable energy (EVER). Monaco, March, pp 26–29
21. Moubray J (1997) Reliability-centered maintenance. Industrial Press Inc., USA
22. Rausand M, Høyland A (2004) System reliability theory: models, statistical methods, and applications, vol 396. Wiley
23. Campbell J, Jardine A (2001) Maintenance excellence: optimising equipment life-cycle decisions. Marcel Dekker, New York, NY
24. Crespo Márquez A (2007) The maintenance management framework: models and methods for complex systems maintenance. Springer, London
25. Pham H, Wang H (1996) Imperfect maintenance. Eur J Oper Res 94:425–438
26. Mobley K (2002) An introduction to predictive maintenance. Elsevier, Amsterdam
27. Cox DR, Oakes D (1984) Analysis of survival data. Chapman and Hall, London
28. Klein J, Moeschberguer M (1997) Survival analysis techniques for censored and truncated data. Springer, New York Inc
29. Law AM, Kelton WD (1991) Simulation modeling and analysis. McGraw-Hill, New York
30. Lindsey JK (2001) The statistical analysis of stochastic processes in time. Cambridge University Press
31. Smith PJ (2002) Analysis of failure and survival data. Chapman-Hall, New York
32. Hougaard P (2000) Analysis of multivariate survival data. Springer, New York
33. Lee ET (1992) Statistical methods for survival data analysis. Wiley
34. Blischke WR, Murthy DNP (2000) Reliability modelling, prediction and optimization. Wiley, New York
35. Mitchell E, Robson A, Prabhu VB (2002) The impact of maintenance practices on operational and business performance. Manag Auditing J 11(1):25–39
36. Stewart TT (1992) A critical survey on the status of multiple criteria decision-making theory and practice. OMEGA Int J Manag Sci 20(5/6):569–586
37. Cox DR (1972) Regression models and life-tables. J Roy Stat Soc Ser 34:187–220
38. Ohno-Machado L (2001) Modeling medical prognosis: survival analysis techniques. J Biomed Inform 34:428–439
39. Xianga A, Lapuerta P, Ryutova A, Buckley J, Azena St (2000) Comparison of the performance of neural network methods and Cox regression for censored survival data. Comput Stat Data Anal 34(2):243–257
40. Biglarian A, Bakhshi E, Baghestani AR, Gohari MR, Rahgozar M, Karimloo M (2013) Nonlinear survival regression using artificial neural network. Hindawi Publishing Corporation. J Probab Stat 2013. Article ID 753930
41. Wilson RL (1986) Operations and support cost model for new product concept development. Comput Ind Eng 11:128–131
42. McClelland JL, Rumelhart DE, PR Group (1986) Parallel distributed processing. In: Explorations in the microstructure of cognition, vol 2
43. Blanes JM, Toledo FJ, Montero S, Garrigos A (2013) In-site real-time photovoltaic I–V curves and maximum power point estimator. IEEE Trans Power Electron 28:1234–1240
44. Faraggi D, Simon R (1995) A neural network model for survival data. Stat Med 14:73–82
45. Liestol K, Andersen PK, Andersen U (1994) Survival analysis and neural nets. Stat Med 13:1189–1200

46. Brown SF, Branford A, Moran W (1997) On the use of artificial neural networks for the analysis of survival data. IEEE Trans Neural Netw 8:1071–1077
47. Ravdin PM, Clark GM (1992) A practical application of neural network analysis for predicting outcome of individual breast cancer patients. Breast Cancer Res Treat 22:285–293
48. Biganzoli E, Boracchi P, Mariani L, Marubini E (1998) Feed forward neural networks for the analysis of censored survival data: a partial logistic regression approach. Stat Med 17:1169–1186

Author Biographies

Fernando Agustín Olivencia Polo is Ph.D. in Technology and Engineering, Telecommunications Engineer and International Commerce MBA. He has more than 20 years of experience working for utilities in different sectors: telco, energy, water and railways, and has a sound knowledge of business and operation Information systems for this kind of companies. He has combined his business activity with academic and research life as an associate professor (PSI) for the University of Córdoba, Telefónica I+D and MAGTEL, respectively.

Jesús Ferrero Bermejo is currently Ph.D. candidate in Industrial Engineering, and his main academic education is Statistician and with a Master in Industrial Organization. He is the author of several national and international publications in journals, books and conferences. He works in Magtel Operations as Outsourcing Manager in the System's Division, leading several projects in relation to TIC sector for public companies. He has participated in different projects of consultancy in national and international companies, with a high technical and managerial specialization in different areas about maintenance and with a wide experience in Photovoltaic Installations. He has collaborated in diverse R+D+i projects for Magtel, being awarded with the Best Record in both of them. He was also awarded with the Best Master Thesis by AEC (Spanish Association for Quality) and AEIPRO (Spanish Engineering and Projects Management Association).

Juan Francisco Gómez Fernández is Ph.D. in Industrial Management and Executive MBA. He is currently part of the Spanish Research & Development Group in Industrial Management of the Seville University and a member in knowledge sharing networks about Dependability and Service Quality. He has authored publications and collaborations in journals, books and conferences, nationally and internationally. In relation to the practical application and experience, he has managed network maintenance and deployment departments in various national distribution network companies, both from private and public sector. He has conduced and participated in engineering and consulting projects for different international companies, related to Information and Communications Technologies, Maintenance and Asset Management, Reliability Assessment, and Outsourcing services in Utilities companies. He has combined his business activity with academic life as a associate professor (PSI) in Seville University, being awarded as Best Thesis and Master Thesis on Dependability by National and International Associations such as EFNSM (European Federation of National Maintenance Societies) and Spanish Association for Quality.

Adolfo Crespo Márquez is currently Full Professor at the School of Engineering of the University of Seville, and Head of the Department of Industrial Management. He holds a Ph.D. in Industrial Engineering from this same University. His research works have been published in journals such as the International Journal of Production Research, International Journal of Production Economics, European Journal of Operations Research, Journal of Purchasing and Supply Management, International Journal of Agile Manufacturing, Omega, Journal of Quality in

Maintenance Engineering, Decision Support Systems, Computers in Industry, Reliability Engineering and System Safety, and International Journal of Simulation and Process Modeling, among others. Prof. Crespo is the author of seven books, the last four with Springer-Verlag in 2007, 2010, 2012, and 2014 about maintenance, warranty, and supply chain management. Prof. Crespo leads the Spanish Research Network on Dependability Management and the Spanish Committee for Maintenance Standardization (1995–2003). He also leads a research team related to maintenance and dependability management currently with 5 Ph.D. students and 4 researchers. He has extensively participated in many engineering and consulting projects for different companies, for the Spanish Departments of Defense, Science and Education as well as for the European Commission (IPTS). He is the President of INGEMAN (a National Association for the Development of Maintenance Engineering in Spain) since 2002.

Expected Impact Quantification Based on Reliability Assessment

Fredy Kristjanpoller Rodríguez, Adolfo Crespo Márquez,
Pablo Viveros Gunckel and Luis Barberá Martínez

Abstract Currently, a lack of interpretation tools and methodologies hinders the ability to assess the performance of a single piece of equipment or a total system. Therefore, a reliability, availability and maintenance (RAM) analysis must be combined with a quantitative reliability impact analysis to interpret the actual performance and to identify bottlenecks and improvement opportunities. This paper proposes a novel methodology that uses RAM analysis to quantify the expected impact. The strengths of the failure-expected impact methodology include its ability to systematically and quantitatively assess the expected impact in terms of RAM indicators and the logical configuration of subsystems and individual equipment, which show the direct effects of each element on the total system. This proposed analysis complements plant modelling and analysis. Determining the operational effectiveness impact, as the final result of the computation process, enables the quantitative and unequivocal prioritization of the system elements by assessing the associated loss as a "production loss" regarding its unavailability and effect on the system process. The Chilean Copper Smelting Process study provides useful results for developing a hierarchization that enables an analysis of improvement actions that are aligned with the best opportunities.

F. Kristjanpoller Rodríguez (✉) · A. Crespo Márquez · P. Viveros Gunckel ·
L. Barberá Martínez
Department of Industrial Management, School of Engineering,
University of Seville, Seville, Spain
e-mail: fredy.kristjanpoller@usm.cl

A. Crespo Márquez
e-mail: adolfo@us.es

P. Viveros Gunckel
e-mail: pablo.viveros@usm.cl

L. Barberá Martínez
e-mail: lubarmar@us.es

F. Kristjanpoller Rodríguez · P. Viveros Gunckel
Department of Industrial Engineering, Universidad Técnica Federico Santa María,
Av. España 1680, Valparaíso, Chile

© Springer International Publishing AG 2018
A. Crespo Márquez et al. (eds.), *Advanced Maintenance Modelling for Asset Management*, DOI 10.1007/978-3-319-58045-6_18

Keywords Reliability analysis · Failure · Reliability engineering · Maintenance · Industrial engineering · Life cycle assessment

Glossary

α	Scale Parameter of Weibull Distribution
β	Form Parameter of Weibull Distribution
λ	Failure Rate
Γ	Gamma Function
A	Availability
CAPEX	Capital Expenditures
E-DFP	Expected Downtime Factor Propagation
E-OCI	Expected Operational Criticality Impact
ERP	Enterprise Resource Planning
FEI	Failure Expected Impact
FTA	FailureTree Analysis
GRP	Generalized Renewal Process
HPP	Homogeneous Poisson Process
KPI	Key Performance Indicators
LCC	Life Cycle Cost
MTTF	Mean Time to Failure
MTTR	Mean Time to Repair
NHPP	Non-Homogeneous Poisson Process
OPEX	Operational Expenditures
PLP	Power Law Process
PRP	Perfect Renewal Process
RAM	Reliability Availability and Maintainability
RBD	Reliability Block Diagram
SAP-PM	Plant Maintenance Module of SAP Enterprise Resource Planning software
TTR	Time To Repair
UGF	Universal Generating Function

1 Introduction and Background

1.1 Literature Review

The effectiveness of production processes and their associated equipment is an important tool for assessing total system effectiveness [1], which is generally measured according to the results of reliability and availability indicators and life cycle economic analysis [2]. The total equipment effectiveness indicator measures the productive efficiency using the control parameters as a basis for calculating the fundamentals of industrial production: availability, efficiency and quality [3].

In the current literature, several investigations have been performed to identify the principal factors that directly affect the maximization of economic benefits; these factors converge at the empirical consideration of reliability, maintainability and availability (RAM) indicators. The traditional reliability analyses that are based on logical and probabilistic modelling contribute to the improved key performance indicators (KPI) of a system [4], which directly influence optimal operation designs [5]. However, many alternatives are available for system reliability analyses that employ analytical techniques, such as Markov models [6], Poisson models [7], Universal Generating Function (UGF) and decision diagram [8]. This systematic study is based on techniques such as reliability block diagrams (RBDs) [9, 10], fault trees (FTs) [11], reliability graphics (RGs) [12] and Petri nets (PNs) [13]; these techniques can be used to determine logical relationships that underlie the behaviour or dynamics of a process.

As an example, the productive processes in the mining industry have numerous equipment and systems, rendering a systematic analysis of the plant more difficult [4]. Different analysis methodologies, such as the RBD methodology [9, 10], have been developed and extensively applied in the mining sector due to their adaptability in representing complex arrangements and environments with large amounts of equipment, simplifying reliability analysis. For the correct development of a RAM analysis, a complete scan of the data must be performed to fit the data to a statistical model and to then obtain key indicators [14]. Using a maintenance management support tool, different improvement opportunities can be identified, and recommendations can be offered to develop the most appropriate actions [15].

The Birnbaum importance measure (IM) [16] ranks the components of the system with respect to the impact of their failure on the overall system's performance; however, its application is primarily related to epistemic uncertainties.

Interpretation tools and methodologies for understanding the performance of a single piece of equipment and a total system are lacking; this deficiency is even more pronounced when the selected analysis process has been disaggregated on many levels and each level has been disaggregated across several pieces of equipment. Therefore, a RAM analysis must be combined with a quantitative reliability impact analysis to determine the real performance and identify bottlenecks and opportunities for improvement.

1.2 Motivation of Work

This article proposes an integral and quantitative innovative methodology to analyse the reliability, availability, maintainability and plant failure expected impact. The failure expected impact (FEI) analysis is related specifically to production capacity and the effect of preventive and corrective maintenance intervention on system availability. This proposal designs a novel algorithm to compute an impact index based on the frequency of failures associated, with the reliability and maintainability of the machinery and the expected impact according to different scenarios and configurations. This impact index, based on a probabilistic approach, defines the expected condition of the item in the system from a perspective of evaluation of its possible states (intrinsic behaviour) and related to the logical and functional configuration of the system. This approach enables an overall comparison of elements and the prioritization of those elements, as well as a partial effectiveness assessment.

What is the motivation for applying the FEI methodology?

In the finance area, for example, when a single stock of the NASDAQ index has a variation price of 10% and this stock represents 5% of the index composition, the NASDAQ index increases by only 0.5%. Developing a similar analysis over an "element" failure and determining the system consequences is simple when the "elements" have a serial configuration. If the configuration employs redundancy logic, the result is uncertain and dependent on the reliability and maintainability of the elements that compose the subsystem. The FEI methodology solves this problem by proposing four steps and applying them to a mining process—specifically, a copper smelting process (CSP) in Chile. Related to failure impact methodology, a novel algorithm is proposed to compute a failure impact index for the total availability performance of the system based on the reliability (frequency of failures), maintainability (down time) and availability of the elements. This impact index defines the expected condition of the item in the system by evaluating its possible states (intrinsic behaviour) and the logical and functional configuration of the system. This analysis enables a total comparison of the elements, their prioritization and a failure impact evaluation.

1.3   Scope of Work

The two key indicators of FEI methodology are outlined in Fig. 1, which incorporates general formulas for analysis and calculation. This methodology seeks to explain the level of responsibility for a failure in a single piece of equipment (b3) for system inoperability from a historical perspective. Two important steps are included:

1. Explain the effect of a single failure (downtime Ti of equipment b3) in terms of the single downtime propagation in the system (equivalent downtime ti). This indicator is named the expected downtime factor propagation (E-DFP).
2. Explain the effect of a single downtime propagation (equipment b3) on the total downtime of the system (downtime responsibility). This indicator is named the expected operational criticality index (E-OCI).

A copper smelting plant in Chile is examined as a case study. The smelting process is one of the most important and critical phases in any mineral processing system, especially in copper plants [1]. The selected mining process is divided into four main subsystems: drying, concentrate fusion, conversion and refining.

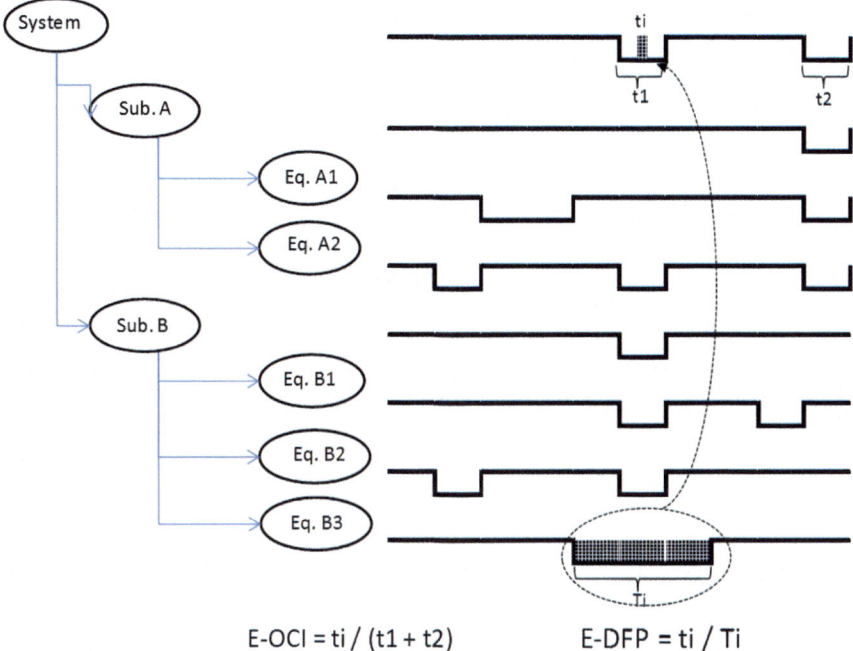

$$E\text{-OCI} = ti / (t1 + t2) \qquad E\text{-DFP} = ti / Ti$$

Fig. 1 Expected impact scheme

The paper is structured as follows: Sect. 2 describes the reliability assessment based on the FEI methodology and presents its conceptual and mathematical basis; Sect. 3 introduces and develops the case study according to the methodology; and Sect. 4 explains the main results and conclusions.

2 FEI Methodology

The FEI methodology should consider the joint processes that are necessary for identifying improvement opportunities and generating maintenance recommendations. These processes can be summarized in four steps: data cleaning, data management, RAM analysis, and FEI quantification (E-OCI and E-DFP) and decision-making.

2.1 Data Cleaning

The first step is to purge and filter the obtained data to improve the data quality (missing values, usefulness records and erroneous data) [17].

2.2 Data Management

To achieve an efficient data management, some methodologies can be considered based on the equipment (historical reparable behaviour). A repairable system after a failure scenario can be restored to its functioning condition (perfect and imperfect) by maintenance actions, with the exception of the replacement [18] (non-repairable item). A reparable system is defined as: "A system that, after failing to perform one or more of its functions satisfactorily, it can be restored to fully satisfactory performance by any method other than replacement of the entire system" [19]. The model and analysis of repairable equipment have high importance, mainly in order to increase the performance oriented to reliability and maintenance as part of the cost reduction in this last item. Depending on the type of maintenance given to equipment it is possible to find 5 cases [20].

- Perfect maintenance or reparation: Maintenance operation that restores the equipment to the condition "as good as new".
- Minimum maintenance or reparation: Maintenance operation that restores the equipment to the condition "as bad as old".
- Imperfect maintenance or reparation: Maintenance operation that restores the equipment to the condition "worse than new but better than old".
- Over-perfect maintenance or reparation: Maintenance operation that restores the equipment to the condition "better than new"
- Destructive Maintenance or reparation: Maintenance operation that restores the equipment to the condition "worse than old".

For a perfect maintenance, the most common developed model corresponds to the Perfect Renewal Process (PRP). In it, we assume that repairing action restores the equipment to a condition as good as new and assumes that times between failures in the equipment are distributed by an identical and independent way. The most used and common model PRP is the Homogeneous Poisson Process (HPP), which considers that the system not ages neither spoils, independently of the previous pattern of failures. That is to say, it is a process without memory. Regarding case (b), "as bad as old" is the opposite case to what happens in case (a) "as good as new", since it is assumed that the equipment will stay after the maintenance intervention in the same state than before each failure. This consideration is based on that the equipment is complex, composed by hundreds of components, with many failure modes and the fact that replacing or repairing a determined component will not affect significantly the global state and age of the equipment. In other words, the system is subject to minimum repairs, which does not cause any change or considerable improvement. The most common model to represent this case is through Non–Homogeneous Poisson Process (NHPP), in this case the most used model to represent NHPP is called "Power Law". In this model, it is assumed a Weibull distribution for the first failure, that later it is modified over time. Although the models HPP and NHPP are used mostly, they have a practical restriction regarding its application, since a more realistic condition after a repairing action is what we find between both: "worse than new but better than old". In order to find a generalization to this situation and not distinguishing between HPP and NHPP it was necessary to create the Generalized Renewal Process (GRP) [21]. The main objective of this stage is to generate parameters evaluation and basic indicators.

2.3 RAM Analysis

2.3.1 Reliability Analysis

Different alternatives have been proposed for the individual and systemic logical-functional representations of processes [4]. Modelling a complex system using RBDs is a well-known method that has been adopted for different applications in system reliability analysis [5, 22]. An RBD is constructed after performing a logical decomposition of a system into subsystems; the RBD is constructed to express reliability logics such as series, redundancy and standby in a network of subsystems. RBDs are considered to be a modelling tool that is consolidated and available for the normal duties of reliability analysis. The RBD analysis methodology [9, 10] is used extensively in the mining sector due to its adaptability for representing complex provisions and environments with large amounts of equipment. An RBD can be applied in addition to other reliability techniques, such as a fault tree analysis (FTA).

2.3.2 Maintainability Analysis

Maintainability performance is defined as "the ability of an item under given conditions of use to be retained in, or restored to, a state in which it can perform a required function, when maintenance is performed under given conditions and using stated procedures and resources" [23]. The "given conditions" refer to the conditions in which the item is used and maintained, e.g. climate conditions, support conditions, human factors and geographical location. The maintainability of equipment can be represented and understood as a probability distribution of the maintenance completion time. Parametric methods have been used to analyse historical time to repair (TTR) data sets in many case studies [24].

2.3.3 Availability Analysis

According to Dhillon [5], availability is the probability that the equipment is available as required. Assuming that the required equipment must always be operating and that maintenance orders are immediately executed after a failure, the expected availability of the determined equipment can be defined [25]:

$$A(t) = \frac{\text{System uptime}}{\text{System uptime} + \text{System downtime}} \approx \frac{\text{MTTF}}{\text{MTTF} + \text{MTTR}} \qquad (1)$$

Availability is probably the most important parameter because it is directly related to the equipment performance, especially in a production environment that is based on volume, such as the mining industry. In this environment, the rate of production, among other variables, determines the level of benefit that is derived from the exercises. Monte Carlo simulation is used as a modelling framework to represent the realistic features of the equipment and the complex behaviour of high-dimensional systems to obtain the performance indicators for the availability [4].

2.4 FEI Quantification and Decision-Making

To implement an efficient asset management, an item classification should be developed based on criteria such as direct and indirect costs, failure rate and operational impact. The objective is to identify the relevant elements in making priority maintenance decisions and efficiently allocating resources. Many techniques exist for asset hierarchy, and each technique has advantages and disadvantages that depend on the operation context [4]. For this proposal, the authors present a novel quantitative methodology for measurement based on reliability impact using previous results of RAM analysis.

2.4.1 FEI Methodology

In this phase, the first indicator is the E-OCI of each element in the system, which is determined by decomposing the global index and each subsystem index [26] on the following levels:

$$\text{Expected Operational Criticality Index E-OCI}_{\text{system } i=0} = 1 \qquad (2)$$

The Eq. 2 represents the start of the process considering the E-OCI as a 100%. Then, the breakdown of each level begins with the following equations:

$$\sum_{j=1}^{n} \text{E-OCI}_{i;j} = \text{E-OCI}_{i-1} \quad \forall i : 1, \ldots, r; \quad \forall j : 1, \ldots, n \qquad (3)$$

In Eq. 3, the E-OCI of a level is distributed in all the elements that composed it, all this to keep the consistence of the impact quantification. To determine the E-OCI of each element of the lower level it is necessary to develop the Eq. 4, which considers the distribution by an unavailability factor. In this sense, the most unavailable element of the lower level, obtains a bigger proportion of the E-OCI from the upper level.

$$\frac{\text{E-OCI}_{i;j}}{\text{E-OCI}_{i;j+1}} = \frac{(1 - A_{i;j})}{(1 - A_{i;j+1})} \forall i : 1, \ldots, r; \quad \forall j : 1, \ldots, n \tag{4}$$

where E-OCI$_{i;j}$ is the E-OCI for element j (from 1 to n) that is included in decomposition level i (from 1 to r) and $A_{i;j}$ is the expected availability for element j (from 1 to n) that is included in decomposition level i (from 1 to r).

In simple terms, the E-OCI shows the final contribution of each element toward mitigating the system's lack of effectiveness based on the production capacity loss. When considering a lower detail level, such as level r, the sum of all E-OCI values is 100% of the system.

$$\sum_{i=1}^{r} \sum_{j=1}^{n} \text{E-OCI}_{i;j} = 1 \tag{5}$$

Once the E-OCI$_{i;j}$ of each item is known, its level of impact is divided into two main aspects: frequency (by the unavailability of the element) and consequence (by the impact of the element). This consequence is E-DFP$_{i;j}$, which represents the effect on system j of element i stopping. The effect of stopping element i may have different results on system j depending on the state of the elements that are also on level i. Particularly, the Eq. 7 is deducted from the definition of E-OCI and the relation between the element and system unavailability (Eq. 6).

$$\text{E-OCI}_{i;j} = \frac{(1 - A_{i;j}) * \text{E-DFP}_{i;j}}{(1 - A_{\text{system}})} \tag{6}$$

$$\text{E-DFP}_{i;j} = \frac{\text{E-OCI}_{i;j} * (1 - A_{\text{system}})}{(1 - A_{i;j})} \tag{7}$$

Figure 2 shows the FEI methodology and each phase of the process.

Table 1 shows a comparative analysis between the main criticality and the operative impact methodologies.

3   Case Study

For the analytical development of this case study, the selected process and equipment in the analysis are presented to develop the logical-functional sequence RBD due to the complexity of the system with a large amount of equipment. Then, the time to failure (TTF) and the time to repair (TTR) are analysed to validate and identify trends and correlations. The parameterization is performed according to the

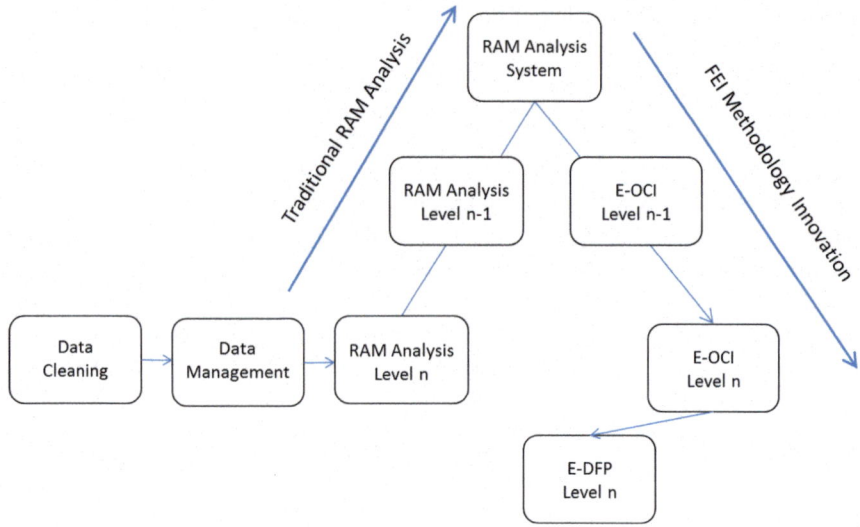

Fig. 2 FEI methodology

Table 1 Comparison of operative impact assessment methodologies

Methodology	Typology	Focus	Failure effect estimation	Flexibility
Failure tree [11]	Qualitative	Individual	Failure analysis	High
US Department [32]	Qualitative	Individual	Failure analysis	High
Crespo Proposal [33]	Quantitative	Systemic	Expert criteria (1–5)	Medium
FEI [26]	Quantitative	Systemic	Probabilistic impact analysis	High

most suitable stochastic model, which represents both the nature of the failure and the process of repairing the equipment. The FEI is individually and systematically developed to obtain the main indicators and to identify the equipment with the highest impacts on the process.

The data were collected over a period of 16 months using the SAP-PM report from the mining industry in Chile. Considering that current automatic capture systems provide rich and complete data with respect to operational parameters, focused on prognosis and health management application, data related to the state of the asset, specific process parameters and downtimes. This data repository is linked with information related to work notifications and work orders. Accordingly, an ERP solution permits the complete integration of information flow from all functional areas by means of a single database that is accessible through a unified interface and channel communication. Hence, it is possible to apply the proposed methodology using this consolidated database validating the quality and quantity of the information.

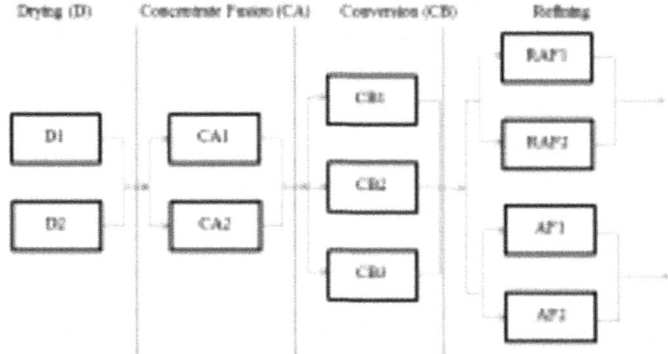

Fig. 3 Case study process diagram

3.1 The CSP Case Study

The case study is the smelting process of a mine in Chile; the first stage of this process is smelting ore, which contains a concentration of approximately 26% copper. The pyro-metallurgical operations, which enable the extraction of metallic copper, are performed in type A converters (the melting process), type B converters (the conversion process), slag cleaning furnaces (the copper recovery process), fire refinement stoves (the refined copper preparation process) and anodic furnaces (the anodic copper preparation process). The nominal production capacity of the smelting process is 60 tons/h. The resulting gases from the fusion conversion process are treated in gas cleaning plants, which also produce sulfuric acid that is primarily marketed in the mining industry in the northern region of Chile. Therefore, the portfolio of products obtained through these processes includes copper plates with 99.7% purity, fire refined copper ingots with 99.9% purity and sulfuric acid derived from the processing of smelter gases.

The analytical development of the case study uses the main operational flow of the concentrate in the smelting process, which is represented by the equipment and subprocesses shown in Fig. 3 (Table 2).

The mining subprocesses that are considered in this paper are drying, concentrate fusion, conversion and refining.

3.2 Modelling the System

The logic behind the process operations can be represented using FT diagrams, which enables the subsequent development of the RBD configuration. According to Table 1, the FT is constructed as shown in Fig. 4.

According to Fig. 2, the smelting process is composed of four subsystems arranged serially: the drying subsystem (DS), which consists of two dryers in full

Table 2 CSP information

Equipment	Label	Basic function	Capacity	
			Nominal [ton/h]	Maximum [ton/h]
Dryer 1	D1	Drying	30	60
Dryer 2	D2		30	60
Type A converter 1	CA1	Conversion type A	30	60
Type A converter 2	CA2		30	60
Type B converter 1	CB1	Conversion type B	20	30
Type B converter 2	CB2		20	30
Type B converter 3	CB3		20	30
RAF furnace 1	RAF1	Refining A	30	60
RAF furnace 2	RAF2		30	60
Anode furnace 1	AF1	Refining B	30	60
Anode furnace 2	AF2		30	60

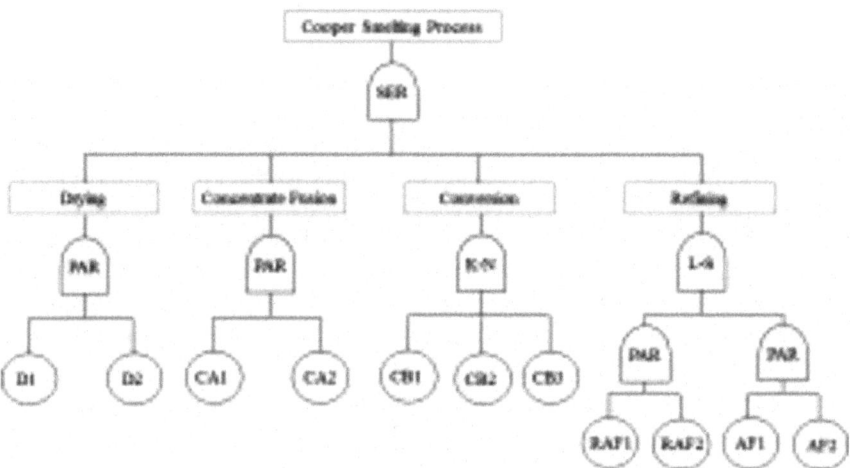

Fig. 4 FT representation of the process

redundancy, the concentrate fusion subsystem (CFS), which consist of two type A converters in full redundancy, the conversion subsystem (CS), which consist of three type B converters in a 3-2 configuration with partial redundancy and the refining subsystem (RS), which consists of two work lines that separate the RAF production from the AF plate production, with a load distribution of 60 and 40%, respectively. Each subprocess is composed of two elements in total redundancy. The classification for the reliability analysis of the CSP is presented in the FT diagram.

The quantitative analysis of the CSP considers the operating conditions, failure data, maintenance times and other functional information needed to estimate the

RAM indicators for each piece of equipment, each subsystem and the total system. Reliability analysis was performed using traditional algorithms [26].

3.3 Data Analysis

The collected data includes time to failure (TTF) and TTR for each equipment. The next step in data management is to determine the nature of the equipment that is involved in the process. In this case, all of the equipment uses the dynamics of serviceable equipment, and its subsequent distribution must be selected using relevant stochastic models [27] according to the behaviours of the data in terms of trend and independence.

3.3.1 Trend and Correlation Analysis

Analysing the data of the equipment that is involved in the process, the independence and trend indicators are calculated. The calculated values of the test statistics for all of the equipment failures and repair data are listed in Table 3. Using the null hypothesis of a homogeneous Poisson process, in which the test statistic U is X^2 distributed with $2(n-1)$ degrees of freedom, the null hypothesis is not rejected at a 5% level of significance in most of the equipment. The statistical results show that the data sets for the majority of the equipment, with the exceptions of the TTF data for CB3, RAF1 and AF1, show no trends or serial correlation. Therefore, the i.i.d. assumption is rejected for these equipment. Identical results were obtained from the graphical trend analysis [14].

3.3.2 Distribution Fitting and Calculation of Basic Indicators

The definition of the probability distributions is commonly used to describe the equipment failure and repair processes. Different types of statistical distributions were examined, and their parameters were estimated using MATLAB [28]. A statistical goodness of fit test was performed to define the distributions of the operating times and TTR. In particular, Kolmogorov–Smirnov tests [25] at a significance level of a = 0.1 were required to be satisfied for p.0.1 at each setting. The null hypothesis H0 is as follows: the data follows a normal, lognormal, exponential or Weibull distribution.

The equipment trend data should be analysed using a stochastic model for repairable elements. The NHPP model used in this study is based on the Power Law Process (PLP). The X^2 test and the Kolmogorov–Smirnov test were classically encountered to validate the best-fit distribution [29]. The parameters that were estimated from the failure data are listed in Table 4.

Table 3 Statistical results for TTF and TTR data

Subsystem	Equip.	Data set	Degrees of freedom	Statistic U	Rejection of null hypothesis at a 5% level of significance
DS	D1	TTF	96	97.25	Not rejected (>76.11)
		TTR	96	82.21	Not rejected (>74.40)
	D2	TTF	66	57.32	Not rejected (>52.34)
		TTR	66	51.31	Not rejected (>47.85)
CFS	CA1	TTF	62	47.56	Not rejected (>39.65)
		TTR	62	45.44	Not rejected (>40.67)
	CA2	TTF	56	46.17	Not rejected (>38.95)
		TTR	54	45.76	Not rejected (>38.65)
CS	CB1	TTF	62	49.13	Not rejected (>33.56)
		TTR	62	50.17	Not rejected (>34.55)
	CB2	TTF	70	54.23	Not rejected (>47.36)
		TTR	70	49.95	Not rejected (>42.72)
	CB3	TTF	92	89.34	Rejected (<92.71)
		TTR	90	87.36	Not rejected (>70.23)
RS	RAF1	TTF	70	52.14	Rejected (<53.11)
		TTR	70	56.22	Not rejected (>52.34)
	RAF2	TTF	68	49.76	Not rejected (>42.57)
		TTR	68	47.66	Not rejected (>43.54)
	AF1	TTF	62	58.27	Rejected (<59.56)
		TTR	62	60.22	Not rejected (>53.33)
	AF2	TTF	54	41.56	Not rejected (>39.44)
		TTR	52	41.75	Not rejected (>37.72)

Table 4 also lists the basic indicator of reliability, which is the mean time to failure (MTTF), the basic indicator of maintainability, which is the mean time to repair (MTTR), and the respective parameters of the fitted curves.

According to the traditional setting (no trend), the reliability parameters of the fitted curves (α, β), which are based on the life cycle theory [29], show that the equipment are in different phases of the bathtub curve. $\beta = 1$ is related to a constant failure rate or "useful" life phase; $\beta > 1$ is the phase in which the failure rate typically increases and additional service and maintenance are needed. In the "wear-out" life phase, a technical and economic assessment is required to determine the need for possible replacement, and in the $\beta < 1$ phase, the component failure rate decreases over time. Early life cycle problems are often due to failures in design, incorrect installation and operation by poorly trained operators. Therefore, the statistical information obtained from the curve fitting should be used to estimate the system performance instead of explaining individual equipment behaviour.

The probability distribution that is commonly used to represent repair times is the normal-logarithmic distribution, which explains the variability of repair by two

Table 4 Fitting distributions of the TTF and TTR data

Equip.	Best fit (TTF data)				
	Distribution	p value (K-S)	Par. 1	Par. 2	MTTFi
D1	Weibull	0.29	$\alpha = 238$	$\beta = 1.13$	228
D2	Weibull	0.17	$\alpha = 224$	$\beta = 1.2$	211
CA1	Exponential	0.14	$\alpha = 253$	$\beta = 1.6$	227
CA2	Weibull	0.43	$\alpha = 330$	$\beta = 1.4$	300
CB1	Exponential	0.32	$\alpha = 256$	$\beta = 1$	256
CB2	Weibull	0.29	$\alpha = 792$	$\beta = 1.34$	722
CB3	NHPP—PLP	0.31	$\alpha = 402$	$\beta = 1.32$	269
RAF1	NHPP—PLP	0.28	$\alpha = 351$	$\beta = 1.27$	211
RAF2	Weibull	0.40	$\alpha = 144$	$\beta = 1.30$	133
AF1	NHPP—PLP	0.23	$\alpha = 324$	$\beta = 1.31$	203
AF2	Exponential	0.61	$\alpha = 315$	$\beta = 1$	315
Equip.	Best fit (TTR data)				
	Distribution	p value (K-S)	Par. 1	Par. 2	MTTRi
D1	Lognormal	0.57	$\mu = 0.478$	$\sigma = 2.24$	4.94
D2	Lognormal	0.24	$\mu = 0.731$	$\sigma = 1.3$	3.97
CA1	Lognormal	0.71	$\mu = 0.940$	$\sigma = 1.8$	6.29
CA2	Normal	0.61	$\mu = 4.8$	$\sigma = 1.2$	4.84
CB1	Lognormal	0.66	$\mu = 1.31$	$\sigma = 1.57$	8.12
CB2	Lognormal	0.35	$\mu = 1.45$	$\sigma = 2.02$	11.71
CB3	Normal	0.69	$\mu = 7.33$	$\sigma = 1.4$	7.33
RAF1	Lognormal	0.21	$\mu = 1.44$	$\sigma = 1.1$	7.31
RAF2	Lognormal	0.61	$\mu = 0.872$	$\sigma = 1.24$	4.41
AF1	Lognormal	0.45	$\mu = 1.31$	$\sigma = 1.45$	7.65
AF2	Lognormal	0.24	$\mu = 0.91$	$\sigma = 1.7$	5.82

phenomena [30]: the variation of the time associated with accidental factors (negative exponential distribution) and the factors related to typical repair (normal distribution).

3.3.3 System Reliability Analysis

CSP is divided into four subsystems, which comprise a logical-functional configuration series, i.e. all subsystems must be operating to ensure that the process performs properly. Subsystem and system reliability is generally calculated using the standard RBD formulas [5, 22].

Table 5 presents the main reliability results for different operating times.

The reliability results for the main subsystem and total system are shown graphically in Fig. 5.

Table 5 Reliability evaluation of the CSP

	R(0)	R(50)	R(100)	R(150)	R(200)	R(250)	R(300)	R(350)	R(400)
System	1.000	0.948	0.785	0.568	0.360	0.196	0.088	0.031	0.009
DS	1.000	0.976	0.901	0.794	0.674	0.556	0.449	0.356	0.278
D1	1.000	0.842	0.687	0.552	0.440	0.347	0.273	0.213	0.166
D2	1.000	0.848	0.684	0.539	0.418	0.320	0.242	0.181	0.135
CFS	**1.000**	**1.000**	**0.999**	**0.987**	**0.943**	**0.833**	**0.643**	**0.412**	**0.218**
CA1	1.000	0.928	0.797	0.648	0.503	0.375	0.269	0.186	0.125
CA2	1.000	1.000	0.993	0.964	0.885	0.732	0.512	0.278	0.106
CS	**1.000**	**0.987**	**0.942**	**0.869**	**0.778**	**0.679**	**0.580**	**0.486**	**0.401**
CB1	1.000	0.823	0.677	0.557	0.458	0.377	0.310	0.255	0.210
CB2	1.000	0.979	0.946	0.907	0.864	0.820	0.773	0.727	0.681
CB3	1.000	0.953	0.882	0.802	0.721	0.640	0.564	0.492	0.426
RS	**1.000**	**0.984**	**0.925**	**0.834**	**0.728**	**0.622**	**0.523**	**0.436**	**0.360**
RAF Subsystem	1.000	0.981	0.911	0.801	0.677	0.556	0.449	0.359	0.284
RAF1	1.000	0.917	0.807	0.694	0.588	0.491	0.405	0.331	0.268
RAF2	1.000	0.777	0.537	0.348	0.216	0.129	0.075	0.042	0.023
AF Subsystem	1.000	0.988	0.948	0.884	0.806	0.721	0.635	0.551	0.473
AF1	1.000	0.917	0.807	0.694	0.588	0.491	0.405	0.331	0.268
AF2	1.000	0.853	0.728	0.621	0.530	0.452	0.386	0.329	0.281

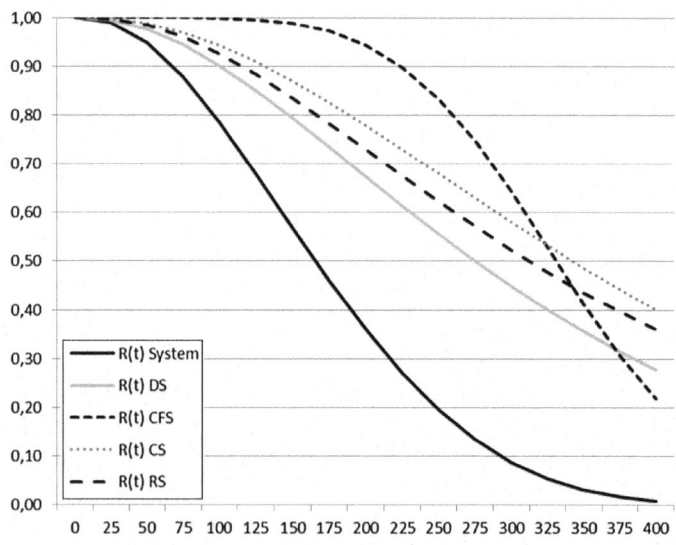

Fig. 5 Reliability curves for the main subsystems in the CSP

Through numerically and graphically analysing the reliability results, the DS and RS subsystems show an accelerated decrease in reliability compared with the other subsystems (exponential behaviour). The CS subsystem presents the best condition due to the redundant configuration of the subsystem and the distinctly reliable behaviour of the equipment. During the first 300 h, the CFS subsystem presents the most reliable behaviour due to its redundancy; after this time, the reliability decreases exponentially (wear-out behaviour).

To identify opportunities to improve the reliability, processes should focus on subsystems with less reliability over time, i.e. DS and RS. This information is the key for defining the intervals for preventive interventions. For example, if the reliability defined by an organization to develop the preventive intervention of critical equipment ranges between 75 and 80%, the planning activities for the DS subsystem should range between 75 and 100 h of operation. However, the examined system presents an important level of redundancy in three of its subsystems, which should also be considered when evaluating and defining future maintenance policies. In practical terms, a corrective policy is generally assumed in redundant systems. However, to reach this conclusion, individual assessment and identification of critical subsystems and equipment as well as the real costs of preventive and corrective interventions are required [5].

3.3.4 System Maintainability Analysis

In terms of individual analyses, the maintainability has an important effect on the equipment and systemic availability results [23]. For each element, TTR and $MTTR_i$ are required for the next analysis. Table 5 lists the parametric information for computing the $MTTR_i$.

3.3.5 System Availability Analysis

With the information obtained from the curve adjustments, the expected availability [5, 22, 31] was analysed.

Table 6 presents the results of the expected availability calculations for the equipment, subsystems and system.

According to the results, the subsystem with the least expected availability is RS, while CS has more availability. For this particular case, the availability results are consistent with the reliability results. Therefore, a practical mechanism is needed for identifying the highest E-OCI and integrating the RAM results.

According to procedure 2.4.1 and Eqs. 2, 3, 4, 5 and 6, the impacts of each piece of equipment, each subsystem and the system are presented in Table 7.

The above table shows that D1 explains 14.57% of the lack of effectiveness of the system, which is expressed as the E-OCI; each D1 failure results in an expected 33.09% loss of production capacity for the total system, which is expressed as the E-DFP. Both impact indices are dependent on the behaviour of the equipment in

Table 6 Expected availability calculation

	MTTF [hr]	MTTR [hr]	Availability %	Subsystem availability %	System availability %
System					**95.17%**
DS				**98.66%**	
D1	227.69	4.94	97.88%		
D2	210.71	3.98	98.15%		
CFS				**98.62%**	
CA1	226.83	6.30	97.30%		
CA2	299.95	4.80	98.42%		
CS				**99.83%**	
CB1	256.00	8.13	96.92%		
CB2	721.85	11.70	98.40%		
CB3	269.38	7.33	97.35%		
RS				**97.99%**	
RAF subsystem				97.77%	
RAF1	211.13	7.32	96.65%		
RAF2	133.00	4.45	96.77%		
AF subsystem				98.31%	
AF1	202.94	7.58	96.40%		
AF2	315.00	5.81	98.19%		

Table 7 Calculation of E-OCI and E-DFP

	E-OCI (%)	E-DFP (%)
System	100.00	100.00
DS	27.27	98.35
D1	14.57	33.09
D2	12.71	33.09
CFS	28.22	98.35
CA1	17.82	31.85
CA2	10.39	31.85
CS	3.47	98.35
CB1	1.46	2.29
CB2	0.76	2.29
CB3	1.26	2.29
RS	41.04	98.35
RAF subsystem	24.62	53.36
RAF1	12.53	18.05
RAF2	12.10	18.05
AF subsystem	16.42	46.78
AF1	10.92	14.65
AF2	5.50	14.65

terms of the RAM results, operational context and logical dependencies as well as the indicators for the other pieces of equipment in their subsystem (immediately higher level). The E-DFP values for each subsystem are equal, which is attributed to the serial configuration of the system.

Using this analysis, the equipment and subsystems that generate the highest impact on the availability or production of the main system can be grouped together. However, the results are not distinct. Therefore, continuity with a dispersion analysis is proposed in which the X-axis corresponds to the unavailability and the Y-axis corresponds to the E-DFP. In this manner, the generated curves correspond to the expected operational criticality iso-impact curves.

First, an analysis is performed at the subsystem level (Fig. 6); DS, CFS, CS and RS have the same level of E-DFP because they are in series. Therefore, the unavailability of each subsystem creates a different E-OCI. The subsystems RAF and AF have a smaller E-DFP because they employ a load-sharing configuration (60% of the load and 40% of the load, respectively) in the RS subsystem.

Figure 7 shows the relatively low E-DFP for the equipment, which is attributed to the redundancy in each of the subsystems. Due to the serial configuration of all subsystems, the E-DFP values are all identical, with the exception of the RAF and AF subsystems, which have different capacities and operational contexts. The order of the subsystems in terms of operational effectiveness is RS, CFS, DS and CS. At the individual level, the order of the equipment is as follows: CA1, D1, D2, RAF1, RAF2, AF1, CA2, AF2, CB1, CB3 and CB2.

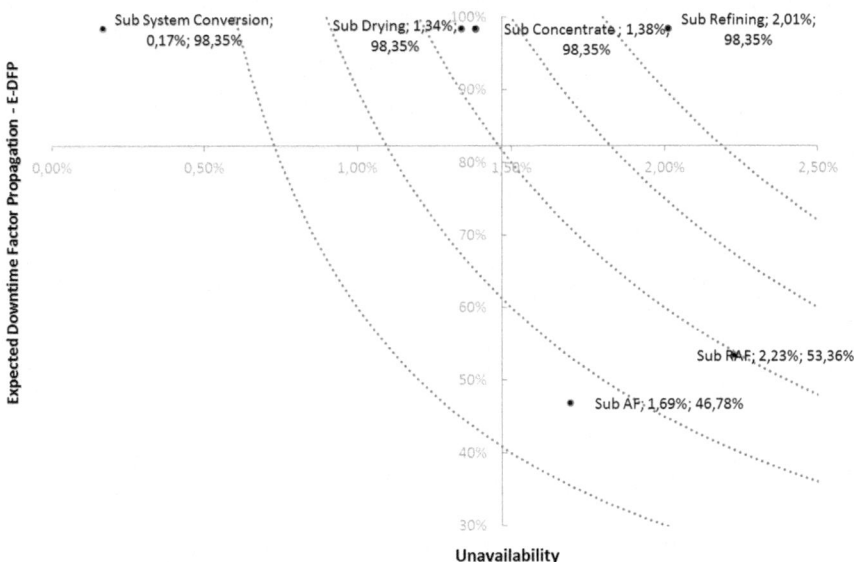

Fig. 6 FEI methodology dispersion analysis for the subsystems

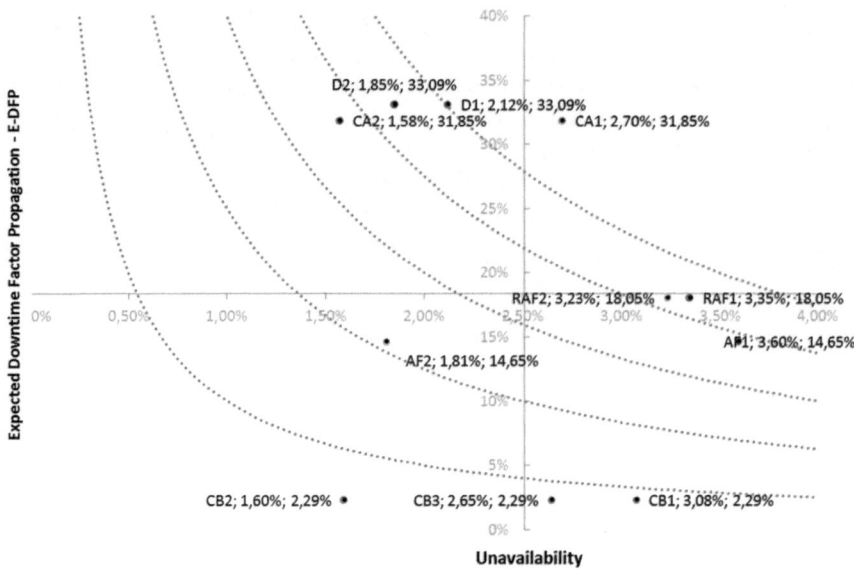

Fig. 7 FEI methodology dispersion analysis for the equipment

4   Conclusions

The reliability impact study is a relevant analysis to develop a decision making process. Considering the standard methodologies, it is possible to establish that there are no formal criteria to identify the impact of each asset and its behaviour or failure. So, it is necessary to define a key performance indicator (KPI) oriented to establish a hierarchy and determine the effectiveness of the KPI's impact on the elements.

A deep reliability analysis requires quality and quantity of data, therefore, this paper shows the importance of the quality of information that is available for analysis; the existing data should be audited to validate the previous analysis.

The FEI methodology has significant potential applications in several engineering problems, industrial realities and productive sectors. FEI is a powerful tool for analysis and decision-making for the various phases of an industrial project via life cycle cost (LCC) for design-oriented operations, such as CAPEX and OPEX, which are associated with the improvement process.

The strengths of the FEI methodology includes its ability to systematically and quantitatively assess the operational criticality in terms of the RAM indicators and the logical configuration of subsystems and individual equipment, which directly affect each element in the total system. This proposal complements plant modelling and analysis, from traditional methodologies. The E-OCI is the final result of the computation process, which enables the quantitative and unequivocal prioritization of the system elements to assess the associated loss as "production loss" regarding its unavailability and effect in the process. The latter concept enables estimating the E-DFP of the equipment to determine the individual effects of the detention and to assess complex and redundant scenarios.

The case study provides useful results for developing a hierarchization that enables an analysis of improvement actions that are aligned with the best opportunities.

Considering the FEI methodology structure, it is possible to conclude that it is replicable in different application fields and can be easily automated. Rankings that are based on the expected impact in the operation are effective, recognize weaknesses and opportunities, and serve as the basis for action plans based on reliability and maintainability.

References

1. Asaki Z (1992) Kinetic studies of copper flash smelting furnace and improvements of its operation in the smelters in Japan. Miner Process Extr Metall 11:163–185
2. Parra C (2009) Desarrollo de modelos de cuantificación económica del factor Fiabilidad en el coste total de ciclo de vida de un sistema de producción. Universidad de Sevilla, Spain, Tesis Doctoral
3. Pham H (2003) Handbook of reliability engineering, 2003rd edn. Springer, New jersey
4. Viveros P, Zio E, Arata A, Kristjanpoller F (2012) Integrated system reliability and productive capacity analysis of a production line. A case study for a Chilean mining process. Proc Inst Mech Eng Part O J Risk Reliab 226:305–317. doi:10.1177/1748006X11408675
5. Dhillon BS (2006) Maintainability, maintenance, and reliability for engineers. Taylor & Francis Group, Boca Raton
6. Welte TM (2009) A rule-based approach for establishing states in a Markov process applied to maintenance modeling. Proc Inst Mech Eng Part O: J Risk Reliab 223:1–12. doi:10.1243/1748006XJRR194
7. Heinrich L (1991) Goodness-of-fit tests for the second moment function of a stationary multidimensional, poisson process. Statistics 22:245–278. doi:10.1080/02331889108802308
8. Peng R, Zhai QQ, Xing LD, Yang J (2014) Reliability of demand-based phased-mission systems subject to fault level coverage. Reliab Eng Syst Saf 121:18–25. doi:10.1016/j.ress.2013.07.013
9. Rausand M, Hoyland A (2003) System reliability theory: models, statistical methods, and applications, 3rd edn. Wiley, New York
10. Guo H, Yang X (2007) A simple reliability block diagram method for safety integrity verification. Reliab Eng Syst Saf 92:1267–1273. doi:10.1016/j.ress.2006.08.002
11. Rauzy AB, Gauthier J, Leduc X (2007) Assessment of large automatically generated fault trees by means of binary decision diagrams. Proc Inst Mech Eng Part O J Risk Reliab 221:95–105. doi:10.1243/1748006XJRR47
12. Distefano S, Puliafito A (2009) Reliability and availability analysis of dependent-dynamic systems with DRBDs. Reliab Eng Syst Saf 94:1381–1393. doi:10.1016/j.ress.2009.02.004
13. Volovoi V (2004) Modeling of system reliability Petri nets with aging tokens. Reliab Eng Syst Saf 84:149–161. doi:10.1016/j.ress.2003.10.013
14. Jiang W, Xie C, Wei B et al (2016) A modified method for risk evaluation in failure modes and effects analysis of aircraft turbine rotor blades. Adv Mech Eng 2016 8(4): first published on April 22, 2016. doi:10.1177/1687814016644579

15. Barberá L, Crespo A, Viveros P, Arata A (2013) The graphical analysis for maintenance management method: a quantitative graphical analysis to support maintenance management decision making. J Qual Reliab Eng Int 29:77–87. doi:10.1002/qre.1296
16. Baraldi P, Compare M, Zio E (2013) Component ranking by Birnaum importance in presence of epistemic uncertainty in failure event probabilities. IEEE Trans Reliab Inst Electr Electron Eng 62:37–48
17. Chapman AD (2005) Principles of data quality, version 1.0. Report for the global biodiversity information facility. GBIF, Copenhagen
18. Gámiz ML, Lindqvist BH (2016) Nonparametric estimation in trend-renewal processes. Reliab Eng Syst Saf 145:38–46. doi:10.1016/j.ress.2015.08.015
19. Ascher H, Feingold H (1984) Repairable systems reliability: modeling, inference, misconceptions and their causes, vol 7. Statistics, New York
20. Veber B, Nagode M, Fajdiga M (2008) Generalized renewal process for repairable systems based on finite Weibull mixture. Reliab Eng Syst Saf 93:1461–1472. doi:10.1016/j.ress.2007.10.003
21. Kijima M, Sumita N (1986) A useful generalization of renewal theory: counting process governed by non-negative markovian increments. J Appl Probab 23:71–88. doi:10.2307/3214117
22. Macchi M, Kristjanpoller F, Arata A, Garetti M, Fumagalli L (2012) Introducing buffer inventories in the RBD analysis of production systems. Reliab Eng Syst Saf 104:84–95. doi:10.1016/j.ress.2012.03.015
23. Moreu P, Gonzalez-Prida V, Barberá L, Crespo A (2012) A practical method for the maintainability assessment in industrial devices using indicators and specific attributes. Reliab Eng Syst Saf 100:84–92. doi:10.1016/j.ress.2011.12.018
24. Barabady J (2005) Reliability and maintainability analysis of crushing plants in Jajarm Bauxite Mine of Iran. In: Proceedings of the annual reliability and maintainability symposium. IEEE, New York, pp 109–115
25. Sundararajan CR (1991) Guide to reliability engineering. Van Nostrand Reinhold, New York
26. Kristjanpoller F, Viveros P, Crespo A, Grubessich T, Stegmaier R (2015) RAM-C: A novel methodology for evaluating the impact and the criticality of assets over systems with complex logical configurations. In: The annual European safety and reliability conference (ESREL), Zurich, Switzerland, 7–10 Sept 2015
27. Viveros P, Crespo A, Kristjanpoller F, Tapia R, González-Prida V (2015) Mathematical and stochastic models for reliability in repairable industrial physical assets. Promoting Sustain Practices through Energy Eng Asset Manage 1:287–310. doi:10.4018/978-1-4666-8222-1.ch012
28. Dogan I (2011) Engineering simulation with MATLAB: improving teaching and learning effectiveness. Procedia Comput Sci 3:853–858. doi:10.1016/j.procs.2010.12.140
29. Levitin G (2007) Block diagram method for analyzing multi-state systems with uncovered failures. Reliab Eng Syst Saf 92:727–734. doi:10.1016/j.ress.2006.02.009
30. Barlow R, Proschan F (1996) Mathematical theory of reliability. Society for Industrial and Applied Mathematics, Philadelphia
31. Lisnianski A (2007) Extended block diagram method for a multi-state system reliability assessment. Reliab Eng Syst Saf 92:1601–1607. doi:10.1016/j.ress.2006.09.013
32. US Department of Defense (1977) Procedures for performing a failure mode and effects analysis, MIL-STD-1629A. Department of Defense, Washington
33. Crespo A, Moreu P, Sola A, Gómez J (2015) Criticality analysis for maintenance purposes: a study for complex in-service engineering assets. Quality and Reliability Engineering International 2015. Published online in Wiley Online Library. doi:10.1002/qre.1769

Author Biographies

Fredy Kristjanpoller Rodríguez is an Industrial Engineer and Master in Asset Management and Maintenance of Federico Santa Maria University (USM, Chile), and doctoral candidate in Industrial Engineering at the University of Seville. Researcher, academic and master program coordinator linked to asset management. He has important scientific papers on indexed journals and international proceedings congress on the following areas: Reliability engineering and Maintenance Strategies. He has developed consultant activities in the main Chilean companies.

Adolfo Crespo Márquez is currently Full Professor at the School of Engineering of the University of Seville, and Head of the Department of Industrial Management. He holds a Ph.D. in Industrial Engineering from this same University. His research works have been published in journals such as the International Journal of Production Research, International Journal of Production Economics, European Journal of Operations Research, Journal of Purchasing and Supply Management, International Journal of Agile Manufacturing, Omega, Journal of Quality in Maintenance Engineering, Decision Support Systems, Computers in Industry, Reliability Engineering and System Safety and International Journal of Simulation and Process Modeling, among others. Prof. Crespo is the author of seven books, the last four with Springer-Verlag in 2007, 2010, 2012 and 2014 about maintenance, warranty and supply chain management. Prof. Crespo leads the Spanish Research Network on Dependability Management and the Spanish Committee for Maintenance Standardization (1995–2003). He also leads a research team related to maintenance and dependability management currently with 5 Ph.D. students and 4 researchers. He has extensively participated in many engineering and consulting projects for different companies, for the Spanish Departments of Defense, Science and Education as well as for the European Commission (IPTS). He is the President of INGEMAN (a National Association for the Development of Maintenance Engineering in Spain) since 2002.

Pablo Viveros Gunckel Researcher and Academic at Technical University Federico Santa María, Chile, has been active in national and international research, both for important journals and conferences. Also he has developed projects in the Chilean industry. Consultant specialist in the area of Reliability, Asset Management, System Modeling and Evaluation of Engineering Projects. He is Industrial Engineer and Master in Asset Management and Maintenance.

Luis Barberá Martínez is Ph.D. (Summa Cum Laude) in industrial Engineering by the University of Seville, and Mining Engineer by UPC (Spain). He is a researcher at the School of Engineering of the University of Seville and author of more than 50 articles, national and main internationals. He has worked at different international Universities: Politecnico di Milano (Italy), CRAN (France), UTFSM (Chile), C-MORE (Canada), EPFL (Switzerland), University of Salford (UK) or FIR (Germany), among others. His line of research is industrial asset management, maintenance optimization and risk management. Currently, he is Spain Operations Manager at MAXAM Europe. He has three Master's degrees: Master of Industrial Organization and Management (2008/2009) (School of Engineering, University of Seville), Master of Economy (2008/2009) (University of Seville) and Master of Prevention Risk Work (2005/2006). He has been honoured with the following awards and recognitions: Extraordinary Prize of Doctorate by the University of Seville; three consecutive awards for Academic Engineering Performance in Spain by AMIC. (2004/2005), (2005/2006) and (2006/2007); honour diploma as a "Ten Competences Student" by the University of Huelva and CEPSA Company (2007); graduated as number one of his class.

Influence of the Input Load on the Reliability of the Grinding Line

Luis Barberá Martínez, Pablo Viveros Gunckel, Rodrigo Mena and Vicente González-Prida Díaz

Abstract The management of physical resources in an organization involves several processes related to innovation and continuous improvement. For this reason, the proper study of the reliability and maintainability analysis is considered essential and is treated as one of the main pillars for decision-making at the tactical and operational levels. This paper proposes a useful support tool for decision-making in the field of maintenance management and reliability analysis, so that such decisions remain aligned with the vision, strategy and economic indicators of the business or industrial organization. This research clearly shows how the variability of different load levels (inflows) on the grinding lines, affects the reliability of a specific sulphur plant (located in Chile), determining after that, what the optimum load should be. The paper identifies the relationship between each line load ranges and the corresponding reliability, all through the development of a real case study conducted in a mining company located in northern Chile.

Keywords Reliability by load range · Grinding line · Management and maintenance optimization · Efficiency and effectiveness in maintenance

1 Introduction and Problem Statement

The development and implementation of tools to support decision-making helps to achieve efficient resource management and physical assets within an industry organization. Even more when there are a large number of devices involved between them and with a functional complexity. In order to obtain real applications based on

L. Barberá Martínez (✉) · P. Viveros Gunckel · V. González-Prida Díaz
Department of Industrial Management, School of Engineering,
University of Seville, Seville, Spain
e-mail: lubarmar@us.es

R. Mena
Department of Industrial Engineering, University Federico Santa María, Valparaíso, Chile

© Springer International Publishing AG 2018
A. Crespo Márquez et al. (eds.), *Advanced Maintenance Modelling for Asset Management*, DOI 10.1007/978-3-319-58045-6_19

analytical models, tools must be generated being practical, functional, simple and innovative, helping to make easily tactical and operational decisions [1, 2].

In the context of mining industry, the variable equipment load is especially important in the behaviour of reliability. This paper presents and analyses a real case study developed in a sulphur plant, where it is considered a priority the operation analysis of grinding lines, all of them subject to changing feeding flows (supply tonnage) and are received by SAG mills. The variability in feeding streams depends mainly on the production goals.

This paper proposes a tool that makes easier the analysis of results supporting a management and supervision framework of different action plans performed in the field of maintenance activities [3]. Throughout this research, it is shown how after a preliminary analysis of stop frequencies in each line, one of them had a lower failure rate coinciding with a lower volume and stable supply.

Based on this observation, it was researched the hypothesis that two lines should be fed in reliable operation ranges. This means that the plant should be operated in load ranges that would achieve real production targets, reducing at the same time the equipment degradation and decreasing operational stops. According to this, this paper tries to achieve a stable operation over the time that does not involve a fast degradation of the line equipment and avoiding, in short term, the occurrence of major faults which can involve stops for periods of time longer than the current averages (Fig. 1).

2   Analysis Methodology

Next, a methodology for analysis and calculation to support decision-making is developed to optimize the reliability of production lines, considering the load range (input supply) required to the equipment [4]. The methodology consists of three stages: operational average load analysis, systemic reliability analysis per working line and, finally, reliability analysis and indicators for load range.

2.1   Operation Average Load Analysis

This first analysis is necessary to estimate the average workloads for two specific moments of operation. Specifically, there is a historical record of 10-month data for grinding lines 1 and 2. The information used has been properly filtered and analysed in terms of trend and correlation quality, ensuring the independence assumption and the lack of trend [6]. This information refers to average input load data per hour of operation. The records belong to failure events, planned activities and operational stops (Table 1).

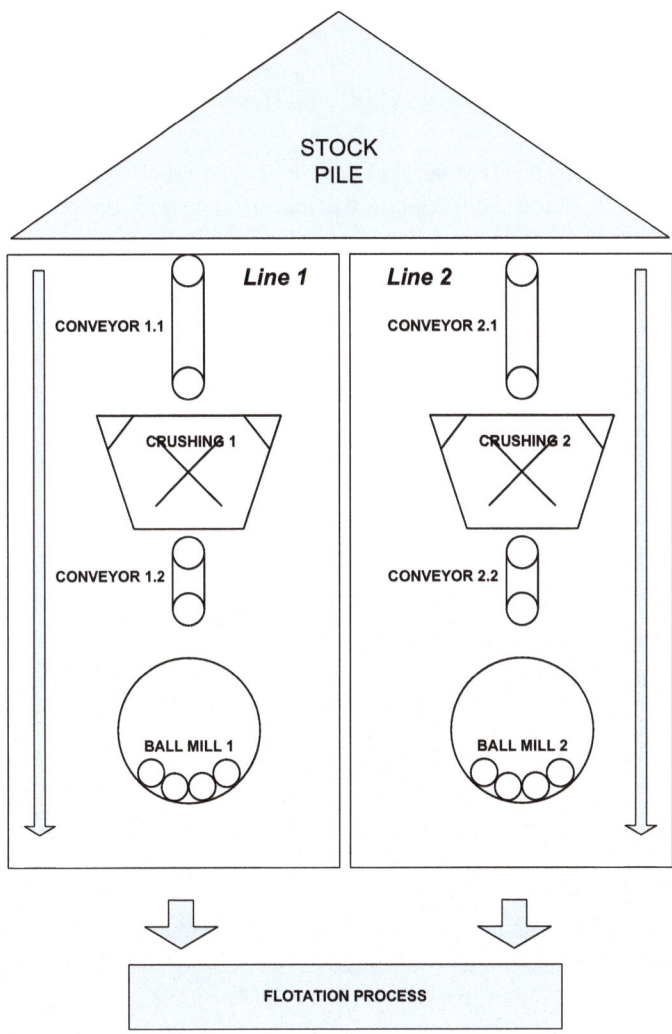

Fig. 1 General representation of grinding lines 1 and 2 in mining plant

Table 1 Historical data representation

Equipment A_j	Event E_i	Event type M_k	Average load $\bar{Q}_i$	T. schedule T_i	T. reparation TR_i	T. Proper function TBF_i
A_1	E_1	M_1	$\bar{Q}_1$	T_1	TR_1	TBF_1
A_5	E_2	M_2	$\bar{Q}_2$	T_2	TR_2	TBF_2
A_1	E_3	M_3	$\bar{Q}_3$	T_3	TR_3	TBF_3
...	...	...	...	...	...	...
A_p	E_n	M_2	$\bar{Q}_n$	T_n	TR_n	TBF_n

$TBF_1 = T_1$ $TBF_2 = [T_2 - T_1 - TR_1]$ $TBF_3 = [T_3 - T_2 - TR_2]$ $TBF_n = [T_n - T_{n-1} - TR_{n-1}]$
$K = 1, 2$ and 3 $j = 1$ to p
$M_1 =$ preventive event $M_2 =$ corrective event $M_3 =$ operational stop

In order to obtain the average load data, it is necessary to perform a graphical analysis (from historical data) relating the loading time with the proper operation time (Figs. 2 and 3).

Figure 2 shows how among failure events (T_1, T_2, ..., T_n), load variations take place (left graph). In order to calculate the average workload supported by the equipment between failure events (in good operation times, TBF), it is performed a graphical equivalence between event areas (total handled load $= A_1 = A_1'$).

Mathematically, the graphic equivalence corresponds to:

$$A_1 = A_1' \quad \& \quad A_2 = A_2'$$

Considering the actual TBF_i from each event, it is possible to estimate the corresponding:

$$\bar{Q}_1; \bar{Q}_2$$

Analysing each pair of events $[E_1; E_{i+1}]$, it is possible to fulfil Table 1, where the type of event M_k, the average load Q_i, the good operation times between TBF_i events and the reparation time TR_i are defined [5].

2.2   Reliability Systemic Analysis by Working Line

In order to perform systemic analysis for each line, there are certain events already known: failure, preventive and operational stop, where the impact I_i that these

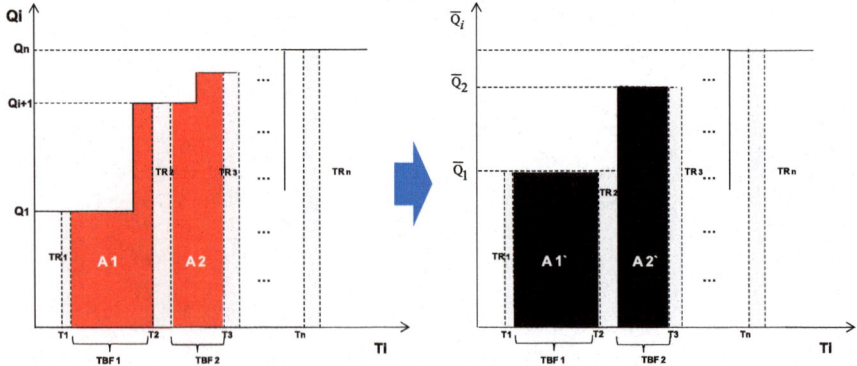

Fig. 2 Graphical representation of average load Q_i by pair of events $[E_i - E_{i+1}]$

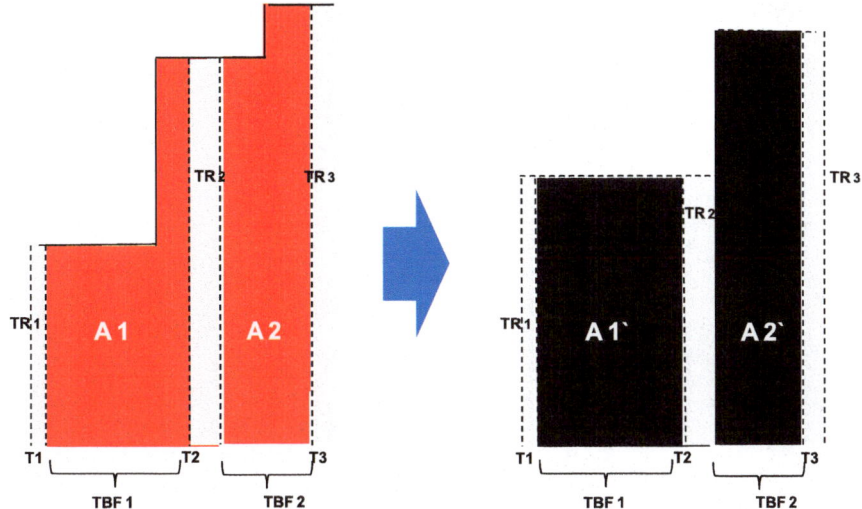

Fig. 3 Graphical approach detail of average load Q_i by pair of events $[E_i; E_{i+1}]$

events generate on mainline load is less than or equal to 100%. It is therefore necessary to establish some criteria for the reliability calculation, regardless the individual impact of failure events on the system:

(a) Any event that has an impact $I_i > 0$, will be considered a visible event at line level. The operational stops have an impact of 100% on the line under analysis. The database available for analysis has an indicator called "conversion factor", equivalent to the impact I_i, which represents the percentage of loss in the main line as a consequence of a single failure.

(b) The line production loss will be considered in proportion to the impact that a single event generates (data obtained from production records).

(c) The system repair times TR_i, for every pair of events $[E_i\_ E_{i+1}]$, will be calculated taking into account the considerations shown in Fig. 4, where the partial failure ($I < 100\%$) is proportionally associated with an fictitious event with impact equal to 100%, i.e. for the reliability calculation is required binary states (system at total operation loss or operating system), for this reason, the events that cause a partial loss of function ($R < 100\%$) should be approximated to an event of total operation loss, both equivalent in loss workload. As a result, the new repair time of the fictitious event (TR'_1) must be lower than the actual repair time (TR_1) to maintain the equivalence of load loss. Graphically, it is represented in Fig. 4.

Where:

$$A_1 = TR_i^*(Q_i - Q_{i+1}) \quad A'_1 = TR_i'^* Q_i$$
$$A'_1 = A_1$$

Fig. 4 Graphical approach detail of system TR_i and TR_i' for a pair of failure event $[E_i; E_{i+1}]$

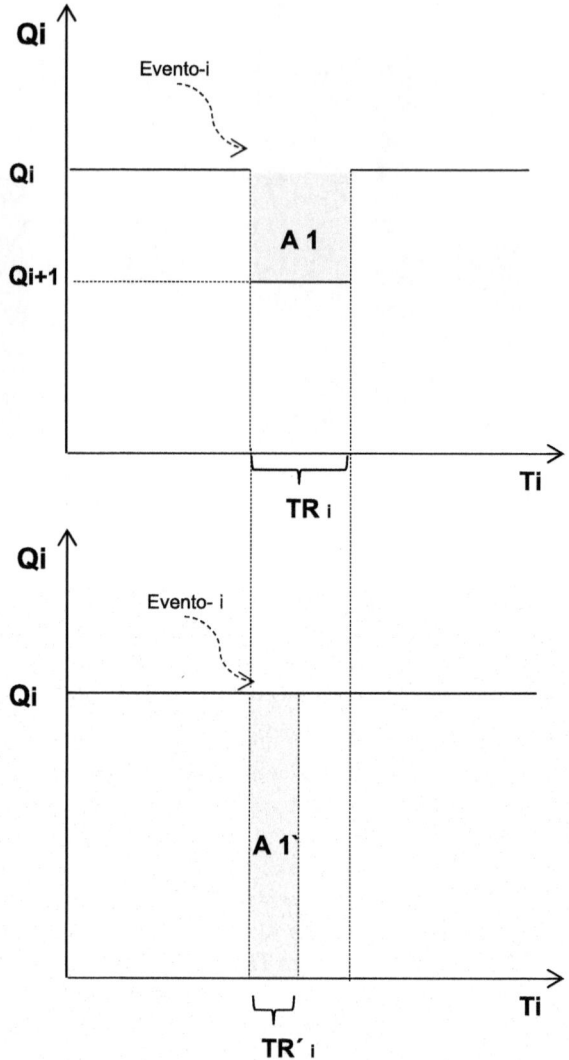

Therefore, the repair time approach of the fictitious event (proportionally equivalent to the real event in terms of loss production on the system) is:

$$TR_i' = \frac{TR_i^*(Q_i - Q_{i+1})}{Q_i}.$$

According to [6], the basic reliability models applied in industry traditionally considered only binary states [0, 1], which cannot be used in this process because

Table 2 Range definition

Range (R_p)	$[Q_A - Q_B]$
Range$_1$	$[Q_1 - Q_2]$
Range$_2$	$[Q_2 - Q_3]$
Range$_3$	$[Q_3 - Q_4]$
...	...
Range$_N$	$[Q_N - Q_{N+1}]$

the failures in the system do not generate a complete loss of production. However, although it is an event with proportional impact, it is a failure event. Once analysed all recorded events, the load ranges must be defined for which the respective reliability curves and indicators will be calculated. These ranges depend on the amount of available information, therefore, the more data is available, the more accurate will be the operating ranges estimation. An example is shown in Table 2.

From $p = 1$ to n, being n the total number of ranges as shown in Table 2. The reliability modelling of these ranges depends on the number of failure events (total or partial) that happens. Subsequently, the information must be crossed in Table 1 associating ranges (R_p) with all the corresponding events E_i, obtaining a considerable number of events, and therefore, functioning times TBF_i, which are necessary for setting the reliability curves. As result, Table 3 is obtained.

In order to perform the reliability analysis in each line, the available information in Table 1 is used to obtain indicators TR_i, TBF_i and the value of average load Q_i for each pair of event $[E_i; E_{i+1}]$ associated to the same equipment. With proper functioning times (TBF_i), the values for reliability curves through the corresponding parameterization are obtained. The calculation methodology is similar to the systemic one, with the difference that it is not necessary to consider the impact on the higher system (Line), and the repair time for each activity is directly obtained without approaches or assumptions. The following considerations must be taken into account:

- The analysis assumes that the data provided to analysts correspond to all events that occur in the process lines.
- Input load records for each grinding line are the arithmetic average for each hour of operation.
- Operational Stops (OS), must conceptually not be due to excessive load on the lines, so OS records associated to these phenomena whose description matches to an "Overload" or "Material overflow" will be evaluated as a line failure.
- Impacts considered for systemic analysis were obtained directly from PI System records, which control the supply to the grinding line. It does not exclude that the analysis can be done considering the functional logic and dynamic effects between equipment and their workloads.

Table 3 Assignment of data to ranges R_i

Range R_i	Events E_i	Typo of events M_i	Average load $\bar{Q}_i$	T. reparation TR_i	T. proper function TBF_i
$[Q_1 - Q_2]$	$E_1; E_2; E_8;$ $E_{45}\ldots E_{150}$	$M_1; M_2; M_8;$ $M_{45}\ldots M_{150}$	$\bar{Q}_1; \bar{Q}_2; \bar{Q}_8;$ $\bar{Q}_{45}\ldots\bar{Q}_{150}$	$TR'_1; TR'_2; TR'_8;$ $TR'_{45}\ldots TR'_{150}$	$TBF_1; TBF_2; TBF_8;$ $TBF_{45}\ldots TBF_{150}$
$[Q_2 - Q_3]$	$E_3; E_5; E_6;$ $E_{43}\ldots E_{211}$	$M_3; M_5; M_6;$ $M_{43}\ldots M_{211}$	$\bar{Q}_3; \bar{Q}_5; \bar{Q}_6;$ $\bar{Q}_{43}\ldots\bar{Q}_{211}$	$TR'_3; TR'_5; TR'_6;$ $TR'_{43}\ldots TR'_{211}$	$TBF_3; TBF_5; TBF_6;$ $TBF_{43}\ldots TBF_{211}$
$[Q_3 - Q_4]$	$E_{11}; E_{17}; E_{21};$ $E_{45}\ldots E_{317}$	$M_{11}; M_{17}; M_{21};$ $M_{45}\ldots M_{317}$	$\bar{Q}_{11}; \bar{Q}_{17}; \bar{Q}_{21};$ $\bar{Q}_{45}\ldots\bar{Q}_{317}$	$TR'_{11}; TR'_{17}; TR'_{21};$ $TR'_{45}\ldots TR'_{317}$	$TBF_{11}; TBF_{17}; TBF_{21};$ $TBF_{45}\ldots TBF_{317}$
$\ldots$				$\ldots$	$\ldots$
$[Q_N - Q_{N+1}]$	$E_7; E_{33}; E_{42};$ $E_{75}\ldots E_{321}$	$M_7; M_{33}; M_{42};$ $M_{75}\ldots M_{321}$	$\bar{Q}_7; \bar{Q}_{33}; \bar{Q}_{42};$ $\bar{Q}_{75}\ldots\bar{Q}_{321}$	$TR'_7; TR'_{33}; TR'_{42};$ $TR'_{75}\ldots TR'_{321}$	$TBF_7; TBF_{33}; TBF_{42};$ $TBF_{75}\ldots TBF_{321}$

2.3 Reliability Analysis and Indicators by Range

For each range R_p, it is recommended to have at least 8 events of corrective nature to perform parameterization, which is a statistical requirement implicit and necessary to parameterize and obtain reliability curves, representative of the system or the lines under study. In the analysis, we consider as failure probability density curves Weibull, Exponential, Normal and Log-normal. For example, Table 4 shows the information to be obtained for each load range.

In reliability modelling, Fig. 5 shows the results according to the load ranges which are, at the same operation time, quite different among the equipment which operates at different load. $R_a(t) > R_b(t) > R_c(t)$.

Some remarks to highlight in the reliability analysis:

(a) As a criterion for the correct parameterization of the data by the evaluators and based on existing studies in the field, if there are no more than 8 corrective data for range R_p, it is used the approach $MTBF_i = 1/\lambda_i$, i.e. failure rate is assumed constant. Reliability is modelled by the formula [7]:

$$R(T) = e^{-\lambda i^{*t}}$$

(b) The curve adjustment using system reliability model assumes that each maintenance event performed, the equipment is as good as new (Perfect

Table 4 Assignment of data to ranges R_i

Range (Q_A–Q_B)	Adjustment	MTBF$_i$ (h)	MTTR$_i$ (h)	Amount of events type M_1	Amount of events type M_2	Amount of events type M_3	Q_i (tons/h)
[5000–5400]	Weibull	85	8	25	45	10	6000
[5400–5600]	Weibull	95	9	35	60	15	6200

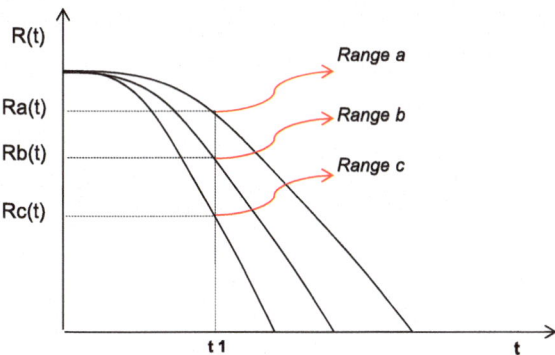

Fig. 5 Reliability *curves* according to load ranges

Maintenance—PRP) [8]. Existing data have been analysed using the Laplace test [6], ruling out the existence of a possible trend.

(c) Q_i Indicator represents the actual load indicator [Ton/hour] for each load range R_i, i.e. this indicator seeks to identify the actual load processed per hour of operation during interventions that occurred for the established load range.

(d) The availability (A) calculation of a system or production line operating at different loads, should be complemented with other analysis since, when there is variability in the workload, availability may conceal important phenomena of analysis. The Indicators: Utilization (U), Production Speed and Outflow Quality compose the global indicator OEE (Overall Equipment effectiveness), as a general indicator of efficiency or loss control [9].

3 Analysis and Results

It should be noted that, based on the proposed methodology, the reliability function $R[a, b](t)$ associated with operating load range [a, b] will be understood as the probability that a failure event does not occur on the system within a time interval t, for which the equipment is operated at an average load within this range [a, b]. In other words, if the instantaneous average load Q^* is monitored, since the last failure event, this operation will determine the range of $Q^* \in [a', b']$, and therefore the reliability function to consider is $R[a', b'](t)$. Specifically, analysis ranges of 100 ton/h are considered. Also, the amount of ranges will vary according to what is deduced from the available historical data. Next, the reliability analysis is presented.

3.1 Failure Mode Analysis in 2 Lines

Figure 6 shows the reliability curves obtained for process line 1. The determined ranges consider operating loads from 1000 to 1900 ton/h. It is relevant to note that the average operating load range [1600, 1700] appears as the most reliable, i.e. if the line operates at an average instantaneous load within this range, the probability of failure events in the line is lower in comparison to any other considered range [10].

Usually, it is expected that at a higher workload, less reliable the systems are. However, this is true in systems where operating load does not present a high variability within the given range for two consecutive failures. In this case, when a failure occurs at a load level Q^*, this is not directly referred to such charge level, since it is necessary to assess the past performance of operating levels until the time of previous failure event, determining average load at which the system was subjected. So, it is possible to assign an occurred failure at a load level Q^* to a different average range.

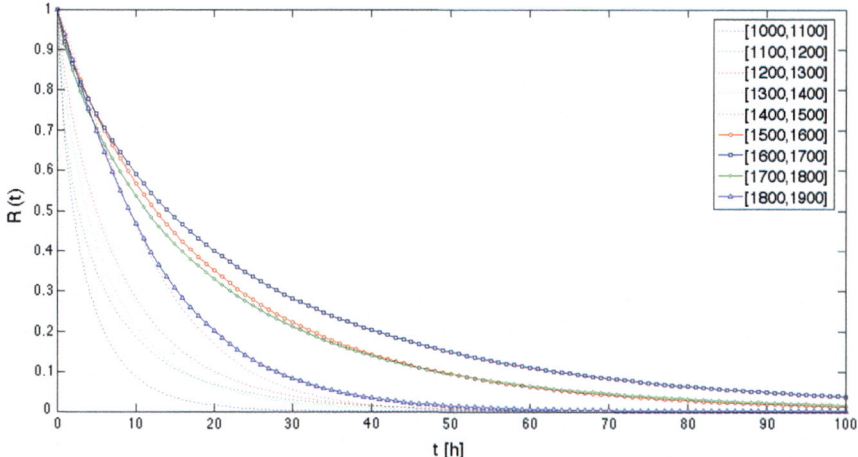

Fig. 6 Reliability functions by load range—line 1

On the other hand, one of the relevant preliminary findings is the reliability sensitivity to an increase in operating load. As noted, the most reliable operation level occurs within the range [1600, 1700]. Furthermore, analysing the graphs of Fig. 6, it can be deduced that the workload must not exceed 1800 tons (upper level), since the reliability curve for that range is considerably lower. The same for loads lower than 1500 tons (Table 5).

The analysis for line 2 gives very similar results to those obtained for line 1 (Fig. 7). However, the new average load range of operation [1700, 1800] presents the highest levels of reliability, so it is the recommended range by analysts. Similar to the case of line 1, line 2 is not recommended to exceed 1800 tons load, since reliability levels immediately decrease (Table 6).

3.2 Remarks and Results

The graphics with the results (reliability by load ranges) for each grinding line (Line 1 and 2) are shown in Figs. 6 and 7.

As a qualitative analysis of information, particularly, if it is referred to existing failure modes records for each failure event, the 100% of the recorded failure modes were revised, and it was observed that there was a confusion between the concepts "symptoms" and "failure mechanisms" [11, 12]. From the total analysed data (382 failure events), only 39.1% had failure modes successfully assigned to failure events. This fact limits the subsequent qualitative analysis, such as a root cause analysis.

Table 5 Reliability and maintainability indicators by ranges—line 1

Line 1

Range (tons/h)	Adjustment	Parameters		MTBF (h)	MTTR (h)	Events amount M_2	Events amount M_1	Q (tons/h)
		α	β					
[1000–1100]	Weibull	3.27	0.8	3.62	1.78	15	2	1056,09
[1100–1200]	Weibull	4.64	0.68	5.85	0.88	10	1	1149,01
[1200–1300]	Weibull	7.59	0.86	8.05	2.21	15	0	1254,91
[1300–1400]	Weibull	5.92	0.83	6.59	1.87	17	0	1351,97
[1400–1500]	Weibull	12.38	1.21	11.59	1.51	12	0	1456,75
[1500–1600]	Weibull	18.97	0.89	17.10	2.85	21	2	1563,78
[1600–1700]	Weibull	22.34	0.80	23.91	3.73	16	1	1646,98
[1700–1800]	Weibull	17.65	0.83	13.38	1.61	14	1	1742,97
[1800–1900]	Weibull	12.86	1.08	12.53	4.02	10	0	1860,16

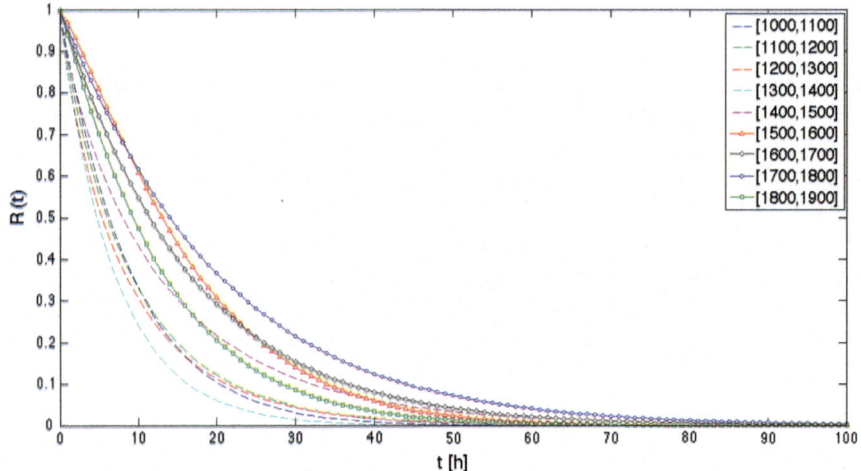

Fig. 7 Reliability function by load ranges—line 2

3.3 Complementary Analysis Methodology Presented

(i) General economic assessment to identify the optimal load range.

Identifying optimal load range, considering overall cost assessment for a longer life period (life cycle), an analysis is required to determine that range which will maximize production and will be economically convenient or profitable. Therefore, it is desirable to sensitize the variables that are affected by the increase of load, among which are known:

Table 6 Reliability and maintainability indicators by ranges—line

Line 2

Range (tons/h)	Adjustment	Parameters		MTBF (h)	MTTR (h)	Events amount M_2	Events amount M_1	Q [tons/h]
		α	β					
[1000–1100]	Weibull	9.15	1.02	8.74	1.61	29	1	1046,34
[1100–1200]	Weibull	9.01	0.92	7.80	2.31	26	3	1150,06
[1200–1300]	Weibull	8.33	0.87	7.60	1.48	34	3	1252,63
[1300–1400]	Weibull	6.97	0.96	7.07	2.42	46	0	1353,45
[1400–1500]	Weibull	12.42	0.86	12.10	1.95	63	4	1455,20
[1500–1600]	Weibull	17.69	1.25	15.34	4.29	40	4	1554,21
[1600–1700]	Weibull	16.38	1.03	13.94	3.83	27	2	1652,91
[1700–1800]	Weibull	20.16	1.05	19.72	2.26	23	0	1745,70
[1800–1900]	Weibull	13.16	1.08	11.58	3.08	14	2	1853,33

- Maintenance Cost (Direct). It must be clear about the real costs of each maintenance event, which requires efficient and effective information systems, as well as adequate "data culture" throughout the organization.
- Cost Inefficiency or Production Loss
- Frequency of Failure (MTBF) and preventive interventions
- Maintenance Budget.

The importance of this analysis lies in the fact that there is no objective measurement of the economic and technical effects that bring an increased load on the nominal design capacity. The uncertainty about the impact on assets will have increased load, specifically in terms of degradation during its life cycle, which is very high. Furthermore, regarding the economic impact caused by failures (mainly catastrophic failures or with a high frequency), direct cost of maintenance and loss production, respectively. For grinding lines analysis, it is only possible to show an increase in the frequency of failure, which directly affects the reliability curves. Therefore, further research should integrate the operating features, maintenance practices, equipment features, among others.

(ii) Complement the analysis with indicators.

Indicators should be useful and applicable to the process and operation environment. Particularly, the implementation of overall indicator of efficiency (OEE) is proposed. The availability indicator at a component or systemic level, in processes with variable load (production rate variable), sometimes conceals disabling phenomena and low performances, showing a partial reality of the process. This situation leads to incorrect decision making, misallocation of resources, unrealistic expectations of production results, among others. It is therefore recommended making future analysis using the Overall Indicator of Efficiency (OEE) and the reliability assessment in each event (M_1, M_2 and M_3).

The main objective of this analysis is to identify the real impact of the reliability on the production, and the corresponding curves according to those load levels required to process (Figs. 6 and 7).

(iii) Maintenance Plan Audit.

A detailed review of the maintenance plans is required, focusing as a first step in their effectiveness, i.e. those preventive or inspection activities that should be aimed at the critical failure modes. This first analysis should also consider: generation of new preventive and inspection activities (potential failure modes according to the new requirements); modification of the current schedule of preventive tasks and inspection; reassessment of the component degradation rates; increment (if necessary) of the maintenance budget, policy review associated to the supply of critical spare parts.

As a second step, it is recommended to check the efficiency of maintenance plans, validating features such as quality of maintenance planning, scheduling, performance of preventive, inspection tasks and review of procedures.

These recommendations should be supported, according to the respective operating context, through tools application as Root Cause Analysis (RCA) [13] and Failure Modes, Effects and Criticality Analysis (FMECA) [11].

Other relevant references in this area are [14–17].

4 Conclusions

The methodology developed and proposed in this paper, aims to support the decision-making process in the field of maintenance management, operation and reliability analysis so that such decisions remain aligned with the vision, strategy and objectives. This article clearly shows how it affects the reliability of a grinding plant variability load levels (input streams) in the lines of the grinding plant and then, it determines the optimal load range would. The paper also identifies the relationship between the load ranges of each line and their respective reliability by developing a real case study carried out in a mining company located in northern Chile, serving as validation of the presented methodology.

In the analysed 2 lines, higher reliability ranges coincide with those whose production is recommended. The fact that in these load ranges there is lower probability of failures occurrence, does not ensure that they cannot occur, but they will happen with less frequency along the time. According to the obtained information, the modelled reliability behaviours match with operating and occurrence conditions of failure events. In this way, theoretical results are empirically validated:

- For Line 1, the range of average operating load [1600, 1700] seams to be the most reliable. That is, if the line operates at an average instantaneous load within this range, the failure occurrence probability in the line is lower, compared to operating in some other considered ranges. Moreover, analysing the graphs of line 1, it can be deduced that the workload should not exceed 1800 tons (upper level), since the reliability curve for that range is considerably lower. The same for loads lower than 1500 tons.
- The analysis for Line 2 shows very similar results to those obtained for line 1. However, the new average load range of operation [1700, 1800] presents the highest levels of reliability, so it is the range recommended by analysts. It is not recommended to exceed an operation average load of 1500 tons.

In order to validate the results of Lines 1 and 2 is necessary to know in depth those specific factors involved in the operation that are not recordable as quantitative data. To improve the analysis, monitoring the identified phenomena by the same methodology is necessary, so it allows the comparison and correlation results under the same parameters.

As a future research in this field, the development of tools to integrate and relate reliability knowledge, workloads and the state in which the equipment or subsystems are operating (failure, preventive or operational stop) is suggested, as shown schematically in Fig. 8. This kind of tool will make easier the analysis and decision making in asset management and maintenance in industrial organizations.

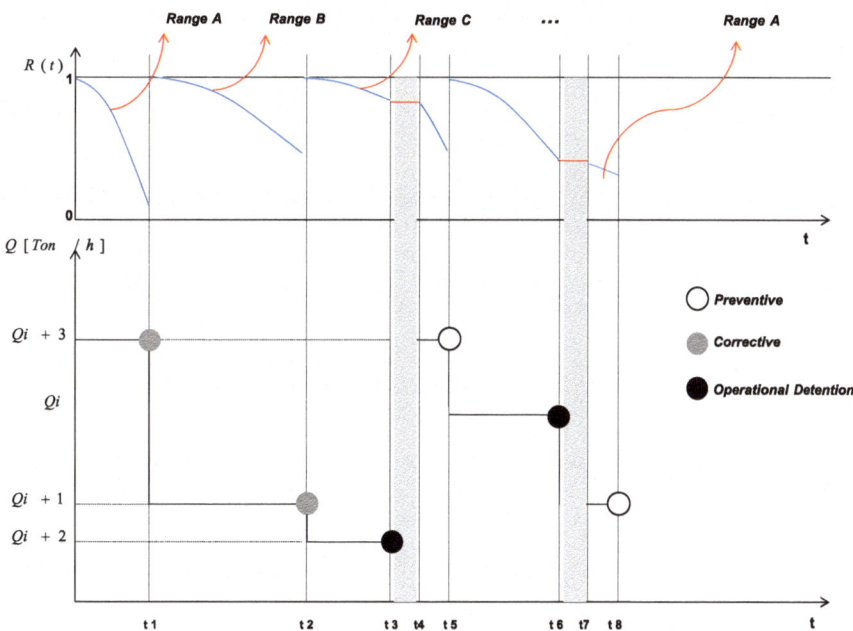

Fig. 8 Simultaneous workload and reliability type analysis

The objective of this proposal is to develop a computer system that monitors online system behaviour, assessing workload in which you have configured the system/equipment, linking it directly with the reliability curve that corresponds to the load. This information is useful for managing maintenance guidelines, control failures and their causes, energizing maintenance plans and to analyse features of the accelerated wear phenomena and its possible causes.

Acknowledgements This research is funded by the Spanish Ministry of Science and Innovation, Project EMAINSYS (DPI2011-22806) "Sistemas Inteligentes de Mantenimiento". Procesos emergentes de E-maintenance parala Sostenibilidad de los Sistemas de Producción, besides FEDER funds. The research work was performed within the context of iMaPla (Integrated Maintenance Planning), an EU-sponsored project by the Marie Curie Action for International research Staff Exchange Scheme (project acronym PIRSES-GA-2008-230814 iMaPla).

References

1. Dhananjay K, Bengt K (1994) Proportional hazards model: a review. Reliab Eng Syst Saf 44 (1994):177–188
2. Barberá L, Crespo A, Viveros P, Arata A (2012) The graphical analysis for maintenance management method: a quantitative graphical analysis to support maintenance management decision making. J Qual Reliab Eng Int. doi:10.1002/qre.1296. Copyright © 2012 John Wiley & Sons, Ltd. (wileyonlinelibrary.com)
3. Wightman DW, Bendell A (1985) The practical application of proportional hazards modelling. Reliab Eng 15(1986):29–53
4. Dhananjay K, Bengt K, Uday K (1992) Reliability analysis of power transmission cables of electric mine loaders using the proportional hazards model. Reliab Eng Syst Saf 37 (1992):217–222
5. Ansell JI, Phillips MJ (1997) Practical aspects of modelling of repairable systems data using proportional hazards models. Reliab Eng Syst Saf 58(1997):165–171
6. Rausand M, Hoyland A (2003) System reliability theory: models, statistical methods, and applications, 3rd edn. Wiley, New York
7. Blischke WR, Murthy DNP (2003) Case studies in reliability and maintenance. Wiley, USA
8. Parra C, Crespo A, Kristjanpoller F, Viveros P (2012) Stochastic model of reliability for use in the evaluation of the economic impact of a failure using life circle cost analysis. Case studies on the rail freight and oil industry. Proc IMechE, Part O: J Risk Reliab 226(4):392–405
9. Pomorski T (1997) Semiconductor fairchild. Manag Overall Equip Effectiveness Optimize Factory Perform, IEEE. doi:10.1109/ISSM.1997.664488
10. Jardine AKS, Anderson PM, Man DS (1987) Application of the weibull proportional hazards model to aircraft and marine engine failure data. Qual Reliab Eng Int 3:77–82
11. Moubray J (1997) Reliability-centred maintenance, 2nd edn.. Industrial Press, Inc., New York, USA. pp 448. ISBN: 0831131462
12. ISO/DIS 14224 (2004) Petroleum, petrochemical and natural gas industries. In: Collection and exchange of reliability and maintenance data for equipment, Oct 2004
13. Mobley RK (1999) Root cause failure analysis. Butterworth-Heinemann
14. Huang SH, Dismukes JP, Shi J, Su Q, Wang G, Rauak MA, Robinson E (2001) Manufacturing system modeling for productivity improvement. J Manuf Syst 21(4)

15. Jardine AKS, Banjevic D, Wiseman M, Buck S, Joseph T (2001) Optimizing a Mine Haul truck wheel motors' condition monitoring program: use of proportional hazards modeling. J Qual Maintenance Eng 7(4):286–301
16. Lugtigheid D, Banjevic D, Jardine AKS (2008) System repairs: when to perform and what to do? Reliab Eng Syst Saf 93(2008):604–615
17. Vlok PJ, Coetzee JL, Banjevic D, Jardine AKS, Makis V (2002) Optimal component replacement decisions using vibration monitoring and the proportional- hazards model. J Operation Res Soc 53(2):193–202, Part Special Issue: The Process of OR (Feb 2002)

Author Biographies

Luis Barberá Martínez is PhD (Summa Cum Laude) in industrial Engineering by the University of Seville, and Mining Engineer by UPC (Spain). He is a researcher at the School of Engineering of the University of Seville and author of more than fifty articles, national and main internationals. He has worked at different international Universities: Politecnico di Milano (Italy), CRAN (France), UTFSM (Chile), C-MORE (Canada), EPFL (Switzerland), University of Salford (UK) or FIR (Germany), among others. His line of research is industrial asset management, maintenance optimization and risk management. Currently, he is Spain Operations Manager at MAXAM Europe. He has three Master's degrees: Master of Industrial Organization and Management (2008/2009) (School of Engineering, University of Seville), Master of Economy (2008/2009) (University of Seville) and Master of Prevention Risk Work (2005/2006). He has been honoured with the following awards and recognitions: Extraordinary Prize of Doctorate by the University of Seville; three consecutive awards for Academic Engineering Performance in Spain by AMIC (2004/2005), (2005/2006) and (2006/2007); honour diploma as a "Ten Competences Student by the University of Huelva and CEPSA Company (2007); graduated as number one of his class.

Pablo Viveros Gunckel Researcher and Academic at Technical University Federico Santa María, Chile, has been active in national and international research, both for important journals and conferences. Also he has developed projects in the Chilean industry. Consultant specialist in the area of Reliability, Asset Management, System Modeling and Evaluation of Engineering Projects. He is Industrial Engineer and Master in Asset Management and Maintenance.

Rodrigo Mena (BS and MSc in Mechanical Engng., Universidad de Chile, 2008, PhD in Industrial Engng., CentraleSupélec, 2015) is senior research consultant at Aramis S.r.l. and post-doc fellow at the Politecnico di Milano since April 2016. His current research is focused on the design and implementation of computational frameworks for modelling, simulation and optimization for the integration of renewable generation technologies onto distribution and sub-transmission power networks. At the basis of his work are Monte Carlo simulation methods and optimal power flow, to model primary renewable energy sources and operating uncer-tainties and assess different economic-reliability-based objective function(s), evolutionary algorithms-heuristic techniques for non-linear combinatorial optimization, clustering techniques to improve computational efficiency and portfolio theory risk measures to formulate diverse optimization strategies. He has also served as RAM engineer in the industrial sector, performing tasks in the fields of physical asset management and reliability and maintenance engineering.

Vicente González-Prida Díaz is PhD (Summa Cum Laude) in Industrial Engineering by the University of Seville, and Executive MBA (First Class Honors) by the Chamber of Commerce. He has been honoured with the following awards and recognitions: Extraordinary Prize of Doctorate by the University of Seville; National Award for PhD Thesis on Dependability by the Spanish Association for Quality; National Award for PhD Thesis on Maintenance by the Spanish Association for Maintenance; Best Nomination from Spain for the Excellence Master Thesis Award bestowed by the EFNSM (European Federation of National Maintenance Societies). Dr. Gonzalez-Prida is member of the Club of Rome (Spanish Chapter) and has written multitude of articles for international conferences and publications. His main interest is related to industrial asset management, specifically the reliability, maintenance and aftersales organization. He currently works as Program Manager in the company General Dynamics—European Land Systems and shares his professional performance with the development of research projects in the Department of Industrial Organization and Management at the University of Seville.

Summary of Results and Conclusions

Adolfo Crespo Márquez, Vicente González-Prida Díaz
and Juan Francisco Gómez Fernández

Abstract In this last chapter of the book, as editors of the work, we would like to summarize for the reader the main contributions included in this manuscript. We want to remark the importance of new and advanced techniques supporting decision-making in different business processes for maintenance and assets management, but mainly we recall the reader the basic need of the adoption of a certain management framework, with a clear processes map and the corresponding IT supporting systems. Framework processes and systems will be the key fundamental enablers for success and for continuous improvement.

Keywords Assets management · Management framework · Management supporting structure · Maintenance engineering techniques

1 Introduction

Framework processes and systems will be the key fundamental enablers for success and for continuous improvement in maintenance and assets management. Our framework will help to define and improve business policies and work procedures for the assets operation and maintenance along their life cycle. Additionally, it improves the technical support to the user, rationalizes decisions concerning spare parts stocking levels, and many other related topic.

As a consequence, it is possible to reach important costs reductions, improvements in the quality of the operations and increasing business profitability and users satisfaction.

Along the book, advanced tools were introduced for the following purposes:

- Reengineering of processes
- Possibility to outsource technical assistance

A. Crespo Márquez (✉) · V.G.-P. Díaz · J.F. Gómez Fernández
Department of Industrial Management, School of Engineering,
University of Seville, Seville, Spain
e-mail: adolfo@us.es

© Springer International Publishing AG 2018
A. Crespo Márquez et al. (eds.), *Advanced Maintenance Modelling for Asset Management*, DOI 10.1007/978-3-319-58045-6_20

- Data analysis
- Analysis, recovery and collaboration with suppliers
- Repair tasks management
- Parts replacement planning and an inventory management policies implementation
- Logic validation of repairs, etc.

The above-mentioned topics are also detailed in other books developed by the SIM Research Groups such as: [1–8].

The benefits of these techniques were solidly proven in many industrial sectors and business environments. Important contributions to safety, health and protection of the environment were appreciated. Beside this, these activities could foster organizational commitment to quality, performance, safety, and did help to mitigate legal, social and environmental risks associated with accidents at industrial facilities. In the following sections, some conclusions regarding each one of these achievements are presented and later, we do also propose possible future lines for a research agenda within this field of assets management.

2 Summary of Results of This Work and Research

The emergence of physical asset management as an important field of action in the industry is an indisputable fact that has given renewed attention to the maintenance of equipment and infrastructure. The main concern is not only to ensure the sustainability of assets in the organizations (it refers that they can continue satisfying their users generating profits in a safe way), but also to guarantee their ecological efficiency as well as maximum interval of their life cycle.

The formulation of global models allows the coherent and integrated use of different techniques, activities, areas of knowledge, tools, etc. They can be applied to a specific engineering problem, proving to be a long-line research field, since it not only allows to design an optimal solution to a particular practical case where this management is necessary, but also, to be used as a support for the development of computer systems and applications in this wide field.

Assets management itself has the handicap of being intensive in the use of information, for which one of the main challenges is the development of information systems that support it. This is why the use of any AMS (Asset Management System) results as a must, as a clear business need, that cannot be avoided when pursuing excellence in this area. Besides the AMS, nowadays, we must welcome new emerging technological trends appearing within the new European initiative named Industry 4.0. The process of permanent transformation of industry by new technologies is here to stay. This is a process of constant evolution whose drivers are competitiveness and sustainability, understood in its broadest sense (environmental, economic and social).

Thanks to the current state of scientific and technological development, this process is facing a new challenge even more important: the transition from discrete technological solutions (which respond to isolated problems), to a global conception where assets, plant, processes and engineering systems are designed and operated as an integrated complex unit. This vision has materialized in the scientific community through the proposal of a series of concepts that serve, in some way, to guide this development: Smart Plants, Cyber-Physical Systems, Factory of the Future or Industry 4.0 are some examples. In our view, this concept of Industry 4.0 clearly illustrates the scope of the qualitative leap that is taking place. The complete integration of Operations and Maintenance (O&M) processes into assets and production systems from its own conception phase is a key matter within this new paradigm. Moreover, this evolution will necessarily cause the emergence of new processes and needs to O&M. That means that O&M will also undergo a deep transformation. In conclusion, this book has been also concerned about the development of innovative models and techniques that may help to address issues of interest, within the field of asset management, such as:

- What is the process to ensure that maintenance can be properly integrated into an overall asset management framework?
- What types of quantitative reliability engineering techniques can be developed to facilitate and improve asset management?
- How can maintenance policies benefit from the development of emerging capabilities?

These are all research topics covered by this book, also becoming the basis for different on-going European R+D+I projects, which are currently under development by the SIM group. Among others, the following projects should be highlighted:

- OptiRail (Innterconecta 2016), which aims to fully develop an Intelligent Asset Management System for application to the management of light trains (metro, tram, etc.)
- Sustain-Owner, Horizon 2020, MSCA-RISE-2014, which addresses the development of models for the calculation of Total Ownership Cost and Total Value of Assets
- O&M 4.0 (Spanish R&D Plan, Challenges 2015). Development of a model for maintenance management including the development of new processes and systems in the Industry 4.0 scenario.

3   Conclusions Extracted from the Different Parts

3.1   A Changing Asset Management Framework

In Chapter "A Maintenence Management Framework based on PAS55", the historical development of maintenance models and frameworks is reviewed. Among all those proposals, PAS 55 standard emerges in 2004 as a first complete framework, not only for maintenance but also for management of the entire life cycle of assets. Desirable characteristics and best practices are in this specification identified as needed for the operation of a modern and efficient maintenance management framework (MMF). At the same time, this chapter describes how the proper exchange of updated maintenance information and the coordination through automatic procedures started to be considered soon as a core activity in order to facilitate knowledge management and proactive decision-making in maintenance. Nowadays, and in line with Industry 4.0, we can start to talk about full digitization, connecting machines, people and processes. Communication among cyber-physical assets of infrastructure systems is now possible and the integration of production and customers feedback with top level business applications, now accessible from any place, can be a reality thanks to cloud computing and the Internet of things.

Chapter "The Integration of Open Reliability, Maintenance and Condition Monitoring Management Systems" illustrates this point, presenting a methodology to develop a systems platform that integrates the database in computerized maintenance management systems (CMMS), with the tools for Reliability Centred Maintenance (RCM) and the results of Condition-Based Maintenance (CBM) tools. The aim of the integration platform developed is to optimize the synergy of the three maintenance tools (CMMS, RCM and CBM) and to create an intelligent system implemented as a demo software. Continuing on the issue, Chapter "Prognostics and Health Management in Advanced Maintenance Systems" remarks that new technical capabilities have been added in maintenance systems, supported by PHM, that open the door to a large room for improvement in terms of competitiveness and profitability, providing the basis for proactive maintenance management. Besides that point, Chapter "A Framework for Effective Management of CBM Programs" discusses deeply about CBM management approaches in complex contexts, proposing a CBM framework with a template to clarify the concepts and to structure and document the knowledge generation for a given condition-based maintenance solution. In these chapters, we can also understand why it is important to integrate at least the CMMS, RCM and CBM systems for decision-making over the interoperability point of view, especially in high-tech complex engineering systems, for full exploitation of their potential. These systems are traditionally supported by isolated software. As a result of this integration, maintenance actions can modify the assets indenture levels and even may render new maintenance actions conveniently, contributing to extend their life.

Consequently, from a pragmatic point of view, Chapters "A Maintenence Management Framework based on PAS55–A Framework for Effective Management of CBM Programs" complement MMF from processes, technical and knowledge management perspectives.

3.2 Pursuing High Management Effectiveness in a Dynamic Environment

Value-based systems models are very demanded to obtain the real value of the assets and the organization. Therefore, the chapters from "Criticality Analysis for Maintenance Purposes to Economic Impact of a Failure Using Life Cycle Cost

Analysis" are focused on this field conceptually and mathematically, researching relevant qualitative and quantitative factors that can contribute to the asset value, but at the same time, improving precision in terms of costs and reliability with adaptation to dynamic changes of the business (Chapters "Criticality Analysis for Maintenance Purposes" and "AHP Method According to a Changing Environment") or the life cycle of the asset (Chapters "Reliability Stochastic Modelling for Repairable Physical Assets" and "Economic Impact of a Failure Using Life Cycle Cost Analysis").

In Chapter "Criticality Analysis for Maintenance Purposes" a process and model for criticality analysis with the idea to easy the maintenance planning process is designed. This chapter is an effort to describe how to turn this process over extensive data into a practical management tool, supported by quantitative formulae and crucial representations for result interpretation.

Chapter "AHP Method According to a Changing Environment", deeps more in this issue with a modified version of the AHP methodology, it describes how this well-known method can be dynamically applied within the area of maintenance. The idea is to support decision-making in questions such as: What information should be relevant to make decisions regarding the maintenance or the post-sales service? how to select a policy for maintenance or warranty assistances in order to improve the profit and the image of the company?

Managers have to identify critical trends of the business behaviour over time and on more relevant variables (alternatives), when strategy is changing dynamically according to emerging market needs. A better selection of possible maintenance policies over time may be reached taking into account the dynamics of the environment in which businesses are embedded.

Both Chapters "Reliability Stochastic Modelling for Repairable Physical Assets" and "Economic Impact of a Failure Using Life Cycle Cost Analysis" translate the strategic perspective of assets to tactical perspective from a reliability point of view, defining mathematically this issue searching to show the asset value over the time. Chapter "Reliability Stochastic Modelling for Repairable Physical Assets" stresses the learning process required for the proper implementation of modelling of repairable industrial assets reliability. This is sometimes forgotten by many researchers concentrated in presenting final results indicating the use of a certain model embedded in a computer tool and without considering the extremely different performance of assets under real process operational conditions instead of described ideal conditions (lab test) from providers.

The study includes the possibility of different degradation levels, where the assets could be placed after specific maintenance activities.

In Chapter "Economic Impact of a Failure Using Life Cycle Cost Analysis", an analysis of the impact of the lack of reliability on asset's life cycle cost is presented. This is many times difficult to visualize and may significantly influence the possibility to extent the useful life of an asset at reasonable costs, and so to the asset value for the business. It is important to mention that for the LCCA, there is room for improvement in the impact of reliability assessment, by using advanced mathematical methods such as:

(a) Stochastic methods;
(b) Advanced maintenance optimization using genetic algorithms;
(c) Monte Carlo simulation techniques;
(d) Advanced reliability distribution analyses;
(e) Markov simulation methods.

These methods will have their particular characteristics and their main objective is to diminish the uncertainty created in the estimation process of the total costs of an asset during its expected useful life cycle.

For these reasons, the chapter remembers that it is not feasible to develop a unique LCCA model, which suits all of the requirements. However, it is possible to develop more elaborated models to address specific needs such as a reliability cost-effective asset development.

This consideration will support the decision-making process. Finally, it is important to mention that the results are developed over real case studies help in understanding and confirm the importance of the procedures that consider the business impacts of reliability in asset life cycle.

3.3 Advanced Methods and Techniques to Improve Management Efficiency

Risk assessment methods are essential to guide the company's continuous improvement towards efficiency. Their proper implementation can contribute to fulfil the enterprise objectives importantly.

There are different implications of this work from a wide variety of perspectives:

(i) From a reliability point of view, activities in the assets can now be improved and reduced by considering asset durability, such as in Chapters "On-line Reliability and Risk to Schedule the Preventive Maintenance in Network Utilities" and "Analysis of Dynamic Reliability Surveillance".
(ii) From a cost point of view, determining a direct relationship among activities execution on the assets and the corresponding implications in the business, such as in Chapter "Customer-Oriented Risk Assessment in Network Utilities".

Taking into account the impacts on risks, these methods allow maintenance management to be accountable and comparable with other organizations or reference standards, from different areas and levels of detail, identifying improvements in reliability and ensuring service quality from customer, business and society perspectives. An explanation of the precise methodologies followed and examples (where all were implemented) of the results obtained are shown. Extensions of this work could be related to the study of maintenance logistics and support improvements when using these risk-based strategies.

Chapter "On-line Reliability and Risk to Schedule the Preventive Maintenance in Network utilities" presents a methodology that can generate suitable estimations of risk per critical failure mode. This can help in the process of releasing PM activities for these assets, minimizing them, as well as minimizing CM activities by increasing reliability. Minimization of maintenance activities can increase asset durability and considerably reduce network technical deterioration. This methodology also offers managers the possibility of controlling maintenance budgets based on asset risk expectations for a certain period, because:

(i) Management of spare parts can now be improved by considering times to deterioration instead of times to failure. Stock levels can therefore be more precisely estimated with considerable savings in inventory holding costs and working capital requirements.
(ii) Asset depreciation can be better defined, according to reliability and original purchasing value. Obviously, the equipment reliability affects ageing equipment depreciating the equipment from this purchase value. Consequently, proper maintenance strategies ensure that the value of asset books is accurate, and that repair or replacement decisions are accountable.

Chapter "Customer-Oriented Risk Assessment in Network Utilities" discusses about the relationship among the level of activities execution on the assets and the corresponding implications in the business. In particular, the relationship between loyalty of customer and service reliability is analyzed. The proposed methodology in this chapter is focused on analyzing the economic implications of maintenance and the occurrence (and reoccurrence) of failures in different scenarios, calculating the economic risk of the failure and comparing it with the costs of prevention. It is a clear result how this assessment should not be only done in terms of budget, but also in terms of real profits directly observable and those due to the avoided damages. It can be appreciated how maintenance performance affects the offered and the perceived quality by the customers, and how investments in maintenance activities (especially preventive/predictive) represent economic advantages, reducing customer's abandonment rates. This methodology can be used as an automated routine to determine the expected reliability of the services for a determined time, and geographical zone, as well as to design marketing and loyalty campaigns appropriate to each customer circumstances.

Chapter "Analysis of Dynamic Reliability Surveillance" explores advanced methodologies in order to improve the dynamic flexibility, learning from conditions changes and implementing this in a practical way in SCADA. It suggests applying an ANN model, showing how these models can ensure longer profitability periods reaching more reliable assets. ANN models' capacity for self-learning using different sources of data (sometimes noised or deprived of communication) is a very powerful feature that can be utilized thanks to reiterative memory. Back-propagation perceptron ANN is recommended for automation developments with real-time utilization. Furthermore, advanced ANN models could be applied when supporting additional variables.

3.4 The Need for Innovation in Assessment and Control

A process-oriented management ensures proper decision-making by controlling process performance as a way to pursue effectiveness and efficiency. Maintenance decision-making will rely importantly on available information and knowledge, to control possible variation of assets performance. That is to say, it is therefore necessary to act in a comprehensive manner on assets; there is a need to evaluate first the causes of variance in activities implementation in order to control and later to define the trend of future developments in the use of the assets. The trend can be automated through the easy use of statistic techniques and diagnostic tools, facilitating its control and proper knowledge management. All the previous aspects are supported through the design of a suitable control panel, a dashboard, verifying the main causes that determine the actual performance of activities on assets as shown in Chapters "A Quantitative Graphical Analysis to Support Maintenance"–"A Graphical Method to Support Operation Performance Assessment".

Successful results of implementation require experience and advanced knowledge about the maintenance management process. Their applicability relies on the quality of the relatively simple database required for the diagrams elaboration. The presented tools provide information that is useful, clear and easy to interpret.

Chapter "A Quantitative Graphical Analysis to Support Maintenance" presents an innovative method named Graphical Analysis for Maintenance Management (GAMM). This method provides useful information regarding the reliability and maintainability of systems in a graphical and very intuitive way. In Chapter "Case Study of Graphical Analysis for Maintenance Management", the method is applied to two slurry pumps comparing the results in order to determine and graphically characterize the deficiencies of the slurry pumps. The GAMM method could also support the decision-making process for the definition of a new maintenance plan.

Following a very similar approach, in Chapter "A Graphical Method to Support Operation Performance Assessment", another graphical method (GAOM) is presented. GAOM integrates the main maintenance and production indicators, and summarizes and complements the decision process to reduce production losses and thus maximizing business benefits. The GAOM method consists of seven input information and one optional production data (production target). Regarding the applicability of this method, the tool allows to process information, generating graphs based on algorithms that are a part of the tool. As future lines of research, there is great development potential of the tool on automation, using more advanced programming languages that enable its implementation in enterprise resource planning systems (ERP).

3.5 Continuous Improvement Through Emergent Process and Technologies

Innovation measures and internal proposals in organizational culture could be established in order to avoid complacency and regression, encouraging a move towards sustainable world-class maintenance. Nowadays, in almost all maintenance tasks, some new technology is to be used and some new concepts need to be mastered. Defining the common bases of knowledge and the support of appropriate technologies based on adequate foundation is critical. In general, the rate of learning emergent processes and technologies are slow initially, according to the levels of expertise. Once they are a part of organizational culture, they can be carried out in a guided and effective form, making easy to promote the exchange and generation of new knowledge.

Chapters "Value-Driven Engineering of e-maintenance Platforms"–"Influence of the Input Load on the Reliability of the Grinding Line" present a variety of technological and innovative tools allowing gaining knowledge and competitive advantage in asset management. Chapter "Value-Driven Engineering of e-maintenance Platforms" presents a starting point to further develop the theory of value assessment of E-maintenance. Missing issues are the analysis of intangible values as well as the complete coverage of technologies and processes target of interest for the development of applied services. A better implementation of the E-maintenance technologies helps to reduce the costs and to facilitate the knowledge generation and dissemination. Chapter "Assistance to Dynamic Maintenance Tasks by Ann Based Models" deals with tools to ensure critical equipment reliability and maximum extension of equipment life cycle in PV plants. For this purpose, this chapter proposes an application of an ANN model, which later allows a very practical implementation in SCADA systems and can easily improve decision-making.

In Chapter "Expected Impact Quantification Based on Reliability Assessment", the FEI methodology is presented, which is a powerful tool which can be used complementing other plant modelling and analysis traditional methodologies and tools. This proposal can be used in various phases of an industrial project via life cycle cost analysis (LCCA) for design-oriented operations. The strengths of the FEI methodology include its ability to systematically and quantitatively assess the operational criticality of the assets and their subsystems. The E-OCI is the final result of the computation process, which enables the quantitative and unequivocal prioritization of the system elements to assess the associated loss as "production loss" regarding its unavailability and total impact on the process.

Another example of an innovative methodology is presented in Chapter "Influence of the Input Load on the Reliability of the Grinding Line": In this case, the tool allows to study the relationship between load ranges of different production lines and their respective reliability. An interesting case study carried out in a mining company located in northern Chile serves as the validation of the

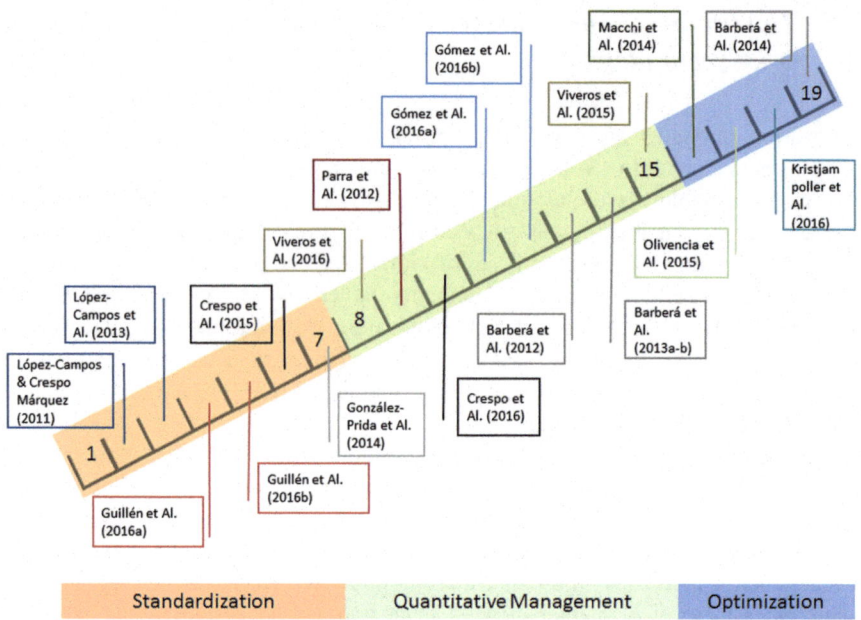

Fig. 1 Chapters according to maturity levels

methodology. This chapter clearly shows how to determine optimal load ranges in a grinding plant to ensure reliability and ultimately production levels.

As a future research in this field, the development of tools to integrate and relate reliability knowledge, workloads and the state in which the equipment or subsystems are operating (failure, preventive or operational stop) is suggested. This kind of tool will make easier the analysis and decision-making in asset management and maintenance in industrial organizations. A computer system that monitors online system behaviour could be developed, assessing workload for which the system/equipment is configured, linking it directly with the reliability curve that corresponds to the load.

4 Synthesis and Possible Future Research Lines

The content of the book can be somehow classified by using three different assets management levels of maturity and excellence.

(I) Standardization level: At this level the organization reaches reasonable performance based on well-defined methodologies, techniques and supporting technologies and using ICT to automate and manage the dispersed, duplicated and unrelated data.

(II) Quantitative management level: This is a more proactive level, where the organization defines anticipatory measures to manage circumstances, guiding the organization for the statistical analysis with the purpose of prediction through the use of the amount of information generated thanks to the ICT evolution.

(III) Optimization level: This level is focused on continuous improvement, based on innovation in a sustainable way over quantitative data and also testing new methodologies, techniques and technologies.

These levels are in accordance with the CMMI-Capability Maturity Model Integration (2007), and are related to the application of the ICTs for asset management purposes. Figure 1 places the different contributions that can be found in this book's Chapters within these three levels. This helps to take into account the evolutionary path that can be followed during a maturity growth.

Our final remark is related to the future to come: the existence of increasingly complex equipment and processes, the increase in the number of assets, the speed of technological change, the need to reduce costs in the modern world, together with increases in the level of excellence of commercial goals such as quality and delivery time, and concern for the safety of workers and the environment, make asset management an important source of benefits and competitive advantages for present and future world-class enterprises. This book has analyzed the above-mentioned factors, and has identified different areas where research lines, connected to assets management topics, could be developed:

- Maintenance policy selection.
- After-sales management.
- Knowledge management.
- Critical asset and infrastructure management.
- Asset life cycle management.
- After-sales maintenance.
- Performance measurement system.
- Sensors and health monitoring systems.
- Reliability centred maintenance.
- Building information modelling.
- Advanced maintenance techniques.
- The set up process.

References

1. Carnero C, González-Prida Vicente (2016) Optimum decision making in asset management. Hershey, PA, IGI Global. doi:10.4018/978-1-5225-0651-5,ISBN:9781522506515
2. Crespo MA (2010) Dynamic modelling for supply chain management. Front-end, back-end and integration issues. Springer Series in Advanced Manufacturing. ISBN: 978-1-84882-680-9

3. Crespo MA (2007) The maintenance management framework. Models and methods for complex systems maintenance. Springer Series in Reliability Engineering. ISBN: 978-1-84628-820-3
4. Gómez F, Juan F, Crespo MA (2012). Maintenance management in network utilities: framework and practical implementation. Springer-Verlag London 2012, ISBN 978-1-4471-2756-7, Springer Series in Reliability Engineering
5. González-Prida V, Raman A (2015) Promoting sustainable practices through energy engineering and asset management. IGI Global, Hershey, PA. doi:10.4018/978-1-4666-8222-1.ISBN:9781466682221
6. Gonzalez-Prida-Diaz V, Crespo-Marquez A (2014) After-sales service of engineering industrial assets. A reference framework for warranty management. Springer Verlag, 2014. 401 P. 1v. ISBN: 978-3-319-03709-7
7. Parra C, Crespo A (2015) "Ingeniería de Mantenimiento y Fiabilidad aplicada en la Gestión de Activos. Desarrollo y aplicación práctica de un Modelo de Gestión del Mantenimiento". Segunda Edición. Editado por INGEMAN, Escuela Superior de Ingenieros Industriales de la Universidad de Sevilla, España
8. Sola Rosique A, Crespo MA (2016) Principios y marcos de referencia de la gestión de activos. AENOR, ISBN: 978-84-8143-924-3; Depósito legal: M-12236-2016

Author Biographies

Adolfo Crespo Márquez is currently Full Professor at the School of Engineering of the University of Seville, and Head of the Department of Industrial Management. He holds a Ph.D. with Honours in Industrial Engineering from this same University. His research works have been published in journals such as Reliability Engineering and System Safety, International Journal of Production Research, International Journal of Production Economics, European Journal of Operations Research, Omega, Decision Support Systems and Computers in Industry, among others. Prof. Crespo is the author of eight books, the last five with Springer Verlag (2007, 2010, 2012, 2014) and Aenor (2016) about maintenance, warranty, supply chain and assets management. Professor Crespo is Fellow of ISEAM (International Society of Engineering Assets Management) and leads the Spanish Research Network on Assets Management and the Spanish Committee for Maintenance Standardization (1995–2003). He also leads the SIM (Sistemas Inteligentes de Mantenimiento) research group related to maintenance and dependability management and has extensively participated in many engineering and consulting projects for different companies, for the Spanish Departments of Defense, Science and Education as well as for the European Commission (IPTS). He is the President of INGEMAN (a National Association for the Development of Maintenance Engineering in Spain) since 2002.

Vicente González-Prida holds a Ph.D. with Honours in Industrial Engineering by the University of Seville, and Executive MBA (First Class Honours) by the Seville Chamber of Commerce. He also has been awarded with the National Award for Ph.D. Thesis on Dependability by the Spanish Association for Quality; the National Award for Ph.D. Thesis on Maintenance by the Spanish Association for Maintenance; and the Best Nomination from Spain for the Excellence Master Thesis Award bestowed by the EFNSM (European Federation of National Maintenance Societies). Dr. Gonzalez-Prida has authored a book with Springer Verlag about Warranty and After-Sales Assets Management (2014) and many other publications in relevant journals, books and conferences, nationally and internationally. His main interest is related to industrial asset management, specifically the reliability, maintenance and life cycle organization. He currently works as Project Manager in the company General Dynamics—European Land Systems and shares his professional performance with the development of research projects within the SIM group in the Department of Industrial Organization and Management at the University of Seville and teaching activities in Spain and Latin-America.

Juan Francisco Gómez Fernández is Ph.D. in Industrial Management and Executive MBA. He is currently part of the SIM research group of the University of Seville and a member in knowledge sharing networks about Dependability and Service Quality. He has authored a book with Springer Verlag about Maintenance Management in Network Utilities (2012) and many other publications in relevant journals, books and conferences, nationally and internationally. In relation to the practical application and experience, he has managed network maintenance and deployment departments in various national distribution network companies, both from private and public sector. He has conduced and participated in engineering and consulting projects for different international companies, related to Information and Communications Technologies, Maintenance and Asset Management, Reliability Assessment and Outsourcing services in Utilities companies. He has combined his professional activity, in telecommunications networks development and maintenance, with academic life as an associate professor (PSI) in Seville University, and has been awarded as Best Master Thesis on Dependability by National and International Associations such as EFNSM (European Federation of National Maintenance Societies) and Spanish Association for Quality.

Printed by Printforce, the Netherlands